# International Symposium on Liquid Metal Processing and Casting 2015 (LMPC 2015)



IOP Conference Series: Materials Science and Engineering Volume 143

Leoben, Austria
20 – 24 September 2015

**Editors:**

**Abdellah Kharicha**
**R. Mark Ward**

ISBN: 978-1-5108-2821-6
ISSN: 1757-8981

**Printed from e-media with permission by:**

Curran Associates, Inc.
57 Morehouse Lane
Red Hook, NY 12571

**Some format issues inherent in the e-media version may also appear in this print version.**



Printed by Curran Associates, Inc. (2016)

For permission requests, please contact the Institute of Physics
at the address below.

Institute of Physics
Dirac House, Temple Back
Bristol BS1 6BE UK

Phone: 44 1 17 929 7481
Fax: 44 1 17 920 0979

techtracking@iop.org

**Additional copies of this publication are available from:**

Curran Associates, Inc.
57 Morehouse Lane
Red Hook, NY 12571 USA
Phone: 845-758-0400
Fax: 845-758-2633
Email: curran@proceedings.com
Web: www.proceedings.com

# TABLE OF CONTENTS

IOP Conf. Series: Materials Science and Engineering **143** (2016) 011001 doi:10.1088/1757-899X/143/1/011001

# Preface: International Symposium on Liquid Metal Processing and casting (LMPC 2015), *20-24 September 2015, Leoben, Austria*

We are delighted to come up with thirty eight (38) contributed research papers selected from the proceedings, as an output of the international Symposium on **Liquid Metal processing and casting** held in Leoben, Austria, from September 20-24, 2015. This event marks the eleventh such event in a highly successful series of symposia devoted to this important technical field. This symposium gathers industrial and scientific communities in the field of melting, refining and casting of special metals and alloys. Contributions cover a wide range of topics including vacuum arc and electro slag remelting, process modeling, innovative casting techniques, segregation, and liquid metal treatment and refining.

Papers from both the presentation and poster sessions are included in this symposium volume. It is divided into the following six topic areas: Electro slag Remelting, Vacuum remelting, Magneto-hydrodynamic applications in metallurgy, ingot casting, continuous casting, and special metallurgy.

More than half of the papers are directly concerned with mathematical modelling and numerical simulation, although experimental and industrial full-scale processing are also included. Steel, superalloys and titanium are the major materials investigated, while the processing of aluminium, silicon, copper and reactive metals are also addressed.

The conference co-chairmen are grateful to the scientific committee and all the reviewers of the papers for their efficient work done in a very short period of time. These proceedings would not have been possible without the tremendous help of Andreas Ludwig from the Montanuniversität Leoben. Thanks to TMS for promoting the conference to their members around the world. Finally, we would like to thank ASMET (*The Austrian Society for Metallurgy and Materials*) for helping with registration and local organization in particular Bruno Hribernik, Melanie Baumgartner and Yvonne Dworak.

**Abdellah Kharicha and R. Mark Ward,**
**Conference Co-chairs and Co-editors**
**E-mail:** abdellah.kharicha@unileoben.ac.at

IOP Conf. Series: Materials Science and Engineering **143** (2016) 011001 doi:10.1088/1757-899X/143/1/011001

# LMPC 2015

## The organisers wish to express their gratitude to the following companies:

special melting INTECO technologies

WACKER

CONSARC
An Inductotherm Group Company

TRANSVALOR

AMG

The Solution
ald

bürkert
FLUID CONTROL SYSTEMS

IOP Conf. Series: Materials Science and Engineering **143** (2016) 011002 doi:10.1088/1757-899X/143/1/011002

# Peer review statement

All papers published in this volume of *IOP Conference Series: Materials Science and Engineering* have been peer reviewed through processes administered by the proceedings Editors. Reviews were conducted by expert referees to the professional and scientific standards expected of a proceedings journal published by IOP Publishing.

IOP Conf. Series: Materials Science and Engineering **143** (2016) 012001 doi:10.1088/1757-899X/143/1/012001

# Electrochemical Aspects of the ESR Process

**Alec Mitchell** [a]
[a]Dept. of Materials Engineering, University of British Columbia, Vancouver, BC, Canada

E-mail: alec.mitchell@ubc.ca

**Abstract.** Current passage in the ESR process is accomplished by at least two Faradaic reaction systems, one at the electrode/slag interface and another at the slag/ingot interface. The nature of these reactions has been investigated and is reported to be a function of the alloy composition, current density and slag composition. In this report, data generated from both DC and line-frequency AC ESR operations is used to relate laboratory investigations on these reactions to results obtained in industrial practice. The operating reaction in the line frequency AC melting of simple alloy steels is found to be principally the reversible anodic oxidation of iron (Fe/Fe2+). In the equivalent DC process there is in addition the cathodic deposition of aluminium, resulting in the formation of substantial amounts of D-type alumina inclusions in the ingot. The Faradaic reactions involve energy exchange which should be taken into account in the process energy balance. They also are likely to change the interface physical characteristics such as interfacial tension. The data is extended to the case of low frequency ESR furnaces in which it is found that the extent of the reactions is strongly dependent on current density. It is concluded that the presence of Faradaic reactions in low frequency furnaces is unlikely to lead to quality problems in the alloys and applications for which these processes are used. This conclusion appears to be supported by industrial practice and leads to a further conclusion that low frequency practices in large ESR furnaces could potentially be replaced by simpler DC systems.

## 1. Introduction

### 1.1 *Electrochemical Reactions*

Both the electrode and ingot in the ESR process are electronic conductors whilst the liquid slag is ionic. For current to pass through the system, Faradaic reactions must take place at each of the slag/metal interfaces. The reactions have been studied by several workers. Beynon [1, 2] reports that the anodic reaction is the corrosion of the base metal, for example iron;

$$Fe = Fe^{2+} + 2e^{-} \quad (1)$$

The cathodic reaction is the deposition of Al or Ca, which dissolve in the base metal. In the case of iron-base alloys the deposition would be to Al; in the case of nickel-base alloys the deposition scheme would include Ca due to its small activity coefficient in liquid nickel;

$$Al^{3+} + 3e^{-} = [Al] \quad (2)$$

The above reaction scheme was confirmed by Etienne [3] and Schwerdtfeger [4]. However, Kawakami [4] postulates that the anodic reaction is the deposition of oxygen into alloy solution;

$$O^{2-} = [O] + 2e^{-} \quad (3)$$

As will be seen below, in the net result of the electrochemical scheme, reactions 1 and 3 are equivalent in terms of final ingot composition. The reaction schemes introduce composition effects into the electroactive surfaces. In the case of the anodic reaction, the formation of Fe2+ on the surface is equivalent to producing FeO in solution in the slag at that surface. The FeO content then produces an equivalent oxygen content in the metal at the surface as has been observed by Beynon [2] and examined by Wcislo [5]. The amount of oxygen introduced depends on the slag composition through the activity of FeO in solution in the slag. Lower oxide ion activity coefficients in the slag lead to lower oxygen potential in the metal [1]. The cathodic reaction also produces a solution of the cathodic deposit in the surface of the metal, as has been reported by Beynon [1] for Al into an iron electrode and Ca into a nickel electrode. This latter reaction has been used industrially to produce "master alloy" compositions of rare earth-nickel alloys in equipment very similar to an ESR furnace, using a slag of $CaF_2$ and RE oxide [7 – 13]. In steels, the solutions of anodic oxygen and cathodic aluminium combine when the two liquids mix in the ingot pool, to form the familiar "D"-type alumina inclusions which are characteristic of ESR processed alloys. The scheme is complicated by thermochemical reactions between the slag and metal in which alumina contained in the slag equilibrates with the alloy being processed in the droplet region which is not electroactive. The net composition of the ingot is hence a combination of the two processes. Bell's results [14] show that the Al and O are not in simple equilibrium with slag (when melting with AC and under an argon atmosphere) and that changes in electrode-to-ingot composition are not in the $Al_2O_3$ stoichiometry. Thomas [15] shows that initial nucleation of alumina is to very small particles but in about 1000s the size range changes to 0.1 – 20 microns. Inclusions nucleated with excess O over Al are round, grow as round particles and eventually agglomerate to collections of round particles as is generally found in ESR ingots.
The characteristics of the processes indicated in 1 and 2 are shown in Figures 1 – 4, where it can be seen that the polarisation extent and polarisation time are dependent on both slag composition and current density. The dependence lies in the solubility of FeO in the slag (as noted above) which in turn affects the formation of the surface layer of FeO formed during the polarisation period [16].

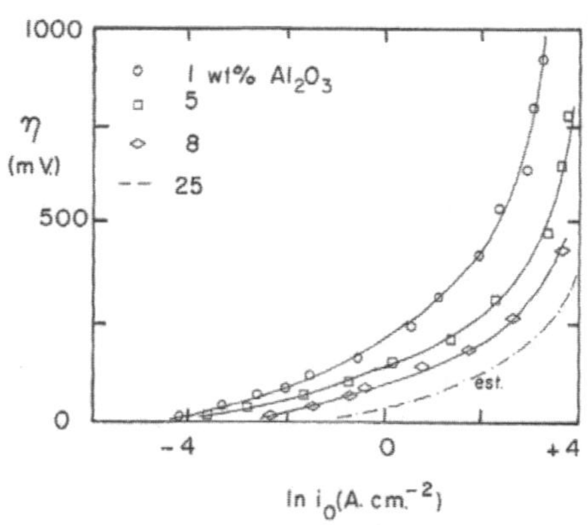

Figure 1

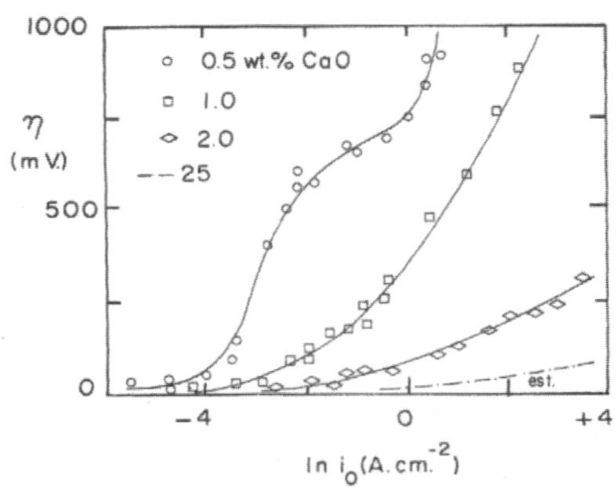

Figure 2

Figure 1. Cathodic polarisation curves for pure iron in $CaF_2 – Al_2O_3$ slags, 1520C.
Figure 2. Anodic polarisation curves for pure iron in $CaF_2 – CaO$ slags, 1520C.

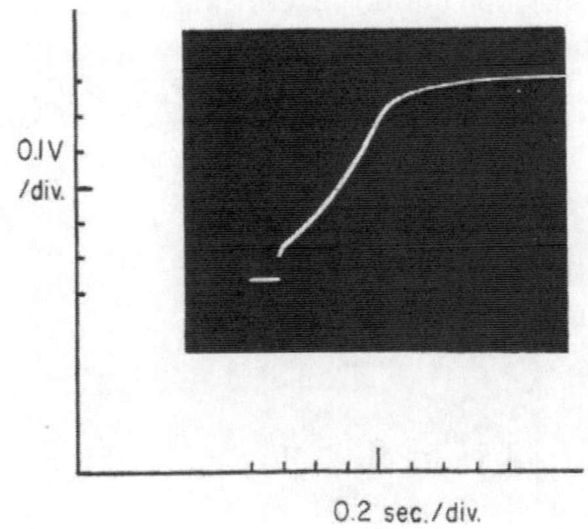


Figure 3

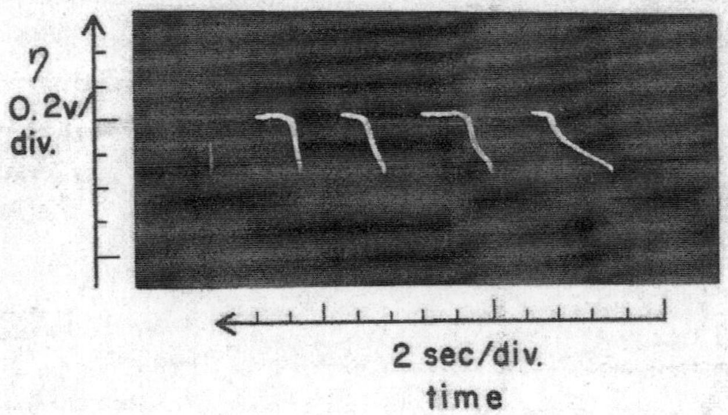


Figure 4

Figure 3. Apparent transition time in the anodic polarisation of pure iron in $CaF_2$ + 5wt% $Al_2O_3$ slag at a current density of $1A/cm^2$, 1520C.

Figure 4. Successive anodic pulses applied to a pure iron anode in $CaF_2$+ 2.5wt% $Al_2O_3$ slag (1520C) showing the gradual disappearance of the transition time when surface saturation is retained between pulses.

The reaction scheme for melting with alternating current has been described by Fraser [17, 18] who indicates that the anodic corrosion reaction 1 is reversible, leading to the system;

metal surface anodic ; $Fe - 2e^- = Fe^{2+}$

metal surface cathodic ; $Fe^{2+} + 2e^- = Fe$ (4)

As is seen from equation 5, at a typical current density found on the electrode of a large ESR furnace, $1A/cm^2$, the polarisation time is approximately 1sec, which implies that when the polarity is reversing at 60Hz, the surface does not reach the saturation polarisation potential. The process follows the laws applicable to a diffusion-controlled reaction in which the polarisation time, t, is a function of current density, $i_0$;

$I_o \times t^{1/2} = 1$ in the range of $0.5 - 2\ A.cm^{-2}$ (5)

The Faradaic reaction at 60Hz takes place over a very small distance across the surface making several atomic distances of iron alternately metallic and ionic. Any resulting bulk changes in composition with respect to anodic or cathodic reactions will be a reflection of the reversibility of this reaction. Further, if the reaction is not 100% reversible, the electrical balance is changed and there would be a resulting asymmetry in the current cycle equivalent to a small DC component in the circuit. The above reaction scheme assumes that current transfer at the electroactive interfaces is entirely Faradaic. In highly reducing slags (e.g.Ca + $CaF_2$) there is sufficient metal solubility to produce an electronic component in the transfer [44]. Such a mechanism would not require chemical changes and substantially change the effective conductivity of the system. It is unlikely that such a mechanism operates in the slags normally used in the ESR process, except in the melting of titanium alloys.

In alloy steels, the alloy elements can be oxidised at the anode according to their electrochemical potentials. An example of this effect is shown in Figures 5 and 6 where the chromium content is found to be significantly depleted in the anodic electrode surface and a corresponding oxygen diffusion into the electrode is observed.

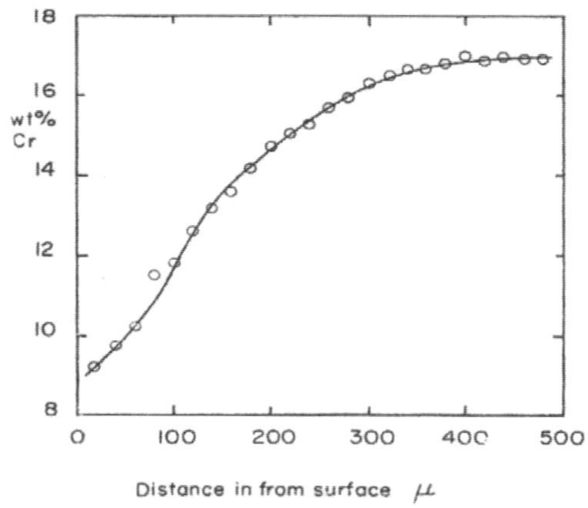


Figure 5

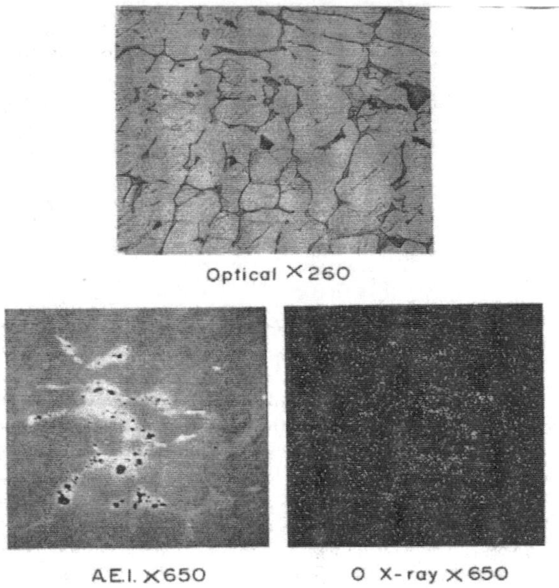


Figure 6

Figure 5. Chromium concentration gradient produced at the surface of an AISI430 stainless steel anode polarised for 1000sec at a current density of 360mA/cm$^2$ in a $CaF_2$+8wt% $Al_2O_3$ slag, 1425C.

Figure 6. Pure iron anode after polarisation for 1000sec, 1A/cm$^2$, $CaF_2$+5wt%$Al_2O_3$ 1520C .

The same effect can also be seen in low alloy steels with the removal of Mn and Si from the surface. The effect is due to the combined reactions of direct electrochemical erosion and the presence of the FeO layer on the anode surface, which raises the local oxygen potential. The effect cannot be seen in electrodes undergoing line frequency AC melting, which suggests that the anodic corrosion of iron is very efficiently reversible at this frequency and hence leads to only a very small residual asymmetry in the reaction.

1.2 *Relation to DC Industrial Processes*

The original ESR furnace, operated by the Firth-Sterling Corp in the USA in the 1940s was a DC furnace. Subsequent developments in the USSR employed AC power, but the reason for the change is unclear, possibly arising in the practical problem of obtaining reliable high-current DC power sources at that time in the USSR as opposed to the ready availability of large transformers with taps for voltage variation. Following that development, almost all ESR furnaces until recently have been operated with line frequency AC. For reasons of electrical efficiency several large ESR furnaces have recently been commissioned with low frequency AC power. It is to be noted that the waveforms in the power supplies used are not simple sine waves. The effect of these waveforms on the Faradaic system is unknown. Although most smaller furnaces have line frequency AC power, several producers have converted VAR furnaces with DC power supplies to produce ESR ingots [19]. Reports of the results from such furnaces are varied, with a variance apparently depending on furnace size. Small experimental DC ESR furnaces produce ingots which have an increased oxygen [20, 21] content relative to line frequency AC ESR, with electrode positive polarity ("DC+") resulting in a lower content than electrode negative polarity ("DC-") [2, 16]. An interesting feature of this research is that there is a marked difference in melting efficiency between the two polarities with DC+ being more efficient than DC-. The reason for this finding lies in the heat effect associated with the respective electrochemical reactions, the anodic reaction being more exothermic than that at the cathode due to the relative electrical resistance of the Faradaic boundary layers. The anode overpotential has been determined to be significantly higher than that at the cathode [2]. The reports of work on large industrial furnaces, however, do not show this marked difference. Dewsnap et al [19] for example report production results in 550mm diameter tool steel ingots showing a small difference in oxygen contents between the modes, but only a modest difference from line frequency AC practice. Similar results were obtained by Paar [22] in which the DC produced ingots (DC+) were found to have a quality level adequate for general sale. Results were reported for the DC+ melting of M250 steel by Udupa et al [21]. Chumanov et al [24], and Starosin et al [25] indicating satisfactory results for the properties of the ingots in industrial production. Kelley [26] reports results from a study of ESR and VAR melting of Hastelloy-X in which the difference between DC and AC melts is small at large ingot diameters and that the difference in oxygen contents between DC+ and DC-noted above is reversed as the ingot diameter is increased. The rationale for the satisfactory property results appears to lie in the fact that the alloys melted were not sensitive to the oxygen content present as very small alumina inclusions. This behaviour has been noted previously in respect of bearing steels, where ESR material with an oxygen content more than double that of VAR material had the same properties (Figure 7).

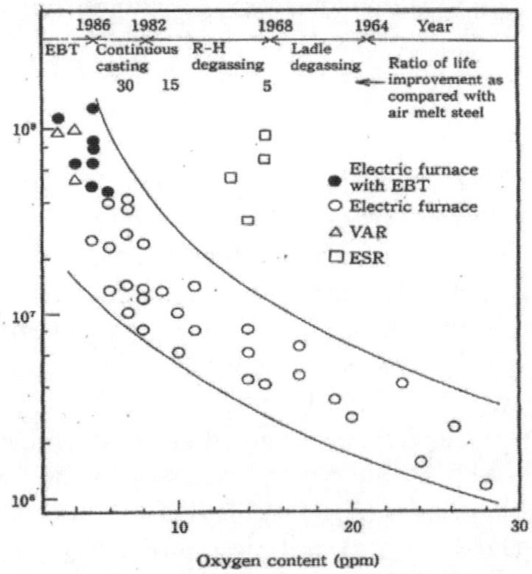


Figure 7. Relationship between the oxygen content and rolling contact fatigue in a bearing steel.

Juvonen [27] also shows that in the case of high strength steels, fatigue cracks are not initiated at inclusion sites which have a characteristic value of (projected area)$^{1/2}$ less than 40 microns implying that inclusions in the size range around 5 microns would not adversely properties in large forging ingots. The effect can be explained by the results of Paar et al [22] who measured inclusion size distributions in tool steel ingots melted by the various modes. They found that although there was a difference in the quantity of small oxide inclusions (<5 microns) between the DC modes and the AC modes, the larger inclusions (>20 microns) (probably coming from the electrode manufacture) were efficiently removed by all three modes of power. This report also indicates that the DC- mode results in a significant pick-up of aluminium in the ingot as might be expected from the cathode reaction.

The results shown in the literature, as reported above, indicate that for steels the changes in oxygen and aluminium cannot be accounted for on the basis of chemical equilibration with the alumina activity of the slag. Thermochemical data for iron solutions on the reaction;

$$Al_2O_3 = 2[Al] + 3[O] \quad (6)$$

show only a small variation of oxygen content with aluminium content in the applicable range at any given temperature but a significant temperature variation. The equilibrium value of oxygen content for a typical Al value in steel (0.03wt%) would be approximately 20ppm for a temperature of 1600C and an alumina activity in the slag of 0.4 (e.g.$40CaF_2;30CaO;30Al_2O_3$ slag). Since the oxygen contents found for DC melting of pure iron [1,2] greatly exceed these values, the Al and O contents of the ingots must have arisen principally by Faradaic reactions and not by chemical equilibration. This finding agrees with published models of the ESR reaction system [18] in which the electro-inactive droplets are a very small part of the overall scheme in spite of their large reaction surface area since their residence time in the slag is very small. The relation between Al and O is complicated in practical furnaces by the possibility of air oxidation. Unfortunately there are no published reports of industrial ingot melting by DC under a protective atmosphere to provide extension of these findings to large ingot practice. Additionally, it is not at present clear how the presence of a large alloy content of Al would influence the reactions, possibly by making the uptake of oxygen from the anodic system more efficient and hence leading to an increased inclusion content and higher total oxygen content in the ingot. As noted above in spite of the high currents used in ESR, the Faradaic reaction does not require the exchange of large amounts of material and the composition changes resulting directly from it are confined to the immediate vicinity of the electro-active surface. Subsequent diffusion and mixing results in the changes described above.

## 2. Applications of Electrochemistry to Processing

There are numerous reports of development in ESR processing by the application of a DC component during line frequency AC re-melting. In respect of sulphur contents, Whittaker et al [28, 29] carried out experiments on a small-scale ESR furnace in which they applied a biased AC supply giving an anodic potential to the electrode relative to the ingot, with the intention of increasing de-sulphurisation of the steel. The system resulted in an increase of 60 – 75% in sulphur removal. Wcislo et al [30] used an ancilliary graphite anode providing a potential between it and the ingot surface which increased the de-sulphurisation of a steel. Nyguyen et al [31] conducted laboratory experimentation in which they found that the sulphur removal process in ESR was an electrochemical system and hence postulated that the removal could be enhanced by an applied potential. Other workers report using ancilliary electrodes to achieve composition variations during ESR. Istomin et al [32, 33] and Chen [34] report using DC power to add V, Nb, and Ti to a steel during remelting, by electrochemical reduction of the respective oxides added to the slag. Goncharov et al [35, 36] report a similar study in which Mg, B, Ti, V, Nb, and Cr were all added during ESR by the same method, at current efficiencies close to 50%. Medovar et al [37] patented a system of AC ESR with ancilliary DC electrodes intended to exploit these reactions, including de-oxidation and removal of hydrogen from the ingot. Mankov et al [38] also studied the electrochemical deposition of elements during AC ESR with added DC current. Chu et al [39] used DC ESR in order to bring about a modification of the graphite structure in SG cast iron by introducing Mg, Ca or rare-earth oxides into the slag. Istomin et al [33] used AC ESR with an added DC+ component to introduce lithium into an alloy steel. They also used a system with a simple asymmetric AC power supply to accomplish the same effect. It is clear from all of the above studies that since the ESR is an electrochemical system, it is possible to change the ingot composition by a suitable biasing of the electrode and ingot. This procedure, however does not, however, appear to have been applied in a routine industrial manner. The reasons for this situation are complex, but principally lie in the necessary electromechanical complexity of the electrode system in the furnace and also in the absence of any suitable non-consumable electrode to supply a bias in the slag. It is evidently easier to accomplish the necessary composition changes by other means, such as suitable processing of the alloy before casting of the electrode or by simple mechanical additions to the slag during melting.

## 3. Implications for Large Ingot Technology

The rationale for using electrical supplies providing power at frequencies lower than the conventional line frequency is twofold. First, to reduce impedence losses in the system due to the necessary mechanical constraints of the furnace design. Second, to use a balanced 3-phase primary input to the furnace and so avoid the penalties of a severely unbalanced electrical load at the input to the site. It appears from both modelling [40] and industrial practice [41] reported at present that there are no penalties in the frequency change with respect to the process heat balance, ingot solidification structure, ingot final minimum temperature or melting efficiency. The chemical situation is, however, not yet clear. The bulk process composition changes (electrode to ingot) appear to be in accordance with existing practice and no undue loss of reactive elements is observed. Details of the ingot inclusion contents have not been reported and oxygen content changes are available in only one report, [42]. In this latter report, the oxygen content in an argon shielded melt (5Hz) of a large ingot is increased during the melting process, to the range of values observed by Paar [22] in DC melting of a similar alloy. In small scale experiments, Chang et al [43] report increasing contents of small (2 – 4 microns) and a corresponding increase in oxygen content in steel ingots as the frequency is decreased from 50Hz to 2.5 Hz when melting under argon. Some of the change in ingot oxygen content may be related to the larger time which is available in the larger ingot pools for the flotation of alumina inclusions.

The current density of the large ingot melts (electrode diameters of approximately 1500mm, ingot diameters of approximately 2000mm) is in the range of 0.8 to 1.5 A/cm2. At this current density, the polarisation time is in the range of 500 to 1500msec depending on the slag composition [1]. The implication is that at a frequency of 1Hz the electro-active surfaces are fully polarising on each cycle. If there was no metal flow through the surfaces, experience with 60Hz melting would suggest that the efficient reversal of the reactions would lead to only a very small residual composition change. Since the boundary layer established during the polarisation is effectively an additional electrical resistance in the overall circuit, a change in the interface heat balance should result. Also, since the formation of the polarised layer involves a significant local composition change, the interfacial surface tension would change periodically, probably leading to significant Marangoni flow in the surface liquids and possibly to interfacial turbulence. Both of these factors should be considered when constructing any mathematical model of the system behaviour. At frequencies lower than 1Hz the system would move progressively towards the DC case and due to diffusion of the polarised layer into the slag, would lead to a progressively reducing efficiency of the back reaction.

The above postulates do not include the effect of metal moving from one electro-active surface to the other, nor the probability that the interfaces are not flat and so can produce a non-uniform metal flow and non-uniform current density. The melting rate for a large ingot is typically 700g/s. The average droplet size is 5 -7g, leading to a rate of approximately 100 drops/s. If the polarisation time at 1Hz is 1s, there is hence a high probability that some of the drops will detach from the electro-active surface at times other than at the zero voltage point when the reactions have been fully reversed. The drops falling into the ingot will therefore be non-uniform in composition, carrying individually varying oxygen and aluminium contents into the ingot pool, to subsequently form the observed alumina inclusions. This effect should become more marked with reducing frequency. At this point in the development of low frequency technology we do not have enough information to clarify the above situation with an accurate model, so preventing us from making the important step of defining the frequency range at which the system behaviour becomes indistinguishable from the DC case. However, appears from the results on large ingots made at low frequencies that the reversal efficiency is high. It is to be noted in this regard that the waveforms of voltage applied in large low frequency power systems are far from the conventional sinewave. The influence of the various modes chosen in relation to polarisation is unknown.

A further point of uncertainty in the potential application of DC melting is the role of the slag cations. In steel melting, the reversal reaction depends on the efficient deposition of aluminium on the cathode surface. In turn, the aluminium must originate as a slag cation and hence in slags which do not contain aluminium the reversal reaction would involve calcium. Due to its low solubility in steel the reversal reaction may be very inefficient and lead to high residual oxygen contents in the ingot. Beynon [2] observes that the polarisation potential is very high in slags without aluminium content, hence increasing the reaction asymmetry even in 60Hz melting.

It has been established that a significant portion of the process current passes through the water-cooled mold. In addition, process variants have been proposed and piloted which utilise a "live mold" in which all of the process current is directed through the mold. In all of these cases, the Faradaic reaction required on the mold/slag interface is unknown, but probably plays a significant part in the composition variations between the electrode and the ingot.

## 4. Conclusions

The ESR process involves Faradaic reactions which can have significant composition effects. The reactions are accompanied by an energy exchange which leads to a thermal effect on the surfaces involved. Both of these points should be considered in chemical or thermal models of the process.

Considering the practical aspects, the above survey indicates that the potential for DC melting as opposed to low frequency practices depends on the alloy, the ingot size being processed and the ultimate property requirements of the product. The ingot content of small inclusions is increased by DC melting as is the oxygen content. However, the large inclusions are efficiently removed by DC melting and so the property requirements sought for large forging ingots should be met with in the same way as for line frequency melting. The same case would probably not hold for the high strength low alloy steels such as M250 which are very sensitive to small inclusions; nor would it likely hold for the nickel base superalloys for the same reasons. The place for DC melting is therefore restricted to large ingots of forging alloys where it would be advantageous to consider the technique in order to reduce equipment complexity.

## 5. Acknowledgement

The author is indebted to numerous colleagues in the industry who have contributed to the ideas expressed above. He is particularly grateful for discussions with Dr K Kajikawa of the Japan Steel Works and to Mr R Roberts of the Consarc Corporation for their many useful comments.

## 6. References

[1] Beynon G, PhD Thesis, University of British Columbia, 1971
[2] Beynon G, and Mitchell A, 1970 Met Trans, 1, pp 1839 - 1844
[3] Etienne M, University PhD Thesis of British Columbia 1969
[4] Schwerdtfeger K, Norbert N and Detief K, 1982 Archiv fur das Eisenhuttenwessen, 53, pp 463 – 468
[5] Wcislo Z, Fedko J and Krucincki M, 1987 Freiberger Forschungshefte, B259, pp 27 - 36
[6] Kawakami M et al, 1977 Tetsu-to-Hagane, 63, pp 2161 - 2171
[7] Cheng T et al, 2010, Xiyou Jinshu Yu Yingzhi Hejn, 38, pp30 – 33
[8] Takeda K, Kensuke H, Ishizaka K, 2009, Jap. Pat 2029228038 A20091008,
[9] Moriyasu T, Masashi K, Nozomi K, Hitoshi M, 1990 Jap Pat. Application 02-071669
[10] Katsuhisa I et al, 1986 Jap Pat JPS61291988(A)
[11] Katsuhisa I et al, 1985 Jap Pat Application, 60-112652
[12] Gocharov A E, 1990 Probl. Spets Elektromet, 4, pp32 – 35

[13] Kawakami M et al, 2007"Innovations in Titanium Technology", eds M N Gugor et al, publ TMS-AIME, 2007, pp 137 – 145
[14] Bell M and Mitchell A, 1972 Can Met Q, 11, pp 363 - 368
[15] Zhang L, Pluschkell W and Thomas B G, Steelmaking Conference Proc. 2002, 85, pp 463 – 476
[16] Kato M et al, 1983 Trans ISIJ, 23, pp 618 - 627
[17] Fraser M E and Mitchell A, 1975 "Metal-Slag-Gas Reactions and Processes" eds Z A Fouroulis et al, publ. Electrochemical Soc., pp 199 - 205
[18] Fraser M E, PhD Thesis University of British Columbia, 1974
[19] Dewsnap P and Schlatter R, 1975 Proc. 5th International Symp on Electroslag, ed GK Bhat and A Simkovitch, publ AVS New York, 1975, pp 91 – 114.
[20] Kawakami M, Takenaka T and Ishikawa M, 2002 J Ironmaking and Steelmaking,29 pp 287 – 292
[21] Udupa K R, Sastry D H and Iyengar G K N, 1990 Bull Mater Sci, 13, pp 351 – 364
[22] Paar A et al, 2013 Proc. LMPC 2013, pp 29 – 36
[23] Paar A et al, 2014 Steel Research Intl., 2014, 85, pp 570 – 578
[24] Chumanov I V and Pyatygin D A, 2006 Izvestiya Vysshikh Zavedenii Chernaya Metallurgiya, 3, pp 22 – 25
[25] Starosin B M et al, 1998 Stal' 3, pp 26 – 28
[26] Kelly T N, Junker D A and Chen R C H, 1978 Proc.3rd Internatioonal Symp. On ESR, Pittsburgh Pa, ed G K Bhat and A Simcovitch, publ Mellon Institute, pp 125 – 141.
[27] Juvonen H PhD thesis, Technical University of Helsinki 2004
[28] Whittaker D A, Ward R G and Purdy G R, 1971 J.Elektroshlak Pereplavov, pp 183 – 202
[29] D A Whittaker, PhD Thesis, McMaster University 1972
[30] Wcislo Z and Krucinski M, 1985 Hutnik, 52, pp 276 – 281
[31] Nguyen Q M and King T B,1979 Met. Trans B, 10B, 623 – 629.
[32] Istomin S A et al, 1989 Problemy Spetsial'noi Elektrometallurgii,2, pp 8 – 12
[33] Istomin S A et al, ibid, pp 12 14
[34] Chen C X and Gao R F, ibid, pp 42 – 47
[35] Goncharov A E, 1990 Prob. Spets. Elektrometallurgii, 4, pp 32 – 35
[36] Goncharov A E, 1969 Trudy Inst Metallurg. Sverdlovsk, 18, pp 261 - 273
[37] Medovar B I et al, German Pat., 1979 – 2908507, 1979
[38] A I Mankov, A Goncharov E and Lepinskikh B M, 1977 Electrochem. Term. Sostava Ionnykh Rasplavov, ed Yu K Delimarskii, pp 67 – 69.
[39] Chu S, Liu H and Wu S, 1989 Huagong Yeijin, 10, pp 15 – 23
[40] Kajikawa K, Mitchell A and Kelkar K M, 2014 Proc IFM 2014, publ. Japan Steel Castings and Forgings Association, Tokyo, 2014, pp 124 - 128
[41] Choudhury A, Jauch R and Tince F, 1980 Stahl u Eisen, 1980, 100, pp 1012 – 1018
[42] Tezuka M et al, 2014 Proc IFM 2014, publ. Japan Steel Castings and Forgings Association, Tokyo, 2014, pp 90 – 94.
[43] Chang L et al, 2009, J Iron and Steel Research International, 16, pp 7 - 11
[44] Haarberg G M, Thonstad J, Pietrzyk S, Egan J J, 2002, J. Light Metals, pp 1083 -1089

IOP Conf. Series: Materials Science and Engineering **143** (2016) 012002 doi:10.1088/1757-899X/143/1/012002

# Manufacturing of 9CrMoCoB Steel of Large Ingot with Homogeneity by ESR Process

**D S Kim, G J Lee, M B Lee, J I Hur and J W Lee**

Casting & Forging BG, Doosan Heavy Industries & Construction,
22, DoosanVolvo-ro, Seongsan-gu, Changwon, Gyeongnam, 642-792, Korea

E-mail: dong-soo.kim@doosan.com

**Abstract.** In case of 9CrMoCoB (COST FB2) steel, equilibrium relation between [B]/[Si] ratio and $(B_2O_3)/(SiO_2)$ ratio is very important to control [Si] and [B] in optimum range. Therefore, in this work, to investigate the thermodynamic equilibrium relation between [B]/[Si] ratio and $(B_2O_3)/(SiO_2)$ ratio, pilot ESR experiments of 9CrMoCoB steel were carried out using the $CaF_2$-$CaO$-$Al_2O_3$-$SiO_2$-$B_2O_3$ slag system according to change of Si content in electrode and $B_2O_3$ content in the slag. Furthermore, through the test melting of the 20ton-class ESR ingot, the merits and demerits of soft arcing were investigated. From these results, it is concluded that oxygen content in the ESR ingot decrease with decreasing $SiO_2$ content in the slag, relation function between [B]/[Si] ratio and $(B_2O_3)/(SiO_2)$ ratio derived by Pilot ESR test shows a good agreement as compared to the calculated line with a same slope and soft arcing makes interior and surface quality of ingot worse. With the optimized ESR conditions obtained from the present study, a 1000mm diameter (20 tons) and 2200mm diameter (120ton) 9CrMoCoB steel of the ESR ingot were successfully manufactured with good homogeneity by the ESR process.

## 1. Introduction

For the USC(Ultra Super Critical) HIP(High and Intermediate Pressure) rotor, the high cleanliness and homogeneous ingots without interior defects such as non-metallic inclusion, porosity and segregation are required. For these demands, most turbine manufacturers have developed and applied an ESR (Electro Slag Re-melting) process to produce 9~12% Cr heat resistance steels.

Among ESR conditions, slag composition and melt rate are the important factors because slag composition gives the remarkable influence on the content of major elements and cleanliness and melt rate also has a great effect on segregation. Therefore it needs to decide carefully slag compositions and melt rate [1].

For the 9CrMoCoB steel(COST FB2) used up to as 630°C class USC power plant materials, it is very important to control the solute element such as Si, Nb and B in homogeneous and optimum range because the exchange reaction of the solutes between slag and molten metal occurs due to their thermodynamic equilibrium during ESR process.

Furthermore, for high cleanliness in the ESR ingot, especially to minimize [Al], they have applied argon gas to prevent slag from oxidizing not to use Al deoxidation during ESR. However, recently, the effect of soft arcing created by ionization of argon gas on the quality of the ESR ingot also has been discussed particularly in the case of large ESR ingot manufacturing. Therefore, in this paper, relation function between [B]/[Si] ratio and $(B_2O_3)/(SiO_2)$ ratio derived by Pilot ESR test is introduced and verified by 20~120 ton ESR melting. Furthermore, the effect of soft arcing on the surface and the inner quality of the ingot is discussed.

### *1.1. Pilot ESR experiments*

The re-melting conditions of Pilot ESR experiments and the composition of the ESR electrode are shown in Tables 1 and 2, respectively. The experiments were carried out using the $CaF_2$-CaO-$Al_2O_3$-$SiO_2$-$B_2O_3$ slag system with varying $B_2O_3$ and $SiO_2$ content to investigate the effect of $SiO_2$ and $B_2O_3$ content on cleanliness, behavior of major element content. The metal samples were taken from the axial center of the ESR ingot and the molten metal and slag samples were taken just before hot-topping stage.

**Table 1.** Re-melting conditions of Pilot ESR experiments

| Classifications | Detailed Conditions | |
|---|---|---|
| Electrode | Composition | 9CrMoCoB steel |
| | Size/Weight | φ130mm/100kg |
| Slag | System | $CaF_2$-CaO-$Al_2O_3$-$SiO_2$-$B_2O_3$ |
| | Weight | 5kg |
| ESR melting | Starting type | Cold start |
| | Mold dia. | φ180mm |
| | Fill ratio | 0.72 |
| | Atmosphere | Argon gas |

**Table 2.** Chemical composition of Pilot ESR electrode (mass %)

| C | Si | Mn | Ni | Cr | Nb | B |
|---|---|---|---|---|---|---|
| 0.12 | 0.04~0.27 | 0.67 | 0.16 | 9.2 | 0.07 | 0.010 |

### *1.2. 20 Ton ESR Ingot Productions*

To manufacture a 20 ton of 9%CrMoCoB of the ESR ingot, the ESR electrode was manufactured in a 30ton EAF and ladle refining furnace.

Detailed ESR conditions are listed in Table 3. ESR slag composition was selected based on the Pilot ESR experiment to obtain the optimum Si, Nb, B content and a low Al content. The melt rate was kept low and stable as much as possible. Table 4 shows the electrode chemical composition target for the 20 ton of 9%CrMoCo steel of the ESR ingot.

**Table 3.** Re-melting conditions of 20ton ESR operation

| Classifications | Detailed Conditions | |
|---|---|---|
| Electrode | Size/Weight | 800mm D, 20 tons |
| Slag | System | $CaF_2$-CaO-$Al_2O_3$-$SiO_2$-$B_2O_3$ |
| ESR melting | Starting method | Hot start |
| | Voltage swing | 4~6V |
| | Electrode surface | Shot blast |
| | Atmosphere | Argon gas |
| | Fill ratio | 0.6~0.8 |

**Table 4.** Chemical composition of 20ton electrode (mass %)

| | C | Si | Mn | Ni | Cr | Mo | Al | Co | Nb | B(ppm) |
|---|---|---|---|---|---|---|---|---|---|---|
| Min. | 0.12 | - | 0.65 | 0.12 | 9.0 | 1.40 | - | 1.15 | 0.05 | 80 |
| Max. | 0.14 | 0.10 | 0.75 | 0.20 | 9.6 | 1.60 | 0.010 | 1.35 | 0.07 | 120 |

The metal samples were taken from the axial surface, middle and center of the product to check the behavior of main elements such as Si, B and Nb content. Molten slag and metal samples for chemical analysis were taken at the end of hot topping to verify relation function between [B]/[Si] ratio and $(B_2O_3)/(SiO_2)$ ratio derived by Pilot ESR experiment.

*1.3. 120 Ton ESR Ingot Productions*

To manufacture the 120 ton-class 9%CrMoCoB of the ESR ingot, the ESR electrode was manufactured in a 100ton EAF and ladle refining furnace. Detailed ESR conditions are listed in Table 5. Among these conditions, especially, voltage swing set-point and control values for voltage swing control were optimized to keep the immersion depth as shallow as possible and prevent the soft arcing. Chemical composition of electrode is shown in table 6.

**Table 5.** Re-melting conditions of 120ton ESR operation

| Classifications | Detailed Conditions | |
|---|---|---|
| Electrode | Size/Weight | 2200mm D, 120 tons |
| Slag | System | $CaF_2$-CaO-$Al_2O_3$-$SiO_2$-$B_2O_3$ |
| ESR melting | Starting method | Cold start |
| | Voltage swing | 8~10V |
| | Electrode surface | Shot blast |
| | Atmosphere | Argon gas |
| | Fill ratio | 0.6~0.8 |

**Table 6.** Chemical composition of 120ton electrode (mass %)

| | C | Si | Mn | Ni | Cr | Mo | Al | Co | Nb | B(ppm) |
|---|---|---|---|---|---|---|---|---|---|---|
| Min. | 0.12 | - | 0.30 | 0.10 | 9.0 | 1.40 | - | 1.10 | 0.04 | 80 |
| Max. | 0.15 | 0.10 | 0.45 | 0.20 | 9.5 | 1.60 | 0.010 | 1.30 | 0.06 | 110 |

As the same as the method with re-melting process of the 20ton-class ingot, molten slag and metal samples were also taken at the end of hot topping to verify the relation function between [B]/[Si] ratio and $(B_2O_3)/(SiO_2)$ ratio derived by Pilot ESR experiment.

## 2. Results and Discussion

*2.1. Pilot ESR experiments*

In the case of 9%CrMoCoB steel containing Si, Mn, Nb and B, B is ought to oxidize first, followed Si, Mn and Nb. These elements have different chemical affinities for oxygen and their oxides have dissociation pressures. Because of this, during ESR, there may exist some chance of redox reaction taking place [2].

*2.1.1 Effect of Si content in electrode on the behaviour Al, Nb content in ingot*

First of all, the effect of Si content in electrode on Al content in the ingot was investigated. The equilibrium Si between slag and liquid metal pool can be represented as the following reaction (1). In the reaction (1), it can be seen that Al content can be controlled by using $SiO_2$ containing slag system [3].

$$3[Si] + 2(Al_2O_3) = 4[Al] + 3(SiO_2) \qquad (1)$$

The behavior of Al content according to Si content (0.08~0.27%) is shown in figure 1. Al content decreases with decreasing Si content and it can be suggested that Al content can be controlled below 0.005% when Si content is less than 0.1%.

In sequence, change of Nb recovery with Si content is also shown in figure 2. During ESR, equilibrium relation between Si and Nb can be represented as following reaction (2). Recovery of Nb decreases with the decrease in Si content and it can be seen that recovery of Nb becomes below 80% when Si content is less than 0.06%.

$$5(SiO_2) + 4[Nb] = 2(Nb_2O_5) + 5[Si] \quad (2)$$

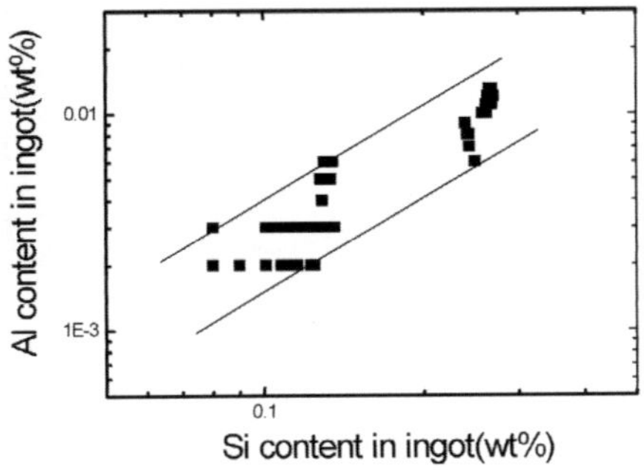


**Figure 1.** The effect of Si content on Al content

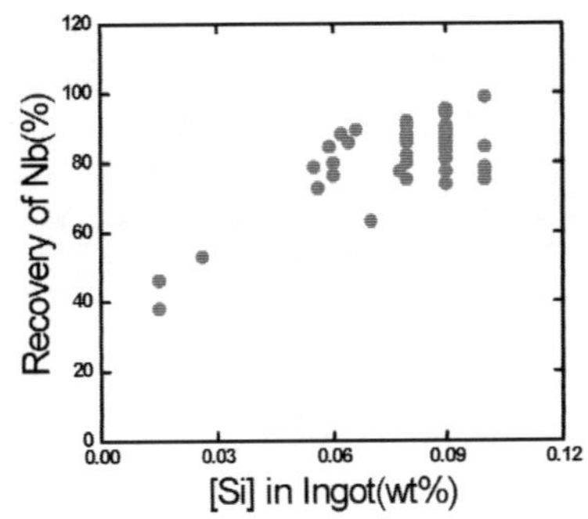


**Figure 2.** The effect of Si content in electrode on Nb recovery in the ingot

*2.1.2 Effect of $SiO_2$ content in slag on ESR Ingot quality*

The effect of $SiO_2$ content in slag on oxygen content in the ingot was investigated and the result is shown in figure 3. With increasing $SiO_2$ content in slag, oxygen content in the ingot increases because equilibrium oxygen content increases with an increase in $SiO_2$ activity as shown in equilibrium reaction (3) and the calculated line shown in figure 3 [4].

Change of cleanliness in the ingot according to slag basicity ($CaO/SiO_2$) is also shown in figure 4. In this figure, cleanliness means area ratio of non-metallic inclusion per unit area. As shown in figure 4, cleanliness improves with an increase in slag basicity. The reason seems to be attributed that equilibrium oxygen content decrease with increasing slag basicity.

$$\begin{aligned} &SiO_2 = \underline{Si} + 2\underline{O} \\ &\log K_3 = \frac{30110}{T} + 11.40 \\ &\log[\%O] = \tfrac{1}{2}(\log a_{SiO_2} - \log f_{Si}[\%Si] - 2\log f_O + \log K_3) \end{aligned} \quad (3)$$

The recovery of major element such as Si, Mn and Nb according to slag basicity was evaluated and the result is shown in figure 5. The equilibrium of the Si, Nb and Mn between slag and liquid metal pool can be represented respectively as reaction (2) ~ (4).

$$(SiO_2) + 2[Mn] = 2(MnO) + [Si] \quad (4)$$

First of all, in the case of Si, it can be seen that the recovery of Si increases with a decrease of slag basicity (increase $SiO_2$ content) in slag and Si pick-up occurs and then the recovery of Si becomes over 100% when the $SiO_2$ content in slag higher than that equilibrated with the Si content in electrode. On the other hand, in contrast to Si recovery, it can be seen that the recovery of Mn and Nb decrease with a decrease of slag basicity in slag and oxidation loss of Mn, Nb occurs and then the recovery of Mn and Nb greatly decreases when the $SiO_2$ content in slag higher than that equilibrated with the Si content in electrode.

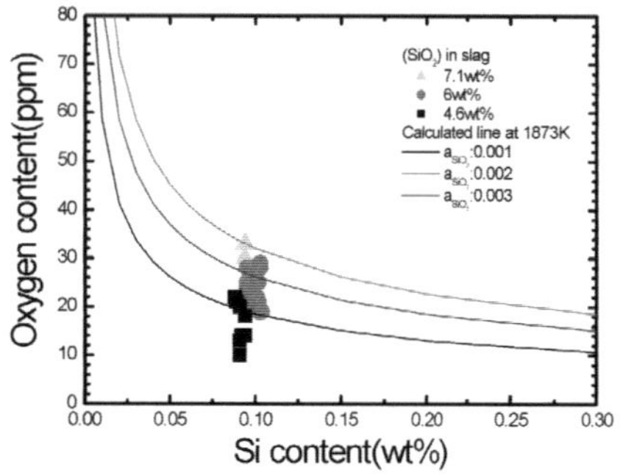


**Figure 3.** Effect of SiO2 content in slag on oxygen content in the ingot

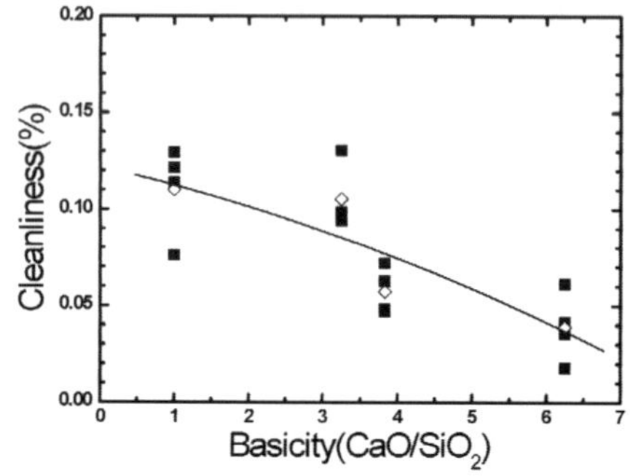


**Figure 4.** Effect of slag basicity on the ingot cleanliness

(Cleanliness(%)= (Total area of inclusion/Unit area) x 100)

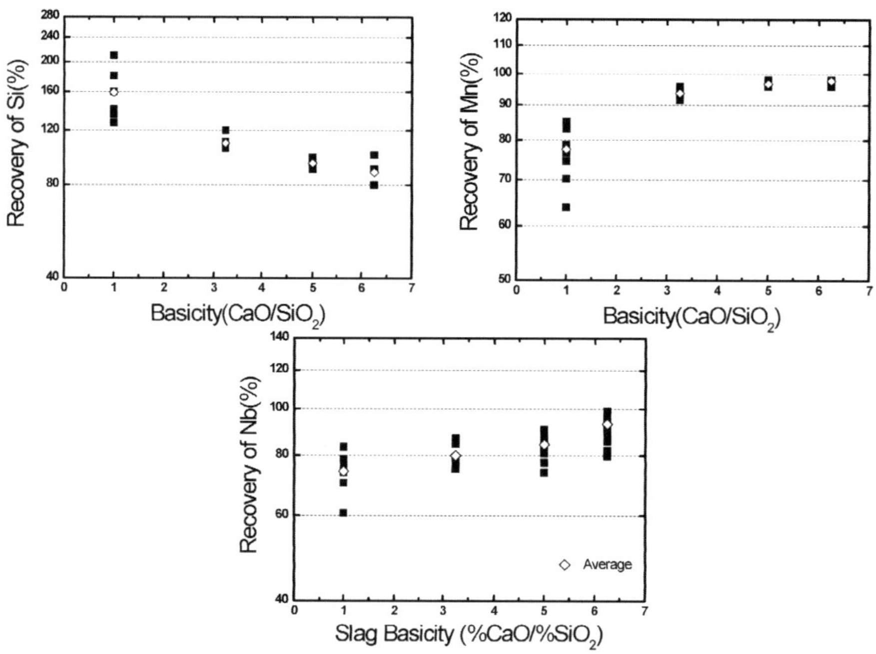


**Figure 5.** Effect of slag basicity on Si, Mn and Nb recovery during ESR

*2.1.3 Equilibrium relation between [B]/[Si] ratio and ($B_2O_3$)/($SiO_2$) ratio during ESR*

The equilibrium B between slag and liquid metal pool can be represented as reaction (5) [2]. In reaction (5), it can be seen that equilibrium B content in molten metal depends on $B_2O_3$ content in slag and Si content in electrode at the constant $SiO_2$ content in slag. Therefore, in order to control B content, the addition of $B_2O_3$ in to the slag is required to establish a thermodynamic equilibrium between the slag and molten metal. At the different Si content (0.08~0.27%) in electrode, the influence of $B_2O_3$ content in the slag on B content in the ingot is represented in figure 6. Equilibrium B content increases with an increase in $B_2O_3$ content in the slag and Si content in electrode.

Reaction (5) can be represented to relation between $log([B]^4/[Si]^3)$ and $log((B_2O_3)^2/(SiO_2)^2)$ as following equation (6).

IOP Conf. Series: Materials Science and Engineering **143** (2016) 012002 doi:10.1088/1757-899X/143/1/012002

$$4[B] + 3(SiO_2) = 2(B_2O_3) + 3[Si]$$
$$logK_5 = \frac{5113.9}{T} + 4.763 \quad (5)$$

$$log\frac{[\%B]^4}{[\%Si]^3} = log\frac{(\%B_2O_3)^2}{(\%SiO_2)^3} - log\frac{f_B^4\gamma_{SiO_2}^3}{f_{Si}^3\gamma_{B_2O_3}^2} - logK_5 \quad (6)$$

Experimental results of Pilot ESR and calculation result using the equation (6) for relation between $log([B]^4/[Si]^3)$ and $log((B_2O_3)^2/(SiO_2)^2)$ are shown in figure 7. As shown in figure 7, experimental result shows a good linearization tendency according to calculated line with a same slope and relation function between $log([B]^4/[Si]^3)$ and $log((B_2O_3)^2/(SiO_2)^2)$ of experimental data can be represented as following equation (7).

$$log(\frac{[\%B]^4}{[\%Si]^3}) = 1.06log(\frac{(B_2O_3)^2}{(SiO_2)^3}) - 2.04 \quad (7)$$

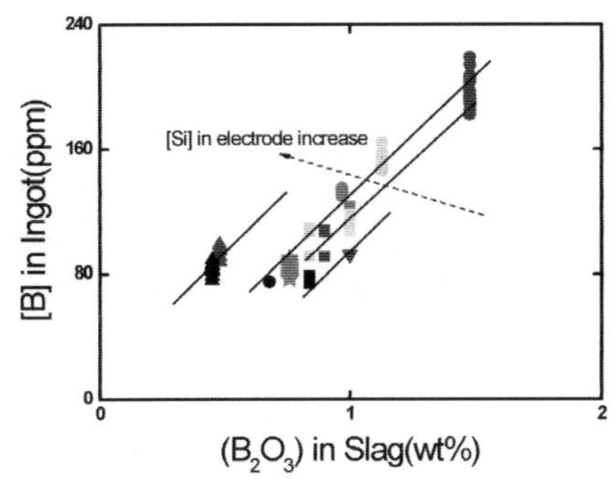


**Figure 6.** Effect of Si content in electrode on B content in ingot with $B_2O_3$ content in slag

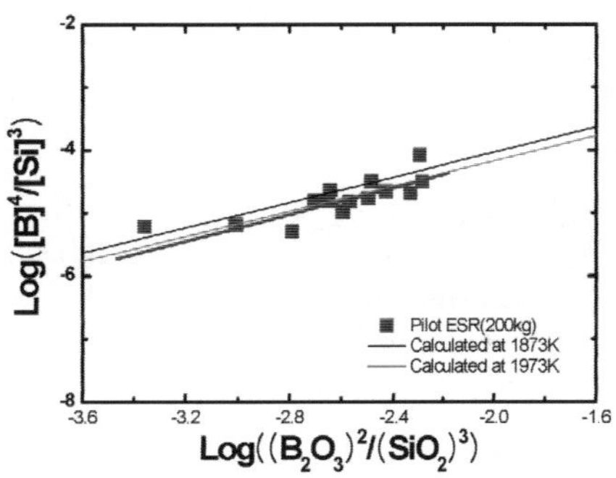


**Figure 7.** Equilibrium relation between [B]/[Si] ratio and $(B_2O_3)/(SiO_2)$ ratio in the experimental and calculated data

From these results obtained by Pilot ESR experiments, Si content in electrode and $SiO_2$, $B_2O_3$ content in the slag are optimized to control major alloy element such as Al, Mn, Nb, B in target range and improve ingot cleanliness.

*2.2. 20 Ton ESR Ingot Productions*

The ESR operation with the conditions shown in Table 3 was conducted. Two electrodes were re-melted but ESR operation results and surface quality were very different. Surface shape of one ingot was relatively smooth but in the other ingot soft arcing was occurred during ESR and deep ripple was generated on the whole surface. It is commonly known that soft arcing is generated between the electrode and the slag under argon gas atmosphere during ESR. The replacement of an air atmosphere with argon gas renders the ESR process more susceptible to instabilities caused soft arcing between the electrode and the slag due to ionization of the argon. During re-melting, ionization of the argon is indicated by the presence of a soft arcing between the electrode and the slag and a dramatic increase in noise from both the arcing and the power supply due to large voltage fluctuations [5].

Therefore, ESR operation results for each melts were compared and interior/exterior quality of the ingot was investigated according to the generation of soft arcing.

Firstly, ESR operation results such as melting current, voltage and melt rate were compared as shown in figure 8 and 9. As shown in figure 8, started melting after 10 minutes later, in the melt generated deep ripple on the ingot surface, melting voltage was abruptly increased, melting current was decreased simultaneously and then melt rate was increased greatly. Furthermore, melting voltage is higher but current is lower than that in other melt even though melt rate is almost same. This abnormal behavior of melting current and voltage was caused by soft arcing due to ionization of the argon between the electrode and the slag. The reason why soft arcing generates is attributed to the high voltage swing set-point in this work.
In additions, melting pattern on the soft arcing appears to resemble the Arc Slag Remelting process developed by the Paton institute [6]. Interestingly, as shown in figure 9, melting power(energy consumption) in melt with soft arcing is about 30% less than that in melt without soft arcing and this is in approximate agreement with results quoted by Paton et al [6].
In sequence, the effect of soft arcing on interior/exterior quality of the ingot was also investigated and the results are shown in figure10~12 respectively. In the case of surface quality of the ingot as shown in figure 10, it can be seen that slag skin becomes thicker and deep ripple is formed on the ingot surface when soft arcing is occurred. Cleanliness in middle part of the ingot height was compared according to generation of soft arcing and the result is shown in figure 11. It can be seen that cleanliness in the ingot without soft arcing is better than that in the ingot with soft arcing.

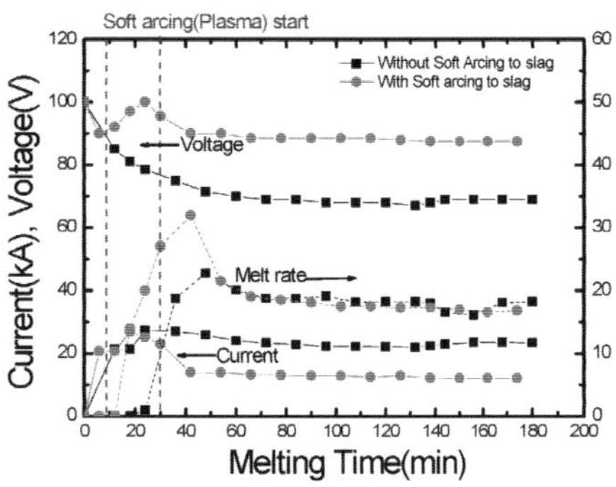


**Figure 8.** Behavior of melting current and voltage according to generation of soft arcing at same melt rate

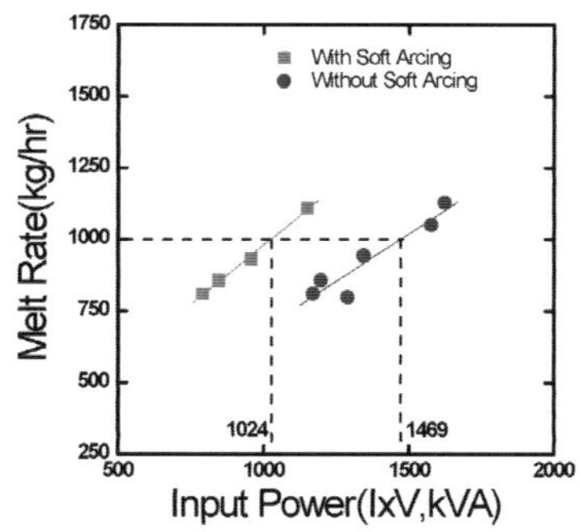


**Figure 9.** Comparison of melting power according to generation of soft arcing

**Figure 10.** Comparison of ingot appearance with the generation of soft arcing

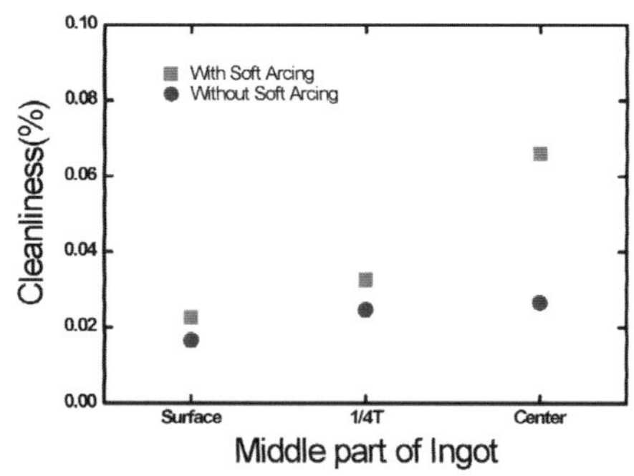


**Figure 11.** Comparison of ingot cleanliness with the generation of soft arcing

To verify the reason that cleanliness of the ingot with soft arcing is worse, specimen taken from the center in middle part of the ingot height with soft arcing was analysed by optical microscope and SEM-EDAX. As shown in figure 12, the reason why cleanliness of the ingot with soft arcing becomes worse seems to be attributed to the non-metallic inclusion and inclusion type is identified as $Al_2O_3$ rich inclusion. Accordingly, it can be estimated that when soft arcing generates, thickness of slag skin becomes thicker and depth of molten metal becomes deeper and then non-metallic inclusion could be caught more at the center part of the ingot.

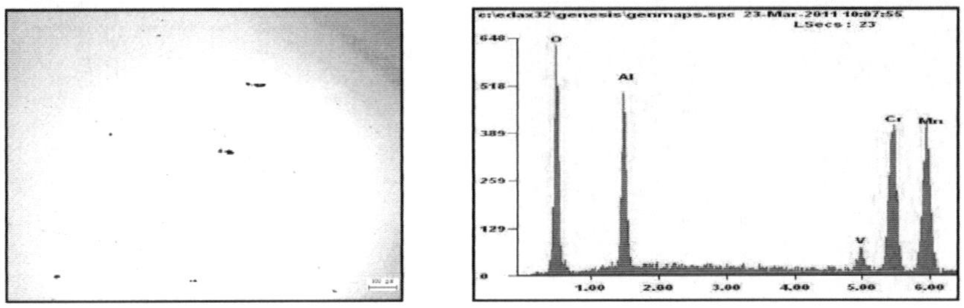


**Figure 12.** Analysis result of specimen taken from the ingot with soft arcing by OM, SEM-EDAX

As mentioned above, it can be conclude that soft arcing has a bad effect on both interior and exterior ingot quality even though melting power can be saved about 30%. Consequently, in order to get a satisfactory ingot quality, it is very important to design the optimum conditions of ESR operation to prevent soft arcing under argon gas atmosphere.

Finally, behavior of major elements in the ingot without soft arcing was analysed and analysis results are shown in figure 13. As shown in figure 13, it can be seen that the content of major elements in the ingot is uniform and nearly similar to alloy content in the electrode. Therefore, it can be concluded that slag composition was nearly close to the equilibrated composition with that of the electrode and very suitable for obtaining the optimum Si, Nb, Mn, B content and a low Al content less than 0.005%.

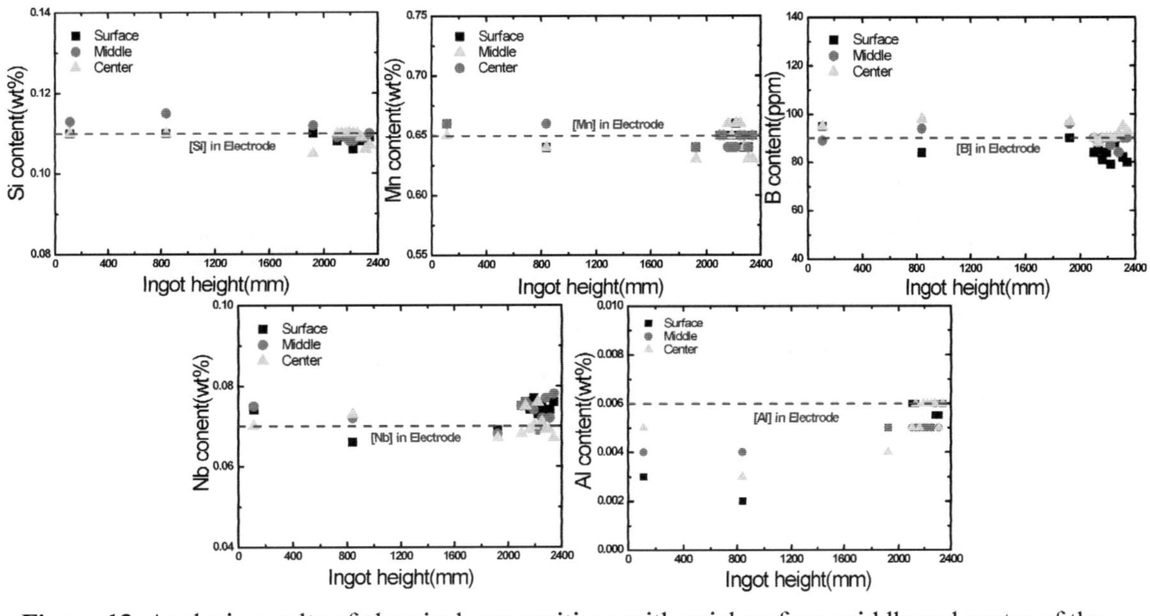


**Figure 13.** Analysis results of chemical compositions with axial surface, middle and center of the 20ton 9CrMoCoB Steel ESR ingot

*2.3. 120 Ton ESR Ingot Productions*

The ESR operation with the conditions shown in Table 5 was conducted. Especially, voltage swing set-point was optimized and voltage swing was controlled as target value to prevent soft arcing and obtain a smooth ingot surface. As shown in figure 14, 120ton 9CrMoCoB Steel ingot was successfully re-melted with very smooth surface and flat hot top. Metal samples for analysis of chemical composition were also taken from the ingot as shown in figure 15.

**Figure 14.** Appearance of 120ton-class 9CrMoCoB Steel of ESR ingot

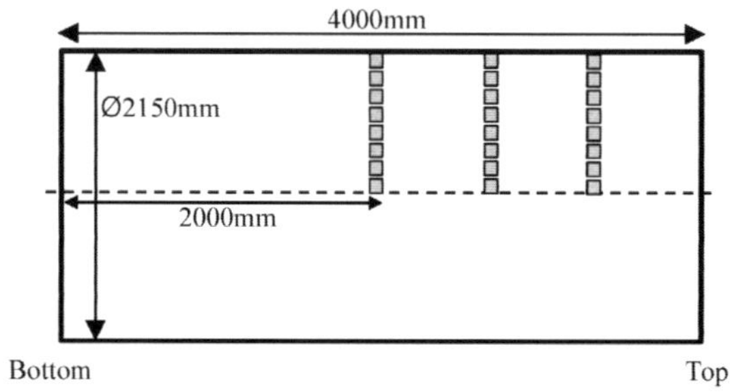


**Figure 15.** Point of chemical analysis of the 120ton-class 9CrMoCoB Steel of ESR ingot

Behaviors of chemical composition of major elements in the ingot are shown in figure 16. As shown in figure 16, it can be seen that the content of Si, Mn, Nb and B is nearly constant and similar to the content of elements in the electrode. Therefore, it can be concluded that slag composition was nearly close to the equilibrated composition with that of the electrode and very suitable for obtaining the optimum Si, Nb, Mn, B content. Furthermore, C, Mn and B having a low distribution coefficient are not segregated and show homogeneity in axis and thickness direction.

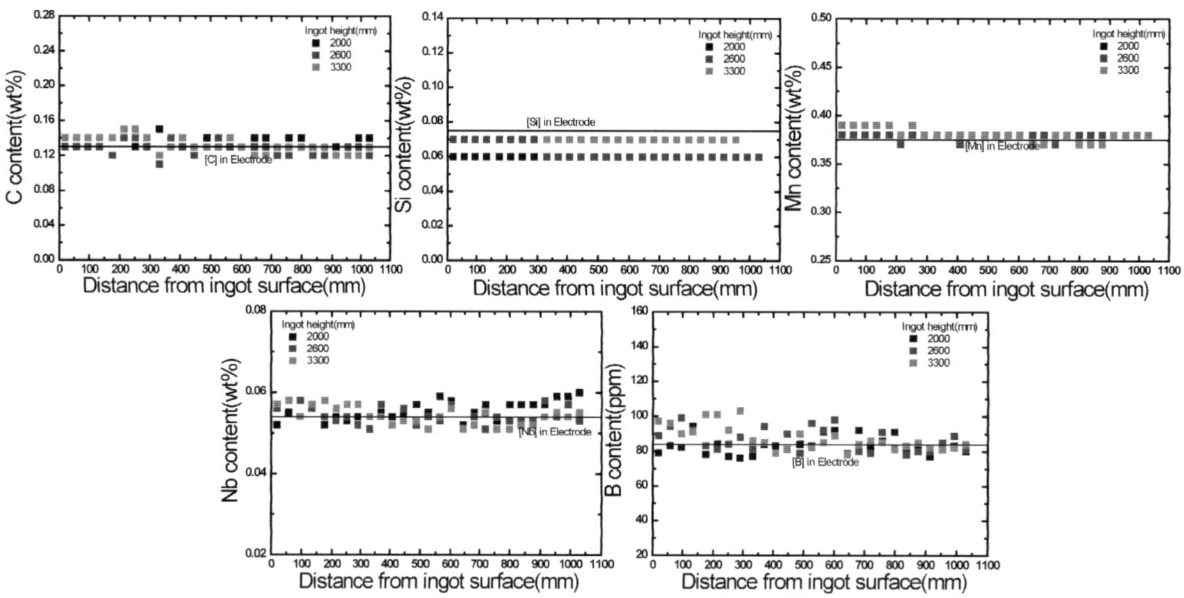


**Figure 16.** Analysis results of chemical compositions of the 120 ton-class 9CrMoCoB Steel of ESR ingot

Finally, content of B, Si and $B_2O_3$, $SiO_2$ in each molten metal and slag taken at the end of hot topping was analysed and the result are plotted on the graph of relation between [B]/[Si] ratio and $(B_2O_3)/(SiO_2)$ ratio derived by Pilot ESR experiment as shown in figure 7 and represented with results of 10ton,

20ton and 70ton re-melting in figure 17. As shown in figure 17, relation between [B]/[Si] ratio and $(B_2O_3)/(SiO_2)$ ratio of 10~120ton re-melting shows a good agreement with relation function derived by Pilot ESR experiment irrespective of ingot weight. Accordingly, it can be suggested that relation function between [B]/[Si] ratio and $(B_2O_3)/(SiO_2)$ ratio as represented in equation (7) is very suitable to control B content in target under inert gas atmosphere during ESR.

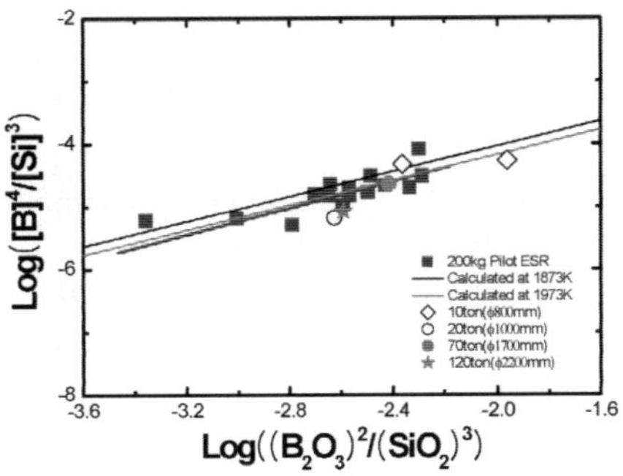


**Figure 17.** Result of 20~120ton re-melting plotted on the graph of relation between [B]/[Si] ratio and $(B_2O_3)/(SiO_2)$ ratio derived by Pilot ESR experiment

## 3. Conclusions

Through the pilot ESR experiments, the effect of Si content in electrode and $SiO_2$ content (slag basicity) in slag on the behavior of major element compositions and cleanliness in 9CrMoCoB Steel ESR ingot were investigated. In addition, equilibrium relation between [B]/[Si] ratio and $(B_2O_3)/(SiO_2)$ ratio was also investigated. Furthermore, in re-melting of 20ton-class ingot, the effect of soft arcing on ingot quality was also discussed.

In the results of Pilot ESR experiments, slag having a high basicity(low $SiO_2$ content) is advantageous to decrease the equilibrium oxygen content and get a good recovery of all oxidative elements such as Si, Mn and Nb simultaneously. Optimizing Si content in electrode enables the Al and Nb content to control in target value. Lastly, relation function between [B]/[Si] ratio and $(B_2O_3)/(SiO_2)$ ratio is also derived from experiments and is verified by re-melting of 10~120ton-class ingot.

In the re-melting of 20ton-class ingot, operation results are compared according to generation of soft arcing. Soft arcing could deteriorate the ingot quality to generate deep ripple on the ingot surface and make an internal cleanliness worse. Accordingly, when ESR is carried out under an argon gas atmosphere, it is extremely necessary to determine optimum conditions of voltage swing control to prevent the soft arcing.

With the optimized ESR conditions obtained from these studies, a 2200mm diameter (120 ton) 9CrMoCoB Steel of ESR ingot was successfully manufactured with good homogeneity in Si, Mn, Nb, B content and no segregation by the ESR process.

## 4. References

[1] D.S.Kim, S.T.Kang, S.K.Jo and J.T.Kim 2007 *International Symposium on Liquid Metal Processing and Casting* 55

[2] J. Fedko and M.Krucinski 1989 *Ironmaking Steelmaking* **16** 116

[3] M.OKamura 1985 *10th International Forging Conferen.* 71

[4] G.K.Sigworth and J.F.Elliott 1974 *Met. Sci.* **8** 298

[5] A.Stewart Ballantyne *Proceeding of the 2002 International Symposium on Electroslag Remelting Technologies(Chicago)*

[6] Paton, E.B., Saenko and V.YA., et al 2001 *Arc-Slag Remelting – Medovar Memorial Symposium* 177

IOP Conf. Series: Materials Science and Engineering **143** (2016) 012003 doi:10.1088/1757-899X/143/1/012003

# Transient melting of an ESR electrode

**A Kharicha[a,b] , E Karimi-Sibaki [a] , J Bohacek[b], M Wu[a,b], and A Ludwig[b]**
[a]Christian-Doppler Lab for Adv. Process Simulation of Solidification & Melting,
[b]Chair of Simulation and Modeling of Metallurgical Processes,
University of Leoben, Franz-Josef-Str. 18, A-8700 Leoben, Austria

E-mail: abdellah.kharicha@unileoben.ac.at

**Abstract.** Melting parameters of ESR process such as melt rate and immersion depth of electrode are of great importance. In this paper, a dynamic mesh based simulation framework is proposed to model melt rate and shape of electrode during the ESR process. Coupling interactions between turbulent flow, temperature, and electromagnetic fields are fully considered. The model is computationally efficient, and enables us to directly calculate melting parameters. Furthermore, dynamic change of electrode shape by melting can be captured. It is necessary to control the feeding velocity of electrode due to melting instabilities in the ESR process. As such, a numerical control is implemented based on the immersion depth of electrode to achieve the steady state in the simulation. Furthermore, the modeling result is evaluated against an experiment.

## 1. Introduction

The electroslag remelting (ESR) process is used for manufacture of superior quality steel and super alloys such as titanium alloys. Within the process, a consumable electrode melts through an electrically resistive slag layer through Joule heating. Then, droplets from the electrode pass through the slag and reach the liquid melt pool. Finally, the melt pool solidifies in a water-cooled mold to build the high-grade ingot. The internal quality of ingot is characterized according to the shape of melt pool especially depth and thickness of mushy zone. A shallow melt pool is favorable to achieve unidirectional solidification. Furthermore, producing an ingot with a defect-free surface is desirable. The electrode melt rate is a key parameter, affecting internal and surface quality of ingot. The influence of melt rate on the pool shape of an ESR ingot was investigated by Mitchell [1]. The depth of melt pool and thickness of the mushy zone were found to increase when the process was operated with higher melt rate. However, Suarez [2] reported that with the increase of melt rate a relatively smooth ingot surface can be obtained.

As reported by Kishida et al. [3], the melt rate, shape and immersion depth of electrode are interdependent parameters that are significantly influenced by the amount of power generated in the process. With the increase of voltage (generated power) in the process, the following melting behaviors were observed: the melt rate increases, the immersion depth decreases, and the shape of electrode tip becomes flatter. Over the last decades, a number of studies were carried out on this topic. For instance, Mitchell et al. modeled the temperature field in the electrode for a laboratory scale ESR process [4]. Their calculated temperature profile agreed with the experimental result. Mendrykowski et al, found that thermal radiation on the surface is ignorable compared to heat conduction along the electrode [5]. Tacke et al. [6] conducted a series of experiments to investigate melting behavior of a laboratory scale ESR electrode. Furthermore, they proposed an approximate formula for the immersion depth which was validated against experiments. By evaluation of thermal fluxes through boundaries, Jardy et al. predicted higher melt rate when buoyancy force is stronger than electromagnetic force [7]. Yanke et al. applied an effective heat transfer coefficient to study the melting behavior of an industrial scale ESR process [8]. Recently, Kharicha et al [9] modeled electrode melting within the framework of multiphase Volume of Fluid (VOF) method. It was stated that the coupling between melt rate and Joule heat is very unstable. Details of flow, temperature, and magnetic field during electrode melting were described. However, the proposed VOF based method is computationally expensive.

In this paper, a dynamic mesh-based simulation framework is proposed to model melting and shape of electrode tip. The proposed method is robust and computationally efficient. Complex interactions between flow, temperature, and electromagnetic fields are considered. In ESR process, it is necessary to continuously track and control immersion depth of electrode due to melting instability. In order to achieve stability in the simulation, a numerical control of immersion depth is implemented. Details of flow, temperature, and electromagnetic fields during remelting are demonstrated. Furthermore, the model is evaluated against experimental results [6].

## 2. A brief description of the model

Previously, the governing equations of flow, heat, and electromagnetic fields together with related boundary conditions as well as simulation parameters such as material properties and operation conditions were presented in details [10]. Here, a short description of the model is given. An axisymmetric computational domain is considered including electrode, slag, air, mold, and liquid zones as shown in Figure 1 (a). The Finite Volume Method (FVM) is used to simulate fluid flow, heat transfer, and electromagnetic fields. The commercial CFD software, FLUENT-ANSYS v.14.5, is the solver that includes a dynamic mesh technique for the simulation of displacements of boundaries. The modeling equations for fields such as electromagnetic field or boundaries (stationary and moving) are implemented using User-Defined Functions (UDF).

The well-known A-$\phi$ formulation is used to compute the electromagnetic field where $\phi$ is the electric scalar potential and $A$ denotes the magnetic vector potential. After calculating the magnetic field and subsequently electric current density distribution, the Lorentz force as well as Joule heating are obtained and added as source terms to momentum and energy equations respectively. Applying A-$\phi$ formulation to calculate electromagnetic field was extensively described by Kharicha [9].

The turbulent flow field in slag and melt pool is calculated considering the effect of Lorentz force and thermal buoyancy (Boussinesq approximation). The turbulence in the system is modeled using SST-k-$\omega$ approach. The essential feature of SST model is an accurate and effective near wall treatment due to insensitivity of the model to grid spacing of near wall cells [11-12]. The enthalpy conservation equation is solved to determine the temperature field taking into accounts Joule heating and latent heat sink ($S_{LH}$) required to melt the electrode [10, 13]. The latter will be described after introducing equations regarding to displacement of boundaries (electrode tip-slag and slag-air interfaces).

Displacements of domain boundaries are modeled using dynamic and deformable meshes in which the mesh nodes can be relocated. The following dynamic mesh schemes are used to handle boundary movement in the system. Firstly, layering scheme is used that involves creation and destruction of cell rows in the vicinity of a moving boundary. Secondly, smoothing scheme is applied in which the grid nodes are considered as a network of interconnected springs. The connectivity between grid nodes remains unchanged, but they can reposition [14]. In the computational domain, slag-air interface (moving) and electrode tip (moving-deforming) boundaries are allowed to move as shown in Figure 1(b). A high resolution mesh with equisized cells is used in the whole domain with the total number of 20000 cell elements (the mesh resolution is shown in melt pool).

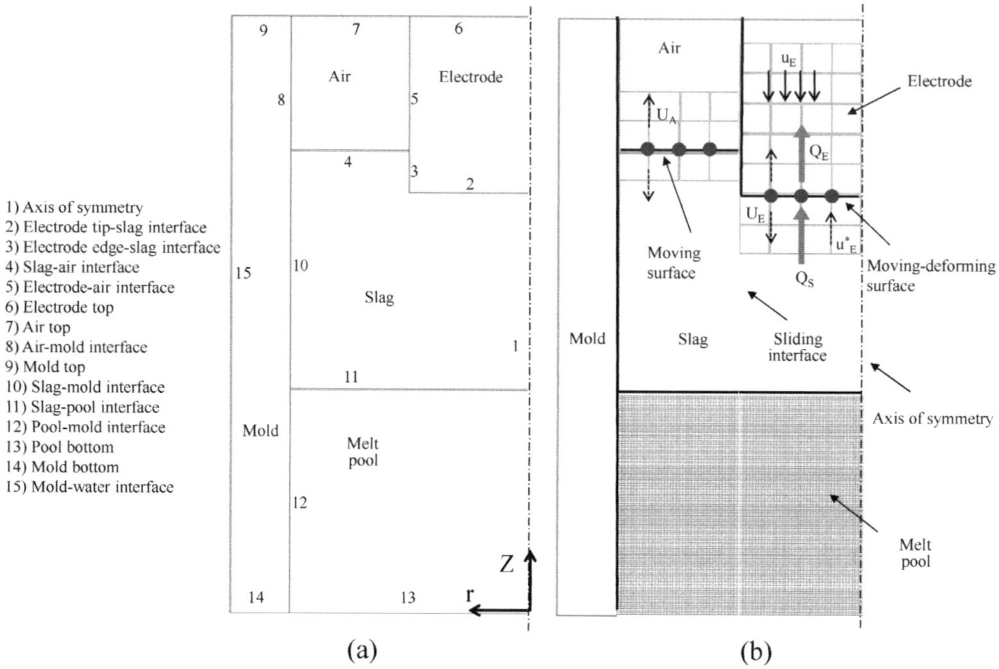


Figure 1. (a) Configuration of computational domain and boundaries, (b) Conceptual demonstration of heat balances across the electrode tip, melting velocity, feeding velocity, total velocity, and slag-air interface velocity (Note that the mesh resolution is very fine with equisized cells in the whole domain that is shown in melt pool, however the mesh is exaggeratedly shown to be coarse near the electrode for illustrative purpose)

Note that, both temperature field of electrode and velocity of melting at electrode tip are unknown. Physically, no dendritic mushy zone was observed during melting process, and the liquid-solid interface remained smooth. As such, melting of electrode can be considered as a Stefan problem where a phase boundary moves with time. The velocity of grid nodes at moving boundaries are computed using a set of balances equations. The velocity of dynamically deforming electrode tip is determined according to the amount of heat provided to the electrode through the slag.

$$U_E = u_E^* - u_E \tag{1}$$

$$Q_S - Q_E = \rho_{metal} \Delta H u_E^* \tag{2}$$

$$\begin{cases} u_E^* \geq 0 & (Q_s > Q_E) \\ u_E^* < 0 & (Q_s < Q_E) \end{cases} \tag{3}$$

According to Eq. (1), the grid node velocity of the electrode tip ($U_E$) is the sum of melting velocity ($u_E^*$) and the electrode feeding velocity ($u_E$). Based on Eq. (2), the melting velocity is dependent on the heat flux provided to the electrode ($Q_S$), heat flux diffused into the electrode ($Q_E$), density of metal ($\rho_{metal}$ ~ 7100 kg.m$^{-3}$), and the latent heat of fusion ($\Delta H$ ~ 260 kJ.kg$^{-1}$). The origin of z-referential is set at the bottom of computational domain. Therefore, the sign (positive or negative) of melting velocity depends on the balance of heat fluxes at the electrode tip as expressed in Eq. (3). Consider that, the right hand side of Eq. (2) describes the amount of latent heat provided to melt the electrode. Therefore, the volumetric source of latent heat ($S_{LH}$) that is absorbed adjacent to electrode tip is given by

$$S_{LH} = \rho_{metal}\ u_E^{*} \Delta H \left(\frac{S_{cell-electrode}}{V_{cell-electrode}}\right) \tag{4}$$

where ($S_{cell\text{-}electrode}$) denotes the face area of a cell that belongs to the electrode tip and ($V_{cell\text{-}electrode}$) is the cell volume.

With the increase of electrode immersion depth, the slag level rises to conserve the total mass of slag. Therefore, the slag-air interface is also modeled as a moving boundary. The velocity of grid nodes ($U_A$) of the slag-air moving boundary is computed as:

$$U_A = \frac{m_0 - m}{\rho_{slag} \Delta t \pi \left(R^2_{mold} - R^2_{electrode}\right)} \tag{5}$$

where ($m_0$) denotes the slag initial mass, ($m$) is the computed mass, ($\rho_{slag} \sim 2700$ kg.m$^{-3}$) is density of slag, ($\Delta t \sim 0.1$ s) is time step size, and ($R_{mold}$) and ($R_{electrode}$) are the radius of the mold and electrode respectively.

The electrode feeding velocity is preferred to be kept constant during industrial operation of an ESR process. It is necessary to adapt feeding velocity ($u_E$) according to the immersion depth to stabilize the electrode position so that the electrode tip neither reaches the melt pool nor the slag free surface [9]. The following restriction is imposed on the feeding velocity of electrode:

$$u_E = u_E^0 \min\left[1, \frac{l_{max} - l}{l_{max} - l_{min}}\right] \tag{6}$$

($l_{min}$) and ($l_{max}$) are minimum and maximum allowable immersion depth of electrode, and ($u_E^0$) is initial feeding velocity which are the input parameters of the process. Furthermore, a uniform velocity field inside the electrode zone is specified which has similar magnitude as the electrode feeding velocity ($u_E$).

The simulation is executed in transient where the immersion depth and shape of electrode tip is continuously tracked. The steady state is achieved once the velocity of grid nodes ($U_E$) converges to zero. The melt rate ($\dot{m} \geq 0$) of electrode can be directly computed as follow:

$$\dot{m} = -\rho_{metal} \iint_{S_{electrode}} \vec{u}_E^{*} \cdot d\vec{S} \tag{7}$$

Following the experiment of Tacke [6] on a laboratory scale ESR process, we used an identical geometry and operating parameters. Table 1 lists the parameters used in the experiment [6] and our simulation. Furthermore, the steady state melting parameters of simulation such as computed immersion depth are listed.

Table 1. Operation conditions and results of a laboratory scale ESR process [6], a comparison is also made with the simulation results (steady state).

| | **Experiment [6]** | **Simulation** |
|---|---|---|
| Electric current mode | DC | DC |
| Electric current (kA) | 1.9 | 1.9 |
| Mold current | Unknown | Yes |
| Electrode radius (cm) | 4 | 4 |
| Mold radius (cm) | 8 | 8 |
| Slag weight (kg) | 4.0 | 4.1 |
| Slag electric cond. ($ohm^{-1}.m^{-1}$) | Unknown | 150 |
| Slag thermal cond.($W.m^{-1}.K^{-1}$) | Unknown | 4 |
| Slag therm. exp. coeff. ($K^{-1}$) | Unknown | $9x10^{-5}$ |
| Voltage (V) | 29 | 28 |
| Power (kW) | 55 | 57 |
| Initial feeding vel. ($cm.s^{-1}$) | Unknown | 0.0153 |
| Steady feeding vel. ($cm.s^{-1}$) | 0.0153 | 0.0076 |
| Melt rate ($g.s^{-1}$) | 5.5 | 3.7 |
| Immersion depth (cm) | 3.8 | 3.56 |
| $L_{min}$ (cm) | Unknown | 2.5 |
| $L_{max}$ (cm) | Unknown | 4.5 |

It must be noted that the following phenomena are not considered in our model. As generally recognized, the electric current is conducted by ions in the slag region. Consequently, difference in melt rate is observed based on the polarity of electrode (positive or negative) due to polarization overpotential [15]. Formation of a liquid film under the electrode and dripping of droplets through the slag are also not included [16-17]. Solidification of liquid pool is also ignored.

## 3. Results and discussions

A constant magnitude of DC current is imposed in the process. Additionally, it is assumed that the current can cross the slag skin flowing through the mold [15, 18-19]. The initial feeding velocity is not reported for the experiment, but the steady feeding velocity is known. Furthermore, the feeding velocity is adjusted adapted by Eq. 6 in the simulation. Figure 2 illustrates the evolution of electrode shape including electric potential, velocity, and temperature fields in slag and submerged part of the electrode. The amplitude of voltage drop and power generated in the system depends on immersion depth (indicated by the distance between electrode tip and isolines of voltage drop). Decreasing immersion depth leads to significant voltage drop and power generation in the system as shown in Figure 2 (c). Therefore, the temperature notably elevates in the bulk of slag. With the increase of immersion depth, the isolines of voltage are shifted and the total voltage in the system decreases. However, the current density increases in the central region of slag and subsequently the Lorentz force become stronger. The latter intensifies the velocity under the shadow of electrode which results in promotion of stirring in the slag zone that is shown in Figure 2 (e). Here, the global energy transport is efficient due to strong turbulence which causes relatively uniform slag temperature. Finally, electrode develops a conical shape at steady state ($t$ = 350 s). The mesh resolution near moving-deforming boundary (electrode tip) is also shown in Figure 2. Accordingly, the quality of mesh is very well preserved.

The variation in melt rate and generated power as well as feeding velocity and immersion depth are plotted during deformation of electrode tip until reaching the steady state which is shown in Figure 3. The results show that variation in generated power is much smoother than melt rate. The peak observed in generated power indicates the situation where the immersion depth is very shallow.

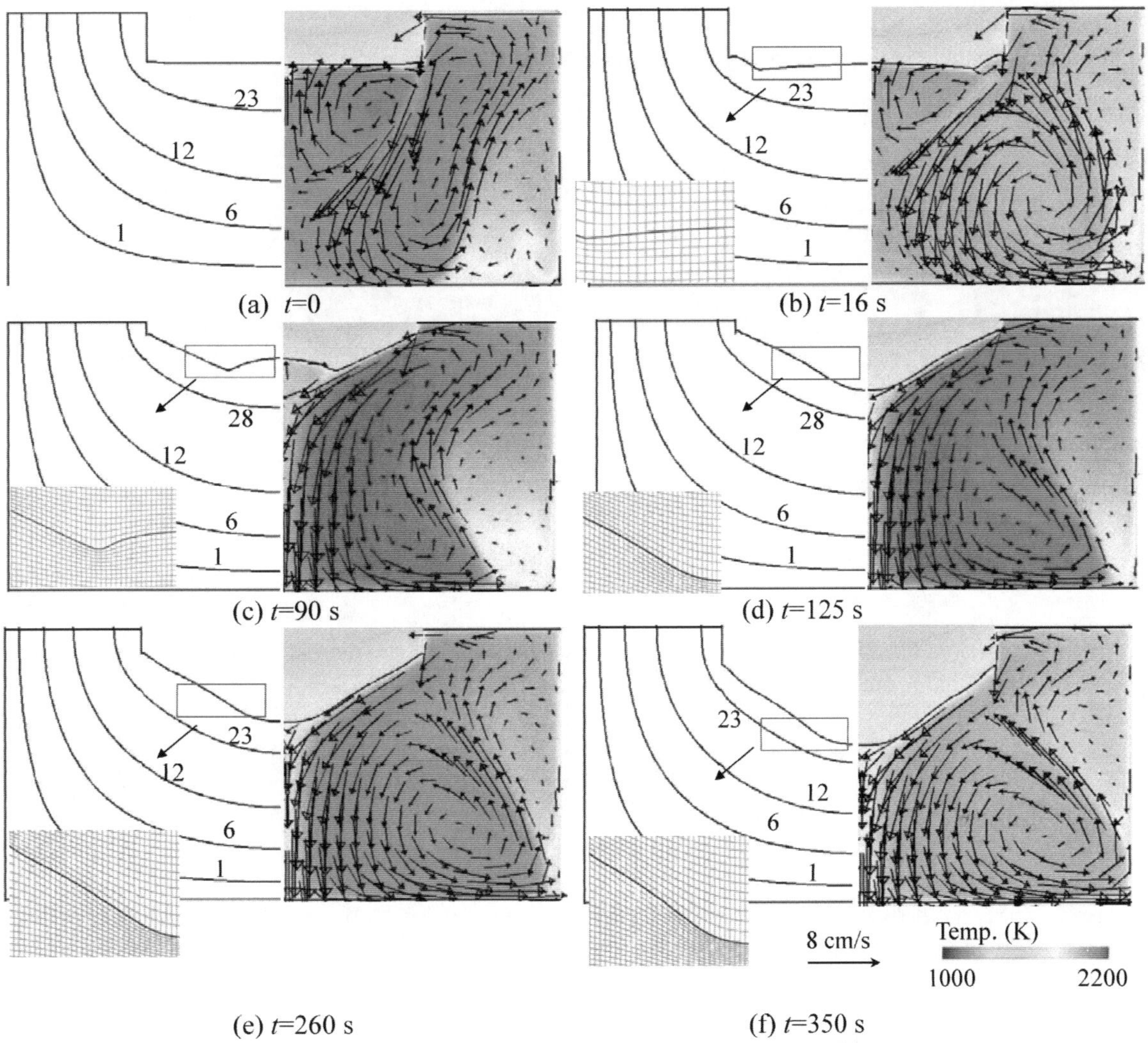


Figure. 2. Evolution of shape of electrode tip: (a) $t$=0, (b) $t$=16 s, (c) $t$=90 s, (d) $t$=125 s, (e) $t$=260 s, (f) $t$=350 s. On left half: isolines of voltage, and the grid node near the moving boundary (electrode tip). On right half: contour of temperature overlaid with the vectors of velocity in the slag region.

Despite gentle variation in generated power, the melt rate can dramatically increase or decrease before reaching the steady state. In fact, the melt rate is governed by the amount of heat provided to the electrode which is highly influenced by complex interactions between flow, temperature, and electromagnetic fields. For example, with decrease of immersion depth, Figure 3 (d), the generated power increases, Figure 3 (a) that leads to hotter slag and increase of melt rate as shown in Figure 2 (c). Furthermore, the constraint imposed on the immersion depth based on Eq. 6 can certainly affect the feeding velocity as well as melt rate of electrode. Further investigation is required to explore the effect of immersion depth control on melting behavior of electrode that is beyond the scope of the current work. Figure 4 compares the experimentally observed shape of electrode with the predicted shape in the simulation at steady state where a relatively good agreement is obtained. In any case, the shape of electrode tip is mainly influenced by temperature and flow distribution which are highly dependent on slag properties. The slag properties are temperature dependent, so a large uncertainty on the reported values exists due to difficulty of measurements at high temperature. For instance, the thermal conductivity of slag was reported to vary between 1 to 5 $W.m^{-1}.K^{-1}$, whereas the electrical conductivity was measured to be between 80 to 300 $ohm^{-1}.m^{-1}$ [20-21].

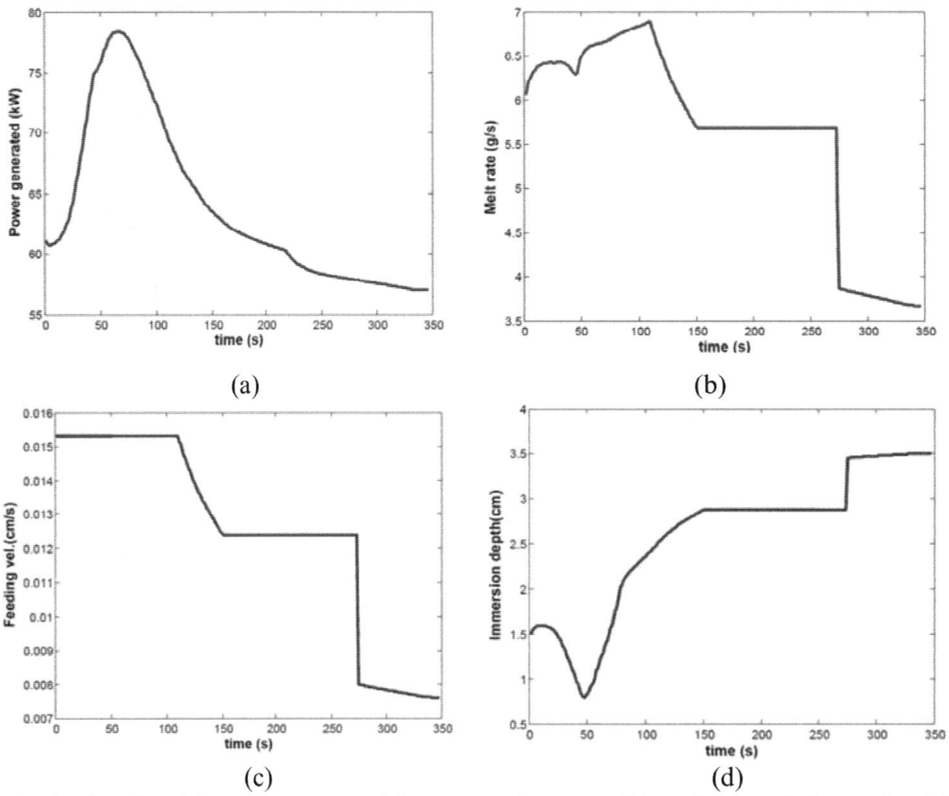


Figure 3. Analysis of melting parameters: (a) generated power, (b) melt rate of electrode, (c) feeding velocity, (d) immersion depth.

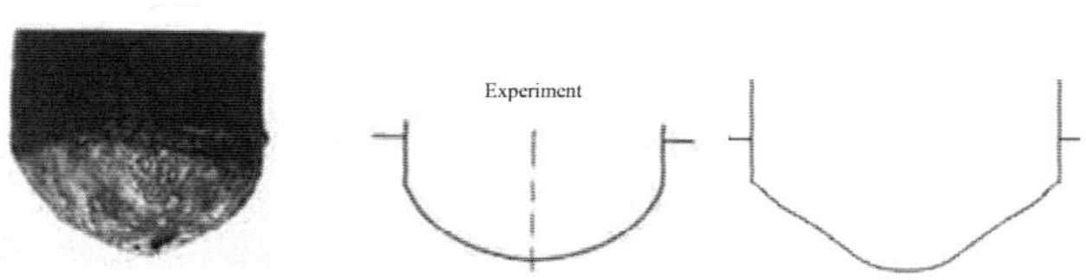


(a) (b)

Figure 4. A comparison of (a) the shape of electrode tip as observed in the experiment [6] with (b) the calculated shape of electrode from simulation.

It must be reported that we ran many simulations to perform a parameter study to investigate the effect of slag physicochemical properties such as electrical and thermal conductivity on the melt rate and shape of electrode. Our investigations were published with details in the previous paper [10]. It must be stated that the predicted melt rate by the simulation is lower than the value reported in experiment [Table 1]. The current model does not take into account effects of electrochemical reactions, formation of liquid film at electrode tip and dripping of droplets through slag. In fact departure and dripping of droplets can create stronger turbulence which enhance global energy transfer and melt rate of electrode [16-17]. The aforementioned phenomena must be included in the future model to improve quantitative accuracy of the model.

## 4. Summary

A dynamic mesh based approach is used to model submerging and remelting of electrode in electroslag remelting (ESR) process. The computational domain is 2D axisymmetric where interactions between turbulent flow, temperature, and electromagnetic fields are fully taken into account. It is shown that the model is capable of predicting the evolution of electrode shape during remelting. In order to achieve steady state, the immersion depth of electrode is continuously tracked and controlled. For the latter, a numerical control of feeding velocity is implemented dependent on the immersion depth. The model is computationally robust that enables us to directly compute melt rate without using any further assumption (e.g. heat transfer coefficient between electrode-slag). The shape of electrode is also validated against the experiment. Melt rate and shape of electrode are highly dependent on physicochemical properties of slag such as thermal and electrical conductivities. The effect of slag properties on melting behavior of electrode is extensively discussed in other paper [10].

## 5. Acknowledgement

The authors acknowledge the financial support by the Austrian Federal Ministry of Economy, Family and Youth and the National Foundation for Research, Technology and Development within the framework of the Christian Doppler Laboratory for Advanced Process Simulation of Solidification and Melting.

## 6. References

[1] Mitchell A 1984 *Conf. on Perspective in Metallurgical Development* (Sheffield) p 89
[2] Suarez F S, Roberts J E, and Schley L.D 1974 *5th Int. Symp. on Electroslag and Other Special Melting Technologies* (Pittsburgh) p 126
[3] Kishida T, Yamaguchi T, Tomioka T, and Ichihara T 1974 *Elect. steel* **45** 219
[4] Mitchell A, Joshi S, and Cameron J 1971 *Metall. Trans.* **2** 561
[5] Mendrykowski J, Poveromo J, Szekely J, and Mitchell A 1972 *Metall. Trans.* **4** 1761
[6] Tacke K H and Schwerdtfeger K 1981 *Arch. Eisenhüttenwesen* **52** 137
[7] Jardy A, Ablitzer D, Wadier J F 1991 *Metall. Trans.* **22** 111
[8] Yanke J, Fezi K, Fahrmann M, and Krane M J M 2013 *Proc. Of LMPC* (New York: Wiley–Interscience)
[9] Kharicha A, Wu M, and Ludwig A 2014 *ISIJ* **54** 1621
[10] Karimi-Sibaki E, Kharicha A, Bohacek J, Wu M, and Ludwig A, *Metall. Mater. Trans. B,* DOI 10.1007/s11663-015-0384-0
[11] Menter F R 1994 *AIAA Journal* **32** 1598
[12] Menter F R, Kuntz M, and Langtry R 2003 *Turb. Heat and Mass Trans.* **4** 625
[13] Karimi-Sibaki E, Kharicha A, Korp J, Wu M, Ludwig A 2014 *Mater. Sci. Forum* **790** 396
[14] Fluent 14.5 user's guide 2012 (Austin: Fluent Inc.)
[15] Kawakami M, Nagata K, Yamamura M, Sakata N, Miyashita Y, and Goto K S 1977 *Testsu-to-Hagane* **63** 220
[16] Kharicha A, Ludwig A, and Wu M 2011 *EPD congress* (San Diego) p 771
[17] Kharicha A, Wu M, Ludwig A, Ramprecht M, Holzgruber H 2012 *CFD modeling and simulation in materials* (New York: Wiley–Interscience) p 139
[18] Kharicha A, Ludwig A, and Wu M. 2005 *Mater Sci. Eng. A* **413** 129
[19] Kharicha A, Schutzenhofer W, Ludwig A, Tanzer R, and Wu M 2008 *Steel Res. Int.* **79** 632
[20] Taylor R, Mills K C 1982 *Arch. Eisenhüttenwesen* **53** 55
[21] Hajduk M and Gammal T E 1979 *Stahl Eisen* **99** 113

IOP Conf. Series: Materials Science and Engineering **143** (2016) 012004 doi:10.1088/1757-899X/143/1/012004

# The influence of the mould cooling temperature on the surface appearance and the internal quality of ESR ingots

**M Kubin[1], B Ofner[1], H Holzgruber[1], R Schneider[2], D Enzenhofer[2], A Filzwieser[3], S Konetschnik[3]**

[1] INTECO special melting technologies GmbH, Bruck/Mur, Austria
[2] University of Applied Science Upper Austria, Wels, Austria
[3] METTOP GmbH, Leoben, Austria

E-Mail: michael.kubin@inteco.at

**Abstract**. One of the main benefits of the ESR process is to obtain an ingot surface which is smooth and allows a subsequent forging operation without any surface dressing. The main influencing factor on surface quality is the precise controlling of the process such as melt rate and electrode immersion depth. However, the relatively strong cooling effect of water as a cooling medium can result in the solidification of the meniscus of the liquid steel on the boundary liquid steel and slag which is most likely the origin of surface defects. The usage of different cooling media like ionic liquids, a salt solution which can be heated up to 250°C operating temperature might diminish the meniscus solidification phenomenon. This paper shows the first results of the usage of an ionic liquid as a mould cooling medium. In doing so, 210mm diameter ESR ingots were produced with the laboratory scale ESR furnace at the university of applied science using an ionic liquid cooling device developed by the company METTOP. For each trial melt different inlet and outlet temperatures of the ionic liquid were chosen and the impact on the surface appearance and internal quality were analyzed. Furthermore the influence on the energy balance is also briefly highlighted. Ultimately, an effect of the usage of ionic liquids as a cooling medium could be determined and these results will be described in detail within the scope of this paper.

## Introduction

Surface quality of as cast ESR ingots is an important precondition in order to further process the ESR ingot in the forging shop without any prior surface dressing. One main key point of a smooth surface appearance, especially for large sized ESR ingots (diameter > 1.600mm), is the precise controlling of the ESR process [1]. However, also the temperature gradient between the mould wall and the ingot has an influence on the appearance of the ingot surface and on the origin of surface defects. The usage of water as a mould coolant in the ESR leads to a very strong cooling effect respectively a very steep temperature gradient which can result in the solidification of the meniscus of the liquid steel on the boundary liquid steel and slag. The overflow of the already solidified meniscus with liquid metal due to local turbulences or an increase of the liquid metal level results in surface defects such as “groves”, “tears” or “metal fins” [2]. The usage of different cooling media like ionic liquids a salt solution which can be heated up to 250°C operating temperature, on the above mentioned temperature gradient and can diminish the meniscus solidification phenomenon.

It has been observed that good surface quality is associated with the presence of a finite depth (few centimetres) of liquid-metal contact on the slag skin [3] also known as “liquid head” or “standing height”. To achieve these requirements, the melting point of the slag must be higher than that of the metal [4]. In most common industrial practice, the problem with poor surface has been dealt with increasing the power input (and so the melt rate), by using the highest stable fill ratio and the lowest electrode immersion depth. Kharicha et al. [5] proposed the hypothesis that the current flowing through the slag/mould and the ingot/mould interfaces is the key factor to explain the empirical correlations gathered by industry on the quality of the ingot surface. It was shown that the final ingot surface quality is controlled by the heat balance at the slag skin. Increasing the power decreases the skin thickness by acting on the heat flux in the slag/ingot side of the skin. Another way consists in acting on the heat flux in mould side by decreasing the amount of heat removed by the coolant liquid. It was indeed suggested by industry that running the process with warmed mould may encourage the formation of smooth skin [6].

Furthermore, the ESR process is a very energy intensive process. According to Mitchell [7] good average power efficiency during the ESR-operation lies in the order of about 1300 kWh/t. The consumed energy cannot be further used as the temperature of the cooling water at the mould outlet reaches only a maximum of approx. 50°C. When using ionic liquids the mould outlet temperature can be increased and the energy extracted from the ESR process can be further used.

These above mentioned considerations were the initial starting point in executing several trial melts at a laboratory ESR furnace to examine the impact on the quality of an ESR ingot as well as on the energy balance when using ionic liquids for the mould cooling.

IOP Conf. Series: Materials Science and Engineering **143** (2016) 012004 doi:10.1088/1757-899X/143/1/012004

## 1. Remelting trials with the laboratory scale ESR furnace

To evaluate the above mentioned influence of the temperature of the mould cooling medium, test trails were performed with the laboratory ESR plant at the University of Applied Science Upper Austria. The test ingots with a diameter of 210 mm were produced by remelting 151,5 mm dia. electrodes using a standard ESR slag (Wacker 2015). The flow rate of the ionic liquid was steadily adjusted during the test melts in order to obtain higher temperatures at the mould outlet (see Table 1). For the first three trials the standard cooling water system was used. For the trails using the ionic liquids an intermediate cooling circuit for the ionic cooling solution was interconnected between the mould and the standard cooling water system. This cooling module including the necessary controlling device was provided by METTOP GmbH.

**Table 1.** Summary of the cooling parameters used for the executed test trials

| Test No. | | Mould inlet temperature | Mould outlet temperature |
|---|---|---|---|
| V082 | Pre- test with water cooling | 30°C | 40°C |
| V083 | Standard test with water cooling | 30°C | 40°C |
| V084 | Repetition Standard test with water cooling | 30°C | 40°C |
| V085 | Trail melt with ionic liquid | 60°C | 110°C |
| V086 | Trail melt with ionic liquid | 60°C | 150°C |
| V087 | Trail melt with ionic liquid | 60°C | 195°C |

For evaluation of the pool profile shape and depth ferro-phosphor and iron sulphide were added just before the hot topping process. All test ingots were evaluated in terms of surface quality, geometrical form and depth of the pool, solidification structure (secondary dendrite arm spacing) and the respective heat transfer balance.

**Table 2.** Summary of the remelting parameters used for the executed test trials

| Test No. | Power [kW] | Steady state melt rate [kg/h] | Current [A] | Immersion depth of electrode [mm] |
|---|---|---|---|---|
| V082 | 105 | - | 4.000 | approx. 20 |
| V083 | 110 | 105,7 | 4.000 | approx. 20 |
| V084 | 105 | 101,6 | 4.000 | approx. 20 |
| V085 | 108 | 105,3 | 4.000 | approx. 20 |
| V086 | 105 | 102,0 | 4.000 | approx. 20 |
| V087 | 110 | 105,5 | 4.000 | approx . 20 |

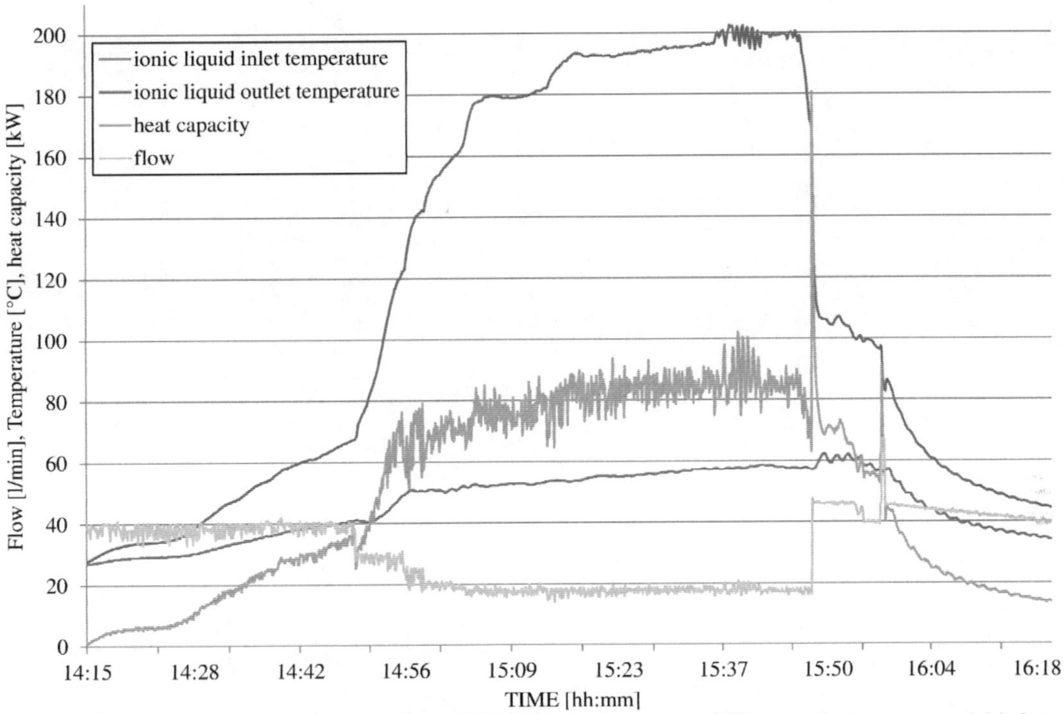


Figure 1: Cooling circuit parameters of test V087. A temperature difference between mould inlet and outlet of 135°C could be obtained

In **Figure 1** a typical trend of the cooling parameters is shown. For the test V087 the highest temperature difference between inlet and outlet was adjusted.

## 2. Quality of produced lab-scale ESR ingots

### *2.1. Surface quality*

A comparison of the achievable surface quality of the trial ingots can be seen in **Figure 2.** As can be seen there is a clear relationship between the temperature of the mould cooling medium and the surface appearance of the ESR ingot. A significant improvement of the surface quality could be determined. Whilst during the test implemented with the conventional cooling method the surface was free of deeper notches but contains irregular depressions. With the implementation of the intermediate cooling circuit and the associated increase of the cooling temperature, the surface quality improved steadily. According to [8] the precondition for achieving a good ingot surface quality is a sufficient overheating of the slag bath. It can be assumed that the higher cooling temperature influences the overheating of the slag bath in a favorable manner so that the effects on the surface quality become quite significant.

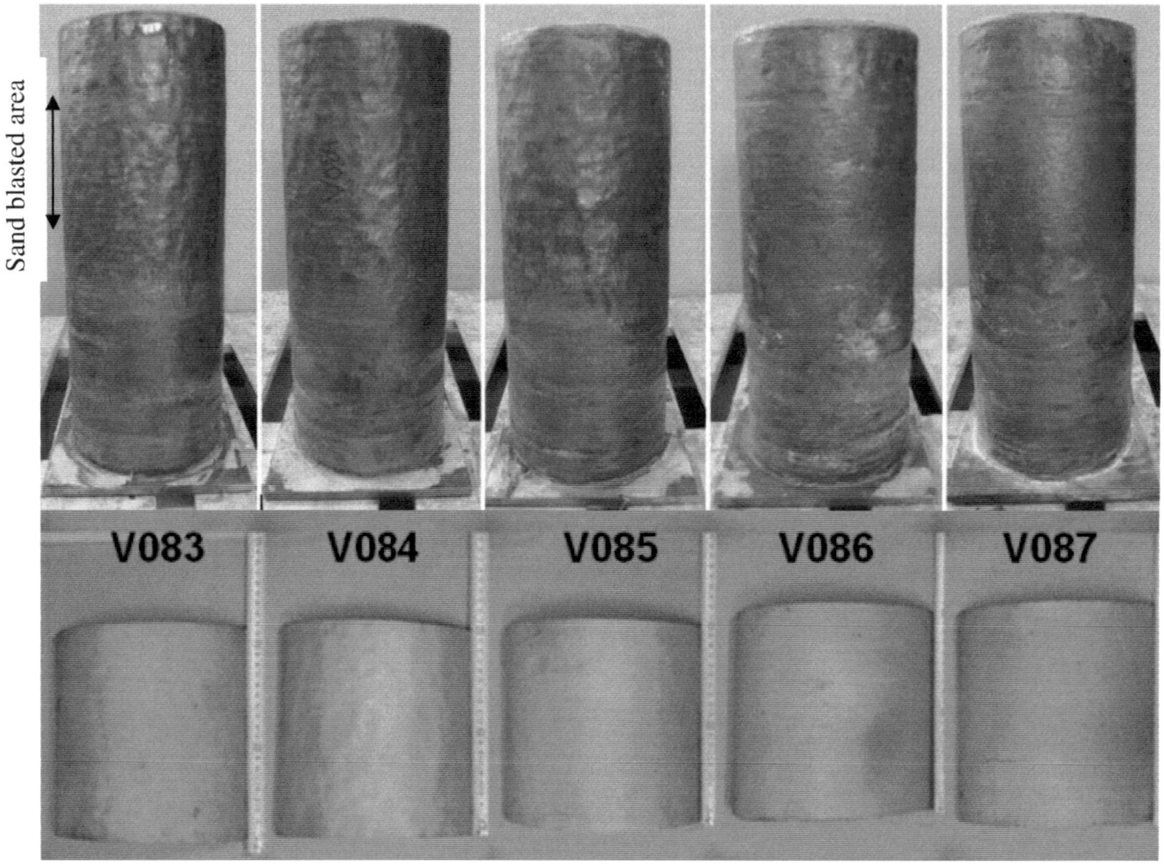


Figure 2: Comparison of the surface appearance (above: entire ESR ingot, below: sandblasted surface)

### 2.2. *Internal quality*

For evaluating the internal quality the pool profile formation (shape and depth of the liquid metal pool) and the secondary dendrite arm spacing were examined. If an ESR ingot is built up at a certain melting rate in the case of ingots with a sufficient length, steady state conditions in terms of thermal conditions are obtained. This means that the melting profile remains approximately steady during this time and with the "ingot building velocity" in the direction of the ingot axis the melt pool moves upwards. The pool profile deepens with the increase of the melting rate [8]. The evaluation of the Baumann print shows a slight increase of the melting pool depth during the test respectively with the rise of the mould cooling liquid outlet temperature as can be seen in **Figure 3**. The shallowest metal pool was achieved at the trail ingots with water cooling with approximately 46 mm. The deepest pool depth was observed at the melt V087 with approximately 59 mm. However it could be explained that the increase in the temperature promotes the overheating of the melt pool and reduces the heat transfer from the ingot to the mould. In addition to the evaluation with the Baumann print, the pool profiles were also investigated with the etching by the Oberhoffer method and the results showed the same tendency. Moreover it can be stated that there is a tendency that the pool profile becomes a more "C-shaped" form with increasing temperature instead of the usual "V-shaped form" in the ESR process. This C-shaped form of the pool profile has certain advantages regarding the formation of macrosegregations [9].

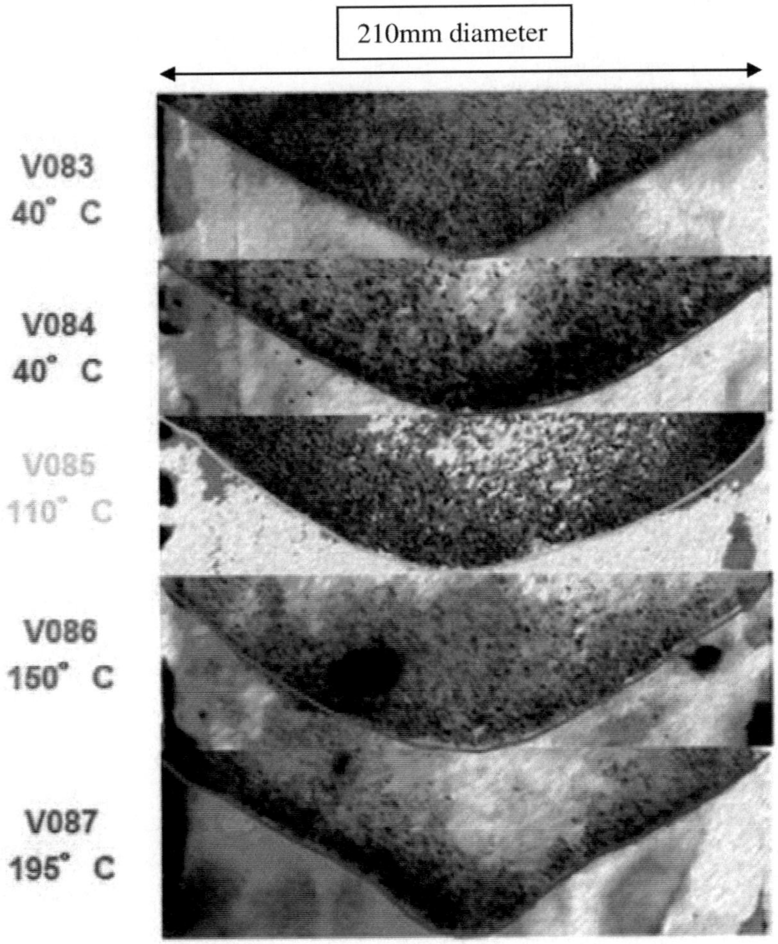


Figure 3: Comparison of the pool shape and depth of the trial ingots

Among the most important parameters of the solidification process affecting the structure and properties of the ingots are the rate of cooling of the metal in the two-phase region and the temperature gradient ahead of the solidification front [10, 11]. With increasing distance from the surface, the rate of cooling of the metal in the two phase zone falls continuously. This reduction along the radius of the ingot is evidently the reason for the increase of the secondary dendrite arm spacing [12]. An appropriate way to evaluate the solidification structure is the analysis of the secondary dendrite arm spacing. Desired microstructures are characterized by short distances between the dendrite arms. During the tests the dendrite arm spacing between the single experiments showed a slight increase of these distances with the rise of the mould cooling liquid outlet temperature (see **Figure 4**). This effect was more significantly detected in the middle of the ingot. This can be explained with the before mentioned effect that the energy remaining in the ingot rose with the increase of the cooling liquid outlet temperature due to the lowered cooling intensity. A higher mould outlet temperature induces a higher rest energy which leads to a lower temperature gradient in the solidification front resulting in higher dendrite arm spacing.

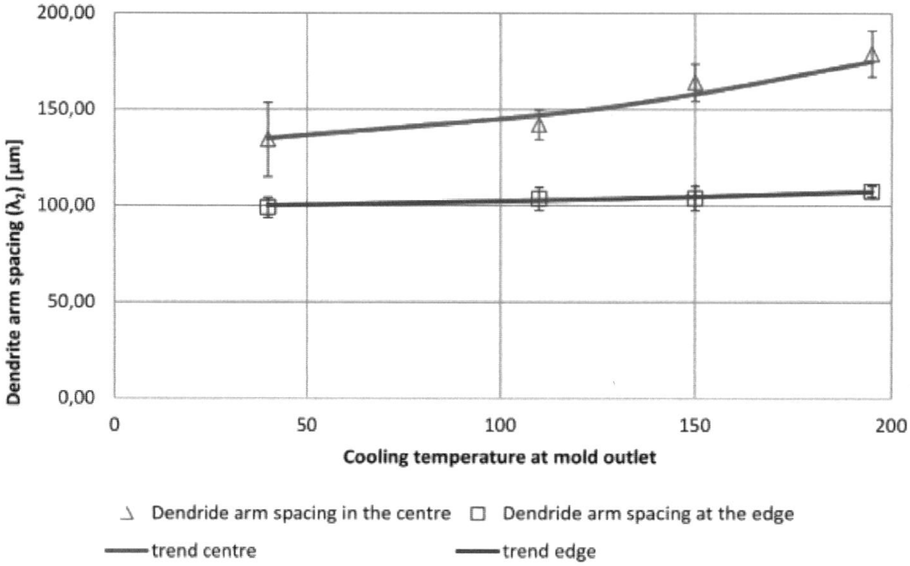


Figure 4: Secondary dendrite arm spacing as a function of the cooling temperature

### *2.3. Heat balance*

According to [7] good average power efficiency during the ESR-operation is in the order of approximately 1300 kWh/t. Concerning these test trials the average power consumption was determined with 1292 kWh/t with a standard deviation of 16 kWh/t. This indicates that the tests all together were implemented on the one hand with very constant parameters and on the other hand in a quite efficient way. With regard to the energy distribution between the cooling energy dispensed to the bottom plate and the mould and the rest energy in the ingot, as can be seen in **Figure 5**, a slight decrease of the cooling energy during the test is recognisable. On the contrary the energy remaining in the ingot got nearly in the same way higher, so that ultimately no significant chance of the whole energy consumption could be noticed.

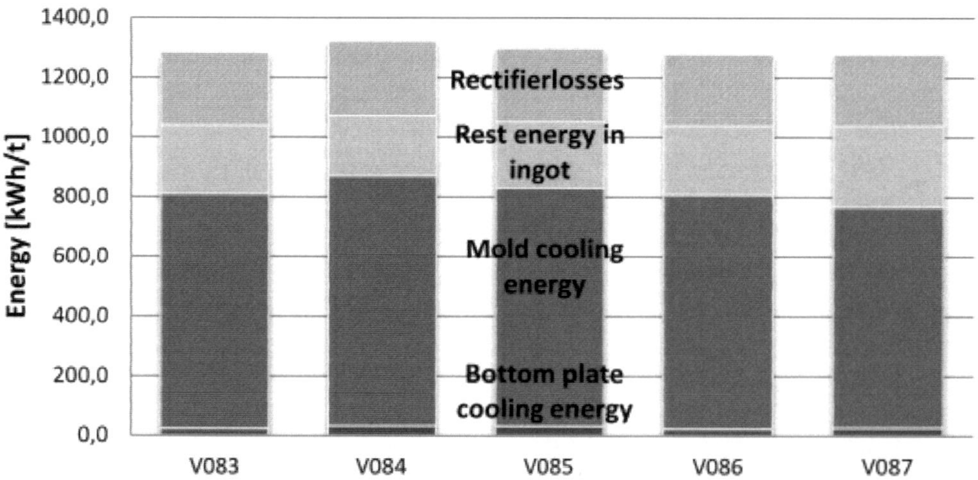


Figure 5: Energy distribution between bottom base plate, mould, ingot and rectifier losses

## 3. Summary, Conclusion and Outlook

The evaluation of the tests demonstrated that the impact of the temperature increase of the cooling media of the mould on the properties to be investigated was apparent. The most significant effects were monitored in terms of the surface quality. Whilst during the tests implemented with water as cooling media the surfaces were in a quite good condition but contained a lot of irregular depressions. With the implementation of the ionic liquid for the cooling and the resulting temperature increase, the depressions got significant fewer so that in the end in case of the last test the surface was smooth all over the entire ingot.
In the case of the melt pool depth, a dependence of the pool profile could be investigated. Higher temperatures lead to slightly deeper pool depth but also a tendency to a more "C-shape" pool profile could be recognized. Concerning the microstructure a statement in respect to the dendrite arm spacing with the cooling temperature could be given. Whereas at the areas closer to the surface of the ingots only a very small increase of the dendrite arm spacing was observed, in the areas in the centre they increased with the rise of the cooling temperature.
Concerning the energy balance it has to be said that the energy needed for the cooling of the mold decreased slightly and in response the remaining rest energy in the ingot increased almost in the same way. This is due to a storing effect of the energy in the ingot and the ionic cooling circuit.

Even though this series of tests displayed a definite answer, to deepen this knowledge a few more tests should be carried out. One aspect could be to decrease the range of the temperature of the ionic cooling liquid between the inlet and outlet of the mold and investigate how this affects the whole process. Also the potential of a possible energy recuperation due to the high temperature of the cooling media at the mould outlet shall be further investigate.

**References**

[1] Kubin M, Scheriau A, Knabl M, Holzgruber H, and Kawakami H 2013 *Proc. of LMPC* (Austin: J. Willey Pub)
[2] Hoyle G 1979 *Proc. of 6th Vacuum Metallurgy Conference*, p. 624
[3] Cameron, J.; Etienne, M.; Mitchell, A. Metallugical Transactions vol. 1 issue 7 July 1970. P. 1839-1844 "Some electrical characteristics of a DC electroslag unit
[4] Bell, M. Mitchell, A. "Some observations on the surface quality of electroslag ingot" Journal of the Iron and Steel Institute, Vol, 209, Part 8, (Aug., 1971), 658-670
[5] Kharicha A., Karimi Sibaki E., Wu M., Ludwig A.: Liquid Metal Processing & Casting Conf. 2013 (LMPC 2013) Sept. 22-25, Austin, Texas (2013) pp. 95-99. "Contribution of the mould current to the ingot surface quality in the electroslag process"
DOI: 10.1002/9781118830857 .ch13
[6] Melberg, P.O., Proc. 4th Int. Symp. On ESR Processes, ISIJ, Tokyo, 1973, p. 13-25. "Temperature distributions in slag and metal during electroslag remelting of ball-bearing steel"
[7] Mitchell A, and Smailer R M 1979 *International Reviews* **5** p. 231
[8] Holzgruber W 1975 *Radex Rundschau* **3** p. 409
[9] K O Yu 1986 *Journals of Metals* **38** 1 p. 46
[10] Kou S, Poirier D, and Flemmings M C 1977 Proc. *of Electric Furnace Conference* **35**, p.221
[11] Thielmann R, and Kreyenberg J 1985 *Proc. of 8 th ICVM Special melting/refining and metallurgical coating under vacuum or controlled atmosphere*, p. 1130
[12] Bordin I P, Goryainov V A Koshman V S, Zaitseva L A, and Tsapaeva L L 1985 *Steel in the USSR* **15** p. 533

# Nitrogen addition using a gas blow in an ESR process

**S. Yamamoto[1], Y. Momoi[2], K. Kajikawa[2]**

[1]Special Melting Shop, Material Manufacturing Department, Muroran Plant, The Japan Steel Works, Ltd., Japan

[2]Muroran Research Laboratory, Research and Development Headquarters, The Japan Steel Works, Ltd., Japan

suguru_yamamoto@jsw.co.jp, koji_kajikawa@jsw.co.jp

**Abstract**. A new nitrogen method for adding in an ESR process using nitrogen gas blown in through the electrode was investigated. Nitrogen gas blown through a center bore of the electrode enabled contact between the nitrogen gas and the molten steel directly underneath the electrode tip. A $\phi$145mm diameter, laboratory-sized PESR furnace was used for the study on the reaction kinetics. Also, we carried out a water-model experiment in order to check the injection depth of the gas blown in the slag. The water model showed that the gas did not reach the upper surface of the molten metal and flowed on the bottom surface of the electrode only. An EPMA was carried out for a droplet remaining on the tip of the electrode after melting. The molten steel from the tip of the electrode shows that nitrogen gas absorption occurred at the tip of the electrode. The mass transfer coefficient was around $1.0\times10^{-2}$ cm/sec in the system. This value is almost the same as the coefficient at the molten steel free surface.

## 1. Introduction

Nitrogen alloyed steels are widely used for drill collars, power-generation components, engine valves, high corrosion-resistance applications, bearings and other applications. Nitrogen alloying has two main benefits. One is strengthening owing to solid solution hardening [1], and the other is improvement of corrosion resistance[2]. Nitrogen-alloyed austenitic stainless steels less are likely to form processing evocation martensite. This enables the production of high-strength austenitic steel.

Several nitrogen addition methods have been developed in steel-making industries. Nitride addition is a popular method for high-nitrogen steel production. This method can increase the nitrogen content in steel effectively. However there is a disadvantage to this method. Nitride addition increases other elements such as Cr or Si. The gas-blow method is commonly used in a ladle furnace. This method has the advantage of eliminating contamination with other elements in nitride alloys. It is difficult to add nitrogen into a metal pool by blowing gas on the slag surface because steel-making slag has almost no nitrogen solubility· Direct contact between the nitrogen gas and the molten steel is required in order to add nitrogen using this method. Although bottom gas blowing can be used for nitrogen addition in a ladle furnace process, it cannot be applied to an electroslag remelting (ESR) process.

There is one solution for adding nitrogen in ESR. Prof. B. I. Medovar *et. al.* had developed Arc Slag Remelting (ASR)[3] as a method allowing direct contact between nitrogen gas and the molten steel in an ESR. However the heating method of ASR is different from that of a conventional ESR.

The electrode is not immersed in the slag and keeps some distance from slag surface in order to make a stable arc. Therefore ASR requires a special control system using feedback with respect to the current and arc voltage in order to stabilize the arc discharge.

A simpler way of nitrogen addition in ESR was investigated in this study. Gas blow through a center hole of the electrode was tried in an ordinary ESR melting condition. In this process, we may not need a special control system and could use a normal ESR control system. We confirmed how much nitrogen addition is possible by gas blow in a conventional ESR system.

## 2. Experimental procedure

Figure 1 shows the schematic of the experiment using a ϕ145mm mold pressurized electroslag remelting (P-ESR) furnace. A ϕ10mm hole was drilled through the center of the ϕ100mm electrode. Nitrogen gas was blown through the center hole to the electrode tip. The gas flow rate was 0.1-2.0 NL/min. The electrode material was SUS304 (AISI304, X5CrNi18-10) steel. The experimental conditions are shown in Table 1. The experiment was run under 0.98MPa pressurized and 0.1MPa atmosphere conditions. The melt voltage was 18V and the current was 2000A. In addition, a 20V-2500A melt was carried out for the 0.98MPa condition. After the ESR melt, the ingots were longitudinally cross-sectioned in order to perform macrostructure observation and chemical composition analysis. 20mm cubic samples for chemical analysis were taken at interval of 50mm from the bottom of the ingot.

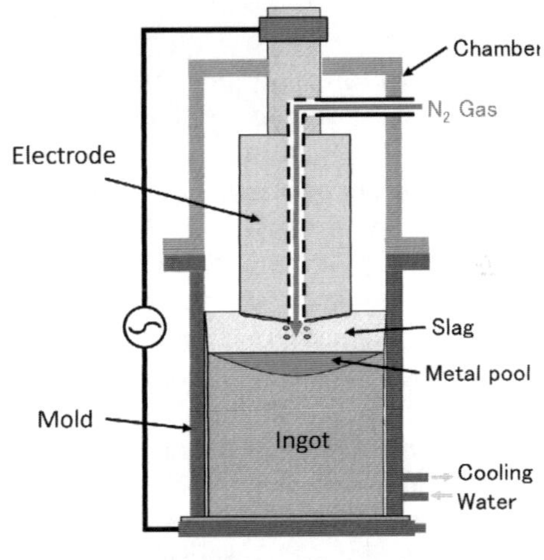


Figure1 Schematic of the Apparatus

Table 1 Experimental condition

| | Atmosphere (MPa) | Voltage (V) | Current (A) | $N_2$ blow rate (NL/min) |
|---|---|---|---|---|
| Ingot 1 | 0.98 | 20 | 2500 | 0.2 – 0.8 |
| Ingot 2 | | 18 – 20 | 2000 | 1.0 – 2.0 |
| Ingot 3 | 0.1 | 18 | | 1 |

In order to confirm the blown gas immersion depth, a water-model experiment was also carried out. The same ϕ145mm mold P-ESR was used for this water model. A transparent plastic container was used for the water bath. It was placed under the electrode in substitution for the water-cooled coper mold. The bubble formation in water was observed from the side and the bottom of the container. The water model was run with gas blow rates of 0.8, 1.6, 3.1, 4.8, and 6.2 NL/min. There are some differences between the water-model and the actual ESR, such as the density of the liquid, the temperature and so on. The gas flow rate for the water-model experiment can be estimated from the modified Froude number $Fr$[4].

$$Fr = \frac{\rho_{g1} Q_{g1}^2}{\rho_{L1} g d_1^5} = \frac{\rho_{g2} Q_{g2}^2}{\rho_{L2} g d_2^5} \quad \cdots \quad (1)$$

where $\rho_g$ is the gas density, $Q_g$ is the gas flow rate, $\rho_L$ is the density of the liquid, $g$ is the standard acceleration due to gravity, and $d$ is the diameter of the nozzle. Equation (1) becomes Equation (2).

$$Q_{g2} = \left( \frac{\rho_{g1} \rho_{L2}}{\rho_{g2} \rho_{L1}} \right)^{1/2} Q_{g1} \quad \cdots \quad (2)$$

The modified Froude number $Fr$, is a dimensionless number that represents the ratio between the forces of inertia and buoyancy. The blow distance can be dependent on the number. The blow rate in the water-model corresponds to blow rate in ϕ145mm SUS304 steel ESR at 1600°C of 0.5, 1.0, 1.8, 2.8, and 3.6 NL/min.

## 3. Result

The melt record of 20V-2000A melt is shown in Figure 2 as an example. The gas blow does little to disturb melting stability. The slag skin thicknesses of the test ingots were 0.5–3.0mm. The thickness values are the same as the ESR melt without gas blow. No clogging occurred in the center hole of the electrode tip after melt and gas blow was uninterrupted during the melt operation. For the blow rate of 0.2 to 1.0NL/min., the voltage swing and current swing value did not vary. However, when the blow rate increased to 2.0NL/min, the voltage and current swing became wider. 2.0NL/min. seemed maximum gas blow rate in the ESR furnace to perform stable melt with gas blow.

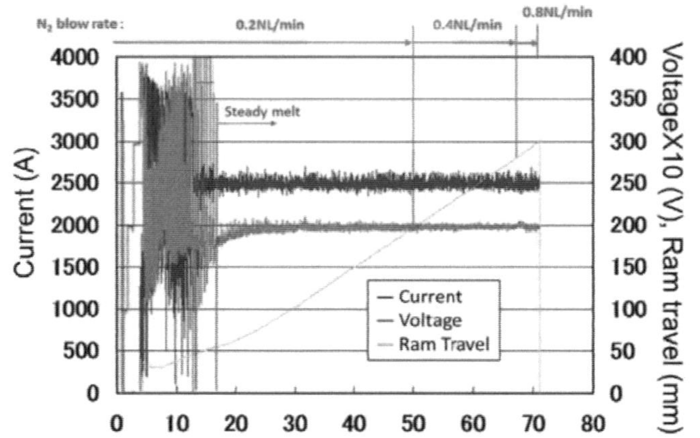


Figure 2 Melt record of Nitrogen blow ESR

The macrostructure of the ingot 1 is shown in Figure 3. An ingot melted without $N_2$ blow is also shown in the figure 3 for comparision. No structural disturbance was observed in the macrostructures. Inclined primary crystals grew from the bottom to the top. There was a slight solidification structure change at the point where the blow rate increased from 0.2NL/min to 0.4NL/min.

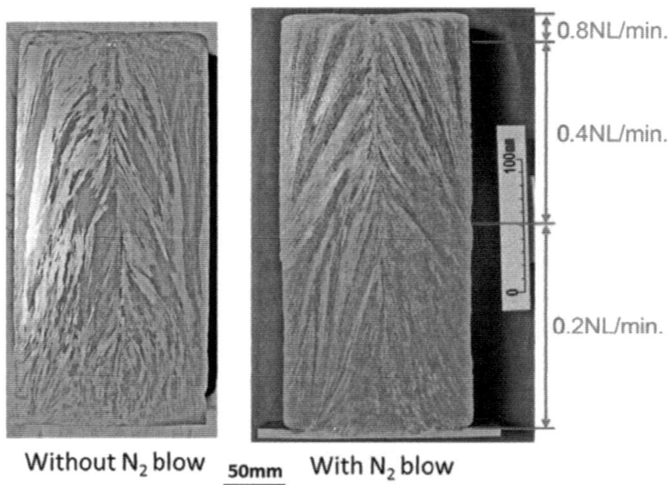


Figure 3 Macrostructure of the ingots

Table 2 shows the chemical composition in the middle of the ESR ingots and the electrodes. The nitrogen increment from the electrode to the ingots is shown in Figure 4. The values of the electrodes were plotted at 0 on the horizontal axis. All the ingots show addition of nitrogen. This figure indicates that nitrogen blowing through an electrode enables the addition of nitrogen into the steel in an ESR process.

In order to investigate the reaction site in this process, test specimens were clipped out from the electrode tip after the ESR melt and from the top of the ESR ingot. The tip of the remaining electrode

Table 2 Chemical composition of test ingots and electrodes (mass%)

| | C | Si | Mn | P | S | Ni | Cr | Cu | Mo | V | Al | N* | O* |
|---|---|---|---|---|---|---|---|---|---|---|---|---|---|
| Electrode1 | 0.049 | 0.44 | 1.33 | 0.03 | 0.0250 | 7.91 | 17.96 | 0.45 | 0.32 | 0.06 | <0.005 | 829 | 117 |
| Ingot1 | 0.051 | 0.41 | 1.32 | 0.03 | 0.0021 | 7.94 | 17.83 | 0.45 | 0.32 | 0.06 | <0.005 | 1600 | 36 |
| Electrode2 | 0.047 | 0.34 | 1.27 | 0.03 | 0.0220 | 7.90 | 18.44 | 0.38 | 0.25 | 0.08 | 0.010 | 636 | 216 |
| Ingot2 | 0.049 | 0.39 | 1.32 | 0.03 | 0.0017 | 8.04 | 18.21 | 0.49 | 0.30 | 0.06 | <0.005 | 1910 | 40 |
| Electrode3 | 0.048 | 0.32 | 1.27 | 0.03 | 0.0200 | 7.87 | 18.50 | 0.39 | 0.25 | 0.08 | 0.013 | 662 | 230 |
| Ingot3 | 0.053 | 0.27 | 1.28 | 0.03 | 0.0017 | 8.04 | 18.44 | 0.40 | 0.26 | 0.08 | <0.005 | 935 | 46 |

*ppm

was clipped out in order to confirm the reaction site of nitrogen absorption. Also a sample from the upper portion of the ingot was clipped out. Those samples were embedded in a resin EPMA mapping

IOP Conf. Series: Materials Science and Engineering **143** (2016) 012005 doi:10.1088/1757-899X/143/1/012005

analysis on the specimen as shown in Figure 5. The left side of the specimen is the top of the ingot and the right side is the tip of the electrode. In mapping for Ni, the columnar structure is shown on the right-hand side of the electrode.

The microstrucure of the left-hand side of the electrode is different from that of the right-hand side. It can be considered that the left-hand side was melted in the ESR process. The N content in this melted portion is the same as that in the ingot. This indicates that the nitrogen absorption reaction in this process occurs at the molten metal film on the electrode.

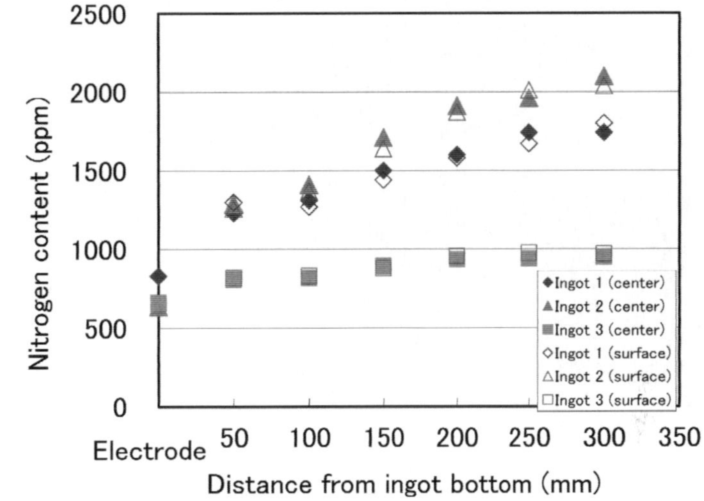

Figure 4 Chemical compositions in test ingots and electrodes

Figure 6 shows gas-bubble formation in the water model as observed from side views and bottom views of the electrode. The gas blow did not spurt out straight into water in all experimental conditions. The gas bubble formed just after the gas came out from the hole. The bubble did not go down because of the buoyancy force and grew underneath the electrode tip. After the bubble reached a certain size, it slid to one side of the electrode and detached from the tip.

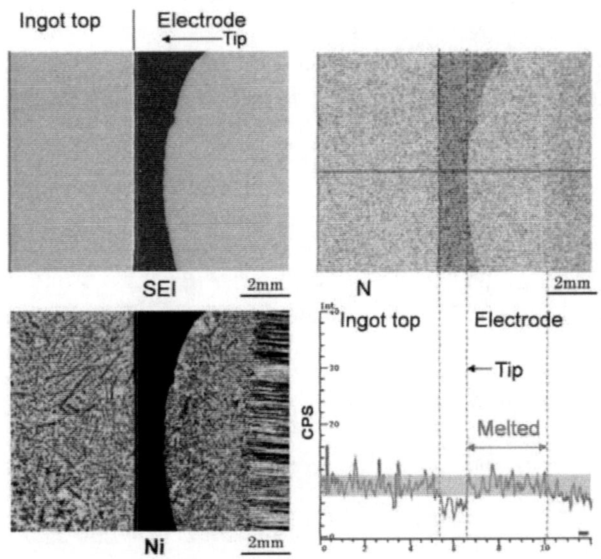

Figure 5 EPMA mapping analysis on the tip of electrode

The relationship between gas blow rate and maximum immersion depth of gas bubble is shown in Figure 7. The maximum immersion depth was around 4.5mm and almost constant for gas blow rates under 4.8NL/min. The maximum immersion depth was about 6.6mm with a blow rate of 6.2NL/min. Nitrogen blow rates in this experiment lower than 2.0NL/min. correspond to rate lower than 3.4NL/min in the water model. The slag weight was 2.2kg. From the slag weight, the slag cap height can be calculated as 50mm. The gas immersion depths around 5mm were much shallower than that of the slag cap height. It is considered that the blown nitrogen bubble did not reach the metal pool interface in this test melt condition.

Direct gas contact with the metal pool surface could not occur in the experimental condition. This observation indicates that this process is different from the one in ASR. Figure 8 shows the time history of the gas bubble projected surface area observation from the bottom of the electrode. The area increased with time and the bubble size reached a certain size. Then the bubble detached from the electrode and a new bubble started to grow again. For each blow rate, the surface area of the gas bubble increased with time in a roughly linear fashion.

Figure 6 Bubble formations in water model

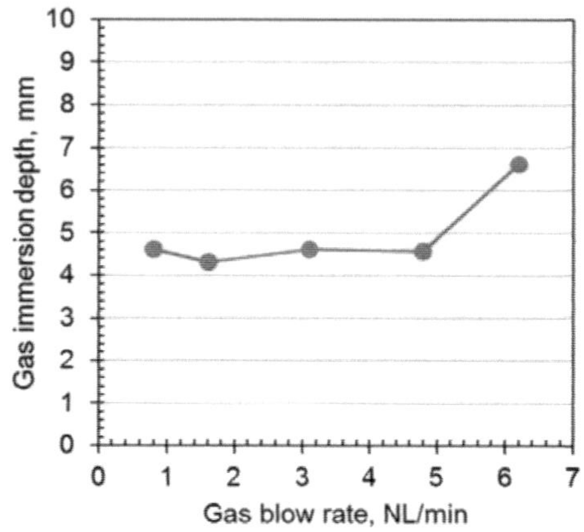


Figure 7 Gas immersion depths in water model

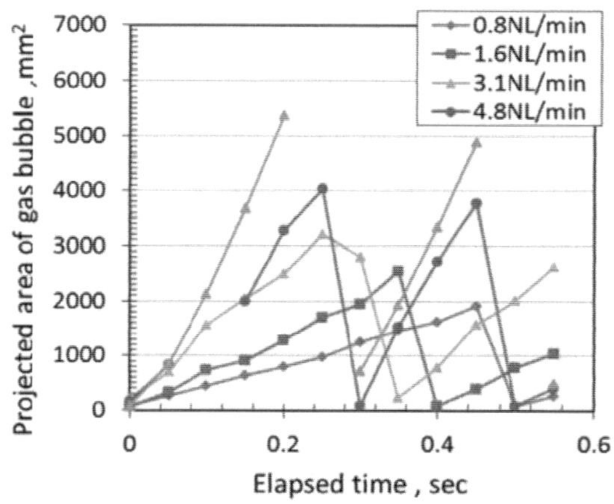


Figure 8 Time history of gas bubble projected area

## 4. Discussion

The apparent mass transfer coefficient was estimated for the gas-absorption reaction under nitrogen-gas blow in the ϕ145mm. P-ESR. The assumed condition for calculation is described here.

(1) The nitrogen-absorption reaction occurs at the molten metal film on the electrode tip. The nitrogen content in the metal droplet remained on the electrode bottom after the ESR. Therefore, it is considered that the nitrogen absorption reaction in this process occurred at the molten metal film on the bottom of the electrode.
(2) The nitrogen-absorption reaction mass transfer rate is limited at the metal side. It obeys the rule of a primary reaction[4].

$$-\frac{dC}{dt} = k_{metal}(\frac{A}{V})(C - C_i) \quad \cdot\cdot\cdot \quad (3)$$

Equation (3) becomes

$$-\ln(\frac{C - C_{eq}}{C_0 - C_{eq}}) = k_{metal}(\frac{A}{V})t \quad \cdot\cdot\cdot \quad (4)$$

Where, $C$ is the nitrogen content in the metal, $C_i$ is the nitrogen content at the gas-metal interface, $C_{eq}$ is the equilibrium nitrogen content for each partial pressure, $C_0$ is the nitrogen content in the metal before nitrogen absorption, $t$ is the reaction time, $A$ is the interfacial area of the metal for reaction, $V$ is the volume of metal for reaction, and $k_{metal}$ is the apparent mass transfer coefficient.
(3) Slag cap/metal pool interfacial temperature at each ingot height
This value was assumed from the Al deoxidation equilibrium value in carbon steel from other experiments using the ϕ145mm. P-ESR.

2000A-20V condition : $T = -0.5112d + 1868.2$
2500A-20V condition : $T = -0.3391d + 1866.3$

Where, $d$ is the distance from the ingot bottom, and $T$ is the slag cap/metal pool interfacial temperature at distance $d$ from ingot bottom.

(4) The temperature vicinity of the electrode bottom (molten metal film)/slag cap for each ingot height is assumed as the slag cap/metal pool interfacial temperature -250°C. This follows the measured slag cap temperature distribution in the ϕ180mm ESR[6].

(5) Metal volume for reaction

This is the volume of molten metal film under the electrode tip. The thickness is assumed as 1mm and the film covered the whole bottom surface of the electrode. This was expected from the EPMA analysis of the electrode tip as shown in Figure 5.

(6) Interfacial area of metal for reaction

This is the average contact area between the nitrogen gas bubble and the molten metal film. The result of the water model is used for this value.

Figure 9 shows the relationship between the distance from the ingot bottom and the apparent mass transfer coefficient $k_{metal}$ for the nitrogen absorption reaction. $k_{metal}$ increases from the bottom to the top of the ingot in either condition. However the increment is not significant and the apparent mass transfer coefficient for the nitrogen absorption reaction did not depend on the ESR ingot height significantly. The value is around $1.0\times10^{-2}$cm/sec.

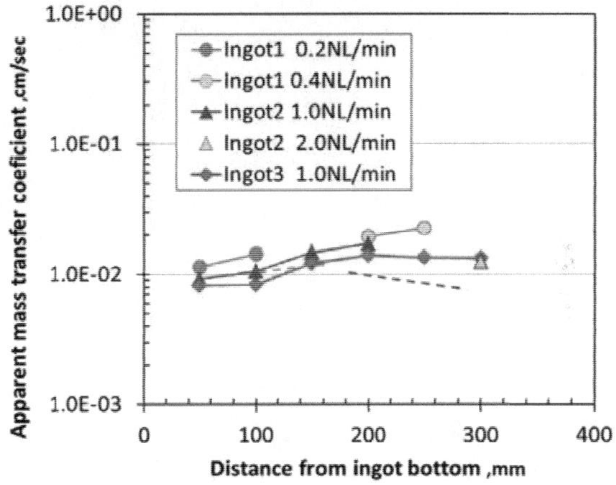


Figure 9 Relationship between distance from ingot bottom and apparent mass transfer coefficient

Table 3 shows the apparent mass transfer coefficients of the nitrogen absorption reaction for the Fe, Fe-Cr, and Fe-Cr-Ni systems at 1600°C[7-11]. These values are for reactions measured at the free surface of the molten steel. The $k_{metal}$ value estimated from this experiment is also given in Table 3. The $k_{metal}$ values estimated from this experiment are consistent with the value reported by Fedorchenko[9]. The estimated value is within the range of the reported values. Nitrogen absorption rates in the longitudinal direction of the ingot could be constant since the $k_{metal}$ values are almost constant as shown in Figure 9. The transition of the nitrogen contents in the ingots are shown in Figure 10. The equilibrium nitrogen contents for each location are also plotted in this figure. These equilibrium nitrogen contents are calculated from the estimated temperature vicinity of the electrode bottom/slag pool. Both nitrogen contents and equilibrium nitrogen contents increased from the bottom to the top of the ingot. Longitudinal variation in the nitrogen contents of the ESR ingots could be due to equilibrium nitrogen contents with metal temperature change in the vicinity of the slag/molten metal film interface. In case of small size ESR such as this apparatus, it may difficult to obtain constant slag temperature. However if the furnace size becomes larger, the temperature change becomes milder. It could be available to obtain near constant temperature condition in a large sized commercial ESR and to make much uniform nitrogen content.

Table 3 Mass transfer coefficients of Nitrogen absorption reaction

| | Alloy type | $k_{metal}$ |
|---|---|---|
| Chou [7] | Fe | 0.0345 |
| | Fe-18Cr | 0.0342 |
| Pehlke [8] | Fe-20Cr | 0.022 |
| Fedorchenko [9] | Fe-18Cr | 0.014 |
| | Fe-18Cr-10Ni | 0.012 |
| Gruszcyk [10] | PN-88 | 0.0065 |
| Takahashi [11] | Fe-10Cr | 0.0088* |
| Present Work | Fe-18Cr-8Ni | 0.0103 |

*) 1823K

## 5. Conclusion

Nitrogen addition by gas blow through a center hole of an electrode was tried with ordinary ESR melting conditions.

- Nitrogen addition by nitrogen gas is possible in a ordinary ESR system using gas blow through an electrode.
- Blown gas from center of the electrode could not penetrate through the slag. There was no direct gas contact on the metal pool surface in the experimental condition.
- The mass transfer coefficient for the nitrogen absorption rate was around $1x10^{-2}$ cm/sec.
- Longitudinal variation in the nitrogen contents of the ESR ingots could be due to equilibrium nitrogen contents with metal temperature change in the vicinity of the slag/molten metal film interface.

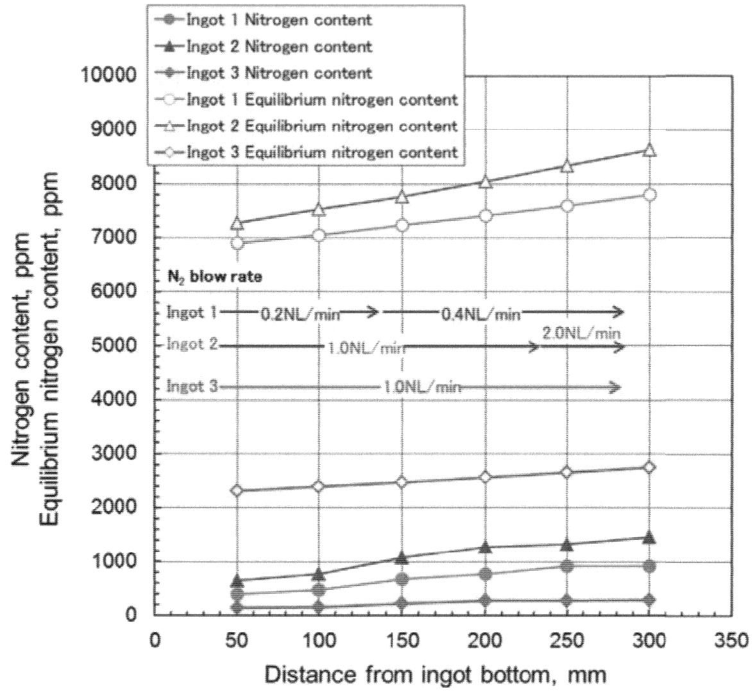


Figure 10 Comparison between nitrogen content and equilibrium nitrogen content calculated from chemical composition.

## 6. Acknowledgments

We greatly thank Prof. Alec Mitchell of the University of British Columbia for useful discussion.

## References

[1] P. G. Uggoweizer and M. Harzermoser: High Nitrogen Steel (1988) p. 174
[2] S. Azuma, H. Miyuki, and T. Kudo: ISIJ int., 36(1996) p.793
[3] B. I. Medovar, V. Ya. Saenko, G. M. Grigorenko, Yu. M. Pomarin, and V. I. Kumysh: “Arc-Slag Remelting of Steel and Alloys”, (1996) Cambridge International Science Publishing.
[4] M. Byrne and G. R. Belton: Metall. Trans. B, 14B(1983) P.441
[5] K. Amano ,K. Ito and H. Sakao: Tetsu-to-Hagane, 62(1976) p.1179
[6] M. Kawakami, K. Nagata, M. Yamamura, N. Sakata, Y. Miyashita, and K. Goto: Tetsu-to-Hagane, 63(1977) No.13, p.2162
[7] T. Choh, T. Yamada, and M. Inoue:Tetsu-to-Hagane, 62(1976) No.3, p334
[8] R. D. Pehlke and J. F. Elliott: Trans. Met. Soc. AIME, 227(1963), p.884
[9] V.I. Fedorchenko V. V. Averin: The Forth Japan-USSR Joint Symposium Physical Chemistry of Metallurgical Process, (1973) p.167
[10] A. Gruszczyk: J. of Achiev. in Manuf. Eng. 26(2008) p.115
[11] F. Takahashi, Y. Momoi, K. Kajikawa, and H. Yamada: Tetsu-to-Hagane, 97(2011) p.525
[12] W. P. Wu and D. Janke: Ironmaking Steelmaking, 23(1996) p.247

IOP Conf. Series: Materials Science and Engineering **143** (2016) 012006 doi:10.1088/1757-899X/143/1/012006

# An attempt to model electrode change during the ESR process

**E Karimi-Sibaki [a], A Kharicha[a,b], M Wu[a,b], A Ludwig[b], H Holzgruber[c], B Ofner[c], A Scheriau[c], M Kubin[c], and M Ramprecht[c]**

[a]Christian-Doppler Lab for Adv. Process Simulation of Solidification & Melting,
[b]Chair of Simulation and Modeling of Metallurgical Processes,
University of Leoben, Franz-Josef-Str. 18, 8700 Leoben, Austria
[c]INTECO Special Melting Technologies GmbH, 8600 Bruck/Mur, Austria

E-mail: abdellah.kharicha@unileoben.ac.at

**Abstract.** The electrode change technology is used to produce very large heavy ingots in which a number of electrodes are remelted one after another during the ESR process. Preparing the new electrode for remelting requires a certain period of time when the electric current is stopped (power off). Here, CFD simulation is used to study the behavior of a large scale ESR process during the electrode change (power off). Firstly, the electromagnetic, temperature, and turbulent flow fields in the process before electrode change are modelled. Mold current and thermal effect due to shrinkage of ingot is considered in the model. Then, a transient simulation is performed and the response of the system to the power off is continuously tracked. It is observed that the pool profile of ingot is preserved before and after electrode change. Details of the flow and temperature distributions during electrode change are presented in the paper.

## 1. Introduction

Nowadays, the demand for very large heavy ingots (> 100 tons) through ESR process is increased especially in chemical, oil and gas industry. Producing a large ingot by ESR process is quite a long process that might take several days. Manufacturing a large ingot requires electrode change technology in which several smaller electrodes are remelted to produce one big ingot. Special care must be taken to avoid internal and surface defects during electrode change due to power interruption in the system.

The entire electrode change procedure is divided to three steps. Firstly, within a very short time (< 5 min), the electric power is turned off, and a new preheated electrode is prepared to replace the remelted last one. Secondly, the electric power is again turned on. Consequently, the temperature of the new electrode rises to reach the melting temperature. Finally, the melt rate increases until reaching the target melt rate. As stated by Holzgruber [1], no traceable change in ingot internal composition was observed during the power interruption. However, Matushkina [2] reported slight defects of surface during electrode change. Jackson [3] monitored thermal field in mold using thermocouples to analyze heat balances across the ingot during power interruption. It was observed that solidification occurred much more rapidly at the pool periphery rather than in the ingot center. It was also found that gross changes in structure and composition of ingot are not expected unless power interruption lasts for too long.

There are only few reports available on this topic. As such, further investigation is required to clarify the influence of power interruption on the process. It is a very first attempt to model electrode change based on CFD simulation. Here, a numerical study is conducted in which the interaction between the turbulent flow, heat, and electromagnetism are taken into account. The model includes the effect of interface motion between molten slag and melt pool. A short collar mold ESR process is studied in which the system behavior is continuously tracked. Formation of surface defects is not explicitly modeled. The thickness of slag skin is kept constant during the whole simulation. This assumption is based on the industrial measurement of skin thickness in plant trial for our case study where a uniformly thin layer is observed along the whole ingot.

Here only results including details of velocity and temperature fields regarding to the first step of electrode change procedure (power off) is presented. The melt pool profile is used as an indicator of the internal quality of ingot. Furthermore, variations in heat fluxes through boundaries are discussed.

## 2. Model

A 2D axisymmetric computational domain with a short collar mold is considered [4]. The computational domain includes electrode, mold, slag, and ingot. The Finite Volume Method (FVM) is used to model turbulent flow, temperature, and electromagnetic fields. The A-ϕ formulation is applied to compute the electromagnetic field in the whole domain where $\phi$ is the electric scalar potential and $A$ denotes the magnetic vector potential ($\nabla \times \vec{A} = \vec{B}$). Details of A-ϕ method applied to solve the electromagnetic field for ESR process were described by Kharicha [5]. Note that, the electric current is allowed to cross the slag skin layer entering into the mold before electrode change (power off) [6-9]. After computing the magnetic field ($B$), the Lorentz force and Joule heating are calculated and added as source terms to momentum and energy equations.

The flow field in slag and melt pool is calculated. The drag resistance of the dendritic mushy zone to the flow is taken into account using the well-known Blake-Kozeny permeability model [10]. Using the previously described model for droplets as mass, momentum, and energy carriers, the impacts of falling droplets are taken into account [11]. Furthermore, the model includes the movement of slag and melt pool interface.

The turbulence is modeled using Scale-Adaptive Simulation (SAS) approach. The advantage of model is the resolution of turbulent spectrum in unstable flow conditions such as interface movement [12]. The temperature field is modeled by solving an enthalpy ($h$) conservation equation:

$$\frac{\partial}{\partial t}(\rho h) + \nabla \cdot (\rho \vec{u} h) = \nabla \cdot (\lambda \nabla T) + Q + S_{LH} \tag{1}$$

Where ($\lambda$) is thermal conductivity, ($\rho$) is density, ($u$) is velocity, ($T$) is temperature, ($Q$) is Joule heating, and ($S_{LH}$) is the source term for treating latent heat released by solidification of the ingot:

$$S_{LH} = -\frac{\partial}{\partial t}(\rho f L) - \rho L \vec{u}_s \nabla f \tag{2}$$

In Eq. 2, ($L$) is latent heat of fusion, ($u_s$) denotes the casting velocity and ($f$) is the liquid fraction. A constant thickness of slag skin (~ 5 mm) is assumed for the whole simulation. A combined convection-radiation condition is considered for slag free surface where the emissivity coefficient is 0.8. Furthermore, formation of air gap between slag skin layer and mold due to shrinkage of ingot is modeled. For the latter, the contact between slag kin and mold wall is empirically assumed to be lost once the temperature at mold wall becomes 100 K lower than alloy solidus temperature. At the contact region the heat is conducted through slag skin to the mold. Otherwise, a combined radiation-convection condition is used to model heat transfer through the air gap. Therefore, a fully coupled simulation is performed including the effect of mold current and the thermal effect of solidifying ingot shrinkage. A summary of parameters used in the simulation is described in Table 1.

Table 1. A summary of parameters used in the simulation.

| Mater. properties | Slag | Steel | Operation parameters | |
|---|---|---|---|---|
| Density (kg.m$^{-3}$) | 2440 | 7000 | Ingot diam. (m) | 1.979 |
| Viscosity (Pa.s) | 0.01 | 0.0062 | Elec. diam. (m) | 1.1 |
| Specific heat (J.kg$^{-1}$) | 1255 | 500-800 | Ingot length (m) | 1.245 |
| Liq. therm. cond. (W.m$^{-1}$.k$^{-1}$) | 1.5-5 | 25-40 | Slag height (m) | 0.307 |
| Sol. ther. cond. (W.m$^{-1}$.k$^{-1}$) | 0.5 | 16 | Melt rate (kg.hr$^{-1}$) | 2200 |
| Ther. exp. Coeff. (K$^{-1}$) | 0.0001 | 0.00011 | RMS current (kA) | 46 |
| Liquidus temp. (K) | 1715 | 1773 | Freq. (Hz) | 0.2 |
| Solidus temp. (K) | 1598 | 1668 | Power (MW) | 2.407 |
| Liq. e. cond. (ohm$^{-1}$.m$^{-1}$) | 180 | 8.8x10$^5$ | Elec. change time (s) | 175 |
| Sol. e. cond. (ohm$^{-1}$.m$^{-1}$) | 15 | 8.8x10$^5$ | | |

First, the process is simulated before electrode change, and temperature, velocity, and electromagnetic fields are obtained. Afterwards, the following conditions are imposed on the system during the first step of electrode change (power off). The electric current is stopped flowing through the system, thus the Lorentz force and Joule heating are vanished. The electrode melting is ceased (no droplets), and subsequently the casting velocity is changed to zero. Additionally, the whole slag surface is considered to be exposed to the air.

## 3. Results and discussions

Figure 1 shows the modeling results of process before electrode change (power interruption) including magnetic flux density, current density, velocity, temperature, turbulence kinetic energy, and turbulence effective thermal conductivity. The generated power in the process is slightly oscillating around the average value (~ 2.4 MW) due to interface movement. As shown in Figure 1(a), a portion of electric current (~ 5%) enters to the mold in slag zone then returns to ingot through the contact region (mold and melt pool). The standing height (liquid metal in direct contact to mold) is influenced by the mold current [11, 13]. The flow and temperature distributions are shown in Figure 1(b). The flow is rotating clock-wisely (converging flow) under the edge of electrode in the slag zone that indicates the Lorentz force is dominant. However, the flow is governed by thermal buoyancy and droplets impact in the bulk of liquid melt pool. Stronger flow under the shadow of electrode is due to the effect of droplets that bring momentum and thermal energy to melt pool. Furthermore, the movement of slag-melt pool interface certainly intensifies the flow in the liquid melt pool [14]. The maximum elevation of interface is observed to be around 7 mm in the simulation. Kharicha [15] reported a maximum elevation of 30 mm for a smaller scale ESR system (Diam. ~ 0.6 m). Nevertheless, with the increase of system size, the current density which is proportional to the inverse of square of diameter decreases in the domain. As such, a smoother movement of interface is expected. Additionally, the temperature distribution and isolines of solid fraction (indicating pool profile) are shown in Figure 1(b). The temperature field is relatively uniform in the melt pool and slag indicating rigorous mixing in the system. The slag is very hot under the shadow of electrode where the current density and subsequently released Joule heating is high. The distributions of turbulence kinetic energy and turbulence effective thermal conductivity are shown in Figure 1 (c). The turbulence kinetic energy is as notably strong in melt pool as in the slag region. Strong turbulence in liquid melt pool has been also reported by Kharicha [16-17] after simulation of the electrode melting in a fully coupled 3D simulation. It must be noted that a region of intense turbulence is created at upper part of melt pool because of interface movement. As shown in Figure 1 (c), the effective thermal conductivity of slag can dramatically increase (~ 150 $W.m^{-1}.K^{-1}$) in the bulk due to strong turbulence. Therefore, the energy transfer is efficiently increased, and it leads to relatively uniform temperature in the slag. However, the turbulence is damped near the walls and slag-melt pool interface where the effective thermal conductivity of slag is in the same magnitude of the molecular thermal conductivity (~ 5 $W.m^{-1}.K^{-1}$). Snapshots of flow, temperature, and solidification during the first step of electrode change (power off) are illustrated in Figure. 2. The pool profile is slightly influenced near the mold wall where the cooling rate is large. However, the flow distribution is totally altered. The clock-wise rotating vortex under the edge of electrode (converging flow) disappears in slag zone. During power interruption, the flow is completely driven by buoyance force results in counter-clock wise rotation of flow (diverging flow) in the slag. Furthermore, the magnitude of velocity decreases in the bulk as the time proceeds whereas the flow is still strong near the mold wall due to thermal buoyancy. The temperature response is observed to be much slower than flow response to the power interruption. Based on the simulation result, a minor impact on the pool profile is observed during power interruption.

The heat loss through the slag free surface and mold normalized by their magnitudes before power interruption is plotted against time as shown in Figure. 3. The amount of heat loss through the mold (~ 0.9 MW) remains almost constant since the thickness of slag skin is invariable. However, the heat loss through slag free surface (slag-air interface) during electrode change is notably larger compared to the value before power interruption (~ 0.8 MW). The heat loss through slag free surface is intensified at the beginning of power interruption due to increase in contact area between slag and air. As time proceeds, the temperature of slag at free surface decreases. Consequently, the convective-radiative heat loss at slag free surface is decreasing. Afterwards, increase in heat loss through slag-air interface is observed once again. Because of thermal buoyancy, the colder liquid slag on the slag-air interface sinks whereas hot liquid slag moves toward the free surface. This behavior of flow is clearly demonstrated in Figure 2(e-f).

According to our investigation, it is even possible to solidify slag at the free surface. Nevertheless, it is highly unlikely that solidified slag layers float on the surface because of having higher density compared to the liquid slag. Most probably, the floating solidified slag pieces sink into the hot slag bath where they remelt once again.

Note that the entire electrode change procedure is not only the first step (power off). The procedure also includes heating of the new electrode until reaching melting temperature (second step) and remelting of the electrode until reaching the target melt rate (third step). Generally, those steps take much longer time (~ 30 min) compared to the first step (power off). During second and third steps, operation conditions of process such as input power or melt rate are dynamically changing. Further investigations are required to explore the behavior of the process during second and third steps.

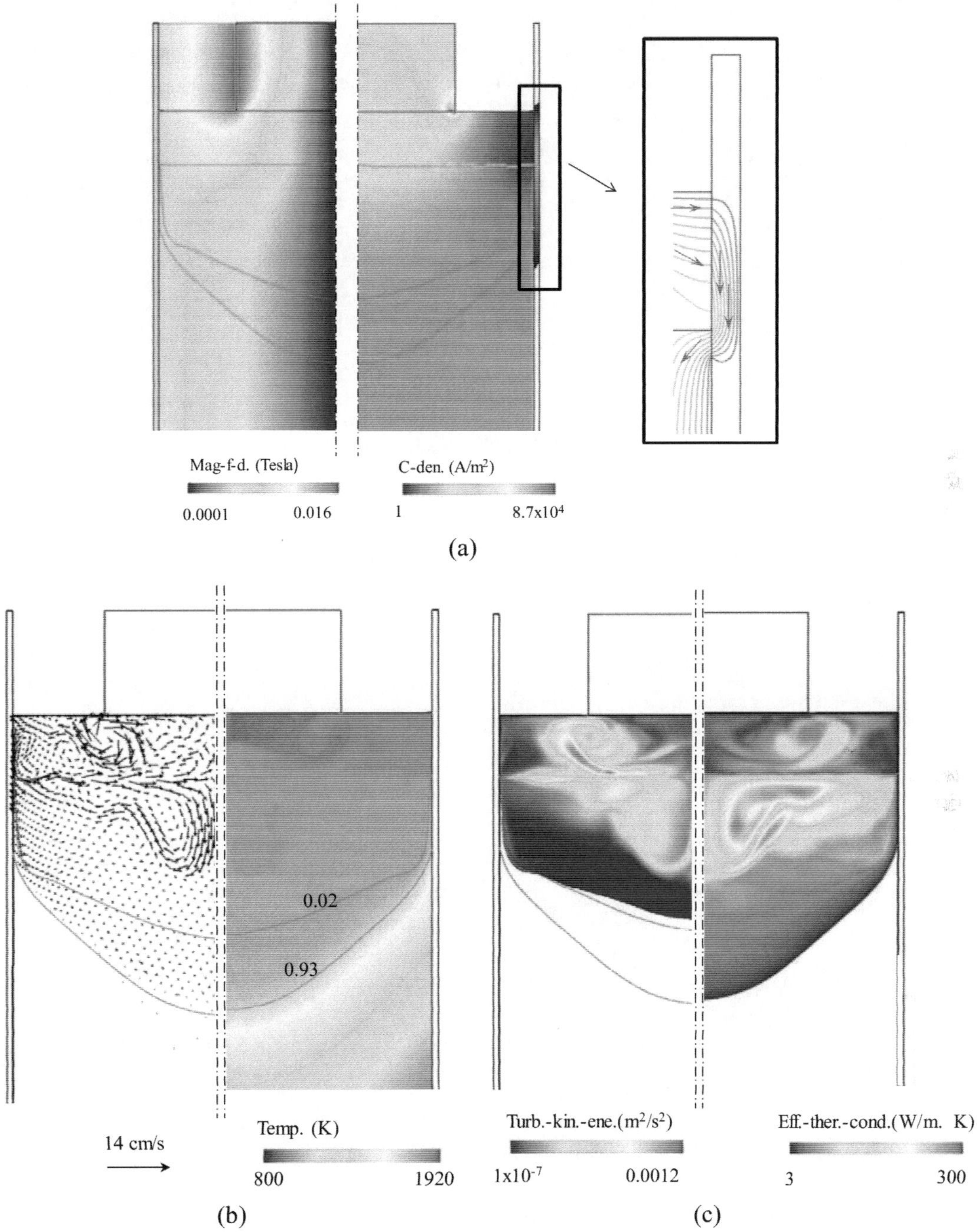


Figure 1. (a) Left: contour of magnetic flux density, Right: contour of electric current density (red arrows show direction of electric current), (b) Left: vectors of velocity, Right: contour of temperature overlaid with isolines of solid fraction, (c) Left: contour of turbulence kinetic energy, Right: contour of turbulence effective thermal conductivity in slag and melt pool.

IOP Conf. Series: Materials Science and Engineering **143** (2016) 012006 doi:10.1088/1757-899X/143/1/012006

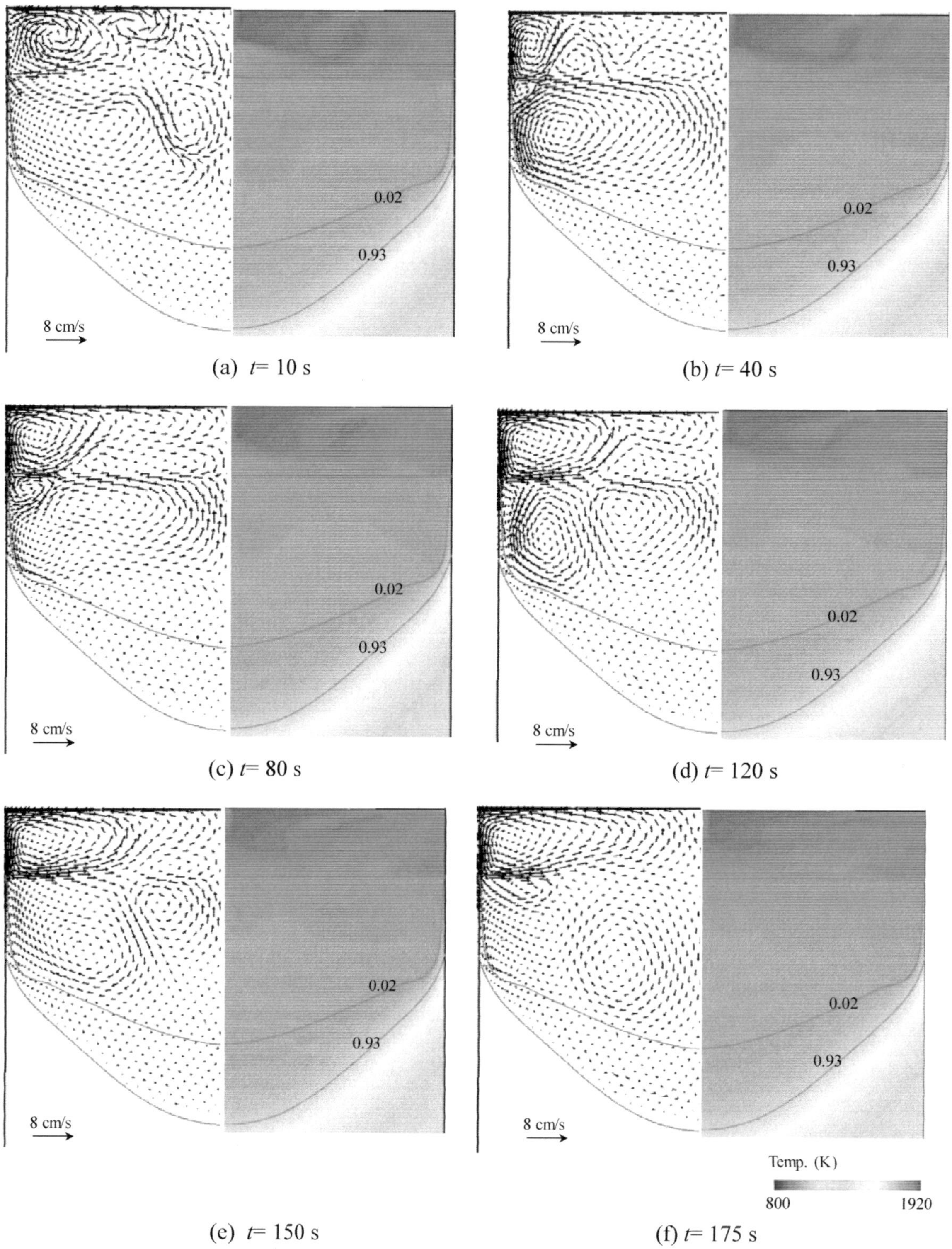


Figure 2. Snapshots of dynamic change of thermal and velocity fields at different times during power off: (a) $t$= 10 s, (b) $t$= 40 s, (c) $t$= 80 s, (d) $t$= 120 s, (e) $t$= 150 s, (f) $t$= 175 s, Left: vectors of velocity, Right: contour of temperature overlaid with isolines of solid fraction.

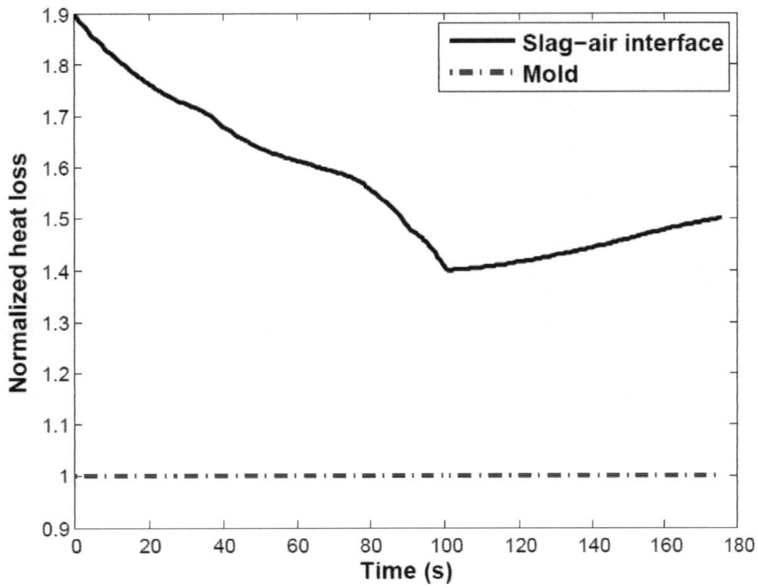


Figure 3. Normalized heat loss through slag-air interface and mold wall is plotted versus time during the first step of electrode change (power off).

## 4. Summary

Producing a large ingot by ESR process requires electrode change technology in which a number of smaller electrodes are remelted one after another. The electrode change procedure is divided into three steps. Firstly, the electric power is turned off (power off), and the end of remelted electrode is replaced by a new preheated electrode. Secondly, the electric power is turned on again, and the new electrode is heated to its melting temperature. Finally, the new electrode is continuously remelted until reaching target melt rate. Here, the influence of power interruption on a large scale ESR process during the first step of electrode change (power off) is studied using CFD simulation tool. A 2D axisymmetric computational domain is considered where the slag-melt pool interface is allowed to move. The interactions between turbulent flow, electromagnetic, and temperature fields are considered. The current is allowed to flow through the short collar mold used in the process. Additionally, the thermal effect of shrinkage of ingot is taken into account in the model. The behavior of system during electrode change including details of flow, and temperature distributions as well as heat loss through boundaries are discussed. Generally, the pool profile is used as an indication of ingot internal quality. Therefore, preserving the original pool profile before and after power interruption is a symptom of maintaining ingot quality. According to simulation results, changes in pool profile of ingot are insignificant during the short period of electrode change (power off). The current study included only the first step of electrode change procedure. Further investigations are required to explore the behavior of process during second and third step of electrode change procedure.

## 5. Acknowledgement

The authors acknowledge the financial support by INTECO GmbH and the Austrian Federal Ministry of Economy, Family and Youth and the National Foundation for Research, Technology and Development within the framework of the Christian Doppler Laboratory for Advanced Process Simulation of Solidification and Melting.

## 6. References

[1] Holzgruber W, Kubisch C, and Jaeger H H 1971 *Neue Hutte* **16** 606
[2] Matushkina L I, Klyuver M, Dedushev L A, Kosyrev L L, Volkov S E, and Sharapov A, 1970 *Sb. Tr. Tsent. Nauch. Issted. Inst Chern. Met.* **75** 167
[3] Jackson R O, Mitchell A, and Luchok J 1972 *J. Vac. Sci. Tech.* **9** 1301
[4] Karimi-Sibaki E, Kharicha A, Korp J, Wu M, Ludwig A 2014 *Mater. Sci. Forum* **790** 396
[5] Kharicha A, Wu M, and Ludwig A 2014 *ISIJ Int.* **54** 1621
[6] Kharicha A, Karimi-Sibaki E, Wu M, and Ludwig A 2013 *Proc. Of LMPC* (New York: Wiley–Interscience) p 95
[7] Karimi-Sibaki E, Kharicha A, Wu M, Ludwig A, Holzgruber H, Ofner B, Ramprecht M 2014 *Proc. of ICRF* (Milan) on data storage device
[8] Kawakami M, Nagata K, Yamamura M, Sakata N, Miyashita Y, and Goto K S 1977 *Testsu-to-Hagane* **63** 220
[9] Holzgruber H, Holzgruber W, Scheriau A, Knabl M, Kubin M, Korp J, Pierer R 2011 *Proc. of LMPC* (Nancy) p 57
[10] Voller V R, and Prakash C 1978 *Int. J. Heat Mass Trans.* **30** 1709
[11] Karimi-Sibaki E, Kharicha A, Wu M, Ludwig A, Holzgruber H, Ofner B, Ramprecht M 2013 *Proc. of LMPC* (New York: Wiley–Interscience) p 13
[12] Menter F R and Egorov Y 2005 *AIAA paper 2005-1095, Reno/NV*
[13] Kharicha A, Schutzenhofer W, Ludwig A, Tanzer R, Wu M 2007 *Proc. of STEEL SIM* (Graz) p 105
[14] Kharicha A, Ludwig A, Wu M 2005 *Mater. Sci. Eng. A* **413-414** 129
[15] Kharicha A, Ludwig A, Wu M 2011 *Proc. of LMPC* (Nancy) p 41
[16] Kharich A, Ludwig A, and Wu M 2011 *EPD congress* (San Diego) p 771
[17] Kharicha A, Wu M, Ludwig A, Ramprecht M, Holzgruber H 2012 *CFD modelling and simulation in materials* (New York: Wiley–Interscience) p 139

IOP Conf. Series: Materials Science and Engineering **143** (2016) 012007 doi:10.1088/1757-899X/143/1/012007

# Characterization of structure and thermophysical properties of three ESR slags

**A. Plotkowski**[1], **J. deBarbadillo**[2], **Matthew J. M. Krane**[1]
[1]Purdue Center for Metal Casting Research, School of Materials Engineering
Purdue University, West Lafayette, Indiana, USA
[2]Special Metals Corporation, Huntington, West Virginia, USA

E-mail: aplotkow@purdue.edu

**Abstract.** The structure and properties of electroslag remelting (ESR) slags were characterized. Slags samples of three compositions were obtained from industrial remelting processes at Special Metals Corporation and from casting in a laboratory vacuum induction melter. The structure of the slag samples was observed using optical and electron microscopy, and phases were identified and their relative amounts quantified using X-ray diffraction. Laser flash thermal diffusivity, density, and differential scanning calorimetry measurements for specific heat were performed to determine the bulk thermal conductivity of the samples. Sample porosity was measured as a function of depth using a serial sectioning technique, and a one-dimensional computational model was developed to estimate the thermal conductivity of the fully dense slags. These results are discussed in context with previous studies, and opportunities for future research are identified. AFRL Case Number: 88ABW-2015-1871.

## 1. Introduction

The scale, cost, and complexity of the ESR process makes it difficult to investigate experimentally. Consequently, a variety of simulation efforts have been undertaken [1–4] to expand the understanding of the effects of process parameters on ingot quality. Slag property data, which are necessary inputs to such models, is sparse in the literature. Mills and coworkers [5-9] have reported solid slag results of various compositions from the ternary $CaF_2$-$Al_2O_3$-CaO and constituent binary systems. They used density, DSC, and laser flash diffusivity measurements to determine thermal conductivity. However, they produced samples via sintering and did not compare the resulting structure to actual slag skins, making it unclear if their results are relevant to ESR simulations. This study seeks to characterize the structure and thermophysical properties of solid slags taken from ESR trials, as well as some produced via solidification in a VIM. Porosity data is also reported, and a simple numerical model was developed to predict the fully dense slag conductivities.

## 2. Experimental Procedure

Slag samples were analyzed either in the as-received state from the ESR process, or after remelting and casting in a VIM at Purdue University. A variety of slag skin compositions were made available from ESR trials at Special Metals Corporation in Burnaugh, KY. Two compositions (Slags A and B) were chosen based on the suitability of their thickness and surface roughness for laser flash diffusivity testing. A third composition, Slag C (Wacker Electroflux ESR2059), was investigated, but ESR samples were not available at the time of property testing, so only VIM samples (produced from virgin slag as received by Special Metals Corporation) were measured. ESR samples of Slag C later became available for microstructural characterization and measurements of phase fractions. The compositions of the three slags are shown in table 1. These values are approximate since reactions with the metal droplets are expected to alter the overall composition during the ESR process. Figure 1 shows the phase diagram of the ternary system and the approximate location of each of the slags.

**Table 1.** Approximate composition (in wt%) of the slags investigated.

| Slag Designation | $CaF_2$ | CaO | $Al_2O_3$ | MgO | $TiO_2$ |
|---|---|---|---|---|---|
| A | 40 | 30 | 30 | - | - |
| B | 60 | 13 | 13 | 5 | 9 |
| C | 49.95 | 19.41 | 21.92 | 5.28 | 2.64 |

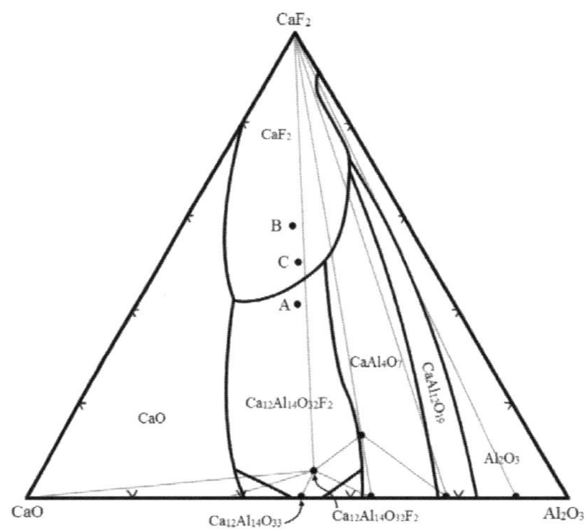


**Figure 1.** A simplified ternary phase diagram showing a projection of the liquidus surface and the approximate location of the three slag compositions (after Busch [10]).

VIM samples were prepared from ESR slag skins with the exception of Slag C in which VIM samples were prepared from virgin slag. VIM samples were melted in a graphite crucible in an argon atmosphere and poured into a steel mold cooled by a water cooled copper bottom plate. The microstructure of samples for the three compositions and each processing route were evaluated by optical and electron microscopy. Phases were identified via XRD and phase fractions were quantified from the XRD spectra by Rietveld refinement [11,12] using the software MAUD [13] and crystallographic data [14-17]. Busch previously found that the phases $Ca_{12}Al_{14}O_{32}F_2$ and $Ca_{12}Al_{14}O_{33}$, shown in figure 1, are practically indistinguishable via XRD, so $Ca_{12}Al_{14}O_{32}F_2$ was assumed where relevant. For electron microscopy, samples were sputter coated with Au-Pd and phase identification was performed using energy dispersive spectroscopy (EDS).

Density, specific heat, and laser flash diffusivity measurements were performed by TPRL, Inc. in West Lafayette, IN. Measurements were performed from room temperature to 1200°C. Once density, specific heat, and thermal diffusivity are measured, the thermal conductivity can be calculated:

$$k = \rho c_p \alpha \quad (1)$$

where $k$ is thermal conductivity, $\rho$ density, $c_p$ constant pressure specific heat, and $\alpha$ thermal diffusivity.

Several samples were prepared for each condition. The slag skins tended to exhibit a layered structure, which may reduce the accuracy of the effective value determine by laser flash measurement.

However, Absi et al. [18] showed that the transient response can still give an accurate effective value of the diffusivity if the phase with the higher diffusivity accounts for less than half the volume fraction of the sample. To assess this criterion, VIM samples of slags A and C were prepared with the thin outer layer (corresponding to rapid solidification at the chill surface) removed by grinding, and their diffusivities were compared to the standard samples. Table 2 shows a summary of the samples that were prepared for thermophysical property measurements.

**Table 2.** Number of samples prepared for thermophysical property measurements.

| Composition | ESR | VIM |
|---|---|---|
| A | 3 | 2 |
| A (1 layer) | 0 | 2 |
| B | 3 | 2 |
| C | 0 | 5 |
| C (1 layer) | 0 | 2 |
| Totals | 6 | 13 |

A significant amount of porosity was observed in the ESR samples, whereas the VIM samples were comparatively sound. These samples were mounted to a steel puck using Crystalbond, and serially sectioned in the solidification direction using a surface grinder, with stereographs taken every 50 μm and pores were analyzed using ImageJ.

## 3. Results

Table 3 shows the phase fractions for each slag and processing route, averaged over the relevant specimens. The differences between ESR and VIM samples for Slags A and B are generally only a few weight percent. However, in Slag C, there is a significant different in $Ca_{12}Al_{14}O_{32}F_2$ content which may be attributed to the difference in composition between the ESR and VIM samples, since the VIM samples were produced from virgin slag that did not undergo reactions with metal during the ESR process.

**Table 3.** Summary of phase fractions (in wt.%) determined by Rietveld refinement of XRD spectra. All values are averages over several samples.

| | Slag A | | Slag B | | Slag C | |
|---|---|---|---|---|---|---|
| Phase | ESR | VIM | ESR | VIM | ESR | VIM |
| $CaF_2$ | 31.7 | 27.7 | 44.5 | 44.4 | 49.9 | 48.3 |
| $Ca_{12}Al_{14}O_{32}F_2$ | 68.3 | 72.3 | 0.9 | 1.9 | 40.0 | 44.5 |
| $MgAl_2O_4$ | - | - | 20.3 | 18.7 | - | - |
| $CaTiO_3$ | - | - | 34.3 | 35.0 | 4.1 | 4.3 |
| MgO | - | - | - | - | 6.0 | 2.9 |

Figure 2 shows optical micrographs comparing the structure of each slag composition between the ESR and VIM samples. Slag A shows similar phase structures between the two processing routes, beginning with a fine combination of two phases near the mold surface (bottom of the micrographs) and coarsening with distance from the mold. EDS results showed that the light phase is primary $Ca_{12}Al_{14}O_{32}F_2$ and the dark phase is $CaF_2$. Note the large pores in the Slag A ESR samples that are not present in the VIM micrograph.

The ESR samples of Slag B consist of three distinct layers. A fine structure near the mold surface which transitions to a middle layer containing two different precipitates in a primarily dendritic matrix with some eutectic material, and an outer layer with a courser dendritic structure containing a high fraction of interdendritic eutectic. The VIM structure is comparatively uniform containing the two precipitate phases and a complex eutectic matrix. The darker precipitates were identified as $CaTiO_3$, while the lighter are $MgAl_2O_4$. The remaining structures were found to contain all the constituent elements, suggesting that either the volume of interaction of the electron beam for EDS was large compared to the phase size, or that one or more of the phases has a high solid solubility for the remaining elements. Similar to Slag A, the ESR samples of Slag B exhibited significant porosity (though generally less than Slag A), while the VIM samples were sound.

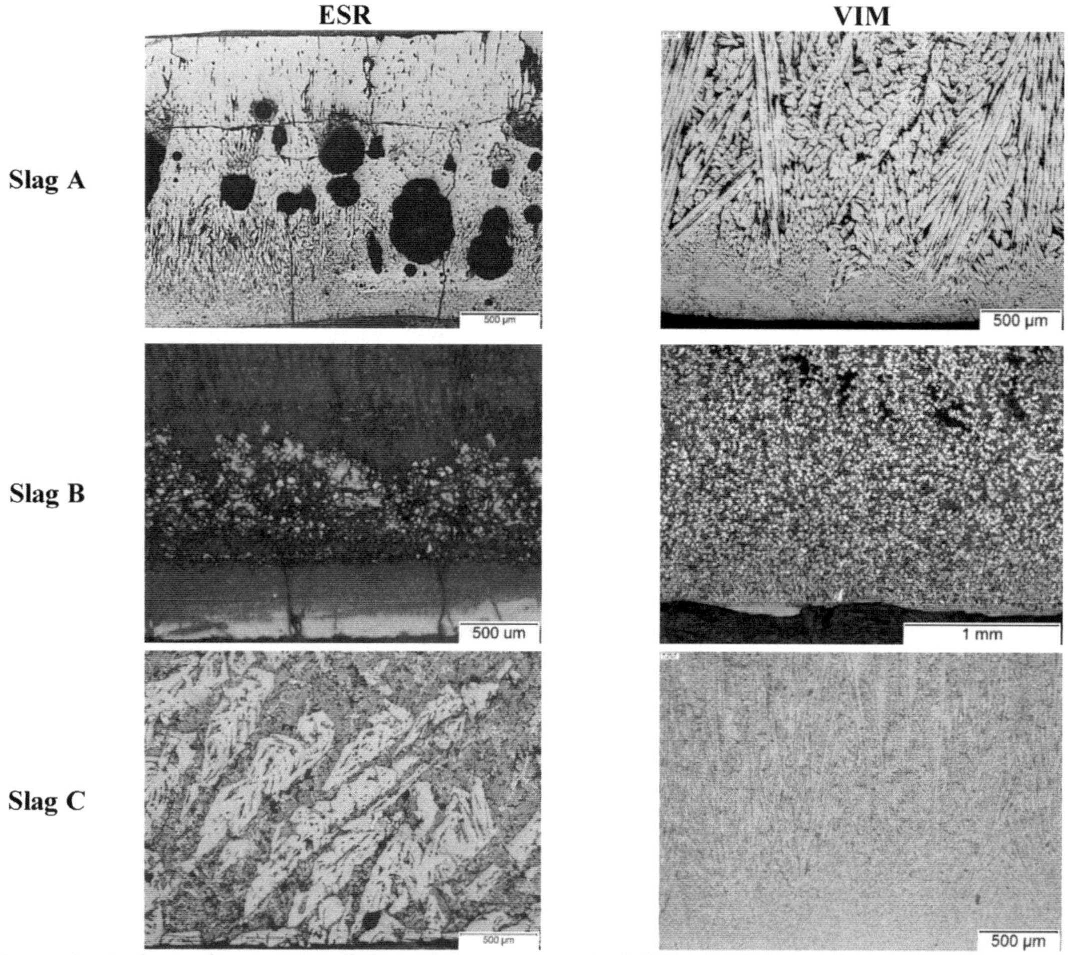


**Figure 2.** Optical micrographs of slag microstructures in ESR and VIM conditions. The bottom of each micrograph is nearest the chill surface.

Micrographs of the ESR and VIM processed Slag C samples show very different structures. The ESR sample shows large, faceted primary $CaF_2$ particles, with smaller scale complex eutectic structures composed of $Ca_{12}Al_{14}O_{32}F_2$, MgO and $CaTiO_3$. The VIM samples show multiphase dendrites, composed of $CaF_2$ and $Ca_{12}Al_{14}O_{32}F_2$ (based on the location in the ternary phase diagram in figure 1 and the dendrites in the ESR samples) with eutectic structures in between. These differences in structure may be attributed to changes in composition in the ESR samples relative to the virgin slag used for the VIM samples, as well as differences in cooling rate and atmosphere during solidification.

Table 4 shows a summary of the densities of the slag samples as measured by TPRL. Each data point represents the average over the relevant samples. A graphical summary of the specific heat and thermal conductivity data are given in figure 3. The single layer results are not shown, but exhibited similar or lower thermal conductivities than the bulk samples, which, according to findings by Absi et al. [18], suggests that the effective values for the multi-layered samples are accurate.

**Table 4.** Averaged room temperature densities for each slag composition and processing route.

| Samples | Density (g/cm³) |
|---|---|
| Slag A - ESR | 2.033 |
| Slag A - VIM | 2.549 |
| Slag A - VIM (L) | 2.672 |
| Slag B – ESR | 2.544 |
| Slag B – VIM | 3.058 |
| Slag C – VIM | 2.745 |
| Slag C - VIM (L) | 2.842 |

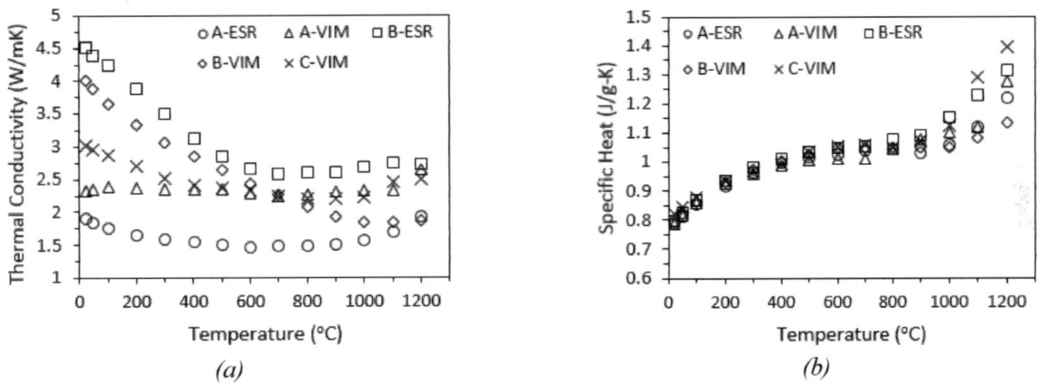


**Figure 3.** (a) Thermal conductivity and (b) specific heat data obtained from TPRL, Inc. All values represent averages of data at each temperature for the multiple samples listed in table 2.

Porosity results are given in figures 4 and 5 for the ESR samples of Slags A and B, respectively, as a function of depth fraction, where zero is the metal ingot surface and one is the mold surface. In general, the pore size tends to increase with distance away from the chilled mold wall, which is consistent with experimental observations [19] of porosity in which the pore size decreases with increasing solidification rate. Additionally, the area fraction porosity tends to be small at either surface, with a maximum in the center of the sample. This result, however, may simply be a consequence of purposefully selecting slag samples with minimal surface texture for maximum suitability for the laser flash diffusivity measurements. In general, Slag A contained much larger, and more numerous pores than Slag B. Slag A also conformed to trends in pore number and size with depth better, but this may be due to increased error in identifying the small pores present in the Slag B.

## 4. Discussion

Comparing the thermal conductivity results for different slags and processes in figure 3 in the context of the slag microstructures reveals some interesting results. For Slag A, the VIM samples had a higher thermal conductivity than the ESR samples. The micrographs show similar phase structures, except for the significant porosity which reduces the effective conductivity of the ESR samples. However, the Slag B ESR samples showed higher conductivity than the VIM samples, despite the increased porosity in the ESR samples. This may be explained by the drastic difference in the phase structure, as seen in the micrographs in figure 2. While the phase fractions, as shown in table 3, are not very different between the two processing routes, the morphology of the phases is very different, resulting in significant changes in the thermal conductivity.

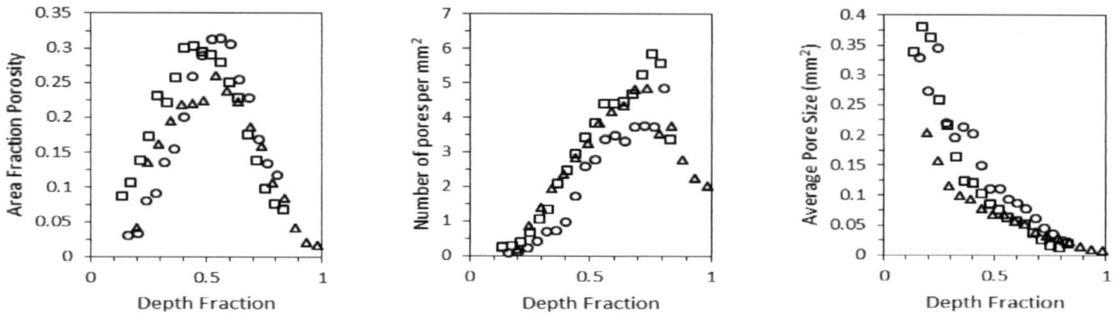


(a) (b) (c)

**Figure 4.** Porosity data for the three ESR samples of Slag A (○=*A-1*, □=*A-2*, and Δ=*A-3* ).
(a) Area fraction porosity, (b) number of pores per unit area, and (c) average pore area are given as a function of depth fraction, where zero is at the ingot surface and one is at the mold surface.

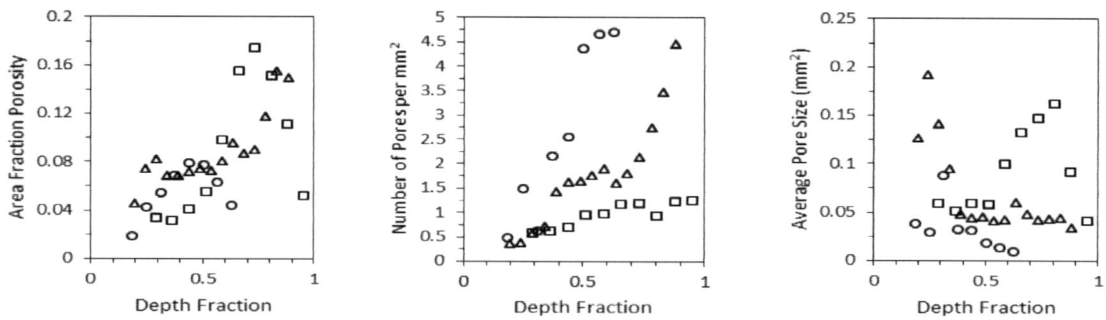


(a) (b) (c)

**Figure 5.** Porosity data for the three ESR samples of Slag B (○=*B-1*, □=*B-2*, and Δ=*B-3*).
(a) Area fraction porosity, (b) number of pores per unit area, and (c) average pore area are given as a function of depth fraction, where zero is at the ingot surface and one is at the mold surface.

In an attempt to predict the fully dense slag thermal conductivity, a simple numerical model of the laser flash diffusivity test was constructed. The model is a control volume based simulation of the laser flash diffusivity test which solves the one-dimensional energy equation for heat conduction:

$$\frac{\partial T}{\partial t} = \alpha \frac{\partial^2 T}{\partial x^2}, \quad (2)$$

where $T$ is temperature, $t$ is time, $x$ is axial position, and $\alpha$ is the thermal diffusivity. The model first assumes a thermal conductivity for the fully dense slag. The porosity distribution data is then used to estimate the density, specific heat, and thermal conductivity of each control volume. The simulation calculates the effective thermal diffusivity based on the ASTM E1461-13 standard using equation 3. That result is compared to the actual TPRL data, and the slag conductivity is adjusted up or down appropriately and the simulation run again. This iteration continues until the simulation diffusivity is within some tolerance of the measured diffusivity:

$$\alpha = 0.13879 L^2 / t_{1/2}, \quad (3)$$

where $L$ is the sample thickness and $t_{1/2}$ the time required for the backside of the sample to reach half of its maximum temperature.

The porosity contributes only negligibly to the mass and specific heat so that these properties may be estimated using the volume fraction porosity directly. The effective thermal conductivity was estimated using a variety of models taken from the literature. The first is the series model or arithmetic mean shown in equation 4. This gives the maximum possible effective thermal conductivity value and minimum slag conductivity [20,21]. The corresponding minimum effective value is a parallel model, or harmonic mean which was found to give unrealistically high slag conductivity predictions. An intermediate between the parallel and series models is the geometric mean [20] as shown in equation 5.

Maxwell's model for randomly dispersed, non-interacting spheres, derived for electrical conductivity, is shown in equation 6 [22], and has shown good agreement with experiments for effective thermal conductivities [23]. Bruggeman's model [24] in equation 7 was derived in a similar manner to Maxwell's model, but with different assumptions about the permeability and electric field strength, and has been found to be more accurate than Maxwell's model at predicting effective thermal conductivities at high volume fractions of the dispersed phase [25,26]. Russell [27] proposed a model (equation 8) based on cubic pores using the laws for parallel and series resistances, which has been found to be accurate within 10% for porous materials [28]. All of the models are summarized in table 5, where $k_m$ is the effective sample conductivity, $k_s$ the slag conductivity, $k_p$ the pore conductivity, and $\varphi$ the volume fraction porosity.

The results of the numerical model are shown in figure 6. It is clear that the geometric relationship over predicts the slag conductivity, while the remaining four models produce similar results. From figure 6a, it appears that porosity alone cannot explain the differences between the ESR and VIM samples for Slag A. It may be concluded that there is some contribution from differences in phase structure or experimental error. Predictions for Slag B in figure 6b reinforce the effect of the phase structure on the effective properties, since the effect of correcting the ESR thermal conductivity values for porosity produces higher values than the fully dense VIM samples.

**Table 5.** Summary of effective thermal conductivity models.

| | | |
|---|---|---|
| *Series Model:* | $k_m = (1-\phi)k_s + \phi k_p$ | *(4)* |
| *Geometric Mean:* | $k_m = k_s^{(1-\phi)} k_p^{\phi}$ | *(5)* |
| *Maxwell:* | $k_m = \dfrac{k_s\left[k_p + 2k_s - 2\phi(k_s - k_p)\right]}{k_p + 2k_s + \phi(k_s - k_p)}$ | *(6)* |
| *Bruggeman:* | $\dfrac{k_m - k_p}{k_s - k_p}\left(\dfrac{k_s}{k_m}\right)^{1/3} = 1 - \phi$ | *(7)* |
| *Russell:* | $k_m = k_s \dfrac{\phi^{2/3} + \frac{k_s}{k_p}(1 - \phi^{2/3})}{\phi^{2/3} - \phi + \frac{k_s}{k_p}(1 - \phi^{2/3} + \phi)}$ | *(8)* |

## 5. Conclusions

The microstructures, phase fractions, and thermophysical properties of three ESR slag compositions were characterized for two different processing routes. It was found that the phase content, phase fractions, porosity content, and phase morphology all contributed to the effective thermophysical properties. A numerical model of the laser flash diffusivity test was designed to predict the fully dense slag thermal conductivity as a function of porosity. These results reinforced the conclusion that the combination of phase structure and porosity content affect the thermophysical properties of slag skins.

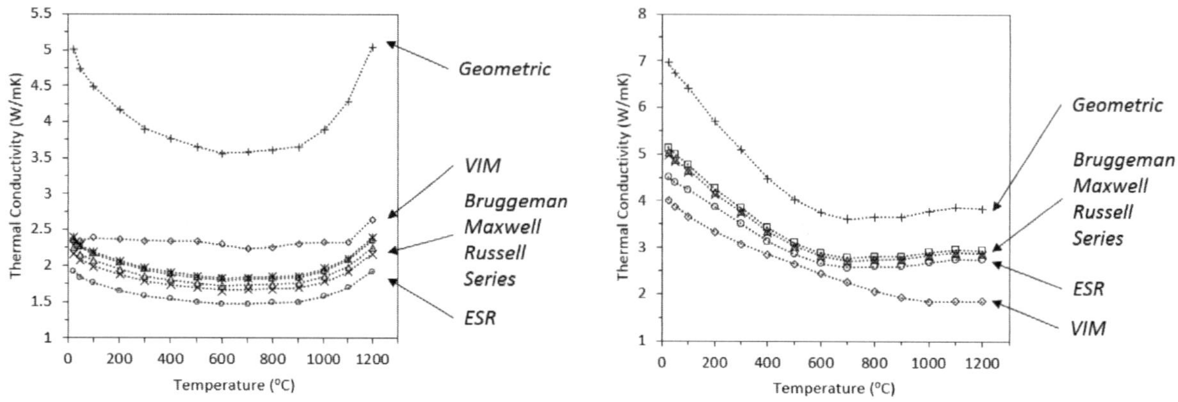


**Figure 6.** Predicted fully dense slag thermal conductivities for (a) Slag A and (b) Slag B.

## 6. References

[1] Kharicha A, Mackenbrock A, Ludwig A, Schützenhöfer W and Maronnier V 2007 *Int. Symp. Liq. Met. Process. Cast.* 113-20
[2] Weber V, Jardy A, Dussoubs B, Ablitzer D, Rybéron S, Schmitt V, Hans S and Poisson H 2009 *Metall. Mater. Trans. B* **40** 271-80
[3] Kelkar K M *et al.* 2013 *Proc. 2013 Int. Symp. Liq. Met. Process. Cast.* 3–12.
[4] Yanke J, Fezi K, Trice R and Krane M J M 2014 *Numer. Heat Transf. Part A Appl.* **67** 268–92
[5] Mills K C, Powell J S, Bryant J W and Keene B J 0981 *Can. Metall. Q.* **20**, 93–9
[6] Keene B J and Mills K C 1981 *Arch. Fur Das Eis.* **52** 311–15
[7] Gohil D D and Mills K C 1981 *Arch. Fur Das Eis.* **52** 335–40
[8] Morrell R and Mills K C 1981 *Arch. Fur Das Eis.* **52** 381–3
[9] Taylor R and Mills K C 1982 *Arch. Fur Das Eis.* **53** 55–63
[10] Busch J D 2012 M.S. thesis, Purdue University
[11] Rietveld H M 1967 *Acta Crystallogr.* **22** 151–2
[12] Bish D L and Howard S A 1988 *J. Appl. Crystallogr.* **21** 86–91
[13] Lutterotti L, Matthies S and Wenk H R 1999 *IUCr Newsl. CPD* **21** 14–5
[14] Gražulis S *et al.* 2009 *J. Appl. Crystallogr.* **42** 726–9
[15] Gražulis S *et al.* 2012 *Nucleic Acids Res.* **40** 420–7
[16] Bergerhoff G and Brown I D 1987 *Int. Union Crystallogr. Chester* 77–95
[17] Belsky A, Hellenbrandt M, Karen V L and Luksch P 2002 *Acta Crystallogr. Sect. B* **58** 364–9
[18] Absi J, Smith D, Naït-Ali B, Grandjean S and Berjonnaux J 2005 *J. Eur. Cer. Soc.* **25** 367–73
[19] Lee P, Hunt J 1997 *Scr. Mat.* **36** 399-404
[20] Woodside W and Messmer J H 1961 *J. Appl. Phys.* **32** 1688–99
[21] Progelhof R C, Throne J L and Ruetsch R R 1976 *Polym. Eng. Sci.* **16** 615–25
[22] Maxwell J C 1904 *A Treatise on Electricity and Magnetism* (Oxford: Clarendon Press)
[23] Sakiyama T, Matsushita Y, Shiinoki Y, and Yano T 1990 *J. Food Eng.* **11** 317–31
[24] Bruggeman V 1935 *Ann. Der Phys.* **5** 636–64
[25] Wang B, Zhou L and Peng X 2003 *Int. J. Heat Mass Transf.* **46** 2665–72
[26] Lee T Y R and Taylor R E 1978 *J. Heat Transfer* **100** 720–4
[27] Russell H W 1935 *J. Am. Ceram. Soc.* **18** 1–5
[28] Cheng S C and Vachon R I 1970 *Int. J. Heat Mass Transf.* **13** 537–46

**Acknowledgements**
The authors would like to acknowledge the support of the Metals Affordability Initiative, AFRL Case Number: 88ABW-2015-1871, and Special Metals Corporation for supplying samples.

IOP Conf. Series: Materials Science and Engineering **143** (2016) 012008 doi:10.1088/1757-899X/143/1/012008

# A numerical study on electrochemical transport of ions in calcium fluoride slag

**E Karimi-Sibaki [a], A Kharicha[a,b], M Wu[a,b], and A Ludwig[b]**
[a]Christian-Doppler Lab for Adv. Process Simulation of Solidification & Melting,
[b]Chair of Simulation and Modeling of Metallurgical Processes,
University of Leoben, Franz-Josef-Str. 18, A-8700 Leoben, Austria

E-mail: abdellah.kharicha@unileoben.ac.at

**Abstract.** Electrically resistive $CaF_2$-based slags are widely used in electroslag remelting (ESR) process to generate Joule heat for the melting of electrode. The electric current is conducted by ions (electrolyte) such as $Ca^{+2}$ or $F^-$, thus it is necessary to establish electrochemical models to study electrical behavior of slag. This paper presents a numerical model on electrochemical transport of ions in an arbitrary symmetrical (ZZ) and non-symmetrical ($CaF_2$) stagnant electrolytes blocked by two parallel, planar electrodes. The dimensionless Poisson-Nernst-Planck (PNP) equations are solved to model electro-migration and diffusion of ions. The ions are considered to be inert that no Faradic reactions occur. Spatial variations of concentrations of ions, charge density and electric potential across the electrolyte are analyzed. It is shown that the applied potential has significant influence on the system response. At high applied voltage, the anodic potential drop near the electrode is significantly larger than cathodic potential drop in fully dissociated $CaF_2$ electrolyte.

## 1. Introduction

In many metallurgical processes such as electroslag remelting (ESR), the slag is core compound served to clean the metal by removing sulfur and non-metallic inclusions. In general, $CaF_2$-based slags are used since they have high degree of chemical reactivity and low electrical conductivity. The heat required to supply into the process is obtained by passing electric current through the resistive slag layer. The mode of current conduction in the slag is either ionic or a mixed ionic/electronic process [1]. Mitchell et al used a galvanostatic pulsing technique to measure magnitudes of overpotentials resulting from concentration polarization at the interface between iron and $CaF_2$-based slags in an electrolytic cell [2]. They found that potentials observed on the ESR electrode agreed with the results from the electrolytic cell. Furthermore, addition of a relatively small amount of $Al_2O_3$ to $CaF_2$ substantially decreases electric conductivity of liquid slag [3]. As a consequence of formation of complex ions such as $AlOF^-_2$, the mobility of $F^-$ reduces [3]. Kojima studied the effect of mode of current (DC or AC) on electrochemical behavior of oxygen in a small scale ESR unit [4]. The distribution of oxygen in ingot was influenced by mode of current and electrode polarity. Kharicha et al [5] addressed possible effects of electrochemical reactions on formation of droplets at an ESR electrode. Metallic droplets which are surrounded by resistive slag can conduct an electric current when the potential drop at their extremities exceeds the sum of anodic and cathodic overpotential. As a consequence of ionic conduction of slag, dependences of melt rate of electrode and ingot composition on the electrode polarity were observed in DC operation of the ESR process [4]. Kawakami et al reported that electrode with positive polarity has higher melt rate compared to electrode with negative polarity [6]. The roles of electrochemical reactions occur in the slag-metal interface are well known. Loss of alloy elements such as Ti and Al by oxidation is reported as a significant issue during DC operation of ESR process [7-8]. Furthermore, the efficiency of sulphur removal in ESR process was reported to be dependent on the condition of applied electric current (AC or DC) [9]. In fact, the aforementioned phenomena are as a result of the ionic character of slag. As such, it is essential to develop an electrochemical model to investigate the electrical behavior of slag. The model must take into account ions response to the applied electric field.

In the current study, a typical electrochemical cell is considered including two parallel, planar electrodes and a completely dissociated electrolyte. The Finite Volume Method (FVM) is used to model the ions transfer under the applied electric potential. No Faradaic reaction is modeled. Mass conservations of inert ions are automatically satisfied that is the advantage of using FVM method. First, an arbitrary binary symmetric electrolyte ($Z^+Z^-$) is investigated in which diffusion coefficients and concentrations of $Z^+$ and $Z^-$ ions are equal. Then, $CaF_2$ is considered as a fully dissociated electrolyte consisted of $Ca^{2+}$ and $F^-$ ions where the diffusion coefficients and concentrations are different. The effect of applied potential and electrolyte properties such as concentrations of ions on the system response is discussed. The goal is to achieve some fundamental understanding of electrochemical behavior of $CaF_2$ slag under the influence of applied electric field.

## 2. Transport of ions

For an electrolyte subjected to an electric field, the Poisson-Nernst-Planck (PNP) equations describe the flux of ions through the electrolyte.

$$-\nabla\cdot(\varepsilon_s\nabla\phi)=F\sum z_i c_i=\rho \tag{1}$$

$$\frac{\partial c_i}{\partial t}=-\nabla\cdot\vec{N}_i \tag{2}$$

$$\vec{N}_i=\vec{u}c_i-D_i\nabla c_i-\frac{Fz_iD_i\nabla\phi}{RT}c_i \tag{3}$$

The net flux of ions ($N_i$) including convection, diffusion, and electro-migration must be conserved as described in Eq. (2) where ($c_i$) is the concentration of *i*-th ion. According to Eq. (3), the convection and diffusion fluxes are dependent on velocity ($u$) and diffusion respectively where ($D_i$) is diffusion coefficient of *i*-th ion. Furthermore, the electro-migration flux of ions is included in Eq. (3) where ($z_i$) is the charge number of *i*-th ion, ($F$) is Faraday constant, ($R$) is the universal gas constant, ($T$) is temperature, and ($\phi$) is electric potential. The Poisson equation, Eq. (1), relates the electric potential to the charge density ($\rho$) of system where ($\varepsilon_s$) is the electric permittivity constant of the electrolyte.
The steady state solution of the PNP equations for a symmetrical binary electrolyte was given in terms of Jacobi elliptic functions by Golovnev [10]. It was shown that the net charge density exists in a very thin boundary layer (diffuse charge layer) near the electrode (~ 1-100 nm). The length of diffuse charge layer is called Debye screening length ($\lambda_D$) in electrochemical systems.

$$\lambda_D=\sqrt{\frac{\varepsilon_s RT}{2z^2F^2C_0}} \tag{4}$$

where ($C_0$) is the bulk concentration of electrolyte.

The specific interfacial region formed near the electrode where the charge is separated is called electric double layer. As illustrated in Fig. 1, the double layer is composed of a diffusion layer, diffuse charge layer, and the compact stern layer (~ 0.5 nm) where ions are attracted to the surface. The diffuse charge layer is not electrically neutral as it has a net charge. However, the diffusion layer is electrically neutral where a non-uniform concentration of electrolyte may exist [11]. The definition of thickness of diffusion layer is arbitrary because the concentration reaches asymptotically to the bulk concentration [12]. For instance in cyclic voltammetry, the thickness is considered as: $(D_i\, t)^{0.5}$ where $(1/t)$ is the scan rate. As such, the thickness can be large (order of micrometer) at slow scan rate or small (order of nanometer) at fast scan rate [12].

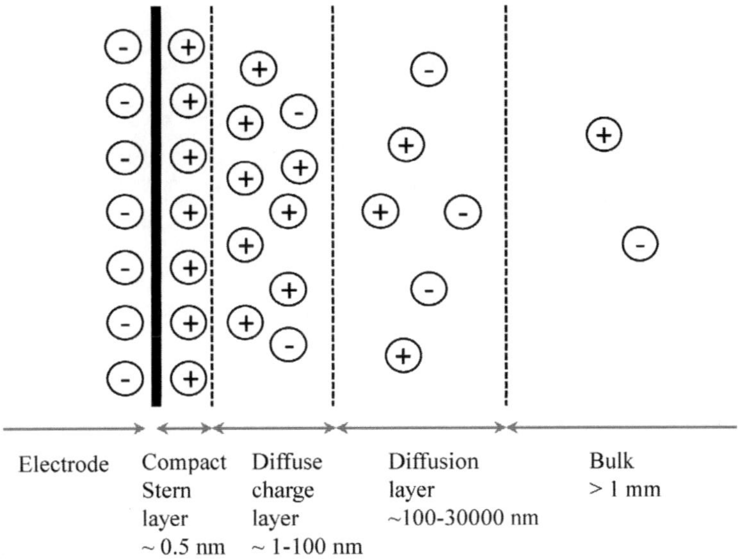


Fig.1. Structure of electric double layer.

## 3. Modeling

Configuration of the computational domain including liquid electrolyte blocked by electrodes (anode and cathode) is illustrated in Fig 2. A variable sized mesh is used applying a very fine mesh near the electrodes to resolve the boundary layer where the gradient is large. The mesh element size is gradually increased away from boundaries. The domain size is 1 mm and the total number of mesh elements is 1000. The simulation is performed with the commercial CFD software FLUENT-ANSYS v.14.5. The required transport equations are specified using User-Defined Scalars (UDS).

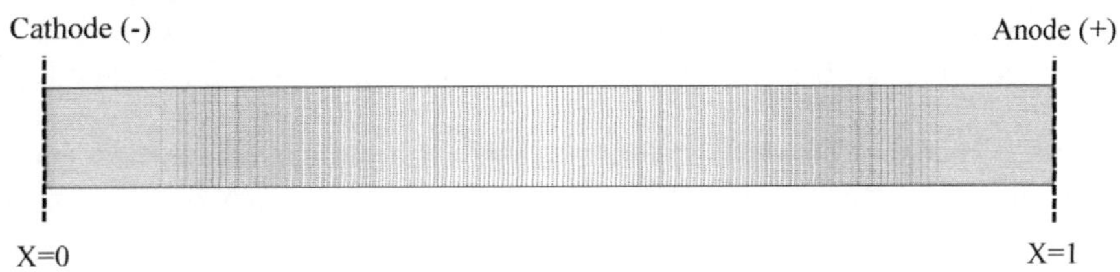


Fig.2. Configuration of computational domain and the mesh.

Applying numerical methods to model the problem requires extremely fine mesh to resolve the boundary layer adjacent to the electrode. As stated by Bazant [13], the investigation of electrochemical systems is significantly simplified using dimensionless formulations of PNP equations. This is the main step to study the asymptotic behavior of an electrochemical system in order to get physical insight into the problem.

Here, a one dimensional model is considered in which the parameter variation is only in one direction across the electrolyte. Furthermore, the electrolyte is assumed to be stagnant ($u=0$). Resultantly, the dimensionless form of PNP equations for the binary electrolyte is expressed as:

$$-\nabla\cdot(\xi^2\,\nabla\phi^*)=z_+C_+ + z_-C_- \tag{5}$$

$$\frac{\partial C_+}{\partial t}=\nabla\cdot(D_+\nabla C_+ + z_+D_+\nabla\phi^*C_+) \tag{6}$$

$$\frac{\partial C_-}{\partial t}=\nabla\cdot(D_-\nabla C_- + z_-D_-\nabla\phi^*C_-) \tag{7}$$

The basic variables in PNP equations are scaled as follow:

$$\phi^* = \frac{F\phi}{RT},\, C_\pm = \frac{c_\pm}{C_{ref}},\, \xi = \frac{\lambda_D}{L_{ref}}$$

Where $C_\pm$ denotes the dimensionless concentration of cation (+) or anion (-), and $C_{\mathrm{ref}}$ is the reference concentration which is typically in the same order as bulk concentration of ions. The electric potential is scaled using thermal voltage ($RT/F$) since diffusion and chemical reactions are thermally activated processes. Furthermore, $\xi$ is the ratio of Debye screening length ($\lambda_{\mathrm{D}}$) to the reference length ($L_{\mathrm{ref}}$) which is typically considered as the distance between electrodes or the size of electrode.
Ions are considered to be inert and subsequently no faradaic reactions occur in our system. Thus the boundary conditions for Eq. (6) and Eq. (7) are zero total flux of ion (including electro-migration and diffusion fluxes). Additionally, Eq. (5) is used to determine the electric potential in the system. For the latter, a fixed value ($\pm FV/2RT$) is considered at the anode (+) and cathode (-) where ($V$) is the operating electric potential of the electrolytic cell. As described by Bazant [13], it is impossible to satisfy all the boundary conditions assuming a negligible screening length compared to the system size ($\xi \sim 0$). Physically, this assumption is corresponding to electro neutrality ($\rho=0$) which is always violated at electrochemical interfaces. A numerical solution can only be achieved using a rather larger value of $\xi$ to emphasize the importance of diffuse charge near the electrodes where the electro neutrality condition is not valid. For instance, the value of $\xi$ can be chosen as large as 0.05 for a micro electrochemical system [13]. In this way, the asymptotic behavior of an electrochemical system can be captured.

Simulations are executed unsteadily, but only final steady state results are analyzed. Initial concentrations of ions are calculated assuming a fully dissociated electrolyte. The concentration of $CaF_2$ is the ratio of density (~ 2600 kg.m$^{-3}$) to molecular weight (~ 78.07 g.mol$^{-1}$). The fully dissociated $CaF_2$ electrolyte ($CaF_2 \leftrightarrow Ca^{2+} + 2F^-$) is consisted of $Ca^{2+}$ and $F^-$ ions in which the concentration of $F^-$ is twice larger than the concentration of $Ca^{2+}$. The electric permittivity of electrolyte is assumed to be the same as permittivity of free space since that is unknown for $CaF_2$ at elevated temperature. Table 1 lists all physical properties required for our calculations.

Table 1. Properties of the system.

| | |
|---|---|
| $T$: Bulk temperature (K) | 1773 |
| $R$: Universal gas constant ($J \cdot mol^{-1} \cdot K^{-1}$) | 8.314546 |
| $F$: Faraday constant ($Amp \cdot s \cdot mol^{-1}$) | 96485 |
| $\varepsilon_s$: Electric permittivity ($F \cdot m^{-1}$) | $8.85418 \times 10^{-12}$ |
| $V$: Low applied voltage (V) | 0.005 |
| $V$: High applied voltage (V) | 2 |
| $C_{ref}$: Reference concentration ($mol \cdot m^{-3}$) | $1 \times 10^{5}$ |
| $L_{ref}$: Reference length (m) | $1 \times 10^{-6}$ |
| $\lambda_D$: Debye length (m) | $1 \times 10^{-9}$ |
| $\zeta$ | 0.01 |
| **ZZ electrolyte** | |
| $Z^+$: Concentration ($mol \cdot m^{-3}$) | $0.333 \times 10^{5}$ |
| $Z^+$: Diffusion coefficient ($m^2 \cdot s^{-1}$) | $4 \times 10^{-9}$ |
| $Z^+$: Charge number | +1 |
| $Z^-$: Concentration ($mol \cdot m^{-3}$) | $0.333 \times 10^{5}$ |
| $Z^-$: Diffusion coefficient ($m^2 \cdot s^{-1}$) | $4 \times 10^{-9}$ |
| $Z^-$: Charge number | -1 |
| **$CaF_2$ electrolyte** | |
| $Ca^{2+}$: Concentration ($mol \cdot m^{-3}$) | $0.333 \times 10^{5}$ |
| $Ca^{2+}$: Diffusion coefficient ($m^2 \cdot s^{-1}$) | $5.66 \times 10^{-9}$ [14] |
| $Ca^{2+}$: Charge number | +2 |
| $F^-$: Concentration ($mol \cdot m^{-3}$) | $0.666 \times 10^{5}$ |
| $F^-$: Diffusion coefficient ($m^2 \cdot s^{-1}$) | $4.12 \times 10^{-9}$ [14] |
| $F^-$: Charge number | -1 |

## 4. Results and discussions

A number of simulations were performed by considering two different operating applied potentials. Additionally, two different electrolytes are investigated. Firstly, an arbitrary binary symmetric electrolyte (*ZZ*) is studied in which diffusion coefficients, concentrations, and charge numbers of ions are identical. Secondly, the fully dissociated $CaF_2$ which is a non-symmetrical binary electrolyte is investigated. Furthermore, responses of electrolytes to the applied potential are compared. Fig. 3 shows the normalized concentration profile of cations ($Z^+$ or $Ca^{2+}$) and anions ($Z^-$ or $F^-$) across the electrolyte at low applied voltage. Note that, initial concentrations of anions are not the same in Fig. 3(b). As expected, cations move toward the cathode ($X=0$) and anions accumulate near the anode ($X=1$). Once the voltage is applied, ions move to screen the electric field originated from electrodes. Therefore counter-ions (relative to electrode) entering to double layers by electro-migration from the bulk. Concurrently, co-ions leave the double layer entering to the bulk. Due to excess of counter-ions in the double layer, diffusion competes against electro-migration until a uniform equilibrium state is reached. In this way, diffusion layers spread across the electrolytic cell from electrodes. As illustrated in Fig. 3 (c-d), the thickness of diffusion layer can significantly decrease by applying a larger voltage in the system. Here, the electro-migration is dominant that results in double layers filled only with counter-ions. In other words, bulk concentrations of ions significantly decrease so that co-ions completely leave diffusion layers.

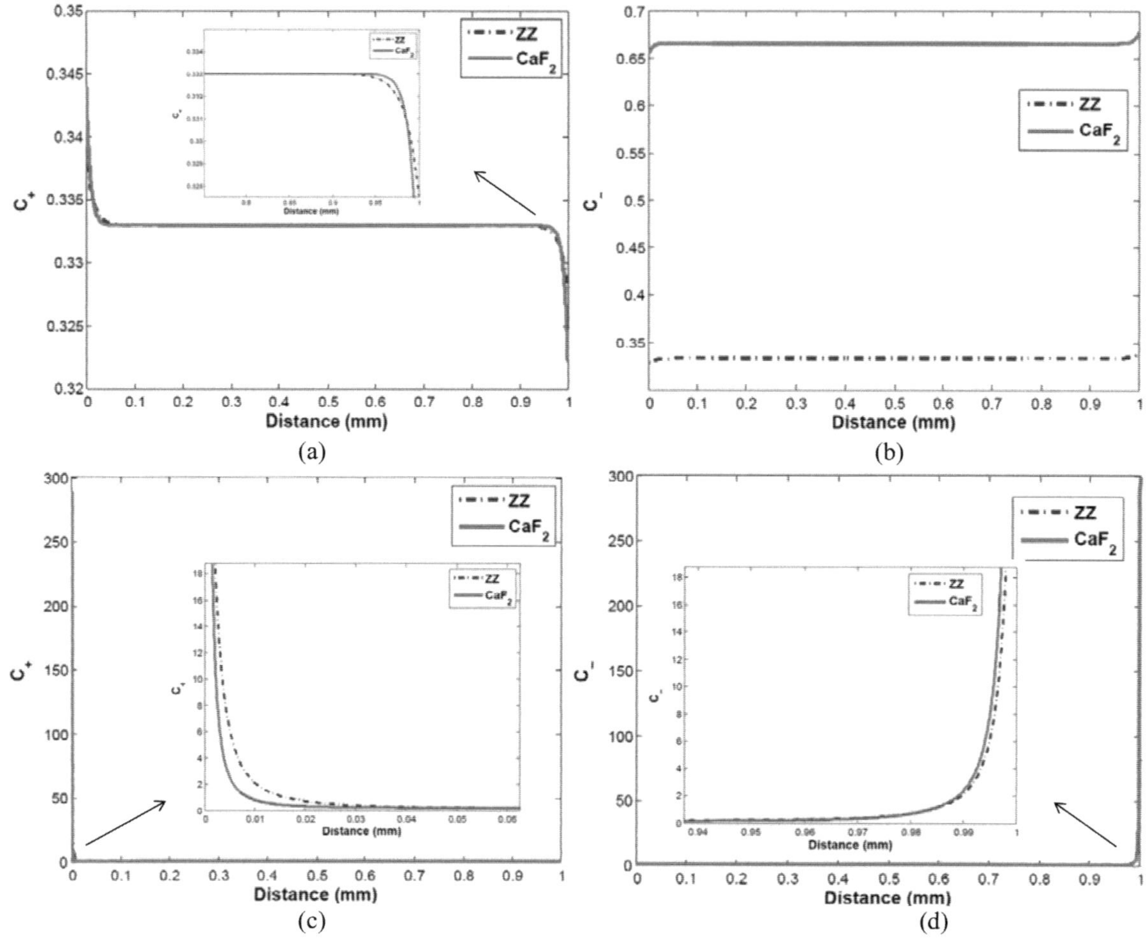


Fig.3. Normalized concentrations of ions are plotted across the electrolyte blocked by cathode ($x=0$) and anode ($x=1$), (a) Cations ($Z^+$ and $Ca^{2+}$) at low applied potential (0.005 V), (b) Anions ($Z^-$ and $F^-$) at low applied potential (0.005 V), (c) Cations ($Z^+$ and $Ca^{2+}$) at high applied potential (2 V), (d) Anions ($Z^-$ and $F^-$) at high applied potential (2 V).

As illustrated in Fig. 4 (a-b), charge density is divided in three regions: neutral region in the bulk and two electrically charged boundary layers near electrodes. The net bulk charge density remains very small ($\rho \sim 0$) that is known as electro-neutrality condition. Most often, electro-neutrality assumption is used for large scale electrochemical systems. This is an excellent assumption that can significantly simplify investigation of large scale systems. However, it is essential to incorporate the effect of double layers into boundary conditions [12]. As shown in Fig. 4 (b), the thickness of diffusion layer is significantly decreased by applying larger voltage which is consistent with Gouy-Chapman theory [12, 15]. According to the theory, counter-ions have higher tendency to be absorbed than co-ions to be pushed out of the double layer once the electric field is applied. As such, the total charge density of ions increases near the electrode and depletes in the bulk [12]. Furthermore, the charge density remains symmetrical for both *ZZ* and $CaF_2$ at low applied voltage as indicated in Fig. 4(a). However, the symmetry breaks strongly applying large voltage for $CaF_2$ electrolyte due to difference in initial concentrations of ions ($F^-$ and $Ca^{2+}$) as shown in Fig. 4(b).

The spatial variation of electric potential is controlled by charge density distribution through Poisson's equation (Eq. 1). Nevertheless, changes in bulk concentration of ions influence the electric potential through the whole electrolyte. In other words, variation in electric potential is not limited near electrodes due to evolution of diffusion layers. The thicknesses of diffusion layers are dependent on the applied potential since electro-migration and diffusion are competing. As shown in Fig. 4(c), the diffusion layer is a narrow region near diffuse charge layer at low applied voltage. Consequently, the potential drop is only near electrodes and the bulk variation of electric potential is negligible. Furthermore, the symmetry of electric potential through the electrolyte is kept independent of electrolyte whether *ZZ* or $CaF_2$. In contrast, strong breaking of symmetry in electric potential profile is observed for $CaF_2$ at large applied potential as illustrated in Fig. 4(d). Additionally, the spatial variation of electric potential is extended to the bulk. The interfacial potential difference near the anode is significantly larger than near the cathode for the $CaF_2$ electrolyte.

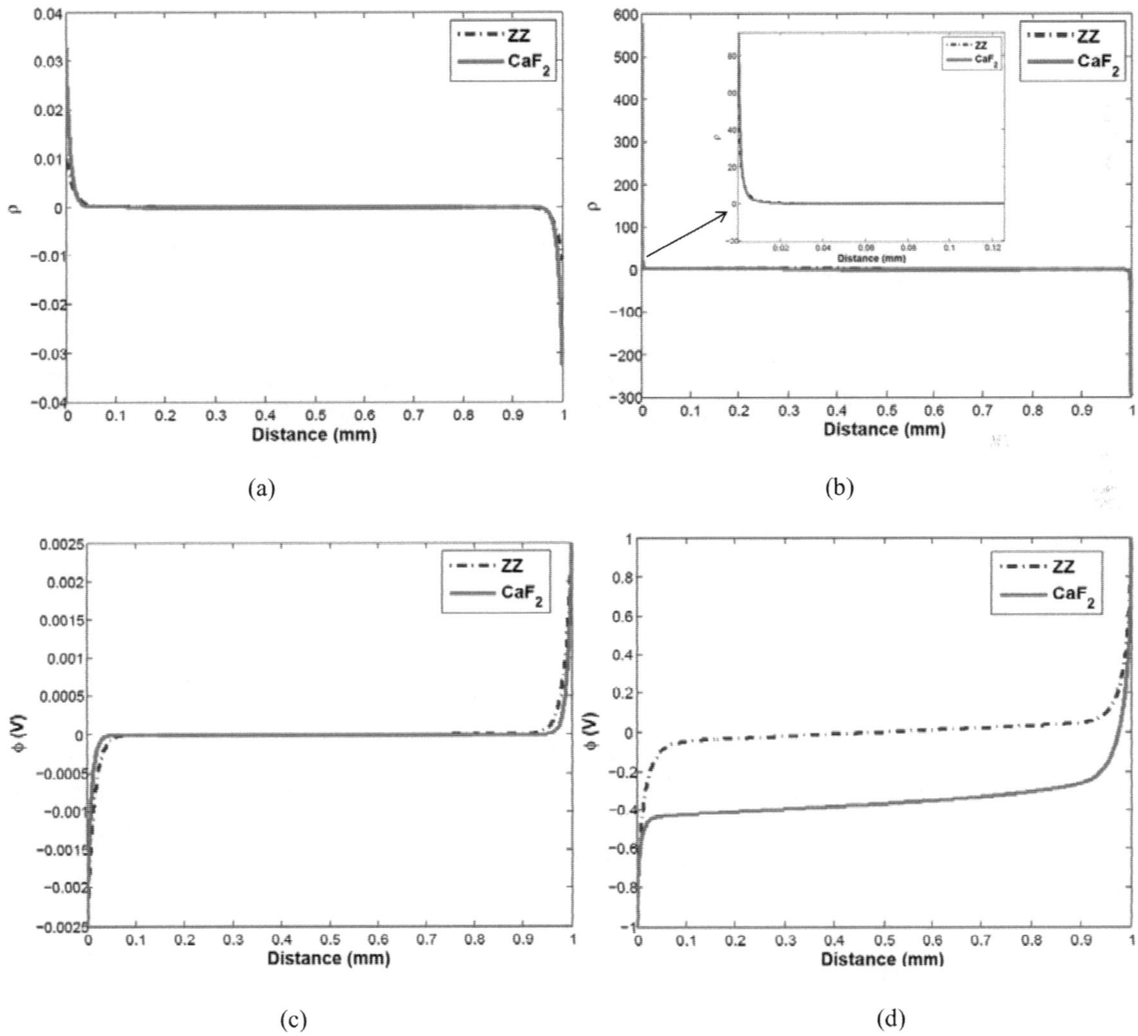


Fig.4. The following parameters are plotted across the electrolytes (*ZZ* or $CaF_2$) blocked by cathode ($x=0$) and anode ($x=1$), (a) Normalized charge density at low applied voltage (0.005 V), (b) Normalized charge density at high applied voltage (2 V), (c) Electric potential at low applied voltage (0.005 V), (d) Electric potential at high applied voltage (2 V).

It must be noted that a conventional slag is typically composed of $CaF_2$, $Al_2O_3$, $CaO$, $SiO_2$, and $MgO$. Therefore, a number of other ions such as $O^{2-}$, $Al^{3+}$, $Mg^{2+}$, and $Si^{4+}$ can play role in electrochemical behavior of the slag. Furthermore, they can participate in Faradaic reactions at electrode surface. Therefore, the model must be further developed to take into account effects of other ions as well as Faradaic reactions on electrical behavior of a conventional slag.

## 5. Summary

A numerical study is performed to investigate electrochemical transport of ions in $CaF_2$ slag as a non-symmetrical binary electrolyte. The behavior of system is compared to an arbitrary binary symmetrical electrolyte (*ZZ*) in which concentration, charge number and diffusion coefficients of ions are identical. Furthermore, the influence of applied potential is studied. The Poisson-Nernst-Planck (PNP) equations considering convection, diffusion, and electro-migration of ions through the electrolyte are solved. The effect of Faradic reactions is not included in the current study. Spatial variations of concentrations of ions, charge density, and electric potential across the electrolyte are compared for *ZZ* and $CaF_2$ electrolytes. At low applied voltage, the interfacial potential difference near electrodes remains just about the same. Additionally, the diffusion layer is thin resulting in uniform concentration of ions in the bulk. In contrast, the diffusion layer is extended to the bulk of electrolytes at high applied voltage. The symmetrical electrolyte (*ZZ*) has identical potential drop near electrodes independent of magnitude of operating potential. However, significant difference of interfacial potential drop is observed for non-symmetrical electrolyte ($CaF_2$). For the latter, the anodic interfacial potential drop is much larger than the cathodic one.

## 6. Acknowledgement

The authors acknowledge the financial support by the Austrian Federal Ministry of Economy, Family and Youth and the National Foundation for Research, Technology and Development within the framework of the Christian Doppler Laboratory for Advanced Process Simulation of Solidification and Melting.

## 7. References

[1] Peover M E 1972 *J. Inst. Metals* **100** 97
[2] Mitchell A and Beynon G 1971 *Metall. Trans.* **2** 3333
[3] Mitchell A and Cameron J 1971 *Metall. Trans.* **2** 3361
[4] Kojima Y, Kato M, Toyoda T, and Inouye M 1975 *Trans. ISIJ* **15** 397
[5] Kharicha A, Wu M, Ludwig A, Ramprecht A, and Holzgruber H 2012 *CFD modeling and simulation in materials processing* (New York: Wiley–Interscience) p 139
[6] Kawakami M, Nagata K, Yamamura M, Sakata N, Miyashita Y, and Goto K S 1977 *Testsu-to-Hagane*, **63** 220
[7] Etienne M 1970 *Ph.D. Thesis* University of British Columbia
[8] Bhat G K and Mitchell A 1969 *Proc. Sec. Symp. On ESR technology*
[9] Whittaker D A, 1968 *Ph.D. Thesis* McMaster University
[10] Golovnev A, Trimper S 2010 *Phys. Lett. A* **374** 2886
[11] Newmann J 2004 *Electrochemical systems*, (New York: Wiley–Interscience)
[12] McNaught A D, and Wilkinson A 1997 *IUPAC, compendium of chemical terminology* (Oxford-Blackwell scientific publication)
[13] Bazant M Z, Thornton K, and Ajdari A 2004 *Phys. Rev. E* **70** 1
[14] Chiho W, and Shunhua X 1993 *ISIJ Int.* **33** 239
[15] Grahame D C 1947 *Chem. Rev.*, **41** 441

IOP Conf. Series: Materials Science and Engineering **143** (2016) 012009 doi:10.1088/1757-899X/143/1/012009

# Deoxidation Limits of Titanium Alloys during Pressure Electro Slag Remelting

**M. Bartosinski[1], S. Hassan-Pour[1], B. Friedrich[1], S. Ratiev[2], A. Ryabtsev[2]**

[1]IME Process Metallurgy and Metal Recycling, RWTH Aachen University
[2]Department of Electrometallurgy, Donetsk National Technical University

mbartosinski@ime-aachen.de

**Abstract**. This paper focuses on deoxidation of titanium alloys produced by aluminothermic reduction (ATR) and subsequent homogenizing and alloying by vacuum induction melting (VIM). The main goal of the performed research work is to outline the deoxidation limit during pressure electro slag remelting (PESR) of the described material.
To obtain electrodes for deoxidation, a Ti-24Al-16V masteralloy was produced by ATR and afterwards melted in a 0.5 litre calcium- zirconate (lab scale) or 14 litres high purity calcia (pilot scale) crucibles with continuous addition of Ti-sponge after reaching liquid state in order to obtain a final Ti-6Al-4V alloy. During melting, in both cases evaporation of calcium was noticed. The cast ingots were analysed for oxygen using inert gas fusion method, matrix and alloying elements were analysed by XRF.
Results show oxygen levels between 0.5 and 0.95 wt.-% for the ingots which were melted in calcium-zirconate crucibles and approx. 1 – 1.2 wt.-% for the material produced by utilization of calcia crucibles. The subsequent deoxidation was carried out in lab and pilot scale electroslag remelting furnaces using a commercially pure calcium fluoride slag and metallic calcium as deoxidation agent. It could be shown, that deoxidation of the highly contaminated material is possible applying this method to a certain limit. Pilot scale trials showed a reduction of oxygen contents by 1500 – 3500 ppm. Oxygen levels in lab scale trials showed weaker deoxidation effects.
In order to describe the achieved deoxidation effects in a quantitative way, the analyzed oxygen contents of the obtained ingots are compared with calculated data resulting from a mathematical kinetic model. The modelled datasets are in good agreement with experimental oxygen values.

## 1. Introduction

Titanium and its alloys combine several attractive properties such as high strength, low density, corrosion resistance and excellent biocompatibility. Despite these characteristics, the wider application of titanium and titanium alloys is limited due to the high manufacturing costs.
Therefore, since approximately 10 years the IME, RWTH Aachen University, is working on the application of a new synthesis and recycling process route for titanium alloys with special regard on decreasing the level of oxygen due to its negative influence on the ductility. Detailed investigations on each process step were performed by Lochbichler [1] Stoephasius [2], and Reitz [3]. An important aspect of the process route is to resort to conventional and established metallurgical processes such as Vacuum Induction Melting (VIM), Electroslag Remelting (ESR) or Vacuum Arc Remelting (VAR).

According to the type of material (scrap, raw materials) and the composition, different process steps are combined to achieve the desired product quality. Main principle to control the oxygen level is the introduction of metallic calcium to the liquid slag or melt as a deoxidant. Especially for the material class of $\gamma$ -TiAl it could be shown, that deoxidation of these alloys from investment casting scraps during ESR is feasible resulting in final oxygen contents below 500 ppm in the product [3]. The research presented in this paper focuses on an alternative process route for Ti6Al4V alloys with special regard on the PESR deoxidation.

## 2. Fundamentals

One of the most challenging tasks in the refining process of titanium alloys is the removal of oxygen present in titanium as dissolved TiO. Besides the high stability of TiO ($\Delta G_f$ = -357 kj/mol @ 1750°C) also the oxygen solubility of up to 33 at% in the system Ti-O indicates that the removal is difficult. The deoxidation by vacuum distillation is not feasible due to the necessary oxygen partial pressure in the furnace atmosphere ($p_o < 10^{\wedge 20}$ bar for 1000ppm in a cp-Ti melt) which is so low, that even present oxygen from the gas phase will be nearly fully dissolved. Thus, the use of deoxidation agents becomes evident.

Regarding the Vacuum Induction Melting of titanium and titanium alloys, the choice of the refractory material is crucial. According to standard Gibbs free energy calculations of oxides, only CaO, $Y_2O_3$ and $ZrO_2$ seem to be sufficiently stable against titanium, whereas CaO should be most stable. Tsukihashi et al. [4] investigated the calcium and oxygen uptake in c.p. titanium and TiAl melts in equilibrium with solid CaO. The experiments were conducted in a closed system by using a lid to avoid Ca evaporation. His research shows, that the reaction according to (1) is causal between for the equilibrium between calcium and oxygen in titanium melts, which depends strongly on temperature. [4] [5]

$$[TiO]_{Ti} + [Ca]_{Ti} = [CaO]_{slag} + Ti \tag{1}$$

$[A]_B$: A dissolved in B

Unfortunately, Tsukihashis results do not allow a prediction of the reachable final oxygen contents in the melt during ESR since the distribution coefficient between calcium in the metal and calcium in the slag is not known. Extending Tsukihashis work, Okabe et al. [6] equiblirated Ti melts with CaO and Ca ($a_{Ca} = a_{CaO} = 1$) and calculated the activity coeffictient $f_{TiO}$. Based on this work, Stoephasius et al. [5] calculated the deoxidation limits of Ti and TiAl melts with metallic calcium as deoxidant. Since during electroslag remelting, instead of pure CaO a $CaF_2$-(CaO) slag (with metallic Ca as deoxidant) is used, the CaO activity is decreased. Therefore, from a thermodynamic point of view, deoxidation of cp titanium melts below 2000 ppm oxygen should be possible in general. However, major challenge during PESR deoxidation of titanium is to achieve a homogeneous oxygen distribution in the whole ingot. Since the CaO concentration is increasing according to (1) during remelting, the Ca activity has to be continuously increased during the process for compensation. Furthermore, the CaO-activity can be decreased by dilution via addition of $CaF_2$ to the slag. Figure 1 exhibits that the necessary Ca content in the slag to reach specification conform oxygen contents is much higher in cp-Ti melts than in TiAl. Thus, because of the high vapour pressure of Ca, this results into difficult handling of high Ti containing melts.

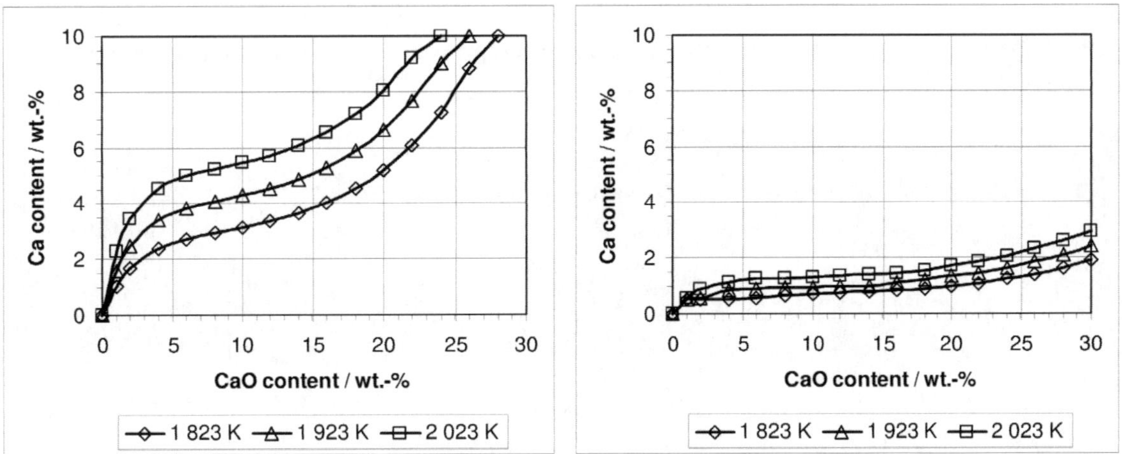


Figure 1: Necessary Ca content in the slag in order to achieve specification conform oxygen contents in titanium and titanium aluminide [5]

## 3. Experimental

### *3.1. Masteralloy synthesis by Aluminothermic Reduction (ATR)*

The feasibility of aluminothermic reduction (ATR) of $TiO_2$ to produce Ti alloys was shown in previous works [7][8]. The main challenge for the ATR process is to produce Ti alloys with high titanium and low aluminium content. However, the high oxygen solubility of titanium does not allow producing Ti alloys with low aluminium and simultaneous low oxygen contents. Because Al is used as alloying element in many Ti materials, residual Al content in the ATR product can be tolerated. The inverse correlation of Al and O content in titanium materials is reported to be 6 wt.-% Al at >10 wt.-% O and 36.-% Al at 1.4 wt.-% O. [9] The typical oxygen content of aluminothermically produced TiAl material varies between 1.3 and 1.8 wt,-% [7][8]. Therefore, a subsequent deoxidation step has to be performed for ATR Ti in order to meet the specifications of 2000 ppm. It is of particular interest to provide aluminothermic reduced titanium for further titanium based alloy production. By co-reduction of $TiO_2$ and other oxides of alloying elements such as vanadium the synthesis of a master alloy is possible. In this respective, Ti-6Al-4V, representing the most utilized titanium alloy, gets relevant.. For the present research the composition of the master alloy is chosen as Ti with 60 wt.-%, Al with 24 wt.-% and V 16 wt.-%. For the subsequent processing by VIM titanium sponge has to be charged with a ratio of 3:1.

The reduction of $TiO_2$ with aluminium requires further heat input to obtain the reaction self-propagated. $KClO_4$ is added to the reaction mixture with the required amount of aluminium. For synthesis of the master alloy, the combined reduction of oxides of titanium and vanadium is performed. There are different oxides available, such as $V_2O_3$, $V_2O_4$, $V_2O_5$ and mixtures of theses oxides whereat $V_2O_5$ provides the highest enthalpy input. In order to decrease the $KClO_4$ amount and hereby decreasing Cl-bearing off-gas $V_2O_5$ is used in every experiment. In order to decrease the liquidus temperature of the alumina slag, lime is added to the ATR mixture as flux. All input materials are added in powder form and blended before charging. Experiments on two different scales (20 L ≙ 6kg of product and 90 L ≙ 30 kg) are performed in the present work. The initial ignition is carried out by a filament on top of the input mixture in all trials. After reaction, the produced metal ingot is analyzed with respect to total oxygen as well as base elements. The ingot is cleaned from residual slag attached to the surface and crushed for the subsequent processing steps.

The cross sections of the ATR products are shown in Figure 2. It can be seen, that a weak segregation occurred in the small scale trail in the middle of metal ingot wherein the melt solidifies at last. On a macroscopic scale, there are no porosities observed within the metal products.

Figure 2: Cross Section of small scale (left) and large scale (right) ATR products

### *3.2. Electrode production by Vacuum Induction Melting (VIM)*

Consolidation, alloying and electrode production were carried out in a 40kW and 150kW vacuum induction furnace at IME. The crucibles used for melting were made of $CaZrO_3$ (lab scale) or CaO (pilot scale) refractory material. Properties of the $CaZrO_3$ crucibles are described elsewhere [10]. At the beginning of each vacuum induction only the ATR material is charged into the crucible. Titanium sponge (ratio 3 Ti: 1 ATR) is added subsequently using a charging system after obtaining a melt in order to keep the titanium activity as low as possible at any time to minimize crucible-melt reactions. During melting, strong evaporation of calcium can be noticed. The melt is hold at a temperature of 1750 °C for 10 min to obtain a good homogenization. Afterwards, the melt is tapped in a water cooled copper mould with a diameter of 50mm and 100mm respectively
After cooling and stripping, the cast ingot is analyzed for oxygen using inert gas fusion method, matrix and alloying elements were analyzed by XRF. Figure 3 shows the cast electrodes.

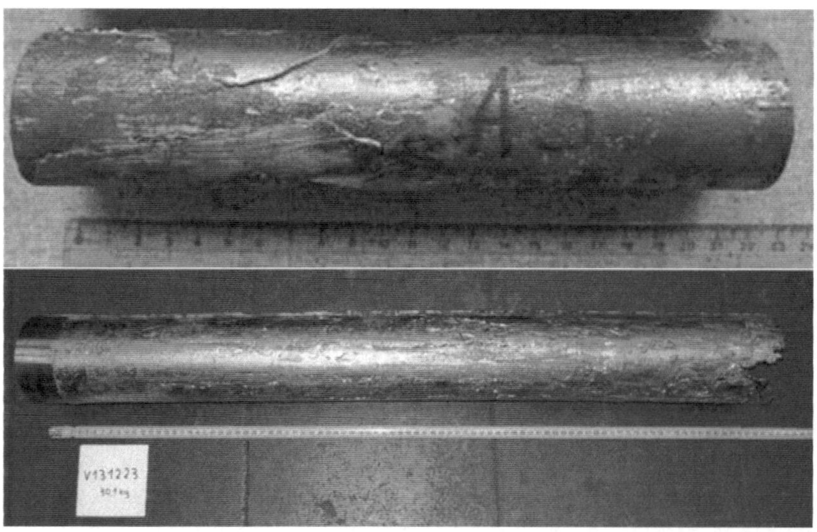


Figure 3: Obtained Electrode after pilot scale VIM, top: lab scale, bottom: pilot scale

### *3.3. Final Deoxidation by (Pressure) Electroslag Remelting (PESR)*

#### *3.3.1. Lab Scale*

Remelting of the lab scale electrodes was performed in a chamber electroslag furnace (ChESR) at Donetsk National Technical University (DonNTU), Ukraine with a maximum operating power of 724kW. The furnace is equipped with a water cooled copper crucible (70 mm diameter, 425 length) and a Chamber for operating under shield gas at slight overpressure. Before remelting, the crucible is filled with 700 g technically pure $CaF_2$ and calcium granules for deoxidation, which were pressed to a tablet. The calcium content varies from 4 to 8 wt.-% of the total slag mass depending on the trial. After this, the chamber is evacuated and backfilled with Argon (99.987%). Starting of the process is performed by arcing between the electrode and calcium tablet. Process control is done by manual adjustment of the current, which is kept by approx. 3 kA. During remelting, the gas pressure in the chamber is maintained at 0.3 -0.4 atm. above atmosphere. The process time varies from 130 to 250 s depending on the electrode length. After cooling, the ingot (Figure 4) is analyzed by OES for metals and inert gas fusion for dissolved oxygen.

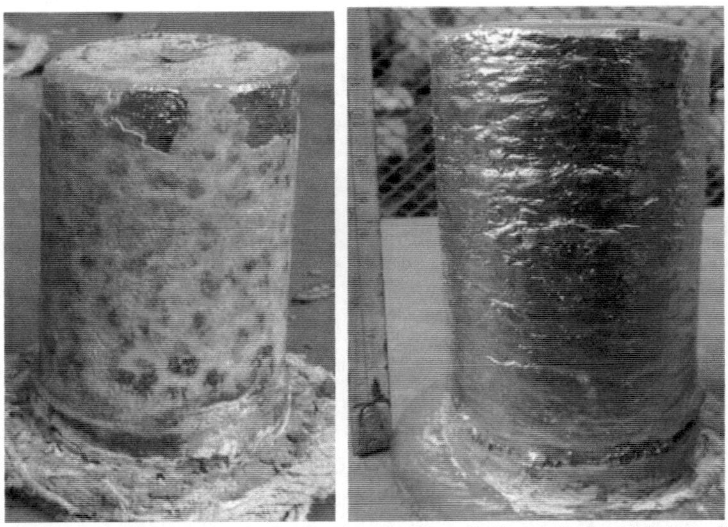

Figure 4: Obtained ingots after electroslag remelting before (left) and after (right) slag skin removal

#### *3.3.2. Pilot Scale*

The experiments for pilot scale deoxidation were performed in the IME pressure electroslag remelting furnace. The furnace is equipped with a charging system, which allows the continuous feeding of additives into the melt during the process at all operating pressures and with adjustable speed. Before the furnace is closed, the electrode is connected to the electrode rod with a titanium stub. A starting plate made of Ti-50Al sputter targets is placed on the crucible bottom to ensure electrical contact during the starting phase of the process. Afterwards, the crucible is filled with 4-4.5 kg of preheated (15 h @ 650 °C) process slag that consists of the technically pure $CaF_2$ slag Wacker Electroflux 2052 (> 97 wt.-% $CaF_2$) as well as 225-300 g of metallic calcium depending on the trial. The bunkers of the feeding system are filled with a $CaF_2$-Ca mixture (ratio 2:1). The furnace is closed and, in order to avoid reactions with oxygen from ambient air, the vessel is evacuated to a pressure of $10^{-1}$ mbar. Consecutively, the vessel is backfilled with Ar gas to the desired process pressure of 20 bar to avoid excessive Ca evaporations. During remelting, constant feeding of the $CaF_2$-Ca mixture was performed in order to compensate Ca-losses and to reduce the CaO activity. After remelting and cooling for several hours, the ingot is stripped, the cap-slag and slag skin are removed, the ingot is sectioned and analyzed by IGF for oxygen.

## 4. Results and Discussion

### *4.1. Masteralloy synthesis*

EDX-analyses exhibit that present inclusions are mainly alumina based oxide particles, spinel type particles of alumina and calcia were detected as well. The EDX-results indicate locally strong variations of the metal composition. The titanium content varies between 51 and 62 wt.-%, Al between 21 and 26 wt.-% and vanadium between 13.6% and 22.4 %. However, the average values for aluminium and titanium match with the results of the ICP-analysis. The metal product of the small scale trial results in lower titanium yields and effects hereby a higher amount of aluminium because not consumed for $TiO_2$ reduction.

Table 1 shows the oxygen contents of the obtained master alloys. With respect to the lower titanium content and higher aluminium content, the lower oxygen amount in the small scale trial is evident and corresponds with literature data. For further processing, the 6kg batch was used for lab scale VIM trials (VIM1a-3a), the 30kg batch for pilot scale (VIM1b-3b) melts.

**Table 1.** Chemical composition of the ATR master alloys (all elements in wt.-%)

| Trial | **Ti** | **Al** | **V** | **Fe** | **Si** | **O** |
|---|---|---|---|---|---|---|
| 6 kg batch | bal. | 31.7 | 12.6 | 2.40 | 0.28 | 0.78 |
| 30 kg batch | bal. | 22.4 | 14.4 | 1.95 | 0.73 | 0.92 |

[a] a – lab scale, b – pilot scale

### *4.2. Vacuum Induction Melting*

Table 2 show the concentration of the major elements in the obtained electrodes from lab and pilot scale tests. Since one lab scale ingot consists of two castings, an oxygen concentration gradient from top to bottom can be measured. In general, the oxygen pickup when using $CaZrO_3$ crucibles is lower, which indicate their stability. On the contrary, a Zr enrichment in the metal can be measured. A detailed description of the $CaZrO_3$-Ti-interaction is given by [10]. Melting in pure lime crucibles leads to an oxygen increase to approx. 1-1.2 wt.-%. All cast electrodes are further processed in the electroslag remelting process.

**Table 2.** Chemical composition of obtained lab- and pilot scale electrodes (all elements in wt.-%)

| Trial | **Ti** | **Al** | **V** | **Zr** | **Fe** | **Si** | $O_{top/bottom}$ |
|---|---|---|---|---|---|---|---|
| VIM 1a | 89.77 | 5.69 | 2.38 | 1.36 | 0.24 | 0.14 | 0.70/0.93 |
| VIM 2a | 90.97 | 5.32 | 2.12 | 0.87 | 0.24 | 0.11 | 0.62/0.72 |
| VIM 3a | 90.51 | 4.87 | 2.47 | 1.43 | 0.25 | 0.12 | 0.74/0.94 |
| VIM 1b | 85.53 | 9.05 | 4.61 | - | 0.36 | 0.20 | 1.05 |
| VIM 2b | 87.00 | 8.20 | 3.91 | - | 0.32 | 0.26 | 1.04 |
| VIM 3b | 85.82 | 8.60 | 4.48 | - | 0.36 | 0.20 | 1.24 |

[a] a – lab scale, b – pilot scale

### *4.3. Electroslag deoxidation*

Table 3 summarizes the main (P)ESR process parameters, final ingot oxygen concentrations as well as the deoxidation efficiency. The trial number refers to the VIM electrode used (see Table 2). Regarding the lab scale trials, the oxygen analyses were performed again in the top and bottom section of the ingot. As results show, deoxidation could be achieved for each ingot in the bottom part. On the contrary, the oxygen content in the top part is equal to or higher than in the initial electrode. An oxygen gradient can be explained by CaO enrichment and Ca loss in the slag. An additional oxygen uptake cannot be explained entirely by considering only dilution effects between top and bottom. Here, more trials are necessary. In respect to the pilot scale tests, a clear deoxidation effect for all Calcium containing trials can be observed. Figure 5 shows the oxygen change after remelting for all pilot scale tests in detail. The oxygen level is significantly decreased, especially trial number PESR number 1b shows a clear tendency which indicates a rising Ca activity in the slag during the melt. Unfortunately, an on-line measurement of the Ca content in the slag is not possible.
The reference melt without Ca-addition shows an oxygen uptake during the whole process which can be explained by dilution of CaO residues in the slag, as well as the dilution of remaining oxide residues from the electrode. In general, a clear tendency between the added Calcium/ingot mass ratio could not be verified and has to be investigated in further trials.

**Table 3.** Experimental parameter and results of lab- and pilot scale deoxidation trials

| Trial | $M_{ingot}$/ kg | melt rate/ kg/min | $m_{CaF2}$/ kg | $Ca_{start}$/ wt.-% | $m_{Ca}/m_{Met}$ g/kg | $O_{top,bottom}$/ wt.-% | $\Delta O_{av}$/ wt.-% |
|---|---|---|---|---|---|---|---|
| ESR 1a | 1.38 | 0.6 | 700 | 8.14 | 44.93 | 0,7/0,78 | 0.00/-0.15 |
| ESR 2a | 1.57 | 0.5 | 700 | 9.20 | 50.96 | 0.78/0.58 | +0.16/-0.15 |
| ESR 3a | 2.73 | 0.7 | 700 | 8.50 | 23.81 | 0.86/0.75 | +0.12/-0.19 |
| PESR 1b | 27.0 | 0.6 | 4500 | 4.76 | 14.81 | 0.73 | -0.32 |
| PESR 2b | 29.6 | 0.6 | 4000 | 6.98 | 22.64 | 0.79 | -0.25 |
| PESR 3b | 30.6 | 0.6 | 4500 | 0 | 1.47 | 1.26 | +0.02 |

[a] a – lab scale, b – pilot scale (20 bar pressure),

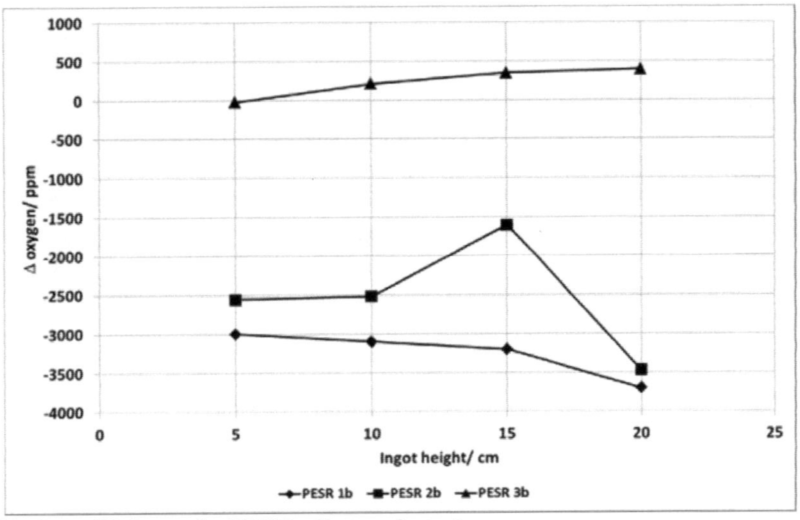


Figure 5: Deoxidation efficiency in PESR pilot scale tests

## 5. Kinetic Model

Fraser [11] investigated various mass transfer aspects of AC ESR of steels (slagging of Mn and desulphurisation) in a very comprehensive thesis work. The findings of his work can be adapted to the conditions for the presented case. General assumption for describing the reaction kinetics during ESR is, that thermochemical equilibrium is only achieved at the metal-slag phase boundary itself, thus it consequently has to be considered that diffusion of Ca and CaO through a finite layer in the slag and O through a layer on the metal side respectively could be rate limiting factors.
In Fraser's model Mn is transported in the metal phase to the phase boundary, is being oxidized by $Fe^{2+}$ ions from the slag and leaves the reaction zone as $Mn^{2+}$ in the slag. The author solves a set of four mass-transfer equations in order to obtain a differential equation for the rate of change of Mn concentration as a function of interfacial area, phase volume, bulk concentration, interfacial concentration at equilibrium, the equilibrium constant K and all transport coefficients $k_i$ in the metal and slag diffusion layer. The adaption of this set of equations to the change rate of oxygen concentration in a titanium melt is presented in [12]

All necessary thermochemical and physical data for calculating the transport coefficients and solving the differential equation was collected from literature or calculated by using different approaches. For a detailed description please refer to [12]. With the necessary data determined, all parameters were put together in MS Excel™ spreadsheets to allow for easy scale up scenario calculations of furnace parameters (e.g. ingot diameter, melt rate), slag and metal composition. The parameters b, c and the differential equation in (2) were numerically solved independently for the three reaction zones at the respective boundary conditions and then merged together in order to calculate the change of the slag composition and the oxygen concentration of the solidifying metal. Based on this principle, the whole remelting process can be modelled in discrete time steps by recursive recalculation of the oxygen concentration.
Figure 6 shows the modelling results of trial PESR 1b. The solid line represents the calculated oxygen content in the melt during the process, the individual data points are oxygen contents in the ingot determined by chemical analysis. Based on the melt rate, for each process time, a corresponding ingot height was allocated. In general, the model is in good agreement with the experimental values and thus indicates a kinetic limit for a deoxidation effect.

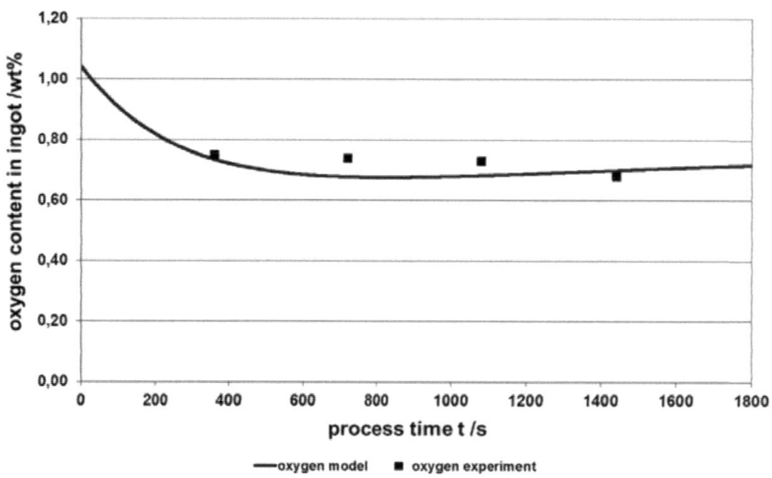


Figure 6: Modelled and experimental oxygen concentrations in the Ti-Al-V ingot (PESR 1b)

In a previous study [13], the authors presented the deoxidation behavior of a standard Ti-6Al-4V alloy with 2000 ppm oxygen. Idea was to enforce a quick drop of the oxygen concentration by rapid Ca-addition to the slag during remelting. Figure 7 shows the obtained results, which predict an oxygen level increase at the beginning of the process. This can be explained by the CaO in a technically pure $CaF_2$ slag, which dissolves in the metal to a level according to the equilibrium constant K of the deoxidation reaction. The further oxygen level tend in the metal calculated by the model can be explained as follows: As the oxygen concentration in the slag reaches its equilibrium state, a dilution of the solidifying metal takes place as drops from the electrode tip enter the liquid metal pool. This leads to a continuous decrease of the oxygen level towards the initial value of 2000ppm (Figure 7, right). The rapid addition of metallic calcium leads to a sharp decrease of oxygen down to 1000 – 1500 ppm. As both figures show, the predicted oxygen trend is mostly confirmed by the experimental trials. Furthermore, the trials show that PESR deoxidation of Ti6Al4V is possible to values below 1000 ppm.

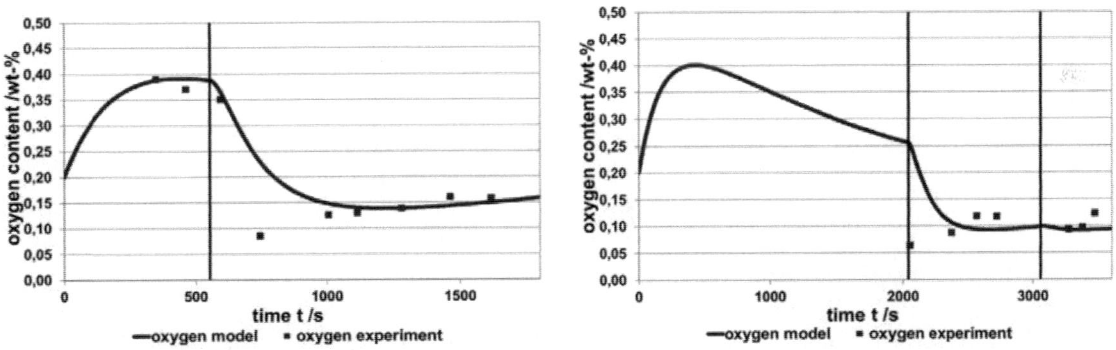


Figure 7: Deoxidation behavior of a standard Ti-6Al-4V alloy with 2000 ppm oxygen

## 6. Conclusions

- Ti-Al-V electrodes containing oxygen contents above specifications were successfully produced by ATR masteralloy synthesis and subsequent alloying via VIM in lab scale ($CaZrO_3$ refractory) and pilot scale (pure CaO refractory) to test the deoxidation limits in the following ESR process
- Deoxidation trials in chamber and pressure ESR units were performed under varying Ca-contents in the slag
- Deoxidation effects vary significantly over the ingot height in lab scale and cannot be explained entirely. Since only slight overpressure (+0.4 bar) was set, Ca losses due to evaporation are possible, since the operating temperature is above the Ca boiling point
- Oxygen levels in PESR trials (20 bars) can be reduced by max. 3500 ppm
- A clear tendency between the calcium/ingot mass ratio could not be verified. Thus, the process chain needs more trials for validation and ensuring reproducibility
- A kinetic model is presented which predicts the oxygen concentration as a function of process time. Modelled and experimental values are in good agreement
- The main challenge remains achieving a homogeneous oxygen level in the whole ingot

## 7. Acknowledgement

The authors would like to thank Deutsche Forschungsgemeinschaft (DFG) for funding this research.

## References


[1] Lochbichler, C.: Design eines prozessstufenarmen Titanrecyclingprozesses, Ph.D. thesis, IME Metallurgische Prozesstechnik und Metallrecycling, RWTH Aachen University, Shaker Verlag, ISBN: 978-3-8440-0514-1, November 2011

[2] Stoephasius, J.: Elektroschlackeraffination aluminothermisch hergestellter gamma-Titanaluminide. Ph.D. thesis, IME Metallurgische Prozesstechnik und Metallrecycling, RWTH Aachen University, Shaker Verlag, ISBN 978-3-8322-5876-4, February 2006.

[3] Reitz, J.: Elektroschlackedesoxidation von Titanaluminiden, Ph.D. thesis, IME Metallurgische Prozesstechnik und Metallrecycling, RWTH Aachen University, Shaker Verlag, ISBN: 978-3-8440-2022-9, July 2013

[4] Tsukihashi, F.; Tawara, E.; Hatta, T.: Thermodynamics of Calcium and Oxygen in Molten Titanium and Titanium-Aluminum Alloy; Metallurgical and Materials Transactions B, Vol. 27B, December 1996, pp. 967-972

[5] Stoephasius, J., Reitz, J. und Friedrich, B.: ESR Refining Potential for Titanium Alloys using a CaF2-based Active Slag. Advanced Engineering Materials, 9, Apr. 2007.

[6] Okabe, T. H. et al.; Thermodynamic Properties of Dilute Titanium-Oxygen Solid Solution in Beta Phase; Materials Transactions, JIM 32 (1991) 5, pp. 485-488

[7] Hammerschmidt, J.: Entwicklung einer Prozessroute zur Herstellung von $\gamma$–TiAl-Legierungen durch Aluminothermie und Schutzgas-Elektroschlackeumschmelzen, D82 (Diss. RWTH Aachen) Shaker Verlag, ISBN 3-8322-1971-4, Aachen 2003Stoephasius

[8] Stoephasius, J.-C.: Modelling of metallothermic reactions - local reaction rates during aluminothermic TiAl-Production, Erzmetall, World of Metallurgy, Band 58, 2004, Nr. 4

[9] Maeda, M.: Aluminothermic Reduction of Titanium Oxide, Materials Transactions, The Japan Institute of Metals, Vol. 34, 1993, No. 7

[10] S. Schafföner, C. G. Aneziris, H. Berek, B. Rotmann und B. Friedrich: Investigating the corrosion resistance of calcium zirconate in contact with titanium alloy melts. In: Journal of the European Ceramic Society 35.1 (2015), S. 259-266

[11] Fraser, M.E.; Mass Transfer Aspects of A.C. Electroslag Remelting; Ph.D. thesis, The University of British Columbia, January 1974

[12] Reitz, J.: Slag-metal interaction during PESR of Titanium alloys; Slags and Fluxes in Modern Metallurgy, Shaker Verlag, 2011, ISBN: 978-3-8440-0481-6

[13] Bartosinski, M., Reitz, J.: Modelling the Oxygen Content of Titanium during Deoxidation in the Pressure Electroslag Remelting (PESR) Process, Proceedings of the EMC European Metallurgical Conference, ISBN: 978-3-940276-51-3, 2013

# In-situ observation of the behaviour of non-metallic inclusions at different interfaces in the system steel-slag-refractory

**S K Michelic[1], U Diéguez Salgado[1] and C Bernhard[1]**

[1] Chair of Ferrous Metallurgy, Montanuniversitaet Leoben, Franz Josef Straße 18, 8700 Leoben, Austria
E-Mail: susanne.michelic@unileoben.ac.at

**Abstract**. The system steel-slag-refractory and reactions of non-metallic inclusions at its interfaces are decisive in numerous metallurgical processes. In order to obtain the desired steel cleanness level, a detailed understanding of the occurring interactions is needed. This study focuses on the in-situ observation of different reactions of non-metallic inclusions in steel and slag. For this purpose a Confocal Scanning Laser Microscope attached to a high temperature furnace is applied which enables the visualization of metallurgical phenomena at steelmaking temperatures up to 1700 °C. Two different examples are discussed: First, the agglomeration of different inclusion types in a Ca-treated steel is described and corresponding attraction forces are calculated. Secondly, the dissolution of $Al_2O_3$ and $MgOAl_2O_3$ in two slags of the system $CaO-Al_2O_3-MgO-SiO_2$ is examined, including the evaluation of the governing dissolution mechanism. Finally, an outlook on further experimental possibilities is given.

## 1. Introduction

Number, size and composition of non-metallic inclusions in the steel matrix directly influence the final product quality for a broad field of steel grades. Thus, a specific adjustment of steel cleanness in secondary metallurgical processes is essential in order to avoid problems in the subsequent casting process, e.g. clogging as well as to achieve the required quality level. Principally, the phases steel, slag and refractory material determine and influence the behavior of non-metallic inclusions. Different reaction sites involving these phases exist. Next to the agglomeration of inclusions in the liquid steel, their separation from the steel to the slag, and their subsequent dissolution in the slag has to be considered. Additionally, inclusions can be modified through the contact with refractory material or slag.

High-temperature Confocal Scanning Laser Microscopy (HT-CSLM) offers a relatively new approach for in-situ investigations of reactions at temperatures relevant to steelmaking processes. Based on the pioneer work done by Japanese researchers [1], this method is applied for the investigation of phase transformations [2, 3] as well as inclusion related questions [4–6]. Within the present study the application of CSLM is shown for two different cases. As a first example, the agglomeration of different inclusion types in a Ca-treated steel is illustrated and the calculation of the corresponding attraction forces between a particle pair is described. Secondly, the dissolution behavior of $Al_2O_3$ and $MgOAl_2O_3$ in two slags of the system $CaO-Al_2O_3-MgO-SiO_2$ is investigated. Absolute dissolution times as well as the dissolution mechanism are evaluated.

## 2. The Method

### *2.1. Experimental Set-Up*

A Confocal Scanning Laser Microscope type VL2000DX from Lasertec and a high temperature furnace type SVF17-SP from Yonekura are used for all investigations. The decisive advantage of HT-CSLM in contrast to conventional microscopes or experimental facilities is the possibility of in-situ observations at temperatures up to 1700 °C in combination with a very good image quality. This is primarily possible due to the fact, that a laser is used as a light source. The laser is characterized by a wavelength of 408 nm and is therewith below the thermal radiation spectrum of the observed samples. So, the contrast of the image can be increased significantly. Heat is provided by a halogen lamp with a maximum power of 1500 W. In combination with a low sample weight, this enables heating rates up to 1200 °C/min and cooling rates of 1000 °C/min. The maximum operating temperature is about 1700 °C at standard configurations and is only limited by the thermocouple type. Since the high temperature furnace is cooled, the thermal strain for the furnace through the radiation source is comparably low. Through the use of an acousto-optic element a very high scan rate can be reached. For the maximum resolution of 1024x1024, 15 images per second are stored. A resolution of 1024x256 enables the recording of 60 images per second [7]. A schematic view of the experimental set-up in the high temperature furnace which shows an elliptic, gold coated inner contour can be found in [7]. The halogen lamp is situated in the bottom focal point. The sample holder with the crucible and sample inside is located in the top focal point of the ellipse. The temperature is measured with a thermocouple fixed at the bottom side of the sample holder. The position of the thermocouple always results in a temperature difference between the sample surface and the thermocouple. Therefore an accurate temperature calibration is necessary before the experiment itself. High-purity argon with a flow rate of 100 $cm^3$/min ensures a neutral atmosphere in the furnace. Additionally, the oxygen content in the furnace is measured for every experiment. The latter typically lies in the range of 10-18 ppm during the experiment.

### *2.2. Experimental Procedure*

Various experiments can be performed with the described set-up. In the present study two different experimental procedures are applied studying the cases of inclusion agglomeration in the liquid steel and inclusion dissolution in a slag phase. Figure 1 gives an overview on the used experimental set-ups for these two types of investigations:

- For type 1 experiments a steel disc with 600μm thickness is placed in an $Al_2O_3$ crucible and subsequently set up on the sample holder of the ellipsoidal furnace. The specimen is heated up following the heating cycle given in Figure 2. An alumina disc is placed under the steel disc in order to avoid the attachment of the steel disc onto the crucible.When the liquidus temperature of the steel is reached, inclusions are starting to emerge from the bulk to the steel surface and their agglomeration behavior can be studied. A posteriori examination of the sample at room temperature is carried out using a Scanning Electron Microscope with Energy Dispersive X-Rays (SEM/EDS). In order to characterize (composition, shape, size) the inclusions observed during the experiment, a fast cooling of the sample is important. In the present study a Ca-treated steel is used (composition given in Table 1).

- For type 2 experiments involving a slag phase a platinum crucible is used instead of $Al_2O_3$. A well-shaped spherical inclusion particle (made by Sandoz Fils, Cugy, Switzerland, details given in ) with a known diameter is placed on the surface of the defined amount of premolten slag in the crucible (compositions given in Table 1). Figure 2 shows a typical heating cycle for dissolution experiments. The whole assembly is heated up to the target temperature using the maximum heating rate. This is important to ensure that the inclusion hardly starts to dissolve before the experimental temperature is reached. At a temperature 50 °C below the final temperature, the heating rate is reduced to 100°C/min in order to avoid overshooting. The experimental temperature can be defined arbitrarily up to 1700 °C. In the present study inclusion dissolution is examined at 1450 °C.

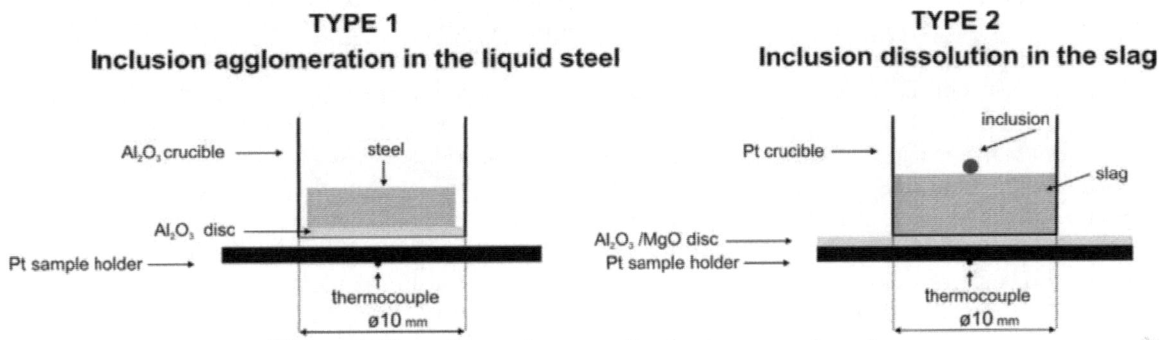


**Figure 1.** Experimental set-ups for the two experiments.

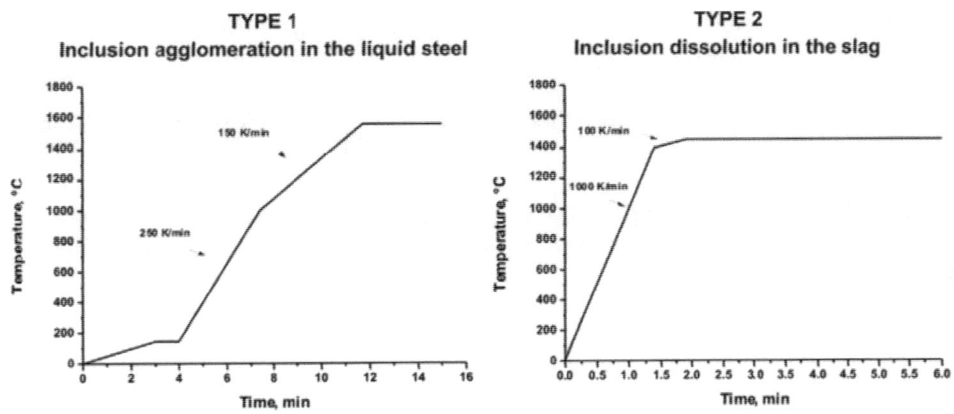


**Figure 2.** Time-temperature profiles of the two experimental types.

**Table 1**: Investigated materials.

| | | | | | | |
|---|---|---|---|---|---|---|
| **Steel** | **C [wt.-%]** | **Si [wt.-%]** | **Mn [wt.-%]** | **Al [wt.-%]** | **Ca [wt.-%]** | **S [wt.-%]** |
| | 0.15 | 0.02 | 1.07 | 0.04 | 0.0012 | 0.004 |
| **Slags** | **CaO [wt.-%]** | **$Al_2O_3$ [wt.-%]** | **MgO [wt.-%]** | **$SiO_2$ [wt.-%]** | | |
| Slag 1 | 40 | 52 | 5 | 3 | | |
| Slag 2 | 33 | 3 | 12 | 52 | | |
| **Inclusion** | $Al_2O_3$<br>$MgOAl_2O_3$ | | | | | |

## 3. In-situ Observations

### *3.1. Inclusion Agglomeration in Liquid Steel*

Inclusion formation during steel processing is unavoidable. The harmful effects of inclusions can be reduced if inclusion chemistries and morphologies are tailored properly and they can even be rendered beneficial in specific cases. In order to optimize steel cleanness in different process steps, a detailed understanding of the factors influencing agglomeration and coalescence of inclusions in molten steel is

necessary. In addition to fluid dynamics, interfacial chemical interactions between particles are important [8–10]. Several research groups already dealt with this topic in the last years especially focusing on the in-situ observation of the agglomeration process and the resultant quantitative evaluation of the obtained data. For example, Yin et al. [6, 11] described the behavior of alumina inclusions at a steel-gas interface. They found out that the non-wettability behavior of the alumina based non-metallic inclusions generate a deformation of the liquid surface, which is called meniscus. If two particles approach one-another, the meniscus in between them is further depressed or drawn downward [12]. This change, which is called capillary attraction, can lead to a difference of capillary pressure between the inside and the outside area of the pair, which will push the two bodies towards each other [11]. The capillary attraction is believed to be the source of the attraction between alumina inclusions at free surfaces and at gas bubbles inside the liquid steel melt [12]. The strength of the capillary attraction force for inclusion particle pairs in molten steel, which was found to be much different for particles with different morphology, can be listed as follows in ascending order [6]: liquid/liquid pair < liquid/semiliquid pair < semiliquid/semi-liquid pair < liquid/solid pair < semiliquid/solid pair and finally solid/solid pair as the strongest. The latter sequence is only valid if only the effect of the particle morphology is considered. But the attraction between particles is also strongly affected by other parameters like for example by the particle size, the contact angle of a particle with steel melt, the particle density, the shape and the interfacial tension [13].
Yin et al. [11] carried out an approximate evaluation of the magnitude of the attraction force from measured acceleration and estimated the mass of particles based on in-situ observations in the confocal microscope. The particle shape was approximated to be an ellipse disk and the lengths of long axis ($d_1$) and short axis ($d_2$) of a particle were observed. The height of all the disks was taken to be 2μm on the basis of confocal observations of the thickness of alumina particles on steel melt surface. Further, the ellipse was approximated to be a round disk of the radius, r, which was taken to be geometric average according to Equation 1.

$$r = \sqrt{d_1 \cdot d_2/4} \tag{1}$$

The acceleration, $a_1$, of the guest particle when the host particle in the particle pair stayed quiescent was determined from the change of the position of the guest particle at defined intervals. The force F was subsequently calculated by

$$F = m_1 a_1 \tag{2}$$

where $m_1$ is the mass of the guest particle (the moving particle toward a bigger one). If the size difference between two particles in a particle pair was big, then the smaller one in the pair (the guest particle) moved accelerated toward the larger one (the host particle), which was almost quiescent while attraction took place. On the contrary, if the pair of particles were almost the same size, then both particles moved accelerative to make a head-on collision. If both inclusions move, a correction factor for the mass $m_1$ was used:

$$\frac{m_2}{m_1+m_2} \tag{3}$$

where $m_2$ is the mass of the largest inclusion. In this force analysis, the friction force that arises from the viscous drag of the particle by liquid steel surface was ignored.

This approach was applied for the investigations in the present study, evaluating the agglomeration behavior of liquid and semiliquid non-metallic inclusions in a Ca-treated steel (composition given in Table 1). Figure 3 exemplarily shows results of inclusion agglomeration experiments (type 1) in a Ca-treated steel. After a large part of the sample surface melted, many inclusion particles emerged from the interior of molten steel bath and floated on the melt surface. After the experiment itself the sample was cooled down as fast as possible in order to analyze the inclusion composition using a Scanning Electron Microscope with Energy Dispersive X-Rays (SEM/EDS). Figure 4a shows a SEM- image of a specimen at room temperature after conducting the HT-CSLM experiment. Almost all the surfaced inclusion particles from the calcium treated steel melt were observed to be in semi-liquid irregular shape and liquid with globular shape. Representative examples of detected inclusions are given in Figure 4b and c. Their chemical composition is summarized in Table 2. The semi-liquid inclusions are composed by two different phases; the calcium sulfide with traces of calcium oxide phase is solid and the calcium aluminate phase is liquid (Figure 4b). The liquid inclusions are composed of calcium aluminate (Figure 4c).

When no local surface flow was observed, the semiliquid and liquid inclusions showed no attraction between them (Figure 3a). When one liquid particle floats up to the melt surface, it pushes neighboring particles until separation is sufficient (Figure 3b and Figure 3c); finally, the original state of dispersion is reached (Figure 3d), they exhibit no relative movement during the experiment, indicating no change of distance between particles, and moving with random direction.

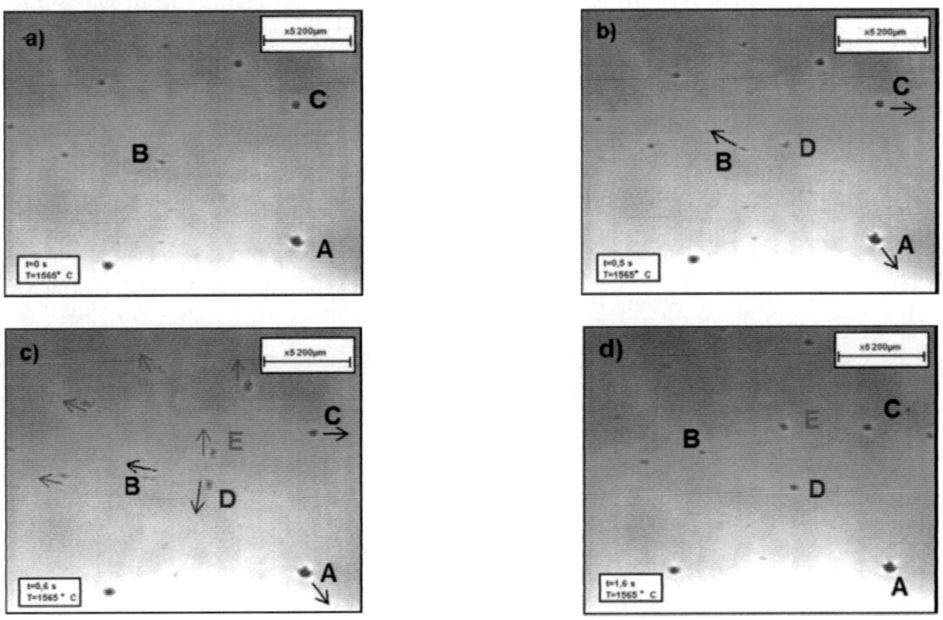


**Figure 3**. (a) Semiliquid particles distributed on the melt surface homogeneously (b) one particle D floats up and pushes the neighboring particles A, B and C (c) a new particle E floats up and again the neighboring particles move (d) the particles are distributed on the melt surface.

Since the contact angle of non-metallic inclusions and steel melt is usually more than 90°, the capillary force will become weak with decreasing density of inclusion particle and contact angle. The contact angle between semiliquid particles and melt is smaller since the liquid layer results in good wetting; this means that the capillary force between semiliquid particles in pairs will be weaker than for solid particles in pairs in agreement with [6,15]. Nevertheless, some interaction between the inclusions is observed close to the solidification front. In Figure 5a some semiliquid as well as liquid inclusions are observed in the pool close to the solidification front; these inclusions seem to be quiet. A single semiliquid inclusion (approx.12μm diameter) appears in the middle of the pool and after 2 second collides with another semiliquid inclusion (approx. 30μm). The trajectory of the particle is observed in Figure 5 as a function of the time. The velocity of the small particle increases when the distance between both particles is decreasing, showing that there is an increasing attraction force between this particle pair which helps them to finally agglomerate.

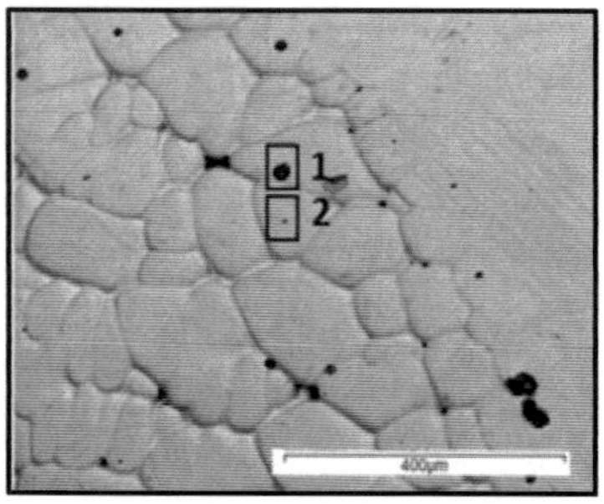


**Figure 4a.** SEM- image of a specimen at room temperature after HT-CSLM experiment.

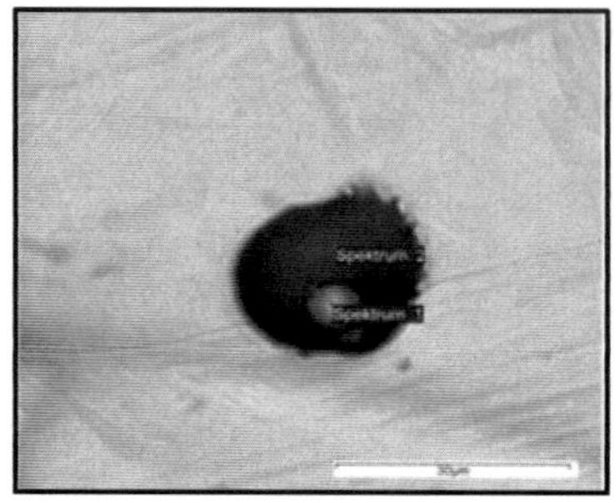

**Figure 4b.** Square 1 from Figure 4a.Semi-liquid inclusion with 22μm diameter, its composition given in Table 2.

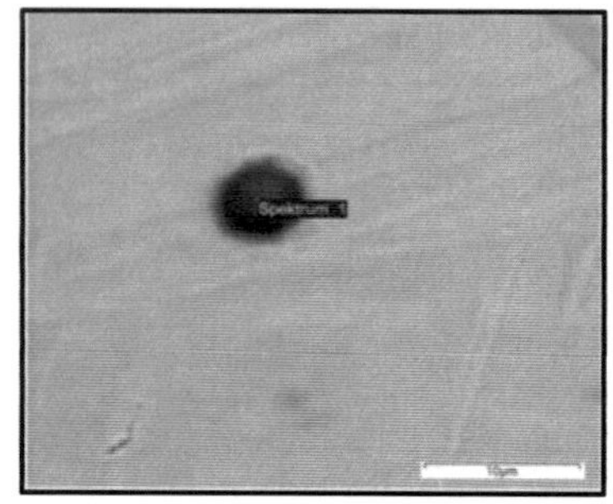

**Figure 4c.** Square 2 from Figure 4a. Liquid globular inclusion with 5μm diameter, its composition given in Table 2.

**Table 2**: SEM/EDS analyses for inclusions illustrated in Figure 4b and c.

| Fig. | Spec. | O [wt.-%] | Al [wt.-%] | S [wt.-%] | Ca [wt.-%] | Mn [wt.-%] | Fe [wt.-%] | Phase |
|---|---|---|---|---|---|---|---|---|
| 4b | 1 | 11.9 | 0.5 | 39.7 | 45.1 | 0.7 | 2.1 | Oxisulfide |
| | 2 | 46.9 | 28.8 | 0.3 | 20.3 | 2.7 | 1.0 | Al-Ca-Oxide |
| 4c | 1 | 43.8 | 30.6 | 1.4 | 16.7 | 1.2 | 6.3 | Al-Ca-Oxide |

Figure 6 displays the results of the measured interparticle distances as a function of time and the hence calculated attraction force for the pair of semiliquid inclusions shown in Figure 5. Principally, the tendency of agglomeration for liquid as well as for semiliquid inclusions is smaller than for solid inclusions. According to literature [6] the attraction force of a pair of semiliquid CA60 particles is approximately $10^{-16}$ -$10^{-15}$ N, which agrees well with the results shown in Figure 6.

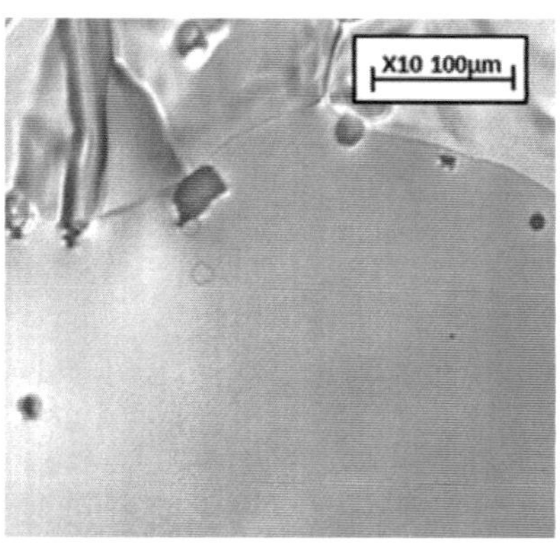


**Figure 5**. Attachment of inclusions at the solidification front in the Ca-treated steel.

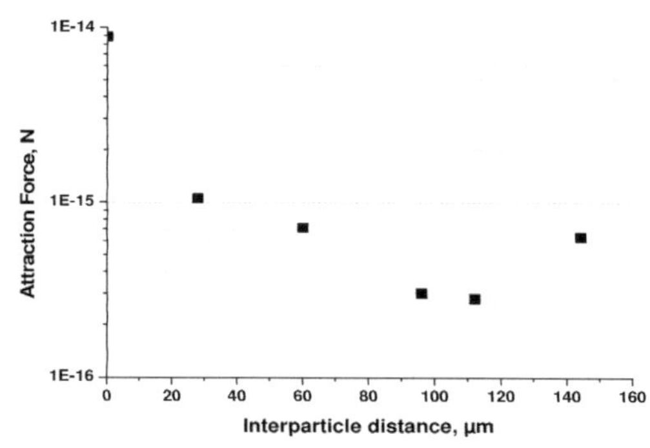


**Figure 6.** Attraction force versus interparticle distance for the semiliquid inclusion pair shown in Figure 5.

### *3.2. Inclusion Dissolution in the Slag Phase*

In the last years the confocal laser scanning microscope frequently replaced classical post-mortem studies [15, 16], dipping or rotating a shaped oxide material into a liquid slag, for investigating the dissolution behavior of various oxides in different slag compositions. The main focus of previous in-situ studies with the CLSM lied on the dissolution of $Al_2O_3$, MgO and $MgAl_2O_4$ in $CaO$-$Al_2O_3$-$SiO_2$ or $CaO$-$Al_2O_3$-$SiO_2$-MgO slags [17–23]. A first output of the dissolution experiments is the absolute dissolution time which is based on the evaluation of decreasing particle diameter with ongoing time. Details regarding the evaluation procedure based on the obtained video from CSLM can be found in [24]. Next to the comparison of absolute dissolution rates of different particles in defined slag composition also the governing dissolution mechanism got in the focus of interest. For this purpose normalized dissolution graphs (transient diameter divided by initial diameter and time divided by total dissolution time) are used. In order to determine the dissolution mechanism different models are proposed in literature: The analytical solution using the Shrinking Core Model (SCM) has been widely applied. [17, 21, 22, 25] In this model either the chemical reaction at the interface between inclusion and slag (linear correlation between particle size and dissolution time) or the mass transport of the dissolving particle in the boundary layer of the slag (parabolic dependence) can be the rate controlling step for the dissolution. Further details of this model are given in literature [26].

Another approach was proposed by Verhaeghe et al. [27–29] using a diffusion equation coupled with Lattice-Boltzmann modeling. Due to the high computational effort needed, it is less applied although its results are usually more accurate. Approximations of the diffusion equation for spherical particles are suited to analyze if the dissolution is governed by diffusion or not. Dissolution controlled by diffusion in a stagnant fluid can generally be described by Fick's first and second law of diffusion. [30] Principally, it has to be distinguished between invariant interface (resulting in an S-shaped normalized dissolution curve) and invariant field approximation (parabolic normalized dissolution curve). Details regarding the two approximations as well as the complete derivation can be found in [24].

A modified diffusion model based on the use of well-shaped spherical particles was introduced by Feichtinger et al. [24] merging the two approaches described above and introducing an additional factor f :

$$\frac{dR}{dt} = -\frac{k \cdot D}{R} - f \cdot k \cdot \sqrt{\frac{D}{\pi \cdot t}}, \tag{4}$$

where R is the radius of the particle, D is an effective binary diffusion coefficient of a particle in the slag, t is time, k is the dimensionless saturation and f is an additional factor ranging between 0 and 1. This approach focuses on the influence of slag viscosity on the governing dissolution mechanism. As equally assumed for all other mentioned models, the formation of a solid layer around the inclusion during dissolution cannot be considered with this approach.

In the present study the dissolution behavior of $Al_2O_3$ and $MgOAl_2O_3$ is examined in two different slag compositions in the system $CaO$-$Al_2O_3$-$SiO_2$-MgO at a temperature of 1450 °C: Slag 1 characterized by a high $Al_2O_3$ content and slag 2 contains $SiO_2$ as the main component. All thermodynamic data (e.g. saturation concentration of $Al_2O_3$ and $MgOAl_2O_3$ in the particular slag) as well as slag viscosities needed for the evaluation were calculated by *FactSage 6.4*. Slag densities were taken from literature [31]. Slag 2 is characterized by a significant higher viscosity (0.58 Pas) than slag 1 (0.23 Pas). Detailed slag compositions are summarized in Table 1. As already described the viscosity can essentially influence the governing dissolution mechanism and therewith the shape of the normalized dissolution curve between S-shape and parabolic shape.

A comparison of absolute dissolution times for $Al_2O_3$ and $MgOAl_2O_3$ in the two investigated slags is given in Figure 7. For each case three experiments were carried out; for the evaluation of the dissolution mechanism the mean out of the three experiments was used. Both inclusion types dissolved significantly faster in slag 1 compared to slag 2. Furthermore, the dissolution of $Al_2O_3$ was always observed to be faster than the $MgOAl_2O_3$ dissolution process. A comparison with data available in literature is difficult due to the at least small deviations in examined slag compositions as well as in experimental temperature, which already can influence the results noticeably.

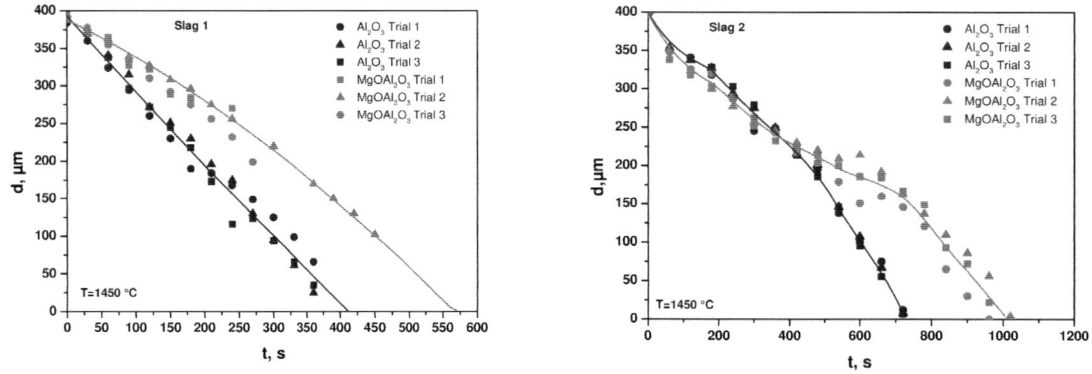


**Figure 7:** Comparison of absolute dissolution rates of $Al_2O_3$ and $MgOAl_2O_3$ in slags 1 and 2.

Figure 8 summarizes the results concerning the evaluation of the governing dissolution mechanism. Figures 8a and 8b compare the normalized dissolution curves to classical models from literature. It can be seen that only one case – the dissolution of $Al_2O_3$ in slag 1 – can be described sufficiently. The latter case shows an almost linear dependence and therefore can be reflected by the reaction rate controlled Shrinking Core Model. The other three cases do not match with the classical approaches. Here, the previously discussed modified diffusion model introduced by Feichtinger et al. [24] was applied. The obtained results are given in Figure 8c, including the used f-factor. The same dependence as already observed in former studies [24] is found: Increasing slag viscosity changes the dissolution pattern from parabolic (characterized by a low f-factor) to S-shaped (characterized by a high f-factor).

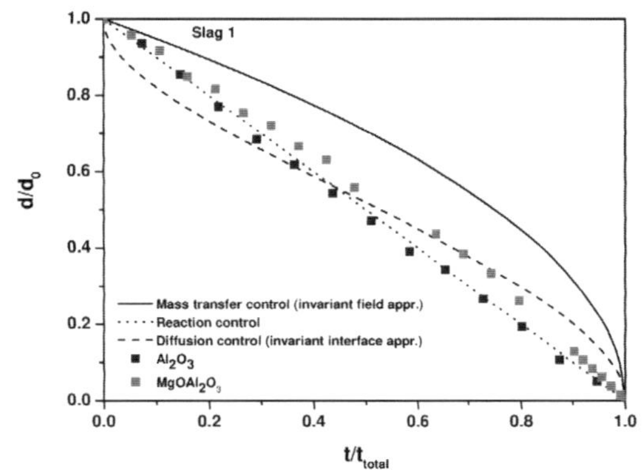


**Figure 8a.** Normalized dissolution curves for $Al_2O_3$ and $MgOAl_2O_3$ in slag 1 including classical dissolution models from literature.

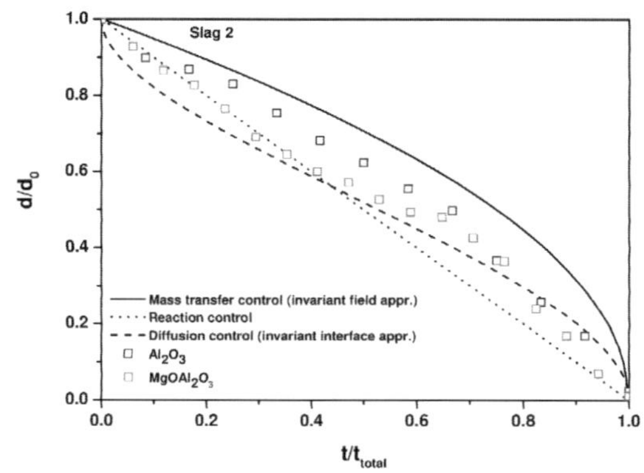


**Figure 8b.** Normalized dissolution curves for $Al_2O_3$ and $MgOAl_2O_3$ in slag 2 including classical dissolution models from literature.

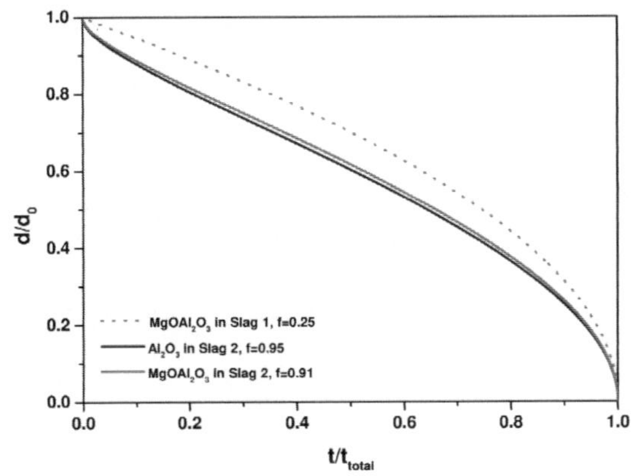


**Figure 8c:** Normalized dissolution patterns for $MgOAl_2O_3$ in slag 1 and $Al_2O_3$ and $MgOAl_2O_3$ in slag 2 calculated using the modified diffusion model introduced by Feichtinger et al.

## 4. Summary and Conclusion

Different reactions and interactions of non-metallic inclusions within in the system steel-slag-refractory can essentially influence the final steel cleanness – in a positive or negative way. Within the present work two types of inclusion reactions are described applying Confocal Scanning Laser Microscopy:

- Inclusion agglomeration in a Ca-treated low carbon steel: The tendency of agglomeration for liquid as well as for semiliquid inclusions is smaller than for solid inclusions. The attraction force of a pair of semiliquid particles was determined with approximately $10^{-16}$ -$10^{-15}$ N, which agrees well with literature.
- Dissolution of $Al_2O_3$ and $MgOAl_2O_3$ in the slag system $CaO$-$Al_2O_3$-$MgO$-$SiO_2$:$Al_2O_3$ proved to dissolve significantly faster than $MgOAl_2O_3$ in the investigated slags. Slag viscosity influences the dissolution mechanism essentially changing the pattern from parabolic to rather S-shaped with increasing slag viscosity.

Confocal Scanning Laser Microscopy does not only enable the in-situ observation of defined reactions on a microscopic scale but also provides important knowledge for processes on industrial scale, especially regarding the improvement of steel cleanness. Next to the two presented experimental types, also the reaction of inclusions at the interface steel-refractory can be studied. For example, these investigations can contribute to understand and evaluate the clogging sensitivity of certain steel-refractory combinations. Relating experiments are currently in progress.

**Acknowledgements**
Financial support by the Austrian Federal Government and the Styrian Provincial Government, represented by Österreichische Forschungsförderungsgesellschaft mbH and by Steirische Wirtschaftsförderungsgesellschaft mbH within the research activities of the K2 Competence Centre on "Integrated Research in Materials, Processing and Product Engineering", operated by the Materials Center Leoben Forschung GmbH in the framework of the Austrian COMET Competence Centre Programme, is gratefully acknowledged.

**References**

[1] Chikama H, Shibata H and Emi T 1996 *JIM Material Transactions* **37** 620–26
[2] Phelan D, Reid M, Stanford N and Dippenaar RJ 2006 *JOM* **58** 67–69
[3] Reid M, Phelan D and Dippenaar RJ 2004 *ISIJ Int.* **44** 565–72
[4] Misra P, Chevrier V, Sridhar S, Cramb AW 2000 *Metall. Mater. Trans. B* **31** 1135–39
[5] Shibata H, Yin H and Emi T 1998 *Phil Trans. R. Soc. Lond.* A **356** 957–66
[6] Yin H, Shibata H, Emi T and Suzuki M 1997 *ISIJ Int.* **37** 946–55
[7] Bernhard C, Schider S, Sormann A, Xia G and Ilie S 2011 *Berg- und Hüttenmännische Monatshefte* **156** 161–67
[8] Aneziris C G, Schroeder C, Emmel M, Schmidt G, Heller H P and Berek H 2013 *Metall. Mater. Trans. B* **44** 954–68
[9] Sasai K 2014 *ISIJ Int.* **54** 2780–89
[10] Coletti B, Vantilt S, Blanpain B and Sridhar S 2003 *Metall. Mater. Trans. B* **35** 533–38
[11] Yin H, Shibata H, Emi T and Suzuki M 1997 *ISIJ Int.* **37** 936–45
[12] Vantilt S, Coletti B, Blanpain B, Fransaer J, Wollants P and Sridhar S 2004 *ISIJ Int.* **44** 1–10
[13] Nakajima K and Mizoguchi S 2001 *Metall. Mater. Trans. B* **32** 629–41
[14] Luo H W 2001 *Scand. J. Metall.* 30 212–19
[15] Taira S, Nakashima K and Mori K 1993 *ISIJ Int.* **33** 116–23
[16] Yu X, Pomfret R J and Coley K S 1997 *Metall. Mater. Trans. B* **28** 275–79
[17] Liu J, Guo M, Jones P T, Verhaeghe F, Blanpain B and Wollants P 2007 *J. Eur. Ceram. Soc.* **27** 1961–72
[18] Monaghan B J and Chen L 2005 *Steel Research Int.* **76** 348–54
[19] Monaghan B J and Chen L 2006 *Ironmaking Steelmaking* **33** 323–30
[20] Monaghan B J, Chen L and Sorbe J 2005 *Ironmaking Steelmaking* **32** 258–64
[21] Sridhar S and Cramb AW 2000 *Metall. Mater. Trans. B* **31** 406–10
[22] Fox A B, Valdez M, Gisby J, Atwood R C, Lee P D and Sridhar S 2004 *ISIJ Int.* **44** 836–45
[23] Verhaeghe F, Blanpain B and Wollants P 2008 *Modell. Simul. Mater. Sci. Eng.* **16** 45007
[24] Feichtinger S, Michelic S K, Kang Y and Bernhard C 2014 *J. Am. Ceram. Soc.* **97** 316–25
[25] Monaghan B J and Chen L 2004 *J. Non-Cryst. Solids* **347** 254–61
[26] Levenspiel O 1999 *Chemical reaction engineering* vol 3 (New York: John Wiley & Sons)
[27] Verhaeghe F, Liu J, Guo M, Arnout S, Blanpain B and Wollants P 2008 *J. Appl. Phys.* **103** 23506
[28] Verhaeghe F, Liu J, Guo M, Arnout S, Blanpain B and Wollants P 2007 *Appl. Phys. Lett.* **91** 124104
[29] Verhaeghe F, Arnout S, Blanpain B and Wollants P 2005 *Phys. Rev. E* **72** 36308
[30] Whelan M J 1969 *Met. Sci.* **3** 95–97
[31] Slag Atlas, 2nd ed., Verlag Stahleisen GmbH, 1995.

IOP Conf. Series: Materials Science and Engineering **143** (2016) 012011 doi:10.1088/1757-899X/143/1/012011

# Characterization of the behaviour of the electric arc during VAR of a Ti alloy

**P Chapelle[1], C Noël[1], A Risacher[1], J Jourdan[1], J Jourdan[2] and A Jardy[1]**
[1]Institut Jean Lamour - UMR 7198 CNRS/Université de Lorraine,
Laboratory of Excellence DAMAS, CS 50840, 54011 Nancy Cedex, France
[2]Timet Savoie, Avenue Paul Girod, 73400 Ugine, France

E-mail: pierre.chapelle@univ-lorraine.fr

**Abstract**. In this paper, we report experimental results based on the direct observation of the electric arc behaviour during vacuum arc remelting of a Ti alloy. These results were obtained in a specifically instrumented industrial furnace using high speed framing camera and optical emission spectroscopy, for a current density level of the order of 10 A/cm$^2$ and a gap length of a few centimetres. It was observed that the arc exhibits a similar operating regime to that described in the literature for the case of Inconel 718 and Zr alloy electrodes. The arc structure corresponds essentially to that of a diffuse metal vapor arc with separate and rapidly moving cathode spots. Several critical parameters of the cathode spots, including their current, size and velocity, and of the interelectrode plasma were evaluated. Also, the interactions between the arc operation and the transfer of metal drops in the interelectrode gap were investigated. Three modes of transfer of the liquid metal drops in the interelectrode gap have been identified depending on the gap length: drop falling, drip short and drop erosion induced by the cathode spots.

## 1. Introduction

The Vacuum Arc Remelting process is an important technology for refining reactive metallic ingots, such as titanium and zirconium alloys. In this process, heating and melting of the metal is achieved with the help of a metal vapor electric arc established between a consumable electrode (cathode) and a molten metal pool that forms on top of an ingot (anode) which solidifies into a water cooled copper mold. To help metal producers to meet new requirements on process performance and ingot quality, it becomes increasingly important to have a more comprehensive understanding of the electric arc behaviour and more generally of all the phenomena taking place in the interelectrode region. In particular, knowledge of the arc behaviour is important for the proper modelling of the heat flux and electric current distributions provided by the arc at the ingot top surface, which determine to a large extent the solidification conditions of the metal, and thus the final ingot quality.

Despite its importance, the information on the arc behaviour are currently very limited due to the experimental constraints caused by the design of VAR furnaces which does not allow to directly observe or easily access with sensors the arc during a melt. Only two works dealing with direct optical investigation of the arc were reported in the literature. The first observations were made by Zanner and his colleagues in the 1980s [1,2] investigating the melts of Ni based superalloys, while, more recently, observations applied to the melt of a Zr alloy were performed by some of the present authors [3]. These investigations were done by installing on an existing furnace a purposely developed module to visualize the arc. Very similar results were obtained for both types of alloy. The main result was that under standard melting practice the arc takes on a so-called diffuse appearance, the arc being composed of multiple fast moving cathode spots distributed over the electrode surface. The sequence of events controlling the transfer of metal in the interelectrode gap was also investigated in these studies, with particular interest in the transient short circuits and arc extinction caused by the metal dropping at short gap lengths.

In the present study, we describe similar experiments carried out while melting a Ti alloy electrode. First, we report observations made with a high-speed video camera of the appearance and evolution of the electric arc, as long as estimations of the main parameters of the cathode spots. A special attention is put on the influence of an external axial magnetic field on the cathode spot distribution. Then, we present the evaluation of some plasma parameters, like excitation temperatures and relative densities, measured using optical emission spectroscopy. Finally, we describe the various modes of transfer of the metal between the electrode and the ingot observed as a function of the gap distance and their interactions with the arc behaviour. We present in particular a new mode of transfer of the metal identified for very small gap distances, which had not been reported in earlier works. A more detailed report on the experiments described in the current work is given in [4].

## 2. Experimental setup

The experiments were performed in cooperation with the CEZUS company using the previously developed setup described in [3]. An industrial VAR furnace was equipped with an additional water-cooled cylindrical chamber, referred to as the module of observation of the electric arc (figure 1). This module, placed between the crucible and the upper vacuum chamber, was equipped with three horizontal (or slightly inclined) viewing tubes that allow to directly observe the arc in a configuration close to that of an actual remelting operation. During an experiment, an electric arc was sustained between an electrode and an initial ingot with a height greater than that of the crucible, so that the arc region was in the horizontal alignment of the viewing ports. The ingot diameter was smaller than that of the crucible to prevent contact between the side of the ingot and the uncooled flanges of the crucible and the observation module. This configuration implied to limit the size of the liquid pool formed at the uncooled ingot top. The duration of an experiment was thus limited to a few minutes.

All experiments were performed with a 160 mm diameter electrode and a 280 mm diameter ingot both in Ti-6Al-4V, except two experiments during which a Ti-10V-2Fe-3Al electrode was used in order to study the influence of the alloy composition. Melting parameters were as follows: the arc voltage was varied between 27 V and 37.1 V, the arc current between 2 kA and 6.1 kA and the gap length between 2.1 cm and 7 cm. Moreover, some experiments were performed applying a vertical magnetic field (the intensity of the field was 15 G) produced by an inductor coil placed around the observation module. Phenomena in the interelectrode region and at the electrode surface were observed with a high-speed video camera (Photron SA5) using frame rates up to $10^5$ frames per second. The high-speed video camera was synchronized with the high frequency (up to 200 kHz) recording of the arc voltage to analyze the electrical signatures associated to the metal transfer mechanisms. Besides arc visualization, we also investigated the plasma emission by optical emission spectroscopy using a Jobin Yvon spectrometer (TRIAX 550) equipped with a 1200 grooves/mm grating blazed at 250 nm and an ICCD detector. The observed spectral domain ranged from 270 nm to 540 nm.

## 3. Results

### *3.1. Arc structure*

Similarly to the observations reported in the case of Ni based and Zr alloy electrodes, the arc appearance during the melt of a titanium alloy appeared to correspond to the diffuse regime, which characterizes "classical" vacuum arcs established between two solid electrode surfaces at low current densities (employed for example in vacuum circuit breakers) [5]. As can be seen in figure 2, the arc was composed of a set of simultaneously operating cathode spots, moving at a high velocity across the electrode bottom end surface and on a fraction of the electrode lateral surface. Each spot was a site of emission of a jet of plasma composed of metal vapor particles originating from the cathode. These plasma jets mixed near the cathode surface to form a diffuse and weakly luminous plasma, which filled approximately homogeneously the interelectrode region. The anode did not play an active role in the discharge.

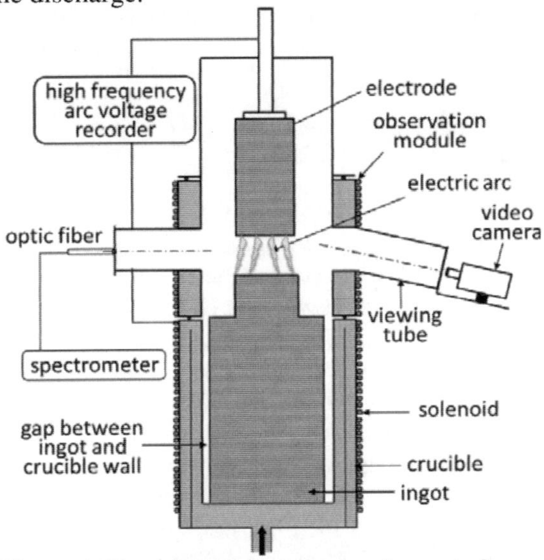


Figure 1. Experimental set-up implemented on an industrial VAR unit.

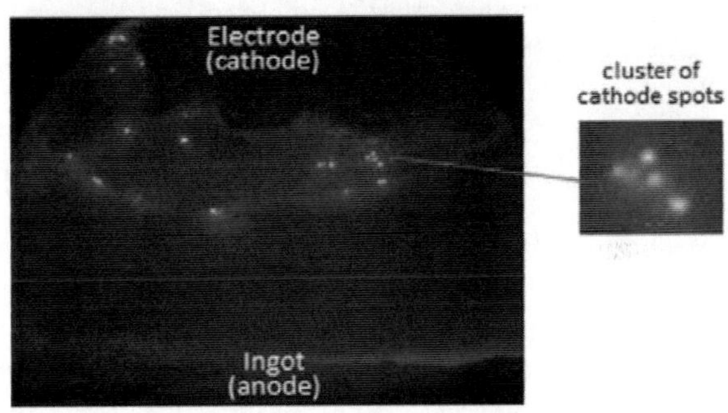


Figure 2. Video image (recorded at 5700 frames/s with an exposure time of 10 μs) of the electric arc.

A more detailed insight into the structure of the arc showed that the arc was composed of at least two structural levels.

- The first level was that of cathode spots. The average current per spot was found to be of the order of 100 A. The size of a spot was estimated to be several hundred micrometers and its apparent velocity over the cathode surface turned out to be of the order of 1 to 10 m/s. As shown in table 1, our results are close to those obtained for the melt of a zirconium alloy and are globally coherent with the spot parameters reported for "classical" vacuum arcs between solid electrode surfaces [6].
- At a larger length scale, a second level of structuration of the arc was observed, corresponding to the formation of clusters of cathode spots. These clusters were most often of small size, with a number of spots per cluster varying between 2 and 13 (for an average value equals to 3.9). The number and the arrangement of the spots inside a cluster were unstable. The clusters might spontaneously extinguish and frequently divide into smaller clusters which tended then to repel each other. It seems that the division of clusters was caused by the increase of the number of the spots inside the cluster. The clusters were characterized by a non-steady and complex dynamic behaviour. They had an apparent random motion which tended to be preferentially oriented towards the edge of the electrode. After reaching the edge, the clusters moved along the side surface of the electrode over a few centimetres, and finally extinguished.

Table 1. Estimated parameters of the cathode spots.

| | Electrode | Properties of the cathode spots | | | Number of spots / cluster |
|---|---|---|---|---|---|
| | | Diameter (μm) | Average current (A) | Apparent velocity (m/s) | |
| Measurements in a VAR furnace | Zy4 [3] | - | 76 - 190 | 1 - 10 | 2 - 5 |
| | Ti-6Al-4V | 629 ± 141 | ~100 | 1 - 10 | 2 - 13 |
| Vacuum arc between two solid surfaces | Ti [6] | 411 ± 66 | 33,5 | 30 | 9 ± 3 |

The influence of an external axial magnetic field on the arc distribution was investigated by following the proportion of spots present at the electrode tip during the melt (i.e. the ratio of the number of spots present at the electrode tip to the total number of spots present both at the electrode tip and on the electrode side surface). This parameter was obtained with the help of an image processing program capable of detecting automatically on each video image the position of all the cathodes spots present on the surface of the electrode. Figure 3 shows that parameter in the form of a distribution function. As shown in this figure, the proportion of spots under the electrode tended to be larger when a magnetic field was applied. Also, the situations where all the spots were established on the lateral surface (no spots under the electrode) were less frequent when a magnetic field was applied. These observations suggest that the application of an axial magnetic field allowed a better confinement of the arc under the electrode, by promoting the location of the spots at the electrode tip instead of the lateral surface. Note that a similar effect of the application of an axial magnetic field is reported in vacuum arc circuit breaker studies [7].

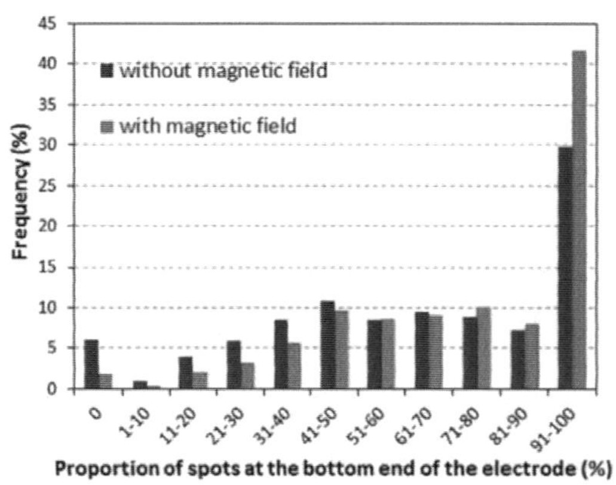


Figure 3. Influence of an external axial magnetic field on the proportion of cathode spots at the electrode tip.

### *3.2. Plasma parameters*

The spectroscopic data provided information on the nature of the excited species (atoms and ions) present in the plasma and has allowed us to determine their excitation temperature. They have also been used to estimate the relative proportion of singly ionized titanium particles with respect to titanium atoms, which gives an indication on the ionization degree of the plasma. Details and equations regarding the processing methods of the spectroscopic data may be found in [4].

The presence of a large number of spectral lines due to singly ionized titanium particles was noted in the emission spectrum of the plasma (suggesting that it was the dominant species in the plasma). Titanium was also present as atoms and (yet to a much lesser extent) in the form of ions $Ti^{2+}$. The emission spectrum contained also spectral lines due to each of the main alloy elements (Al, V and Fe). These elements were present both as atoms and singly charged ions (doubly ionized particles were also observed in the case of vanadium).

Table 2 summarizes the values of the excitation temperature of Ti atoms and $Ti^+$ ions determined for the two investigated alloys. Only the excitation temperature of Ti atoms and $Ti^+$ ions could be determined, as the number of spectral lines observed for the other species was insufficient to perform such a calculation. The temperature of $Ti^+$ ions exceeded (1.5 to 2 times) that of Ti atoms. This difference may be related to differences in the spatial distributions of the two species. Indeed, it is expected that $Ti^+$ ions were mostly produced in the plasma jets emitted by the cathode spots, while Ti atoms originated predominantly from the volatilization of the metal from the liquid film at the base of the cathode and from the surface of the liquid pool on the anode side. No significant influence of the alloy composition on the two calculated temperatures is observed.

Table 2. Estimation from the spectroscopic data of the excitation temperatures of Ti atoms and $Ti^+$ ions and of the density ratio of $Ti^+$ ions and Ti atoms.

| **alloy** | | Ti-6Al-4V | Ti-10V-2Fe-3Al |
|---|---|---|---|
| **Arc voltage (V)** | | 31.2 | 31.4 |
| **Melting current (kA)** | | 3.95 | 4.55 |
| **Gap length (cm)** | | 4.4 | 4.6 |
| **$T_{exc}$ (K)** | **Ti** | **6615** ± 928 | **6963** ± 1741 |
| | **$Ti^+$** | **11387** ± 1509 | **11266** ± 1161 |
| **$n_{Ti+}$ / $n_{Ti}$** | | **24.24** ± 4.14 | **8.81** ± 1.91 |

Table 2 shows also the estimated values of the $Ti^+$ ions to Ti atoms density ratio. It must be pointed out that, contrary to the calculation of the temperatures which relies on the measurement of the intensity of a large number of spectral lines, the calculation of the density ratio relies on the measurement of the intensity of only two spectral lines. This may explain the limited accuracy of the calculation of the density ratio and the important standard deviations obtained. The density of $Ti^+$ was much larger (approximately one order of magnitude greater) than the density of titanium neutral particles. The spectroscopic data did not allow us to calculate the true ionisation rate of the plasma. This would require to take into account all ion and neutral species in the plasma and to determine the excitation temperature of each of them. However, the very large values of the ratio of the densities of $Ti^+$ ions and Ti atoms obtained here and the dominant presence of spectral lines due to these two species in the emission spectrum of the plasma suggest that the plasma was strongly ionized.

### *3.3. Metal transfer*

Three different modes of transfer through the interelectrode gap of the liquid metal drops created at the electrode tip have been observed depending on the gap length. Moreover, the shorter the gap length, the more important the influence of the arc on the transfer mechanisms.

1) For large value of the gap length (greater than about 6 cm), the metal drops were allowed to develop with, in general, negligible activity of the spots on the drop surface. Metal drops grew first by accumulation of metal and then by stretching and eventually ended up falling off into the liquid pool under the effect of gravity. The only interaction with the arc occurred immediately after the drop breakup. A fine filament of metal was left hanging under the electrode, which was quickly eroded under the action of the cathode spots, as the spots tended to accumulate on the filament surface. The total duration of the transfer sequence was typically of the order of several hundred milliseconds.

2) For values of the gap length between 2 cm and 6 cm, the transfer of metal drops involved predominantly a phenomenon called a drip short, which corresponds to the formation of a transient short-circuit caused by the momentary contact of the metal drop with the surface of the metal bath. Such transfer sequence was very similar to that described in the literature for the case of Ni based and zirconium alloy electrodes.

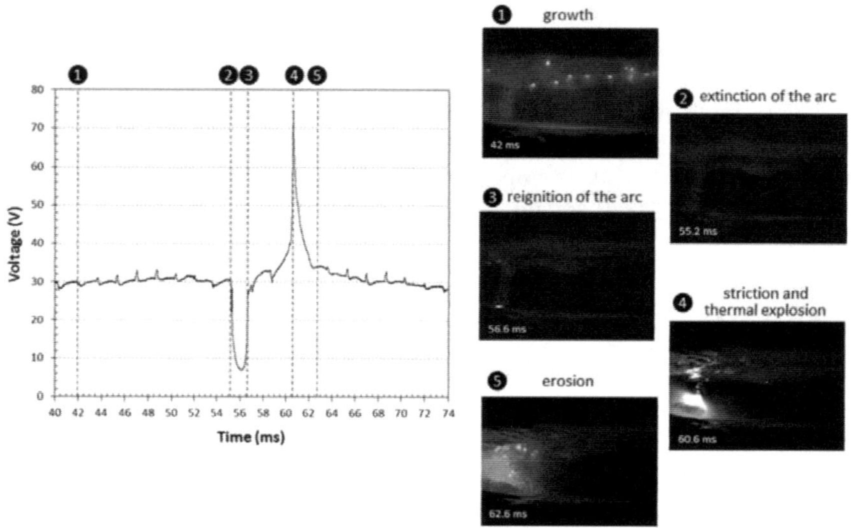


Figure 4. Transfer of a liquid metal drop with a drip-short and the electrical signature associated (gap = 3.6 cm).

Figure 4 shows an example of such a transfer sequence as well as its electrical signature. Note that we describe here the most commonly observed drip short sequence. Slight variations of the sequence are described in [4]. The sequence started with the relatively slow growth of a protuberance of metal, with negligible influence of the arc (image 1). The growing protuberance ended up touching the surface of the ingot metal bath, causing a short-circuit (the voltage dropped consequently to a small value) and the extinction of the arc (image 2). After a short time interval, the bridge of liquid metal connecting the electrode to the ingot broke near the ingot. This was followed immediately by the reignition of the arc between the bottom end of the protuberance and the ingot (image 3). As the cathode spots tended to remain concentrated at the tip of the protuberance resulting in a highly luminous plasma (constricted arc), some local narrowing (pinching effect) of the protuberance was observed. This pinching effect developed until causing a thermal explosion of the metal in the narrow region and the transfer of the bottom part of the protuberance to the ingot (image 4). This phenomenon may be attributed to the fact that an important fraction if not all of the electric current flowed through the protuberance, which induced within the protuberance some intense compressive magnetic forces and Joule heating. The constriction of the arc under the protuberance was responsible for an increase of the voltage level, with a peak value coinciding with the thermal explosion of the protuberance. Eventually, the cathode spots spread over the entire electrode surface and the remaining portion of the protuberance was finally transferred to the ingot by progressive erosion under the action of the cathode spots (image 5).

3) For a gap length value of 2 cm, a specific mode of transfer was observed, which had not been reported in previous studies. The metal drops were generally unable to develop up to a stage where they produce a drip short. Metal transfer was strongly affected by the arc and occurred mostly under the physical action of the cathode spots locating themselves preferentially on the surface of the growing protuberances and causing their progressive erosion. Such erosion took the form of small volumes of metal being torn off from the protuberance and tiny droplets of metal emitted by the spots.

Often the action of the spots was not continuous, but repetitive, with interruptions related to the establishment of the spots on other protuberances.

## 4. Conclusions

Visualization experiments of the arc behaviour and metal transfer mechanisms during the vacuum remelting of titanium alloys have been reported for the first time. Similarly to the results obtained for the melts of Ni based and Zr alloys, the arc was observed to operate in a diffuse mode, characterized by the presence over the electrode surface of a collection of small and fast moving cathode spots. The values of the basic parameters of the cathode spots estimated in this work are comparable to the values measured for cathode spots established on a titanium solid surface. The application of an axial magnetic field has been found to stabilize the arc under the electrode, limiting the establishment of the cathode spots on the side surface of the electrode. The spectroscopic investigation has revealed that the metal vapor plasma is mostly composed of Ti atom and $Ti^+$ ions, the density of the latter being significantly higher than that of atomic titanium. Three distinct modes of transfer of the metal drops have been identified, namely for decreasing gap length: dripping, drip-short and arc controlled erosion of metal drops. This third mode, which had not been described in earlier studies, was observed for very short gap length (about 2 cm)

## 5. Acknowledgments

As a part of the OPERAS (Optimizing Processes based on Electrode Remelting with Arc or Slag) project, this work has been supported by the French National Research Agency (ANR-08-MAPR-0006-04).

## 6. References

[1] Zanner F J 1979 *Proc. Int. Conf. on Special Melting (San Diego, USA)* eds G K Bhat and R Schlatter (American Vacuum Society) pp 417-427
[2] Zanner F J and Bertram L A 1985 *Proc. 8th Int. Conf. on Vacuum Metallurgy (Linz, Austria)* vol 1 pp 512-552
[3] Chapelle P, Bellot J P, Jardy A, Czerwiec T, Robbe X, Champin B and Ablitzer D 2000 An experimental study of the electric arc during vacuum arc remelting *J. High Temp. Mat. Proc.* **4** 493-506
[4] Chapelle P, Noël C, Risacher A, Jourdan J, Jardy A and Jourdan J 2014 Optical investigation of the behavior of the electric arc and the metal transfer during vacuum remelting of a Ti alloy *J. Mat. Proc. Tech.* **214** 2268-2275
[5] Malkin P 1989 The vacuum arc and vacuum interruption *J. Phys. D: Appl. Phys.* **22** 1005-1019
[6] Siemroth P, Schulke T, Witke T and Rackwitz N 1996 *Proc. 7th Int. Symp. on Discharges and Electrical Insulation in Vacuum (Berkeley, USA)* vol 1 (IEEE) pp 226-230
[7] Schade E 2005 Physics of high-current interruption of vacuum circuit breakers *IEEE Trans. Plasma Sci.* **33** 1564-1575

# Faster methods for estimating arc centre position during VAR and results from Ti-6Al-4V and INCONEL 718 alloys

**B G Nair[1], N Winter[2], B Daniel[3], R M Ward[1]**
[1] Metallurgy & Materials, University of Birmingham, Birmingham, B15 2TT, UK
[2] Timet, Witton, Birmingham, UK, [3] Special Metals, Hereford UK

E-mail: r.m.ward@bham.ac.uk

**Abstract.** Direct measurement of the flow of electric current during VAR is extremely difficult due to the aggressive environment as the arc process itself controls the distribution of current. In previous studies the technique of "magnetic source tomography" was presented; this was shown to be effective but it used a computationally intensive iterative method to analyse the distribution of arc centre position. In this paper we present faster computational methods requiring less numerical optimisation to determine the centre position of a single distributed arc both numerically and experimentally. Numerical validation of the algorithms were done on models and experimental validation on measurements based on titanium and nickel alloys (Ti6Al4V and INCONEL 718). The results are used to comment on the effects of process parameters on arc behaviour during VAR.

## 1. Introduction
Presently Vacuum Arc Remelting (VAR) is the most commonly used melting process for the manufacture of high quality segregation-free ingots of titanium and nickel alloys. Due to the stringent targets for the properties of these materials in demanding applications, the process is continuously enhanced to produce these alloys with excellent homogeneity and purity [1]. In VAR the electrode is melted by a plasma arc, however the distributions of heat, current and mass flux within the 'arc' (or 'arcs') are not directly controlled. Studies have shown a time-varying, asymmetry in the distribution of arc current which could affect the velocity and temperature distributions in the melt pool [2]. Ward et al. [3] reported an ensemble arc motion with the time averaged position of maximum current flow, which potentially limits the validity of a 2-D axisymmetric modelling concept of VAR [4]. Due to the aggressive environment some indirect methods of measurement based on voltage gradient on the crucible walls and measurement of magnetic flux surrounding the crucible have been previously proposed [5] to better understand the distribution of current within the process. Woodside et al. [6] reported observing several arc distribution modes during production of Ti6Al4V that could impact solidification times.

The technique of magnetic source tomography (MST) is capable of estimating the locations of multiple centres of current density simultaneously, based on measurements of magnetic flux density outside the crucible. As originally proposed this technique used an optimisation routine based on the Biot Savart Law (BSL-MST algorithm) [7-8]. Due to the computationally intense nature of the BSL-MST algorithm, however, faster methods are appealing. In this paper we present two computationally efficient and faster optimisation algorithms to supplement the BSL-MST method, using Ampere's Law (AL-MST algorithm) and Neural networks (NN-MST), to determine the centre position of a single distributed arc. Numerical validations of the algorithms were done on models developed using a commercial finite element electromagnetic code (Opera 3d).

Experimental studies were based on melts of titanium and nickel alloys (Ti-6Al-4V and INCONEL 718), interpreting measurements of magnetic field, arc voltage and current together with process video. The results from the algorithms are compared and are used to comment on the effects of process parameters on arc behaviour during VAR. In particular, the link between arc behaviour and the growth and re-melting of the solid skin at the ingot edge is explored.

## 2. Theory
Patterns of current distribution during VAR and the expected resultant magnetic field from off-centre current flow are shown in figure 1(*a*)–(*c*) (from [7]). In VAR the current is not deliberately injected at known locations; instead the arc process itself controls the location of the current, and the passive sensors and reconstruction algorithm are used to determine its spatial distribution. Magnetically silent distributions of current (e.g. in which the current flow through the crucible and its return through the ingot are

axisymmetric) do not produce a measurable magnetic field outside the crucible and therefore cannot be reconstructed. However, if either of these flows deviates from axisymmetry, an external field is produced. Such deviations are likely to be caused by drip shorts, by a constricted arc away from the ingot centre line and even by a diffuse arc if not perfectly centred. The term 'arc centre location', often shortened to 'arc location' in this paper, refers to the position of the centre of a locally-axisymmetric distribution of current between the electrode base and ingot top.

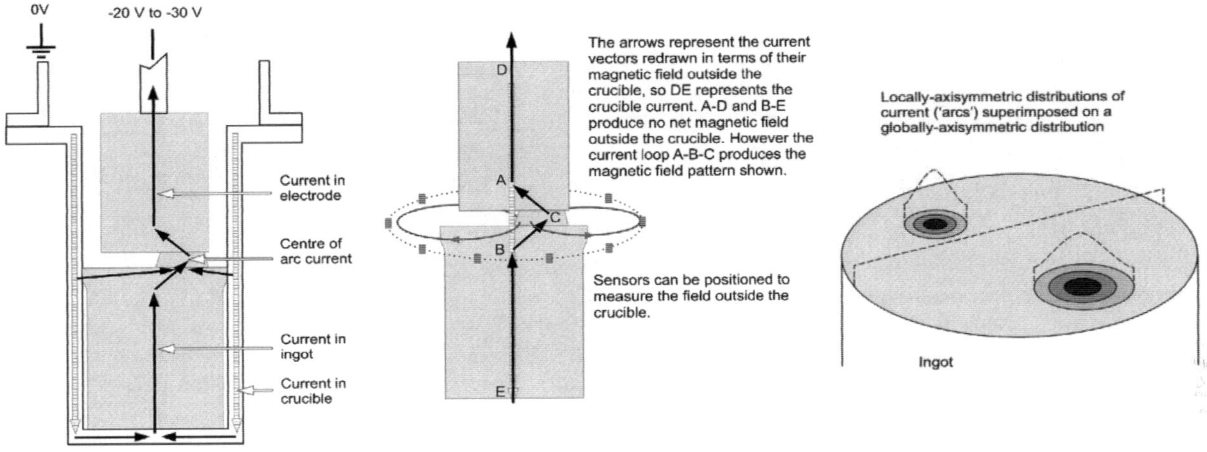


Figure 1. Magnetic fields during VAR. The techniques presented in this paper are to estimate the centre positions and additional currents of off-centre arcs by using measurements of magnetic field outside the crucible. (a) The current paths during VAR and the B field due to an off-centre arc. (b) The current paths redrawn in terms of their field outside the crucible, and the field due to an off-centre arc. (c) Suggested distribution of current during VAR consisting of a globally axisymmetric background current (the diffuse arc), with locally axisymmetric current flows ('arcs') superimposed. Details can be found in [7].

## 3. Magnetic Source Tomography – Three approaches

### *3.1. MST using Biot Savart Law (BSL – MST)*

Assuming that the currents are varying slowly enough that the system can be considered to be magnetostatic, the magnetic flux density $B$ arising from current density $J$ in volumes $dv$ can be calculated using the Biot–Savart law shown in equation (1),

$$B(r) = \frac{\mu_o}{4\pi}\int J(r') \times \frac{r-r'}{|r-r'|^3}dv' \quad (1)$$

where $\mu_0$ is the permeability of free space and also of nonmagnetic objects, $r$ is the vector from the origin to the measurement point, $r'$ is the vector from the origin to field point and $J(r')$ is the current density at $r'$. We considered that the materials used in the important parts of the systems in this paper (Ti6Al4V, INCONEL 718 and copper) had $\mu_r$=1. In BSL-MST, the magnetic field created outside the crucible due to an arc centre at an arbitrary location was represented as a linear interpolation between the $B$ fields resulting from arcs at points on a square grid surrounding the true arc location. Despite the nonlinearity in the underlying equations, given a sufficiently fine grid (1 cm in this case) sufficiently accurate results were obtained. Simulations were performed to make the array of $B$ field values corresponding to arcs at these grid points using the finite element software Opera 3D. Constrained nonlinear optimization was used to estimate the currents and centre locations of a fixed number of arcs using the results from the forward models, so that the predicted B fields outside the furnace matched most closely with the measurements at a given time. A key assumption is that superposition applies when calculating the total field from current split into multiple arcs. Further details of BSL – MST are reported in [6-7].

### *3.2. MST using Amperes Law (AL – MST)*

This method uses 2-D calculations assuming that the furnace can be represented as an infinitely-long crucible and a single off-centre arc, both carrying current and both creating magnetic field which can be measured at sensors outside. Neither of them necessarily carry the total process current, as any current that is co-axial in both of them will produce fields which cancel to zero outside. Each sensor is used to estimate the arc centre position from its $B_{x,\ y}$ readings, $I_C$ , and $I_A$, and the algorithm then chooses $I_C$ and $I_A$ parameters to make all the sensors estimate the arc to be at the same location as far as possible. This approach is also magnetostatic and uses Ampere's law by inverting it so that, given knowledge of $I_C$ , $I_A$ and the position of the crucible current (0,0) , the position of the arc current can be calculated :

$$\oint H.dl = I_{enc} \quad (2)$$

It was found that choosing the $I_C$ and $I_A$ parameters to solely place the arc position estimates as close together as possible led to significant errors in the position estimation; the following cost function was found to be better, which favours solutions that minimise $I_C$ and $I_A$ as well as putting the arc position estimates close together:

$$\overline{D}_i * \sqrt{I_C^2 + I_A^2} \quad (3)$$

where $I_C$ - Crucible current, $I_A$ - Arc current, $D_i$ – distance from the arc position estimate from an individual sensor $i$ to its average from all sensors.

*3.3. MST using Neural Network (NN – MST)*
Various nonlinear connectionist model architectures (in the Matlab Neural Network Toolbox) were tested as a means of predicting the centre location of a single arc based on the external $B$ measurements. Using either modelled or experimental datasets, 70% of the data was used to train the neural network, 15 % was used for validation and the final 15% for testing. A neural network using Bayesian regularisation was finally selected due its performance when testing the network. The network was initially trained using arc centre models positioned on a 2 cm x 2 cm x 2 cm grid, and expected to interpolate at points between these. One advantage of neural networks over e.g. K-nearest neighbours is that once a network has been trained it can predict the arc locations quickly due to being a direct computation, without searching at run-time. However it is important to note that, as this is a purely data-driven approach that doesn't directly contain the process physics, it is likely to be poor at extrapolating beyond the domain of the training data. As an example, it was found necessary to deliberately add noise to the training data in order for the network to be able to function well on noisy test data.

## 4. Experimental methods

*4.1. Ti6Al4V* : Magnetic field measurements were done on a first melt experiment in which an electrode was melted into a crucible of 0.66 m diameter at a higher current (than used for INCONEL 718). To gather the magnetic field data two rings of ten sensor boxes each containing 3 axis Hall effect sensors (measuring azimuthal, radial and axial flux density) were mounted at heights 1.075m and 1.475 m from the base plate on a 1.06 m diameter circle, outside the furnace cooling jacket. The outputs from the sensors were measured using National Instruments differential amplifier and 10 kHz low pass filter at a 1 KHz sampling rate. Video recordings were made for the whole melt using standard video cameras at 25 frames per sec.

*4.2. INCONEL 718* : Magnetic field data were gathered during VAR of an ingot of INCONEL 718. The experiment was done at 6 kA, from 0.42 m / 0.22 m electrode into a 0.508 m diameter mould [3]. Twelve sensor boxes each containing 3 axis Hall effect sensors were mounted in a ring on the crucible inside the water jacket. The outputs from the sensors were measured using National Instruments differential amplifiers and 10 kHz low pass filter at a 5 KHz sampling rate.

In both cases the measurements of magnetic field were compensated using their mean value, owing to the uncertainty in the zero-field output of the Hall sensors used here. Additionally for the Ti-6Al-4V experiment the z-components of measured field were compensated for the effects of the solenoid stirring coil used.

## 5. Results

*5.1. Validation and performance analysis of the algorithms using simulated data*
The algorithms were first tested on simulated data, in which the arc location and B field were known. In all cases the prediction of arc position became more accurate as the amount of simulated noise was reduced, following the trends shown in [7]. The results of trying to locate arcs in the plane of, and 10 cm above, the sensors are shown below (figure 2) for the case when the simulated external B values used had no noise added. (The results for arcs below the sensors were equivalent to those above them). The disks represent cross sections through the plane between the electrode and ingot for a simulation of IN718 VAR, and the colour at each point represents the position error found when attempting to locate a single arc that had been placed at that point.

Good performance was found for BSL-MST, with only a few points on either plane in which the estimated position errors were above 8 cm. AL-MST was found to be effective when the sensors were at the same plane as the arc. However its accuracy decreased strongly when the arc was above or below the sensors, particularly as the arc was closer to the edge of the electrode; this may relate to the truly 3-dimensional nature of the current flow which is being modelled here as purely 2-d. (Although not shown here, AL-MST became increasingly poor at greater vertical displacements, whereas e.g. BSL-MST was able to work at 35 cm vertical difference.) Finally NN-MST was found to be reasonably accurate regardless of the sensor position relative to the arc, and this may be improved by further refinement in the choice of training data and method.

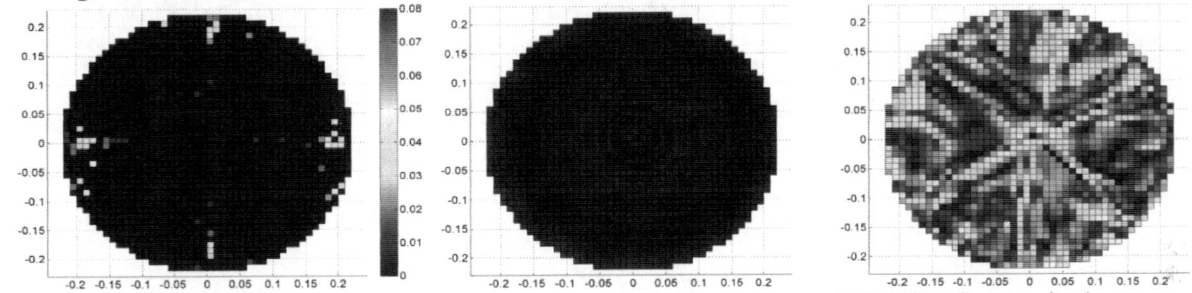

BSL-MST; Sensors in the arc plane — AL-MST; Sensors in the arc plane — NN-MST; Sensors in the arc plane

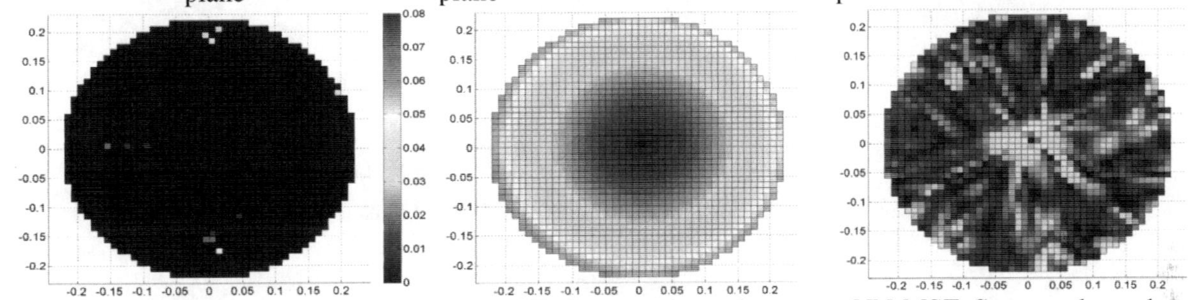

BSL-MST; Sensors above the arc plane — AL-MST; Sensors above the arc plane — NN-MST; Sensors above the arc plane

Figure 2. Arc location error from INCONEL 718 VAR model data. The colour at a point represents the error in m when using a particular algorithm to locate an arc placed there from the simulated magnetic field measurements outside the furnace.

Programmed in Matlab on an Intel Core i7 processor @3.4 GHz, BSL-MST took 1.2 sec to analyse the arc centre position for a single set of B field measurements (probably too slow for real-time use), whereas AL-MST required only 0.017 sec. However NN-MST could analyse an entire melt data set – nearly 300k results from 8 hours at 10 Hz - in less than 2 s in total once it had been trained.

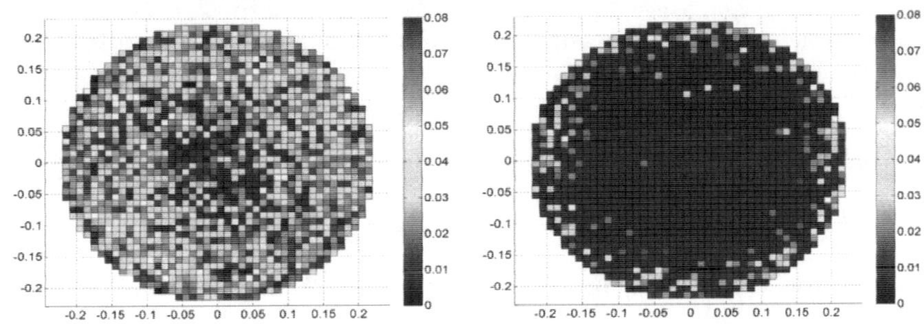

10 sensors — 5 sensors

Figure 3. Ti6Al4V model, Position error computed using BSL – MST algorithm with 10 and 5 sensors

The number of sensors used was also found to be important. In previous work [7] it was suggested that 8 sensors was the minimum needed for accurate arc location given typically noisy measurements. In figure 3 the results of trying to locate arcs in the plane of the sensors in a simulated Ti-6Al-4V melt with

SNR of 40 dB are shown using 10 and 5 sensors; it can be seen that the position errors were large when using only 5 sensors. Ideally, for industrial use, a very small number of sensors would be required (e.g. 2 at each height), but these results imply that more would be preferred.

### *5.2. Physical experiments*

*5.2.1 Determination of Arc centre position.* Figure 4 shows video from a first melt Ti6Al4V trial superimposed with the magnetic measurements. The results from the upper ring are shown with a smaller radius in the figure. The size & direction of the cones represent the magnitude and direction of the B fields measured by each sensor. In this time range the ingot top is closer to the bottom set of sensors; hence the B field is stronger there compared to that at the upper sensors. Note that the numerical values of the B field are not shown but the scaling is constant between frames. The red line in the middle shows the direction of the solenoid current (up=+ve, down=-ve). The arc centre positions computed by the algorithms are shown as 3 dots, using the readings from the lower set of sensors. The yellow and cyan dots are the arc centres computed by BSL-MST and AL-MST ● and the green dot is from the neural network ● . It can be seen that the arc centre positions match well even when computed using different techniques.

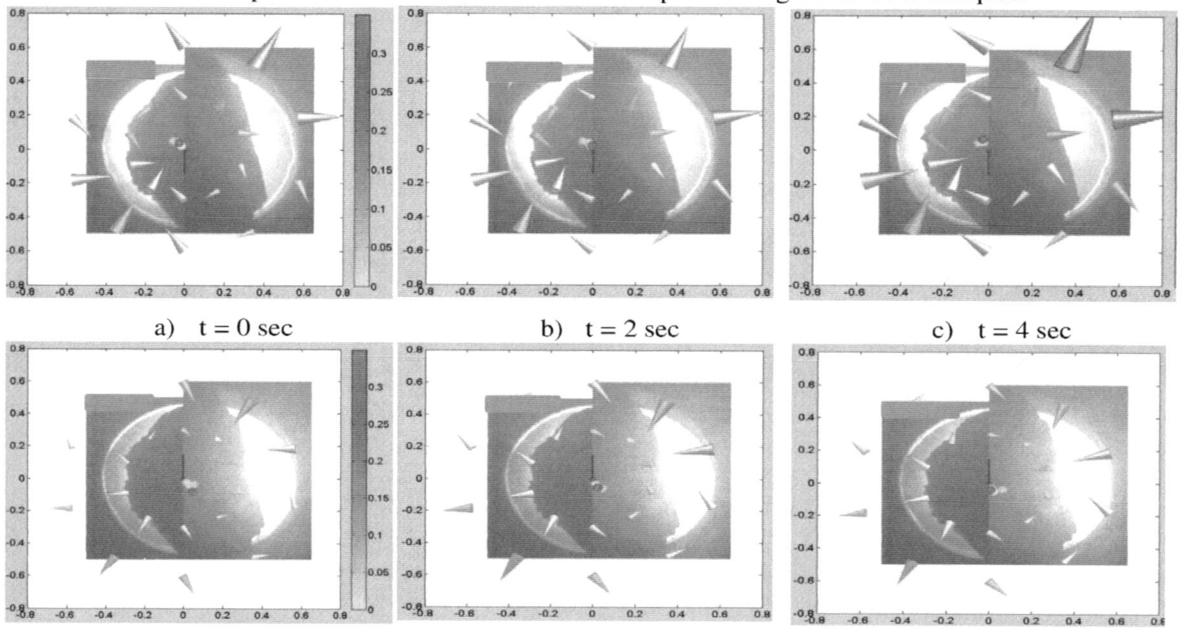

a) t = 0 sec b) t = 2 sec c) t = 4 sec

d) t = 6 sec e) t = 8 sec f) t = 10 sec

Figure 4. Representations of magnetic flux, arc centre and solenoid current at 2 sec intervals during a trial first melt of Ti-6Al-4V

Between figures 4 (a-c) the estimated arc position makes a slight movement corresponding with the variations in the magnetic field. A similar trend is observed in figures 4 (d-f). However when the solenoid current changes direction between fig 4 c and fig 4 d, the estimated arc position makes a larger move corresponding with a large change in the measured magnetic field values. Results from VAR of INCONEL 718 have been shown in previous work e.g. [3, 5].

A difference in arc behaviour during VAR was observed between Ti-6-4 and IN 718. The angle of rotation of the arc centre around the ingot centreline was plotted versus time and is shown in figure 5 for sections of melts of both alloys. It can be seen that for this Ti-6-4 first melt (5 a) the arc position was mostly at around 0 radians when the solenoid current was +ve or +/- 3 radians (180°) when it was –ve. (Although these positions are not fixed; some variation can be seen between 0-200 s and 1300-1500 s). For IN 718 (5 b), however, the arc was observed to rotate fairly continuously around the ingot centreline, reversing direction at random times. This can be seen by the arc position azimuth continuously increasing or decreasing versus time.

The difference between these two VAR experiments can also be seen in figure 6, in which the distributions of current at the estimated arc centre location in the x-y plane over time are shown. Most of the current is carried by arc centres near the edge of the electrode in IN 718 VAR, but by arc centres near the electrode centre in Ti-6-4. (Note that it is not possible to measure the radius of a symmetrical arc current distribution from external measurements, however. So the current could have been almost uniformly distributed over the whole electrode bottom in the Ti-6-4 experiment, with just the centre of the distribution moving slightly in response to the solenoid current). It's not clear whether the arc centre being found close to the ingot centre in Ti-6-4 VAR is compatible with the variations in brightness seen as it moves around; it would be expected that, the further the arc centre from the ingot centre, the greater the variations in brightness that would be observed. However the fact that the algorithm used here is able to estimate the $z$ position of the arc (and hence the ingot top) reasonably accurately suggests that it should also be getting $x$ and $y$ correct.

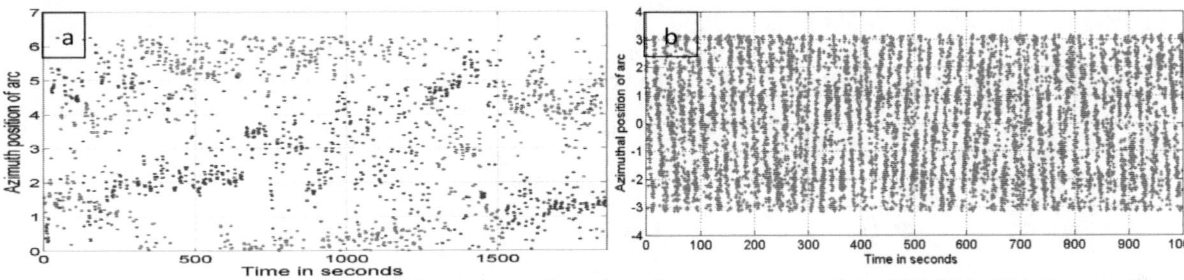


Figure 5. Azimuthal position of the estimated arc location versus time for a)Ti-6Al-4V(where red is used to indicate measurement points for which the solenoid current was positive),b) IN 718

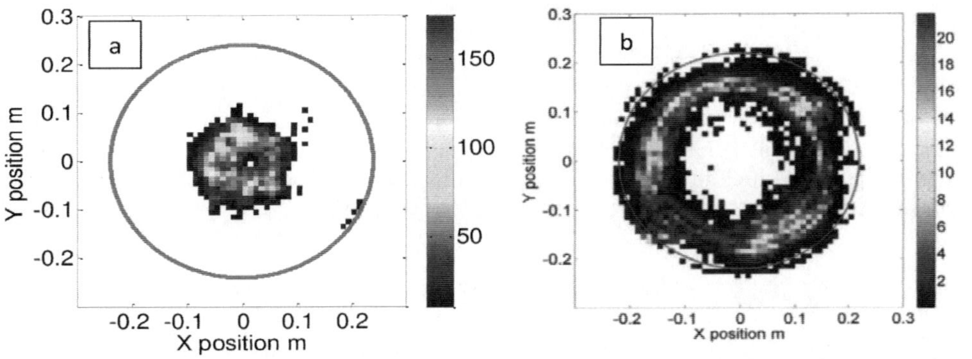


Figure 6. The distribution of current carried by arc centre positions from BSL-MST in terms of A/m$^2$ for a) Ti6Al4V, b) IN718

The differences between IN718 and Ti-6-4 may be related to the drip shorts that occur frequently in typical Ni superalloy VAR, causing arcs continually re-start near the edge of the electrode; these are typically not present in Ti-6-4 VAR due to the higher currents and larger gaps used. The influence of current on drip shorts has been investigated in depth by Zanner and co-authors (e.g. the authors of [9]). However, Woodside et al. [6, fig 7] have interpreted magnetic field data to give final-melt Ti-6-4 VAR arc centre position distributions close to the edge of the electrode as well as near the centre, and they attributed the differences to electrode preparation.

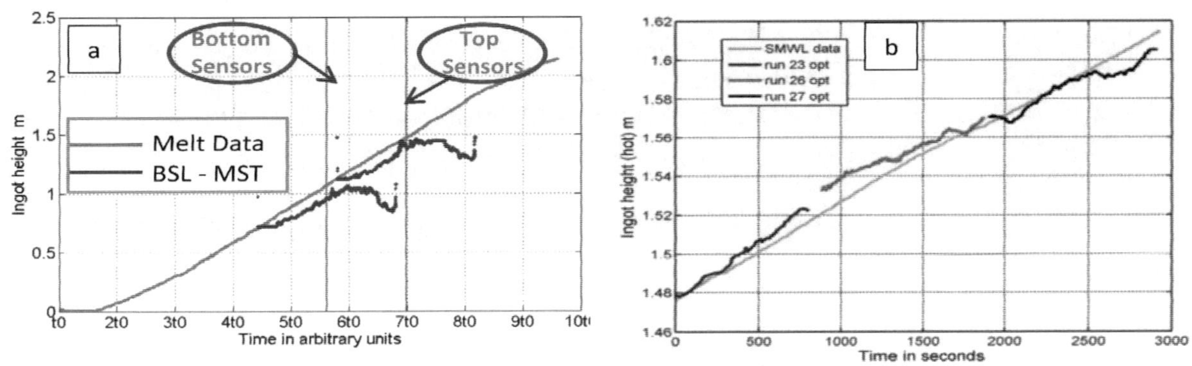


Figure 7. Estimation of the ingot top position using BSL –MST, from a) Ti6Al4V and b) IN 718 melt

IOP Conf. Series: Materials Science and Engineering **143** (2016) 012012 doi:10.1088/1757-899X/143/1/012012

*5.2.2. Estimation of Ingot top position.* The z coordinates from the BSL-MST arc position algorithm were also investigated. They are shown plotted versus time in figure 7, compared with the ingot top position from the melt data. For the titanium alloy example, it is seen that once the ingot top had moved past the sensor rings, the algorithm didn't estimate its position well. However the algorithm behaved reasonably well for the nickel alloy irrespective of whether the ingot top was above or below the sensors.

To investigate this, 2-arc noise-free models of the production of Ti6Al4V and INCONEL718 were produced and the likely estimated arc location errors were calculated using the BSL- MST algorithm. The results are shown in figure 8 (a-b) when the ingot top was 10 cm above the sensors. It can be seen that the errors were greater for the Ti-6-4 furnace simulation. This may be due to the sensors being much closer to the ingot in the IN 718 simulation, corresponding to their physical placement inside the furnace water jacket in this case, as opposed to being outside the cooling jacket for Ti-6-4.

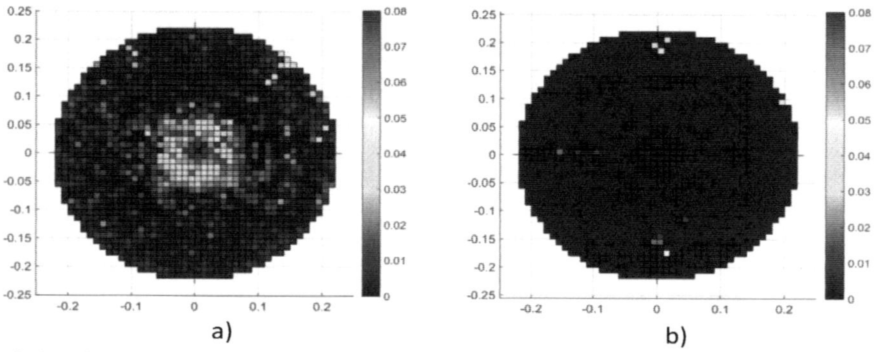


Figure 8. Position error when ingot top is 10 cm above the sensor for models
a) Ti6Al4V,
b) INCONEL 718

*5.2.3. Effects on solidification.* The spatial distribution of heat, current and mass flux on the ingot top will affect the solidification during VAR [9]. The size of the effect may be significant, or not, depending on the conditions; it could influence the thickness of the layer needing to be machined off, the creation of chemical segregation, and the formation of surface irregularities for example.

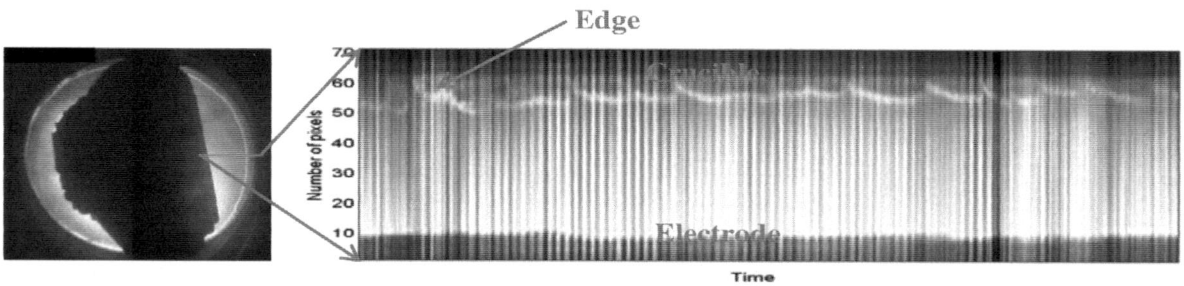


Figure 9: Local brightness and edge thickness over 31 min from a first melt VAR of Ti6Al4V

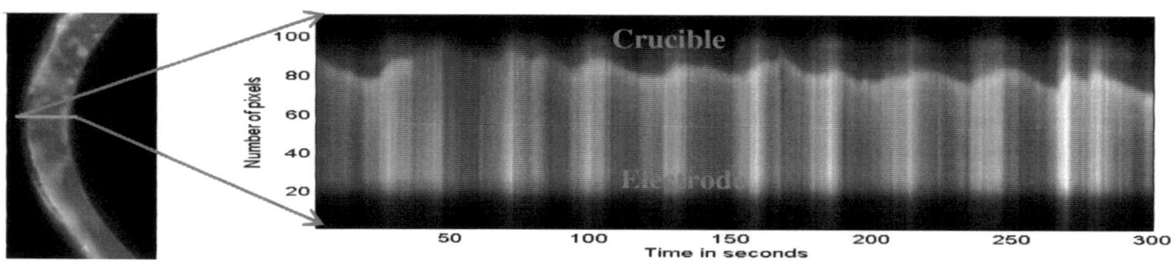


Figure 10: Local brightness and edge thickness over 5 min from a first melt VAR of INCONEL 718

As the arc current flux should correlate with the visible intensity of reflected light, the correlation between the thickness of the solid material at the edge of the ingot and the local brightness was investigated. (The word 'edge' has been used to describe the solid thickness here, as 'shelf' has a specific meaning for Ni superalloys referring to features possibly beneath the top surface.) Some results from the Ti-6-4 first melt are shown in figure 9. It can be seen that the edge thickness does vary over time, but not in direct relation to the local instantaneous brightness. Corresponding results are shown from VAR of an IN718 electrode in figure 10. Here it can be seen that there is significant growth and recession of the solid region at the edge of the ingot during most periods of brighter and darker reflected light, with recession happening when the light is brighter, presumably due to increased heat flux.

The difference between the results from Ti-6-4 and IN718 may be due to the observation that the arc centre location was much closer to the electrode (and ingot) edge during the VAR of IN718 than in Ti-6-4; this would presumably have increased the magnitude of the increase and decrease in local heat and current flux for IN718. The authors thank a referee for the observation that "…for Ti-6Al-4V VAR … At intermediate gaps, it is entirely possible to have the bulk of the arc centered under the electrode but still see arc light intensity rotation in the annulus that correlates with stirring field direction. These variations may not be strong enough to melt back shelf formed at the ingot-crucible interface…". However the ingot edge during Ti-6-4 VAR is still likely to be affected by axisymmetry over time in the arc or ingot/mold heat transfer, coupled with the effects of MHD in the large liquid pool volume at high overall current.

## 6. Conclusion

Faster optimisation algorithms have been developed to estimate the position of the centre of the magnetic centre of the arc current in vacuum arc remelting using magnetic source tomography. The algorithms were validated and compared with numerical models, and then applied to magnetic field measurements from trial VAR melts of titanium and nickel alloys. The slowest algorithm (BSL-MST) was found to perform well on simulated data and also to be able to correctly predict the ingot top position from experimental data. The quicker methods AL-MST and NN-MST were 2-4 orders of magnitude faster than BSL-MST in the implementation here (Matlab), with varying result quality. A strong link between instantaneous arc behaviour and the growth and re-melting of the solid skin at the ingot edge was observed for the nickel superalloy VAR ingot, but for Ti-6-4 the relationship was more complex although growth and recession were still observed; this may be linked to the arc centre position being typically interpreted as close to the electrode centre for the Ti-6-4 melt presented here but close to the electrode edge for IN718.

## 7. Acknowledgements

Dr Ward and Dr Nair are grateful for financial support from TIMET for this work; to Andrew Wilson at TIMET Witton for providing process data; and to Drs Matthew Thomas (Witton) and Ashish Patel (Morgantown) for support.

## 8. References

[1] Zanner F J and Bertram L A 1985 Vacuum Arc Remelting – An Overview Proc.8th Int. Conf. Vac. Metall. 512-522

[2] Shevchenko D M and Ward R M 2007 Liquid metal pool behaviour during the VAR of INCONEL 718 *Proc. Int. Symp. Liquid Metal Processing and Casting* 25-30

[3] Ward R M, Daniel B and, Siddall R J 2005 Ensemble arc motion and solidification during the vacuum arc remelting of a nickel-based superalloy *Proc. Int. Symp. Liq. Met Proc.Cas.*

[4] Koulis P, Djambazov G, Ward R M, Yuan L and Lee P D 2013 A multi-scale 3D model of the vacuum arc remelting process *Metallurgical and Materials Transactions A* **44A** 5365 – 5376

[5] Ward R M and Jacobs M H 2004 Electrical and magnetic techniques for monitoring arc behaviour during VAR of INCONEL 718 : Results from different operating conditions *Journal Mat. Sci.* **39** 7135-7143

[6] Woodside C R, King P E and Nordlund C 2013 Arc distribution during the vacuum arc remelting of T-6Al- 4V *Metallurgical and Materials Transactions B* **44B** 154 -165

[7] Nair B G and Ward R M 2009 An analysis of the use of magnetic source tomography to measure the spatial distribution of electric current during vacuum arc remelting *Meas. Sci. Technol.* **20** 045701

[8] Nair B G and Ward R M 2009 3-D analysis of magnetic field to monitor arc current during vacuum arc remelting *Proc. Int. Symp. Liquid Metal Processing and Casting*

[9] Williamson R L, Schlienger M E, Hysinger C I and Beaman J J 1997 Modern control strategies for vacuum arc remelting of segregation sensitive alloys *Superalloys 718,625,706 and various derivatives* 37 -46

IOP Conf. Series: Materials Science and Engineering **143** (2016) 012013 doi:10.1088/1757-899X/143/1/012013

# Titanium nitride (TiN) precipitation in a maraging steel during the vacuum arc remelting (VAR) process - Inclusions characterization and modeling

**V Descotes[1], J-P Bellot[2], V Perrin-Guérin[1], S Witzke[1] and A Jardy[2]**
[1] Aperam Alloys Imphy, BP 1 Centre de recherches P. Chevenard, F-58160 Imphy, France
[2] Institut Jean Lamour – UMR CNRS 7198 – Université de Lorraine, LabEx Damas, Parc de Saurupt – CS 50840, 54011 Nancy, France

E-mail: vincent.descotes@aperam.com

**Abstract.** Titanium Nitride (TiN) inclusions are commonly observed in a Maraging steel containing Nitrogen and Titanium and remelted in a VAR furnace. They can be easily detected by optical microscopy. A nucleus is observed next to a large number of TiN inclusions. A TEM analysis was carried out on a biphasic nucleus composed of a calcium sulfide (CaS) and a spinel (MgAl2O4), surrounded by a TiN particle. An orientation relationship between these three phases was revealed, which suggests a heterogeneous germination of the TiN particle on the nucleus by epitaxial growth. Based on this observation, on thermodynamic considerations and on previous work, a model has been developed and coupled to a numerical simulation of the VAR process to study the formation and evolution of a TiN distribution in the VAR ingot. Microsegregation is modeled using the lever rule, while the kinetics of precipitation is mainly driven by the supersaturation of the liquid bath. This model highlights the influence of the melt rate on the final size of TiN particles.

## 1. Introduction

The inclusion cleanliness is identified as a key parameter to extend the fatigue life properties of an alloy. In a maraging steel produced by Aperam (Imphy, France), containing Fe, 18%Ni, 9%Co, 5%Mo, 0.45%Ti and some traces of Nitrogen, Titanium nitride inclusions (TiN) are identified as the most harmful defect. This inclusion is very commonly observed in optical metallography [1]. It appears as a squared, yellow-colored particle. The thermodynamics of its formation has been studied by numerous authors, among them Morita and Kunisada [2] and more recently Kim et al. [3]. These studies conclude to the high stability of this kind of inclusion, which forms during the solidification of the alloy.

The production route involves a double remelting into a vacuum induction furnace (VIM) and a vacuum arc furnace (VAR). In this last process, an electric arc is generated between the electrode and the base of the copper mold in a vacuum atmosphere. Once the remelting has begun, the arc is set between the electrode and the growing ingot. The objective of this operation is to better control the solidification of the final ingot, its inclusion population, and possibly take advantage of the vacuum to pursue the steel refining [4].

A study of the TiN precipitation during solidification was carried out by Rocabois et al. [5] and then Lehmann et al. [6], who considered heterogeneous nucleation and diffusion kinetic laws, but no engulfment. Ohta and Suito [7] experimentally characterized the effect of the germs on the TiN sizes and show the particular benefit of MgO particles to increase the numeral density of inclusions and thus reduce their size.

The aim of this paper is to study the formation of the TiN inclusions in VAR ingots. It follows a previous article published on the same subject in 2013 [8].

Industrial and experimental facts are compiled in a first part. Counts of inclusion populations were performed and are presented using log-normal laws.

A model was proposed in previous work [8] to predict the nucleation and growth of a TiN distribution during the solidification of the alloy, which causes segregation of Ti and N in the liquid. This approach is here recalled and coupled to a 2D-axisymetric model of the VAR process (SOLAR code, [9]) in order to quantify the inclusion cleanliness in a whole ingot. The results of the simulation are compared to an experimental remelting. This tool highlights the heterogeneous inclusion size in the VAR ingots and enables one to optimize the process.

## 2. Thermodynamical considerations and experimental observations

### *2.1 Nucleation of TiN inclusions*

The TiN precipitation obeys to the following chemical reaction: $+N = TiN(s)$ . It is characterized by the solubility product $K_{TiN}(T)$ related to the activities of solutes in the steel by:

$$K_{TiN}(T) = a_{Ti}^{eq} \cdot a_N^{eq} \tag{1}$$

Using the software Thermo-Calc [10] and the TCFE7 Steels/Fe-alloys database version 7 [11] to compute the composition at equilibrium, it is found that with 4500ppm of Titanium in the liquid bath, the Nitrogen maximal mass fraction before precipitation starts is 17ppm. Since a Nitrogen mass fraction as low as 10ppm is commonly reached in the electrode, the inclusions that may exist in the bath are unstable and dissolve. As suggested by Young and Mitchell [12], it is assumed that complete dissolution takes place in the VAR liquid pool before solidifying the alloy. At that time, segregation will make the solute concentrations rise above the solubility limit, therefore triggering the precipitation of nitrides.

A detailed characterization of a mixed CaS – $MgAl_2O_4$ - TiN inclusion with a shell-core structure has been presented in reference [13]. An orientation relationship between all phases was found, which suggests that the CaS formed over the spinel, and that the TiN formed itself by epitaxial growth on the two nuclei. This observation strengthens the hypothesis of a heterogeneous nucleation of TiN particles on germs, i.e. inclusions that are stable in the liquid steel above the liquidus temperature. This hypothesis is also evoked by Rocabois et al. [5], Kunze et al. [14] and Lehmann et al. [6].

### *2.2 Optical counts and treatment*

An automated count of all inclusions on a polished sample is performed using an optical microscope. The surface of each inclusion is measured, and then converted into an equivalent radius.

It is noticed that a large amount of small inclusions are not detected by optical microscopy, although they truly exist (they are detected with SEM). Atkinson and Shi [15] report three methods to characterize such an inclusion population and especially choose an indicator of the maximal inclusion size: using a log-normal distribution law, using the extreme value method [16], or using a Pareto distribution law to model the tail.

The modeling of the entire inclusion population using a log-normal distribution law may be more uncertain than only characterizing the tail, as noted by Atkinson and Shi [15], but it enables us to introduce the number density of inclusions in our characterization, which provides fundamental elements to understand the cleanliness differences between samples.

The counting provides a sequence of $n$ inclusion radii that can be sorted in ascending order $r(i)$ ($i$ being the index of each inclusion in the sequence). The distribution of the inclusions will be approximated by a log-normal law with two parameters $\mu$ and $\sigma$. The cumulative distribution function of this law is expressed by:

$$F(r; \mu, \sigma) = \frac{1}{2} + \frac{1}{2}\mathrm{erf}\left(\frac{\ln r - \mu}{\sigma\sqrt{2}}\right) \tag{2}$$

A threshold is set at a radius of 2.5μm. As the total number of inclusions on a sample $n_{tot}$ remains unknown, the model is built with three parameters μ, σ and $n_{tot}$:

$$\forall i \in [\![2; n]\!], \qquad i = n_{tot}(F(r(i); \mu, \sigma) - F(r(1); \mu, \sigma)) \tag{3}$$

where $n_{tot}$ is the theoretical total number of particles that are present on the sample, $n$ the number of particles effectively detected, $i$ the particle index in the radii sequence $r(i)$ sorted in ascent order, and $r(1)$ the smallest radius detected. An optimization on the three parameters is performed using the non-linear least squares "fit" routine of MATLAB and leads to an estimation of $n_{tot}$, $\mu$ and $\sigma$.

A total volume density of particles can be derived, one assuming a uniform spatial distribution of the particles on the sample of surface $S_0$:

$$n_{tot}^{v} = \frac{3}{4}\sqrt{\pi}\left(\frac{n_{tot}}{S_0}\right)^{3/2} \tag{4}$$

An estimation of the upper radius bound with probability 97% is determined using the inverse log-normal cumulative function:

$$r_{97} = \exp\left(\sigma\sqrt{2}\,\mathrm{erf}^{-1}(2 \cdot 0.97 - 1) + \mu\right) \tag{5}$$

where $\mathrm{erf}^{-1}$ is the inverse error function.

Figure 1 presents the diameters $d_{97}^{s} = 2 \cdot r_{97}$ calculated on several industrial samples. It experimentally highlights that:

- the particle diameter $d_{97}^{s}$ increases with the Nitrogen content of the alloy;
- at a given chemical composition, the distribution of nitrides over a large number of nucleation sites decreases the size of the particles.

An effect of the local solidification conditions may explain some of the discrepancies between the samples.

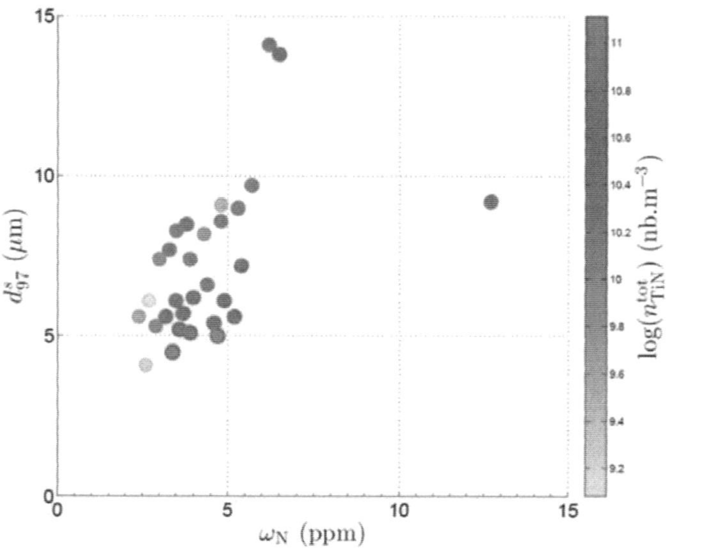


Figure 1: Diameter $d_{97}^{s}$ of the TiN inclusions characterized on several samples, related to the Nitrogen mass fraction in the sample

## 3. Precipitation model and VAR process modeling

Precipitation is thought to happen during solidification thanks to segregation. The VAR ingot growth and solidification was modeled using a 2D - finite volume code. The ingot geometry is discretized into a 2D axisymmetric mesh. The precipitation model developed in ref. [8] is summarized below.

### *3.1 Hypotheses of the inclusion model*

Based on the experimental observations (section 2.1), it is considered that TiN inclusion form on an inert, pre-existing oxide distribution that act as nuclei. Nucleation is therefore considered to be only heterogeneous.

It is also considered that the precipitation reaction stops in the solid phase. This assumption is supported by the fast cooling rate of the ingot during the VAR process, which prevents the establishment of a significant diffusion in the solid state to pursue precipitation.

### *3.2 Inclusion precipitation model*

A competition takes place between the segregation which brings solutes in the liquid phase and the precipitation which consumes solutes. The supersaturation is defined as:

$$S = \frac{a_{Ti} \cdot a_N}{K_{TiN}} \tag{6}$$

The TiN nucleation model is derived from the classic thermodynamics theory developed in several books as in Refs. [17,18]. It considers the Gibbs energy of the chemical reaction of precipitation and an interfacial energy term that acts against the nucleation. Therefore, nucleation can occur only if the supersaturation of the bath is at least higher than unity, the exact threshold depending on the size of the oxide. After nucleation, the growth of an inclusion is described by a first-order kinetic law involving the distance to equilibrium. As stated in [8], a transport equation in the internal coordinate space (inclusion size) for the particle distribution can be derived:

$$\frac{\partial n_{TiN}^l}{\partial t}(r,t) = -k_c(S-1)\frac{\partial n_{TiN}^l}{\partial r}(r,t) \tag{7}$$

where $n_{TiN}^l$ is the distribution of TiN in the liquid phase (density number of inclusions of radius $r$ per unit volume of liquid metal and per unit length of particle radius), $r$ is the radius of an oxide-TiN assembly (m), $S$ the supersaturation, and $k_c$ a kinetic factor which gathers the chemical phenomena at the interface and the Ti and N mass transport from the supersaturated bulk liquid to the TiN particle surface.

When solidification proceeds, the solidification front engulfs some TiN inclusions from the liquid into the solid. Those inclusions will no more be available in the liquid to grow further, so they become inert.

A lever rule model is brought into play to describe the solidification of the alloy, thus assuming infinite diffusion of the solutes within the two phases [19]. The liquid concentration is lowered by the solute consumption from the precipitation reaction, as described in [8].

### *3.3 SOLAR model and coupling*

Simulating the VAR process has been successfully attempted over the past decades. The result of these efforts is collected into several calculation codes like BAR [20], MeltFlow [21] or SOLAR [9].

The SOLAR code has been developed in Nancy during several PhD [9, 22]. It takes as input the electrical current, voltage and the recorded melt rate in order to simulate the VAR process. Fluid dynamics, electromagnetic fields, thermal exchanges, solidification and solute transports are calculated in transient regime using a finite volume SIMPLE numerical scheme [9, 23].

The water cooling of the copper mold and the Helium pressure injected in the gap between the ingot side and the mold are modeled by thermal resistances [22]. The turbulence is computed by a k-ε model. The mesh is refined at the ingot top and side. The growth of the ingot is performed by periodical splitting and continuous growth of cells. It introduces advective source terms in the numerical scheme [24].

A weak coupling is introduced with the precipitation model in each cell: the variation of the solid fraction $g_s$ and associated segregation given by SOLAR determine the precipitation sequence over one time step: germination, engulfment, growth of particles, solute consumption. The coupling between the cells is achieved through the solute transport equation solved in SOLAR.

## 4. Experimental remelting

A full-scale ingot was remelted at Aperam in order to study the effect of remelting parameters on the inclusion cleanliness. It must be stated here that the usual industrial practice largely differs from the remelting conditions that were chosen for this ingot.

### *4.1 Remelting parameters and etching procedure*

The melting rate is suspected to have the main impact on the solidification in VAR process, and therefore on the inclusion cleanliness. Two melt rates were tested, with steep transitions between the two steady-state regimes. A brutal cut of the electrical input was applied at the end of the remelting to freeze the last liquid pool.

A longitudinal slab was extracted from the ingot and cut into four pieces. Each part was polished and chemically etched to reveal the grain structure using FRY7 reactant (HCl + $CuCl_2$). A reconstruction of some liquid pools was performed assuming that the solidification front was perpendicular to the grain columns. Three samples were taken along the axis at different heights and analyzed.

### *4.2 Results - experimental ingot*

The approximation of the TiN counts by a log-normal law lead to the values of table 1. We notice that sizes decrease when the melt rate increases.

Table 1: Results of the three samples taken from ingot W2280

| Height (m) | $r_{97}$(μm) | Comments |
|---|---|---|
| 0.54 | 4.2 | Low melt rate |
| 0.85 | 3.0 | Low melt rate |
| 1.28 | 3.7 | Higher melt rate |

It was observed on the axial section of the ingot that the depth of the liquid pools changed when the melt rate was increased. A deep pool was obtained at the end, materialized by an equiaxed solidification. This structure comes from the abrupt stop of the remelting without wafering the ingot.

## 5. Numerical results

Simulations of the inclusion cleanliness using the coupled model were carried out on the experimental ingot.

### *5.1 Temperature field*

Figure 2 shows the temperature field calculated in the ingot at different steps of the remelting. The mushy zone is materialized by two thick lines. A melt pool depth of around 40cm is achieved at the end, which is what was observed on the etched plate. It can be seen that the mushy zone extent is higher near the end, when a high melt rate is achieved.

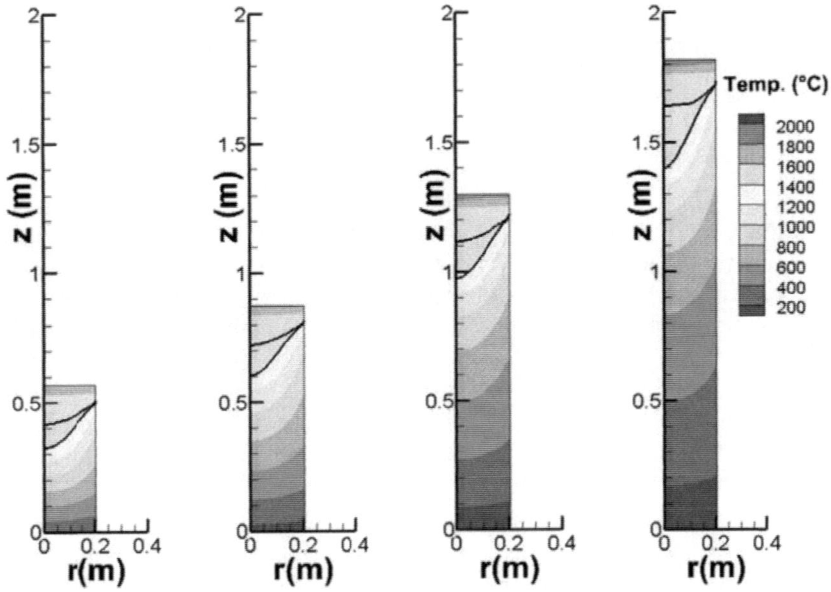


Figure 2: Temperature field at different steps of the remelting

*5.2 Cleanliness aspects*

Cleanliness is characterized by the TiN mass fraction in the solid phase and by the radius $r_{97}$ which represents the upper bound radius of 97% of the TiN distribution in the solid phase (figure 3). The precipitation is influenced by the macrosegregation of N or Ti, which is shown on maps. The end of the solidification after shutting down the power was not simulated.

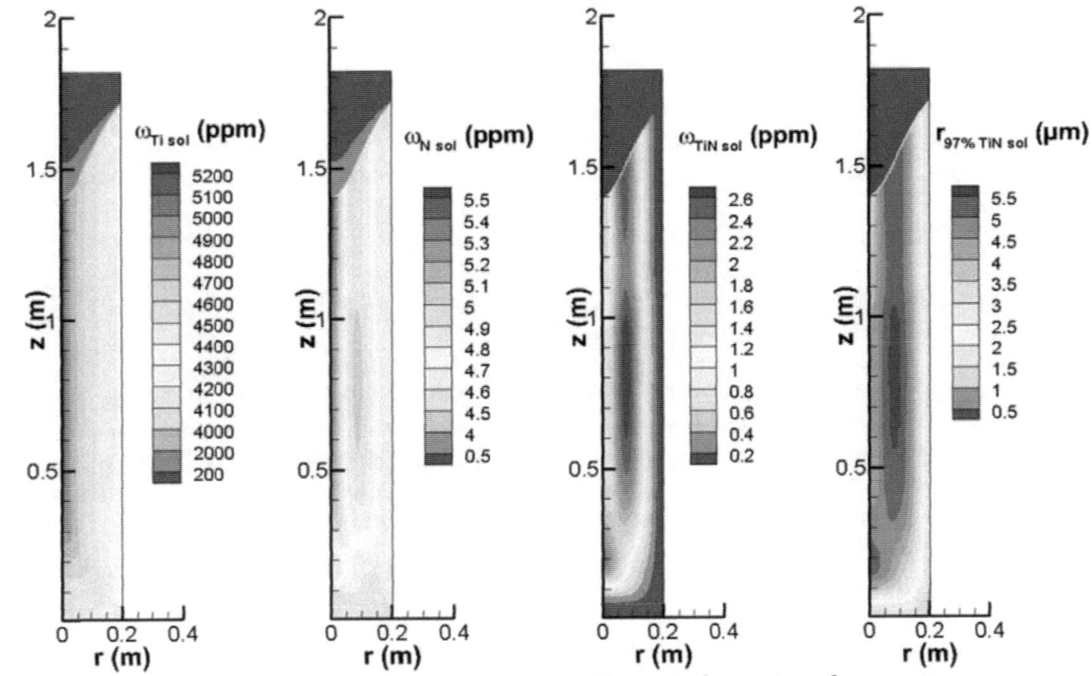


Figure 3: Cleanliness of the remelted ingot ($\omega_{Ti}^{s}$ - $\omega_{N}^{s}$ - $\omega_{TiN}^{s}$ - $r_{97}$)

It is seen that the maximal sizes of TiN particles are reached at mid-radius and rather in the bottom part of the ingot where the melt rate was the smallest. A higher melt rate seems to diminish the inclusion sizes, especially during the transient between the two regimes. Maximal radii of about 5.5μm are achieved. A macrosegregation in Ti and N is visible in the central axis due to the convective movements in the liquid pool.

## 6. Conclusion

Thermodynamic calculation shows that TiN inclusions are unstable in the liquid bath. Segregation in the mushy zone during the solidification raises the Ti and N contents in the interdendritic liquid, making the precipitation reaction possible. Experimental characterization of the inclusions shows that the nucleation is heterogeneous, and that the size distribution is related to the Nitrogen content and to the density of nucleation sites.

A precipitation model was set up in a previous work to study the coupling between these two phenomena [8]. This model was coupled to the SOLAR software to simulate the VAR remelting process and the TiN precipitation in the mushy zone. The results were compared to a full-scale melt with two different remelting conditions: a small melt rate and a high melt rate. It appears that:

- the liquid pool shapes are rather well predicted by the model for this melt ;
- the largest inclusion size can usually be found at mid-radius; a heterogeneity of the inclusion sizes along the radius is clearly seen on the simulations;
- the melt rate may play a certain role on TiN size, by changing the local solidification time.

## 7. Acknowledgments

The authors gratefully acknowledge J. Ratafika (Aperam) who carried out the VAR casting experiment, and G. Pinton and P. Guichard of CRPC Aperam Alloys Imphy who characterized the ingot.

## 8. References

[1] Durand-Charre M 2003 *La microstructure des aciers et des fontes - Génèse et interprétation* (Paris: SIRPE Éditeur)
[2] Morita Z and Kunisada K 1977 *Tetsu-to-Hagané* **63** 1663–71
[3] Kim W Y, Jo J O, Chung T I, Kim D S and Pak J J 2007 *ISIJ International* **47** pp 1082–89
[4] Yu K O 2002 *Modeling for Casting and Solidification Processing* (New York – Basel: Marcel Dekker, Inc.)
[5] Rocabois P, Lehmann J, Gaye H and Wintz M 1999 *Journal of Crystal Growth* **198/199** pp 838–43
[6] Lehmann J, Rocabois P and Gaye H 2001 *Journal of Non-Crystalline Solids* **282** pp 61–71
[7] Ohta H and Suito H 2007 *ISIJ International* **47** pp 197–206.
[8] Descotes V, Bellot J P, Witzke S and Jardy A 2013 Modeling the titanium nitride (TiN) germination and growth during the solidification of a maraging steel *Proc. of the 2013 Int. Symp. on Liquid Metal Processing & Casting* (*Austin, TX, USA, 22-25 September 2013)* ed Krane M J M, Jardy A *et al* (Warrendale: TMS) pp 201–206
[9] Hans S 1995 *Modélisation des transferts couplés de chaleur, de soluté et de quantité de mouvement lors de la refusion à l'arc sous vide (VAR) - Application aux alliages de titane*, Ph.D. thesis, Institut national polytechnique de Lorraine
[10] Andersson J, Helander T, Höglund L, Shi P and Sundman B 2002 *Calphad* **26** pp 273–312
[11] *Thermo-Calc Software, Thermo-Calc Console Mode User Guide* 2013 3.1 edition (Stockholm: Foundation of Computational Thermodynamics)
[12] Young E and Mitchell A 2001 *High Temperature Materials and Processes* **20** pp 79–101
[13] Descotes V, Migot S, Robaut F, Perrin-Guérin V, Witzke S, Bellot J P and Jardy A 2015 *Metallurgical and Materials Transactions A* **46A** pp 2793-5
[14] Kunze J, Spitzer K H, Oswald S and Mickel C 1997 *Zeitschrift für Metallkunde* **88** pp 182–90.

[15] Atkinson H and Shi G 2003 *Progress in Materials Science* **48** pp 457–520
[16] Murakami Y 2002 *Metal fatigue: Effects of Small Defects and Nonmetallic Inclusions* (Amsterdam: Elsevier)
[17] Dantzig J, Rappaz M 2009 *Solidification* (Lausanne: EPFL Press)
[18] Lesoult G 2010 *Thermodynamique des matériaux* (*Traité des matériaux* vol 5) (Lausanne: PPUR)
[19] Kurz W and Fischer D 1998 *Fundamentals of Solidification* (Pfaffikon: Trans Tech publication Ltd)
[20] Bertram L, Schunk P, Kempka S, Spadafora F and Minisandram R 1998 *Journal of Metals* **50** pp 18–21
[21] Kelkar K, Patankar S, Mitchell A, Minisandram R and Patel A 2010 *ASM Handbook* vol 22B ed Furrer D and Semiatin S (ASM International) pp 196–213
[22] Quatravaux T 2004 *Évolution de la modélisation du procédé VAR - Contribution à la description de la dispersion inclusionnaire dans le puits liquide et à la prévention de défauts de solidification*, Ph.D. thesis, Institut national polytechnique de Lorraine
[23] Patankar S 1980 *Numerical Heat Treatment and Fluid Flow* 1st ed (Washington D.C.: Hemisphere Publishing Corp)
[24] Jardy A, Quatravaux T and Ablitzer D 2004 *Multiphase Phenomena and CFD Modeling and Simulation in Materials Processes* (*Charlotte, NC, USA, 14-18 March 2004*) ed Nastaq L and Li B Q (Warrendale: TMS) pp 265–277.

# Controlling pool depth during VAR of Alloy 718

**F Lopez[1], J Beaman[1], R Williamson[1], D Evans[2]**
[1] Department of Mechanical Engineering, University of Texas at Austin, Austin, TX 78705, USA
[2] Special Metals Corporation, New Hartford, NY 13413, USA

E-mail: felipelopez@utexas.edu

**Abstract.** A longtime goal of superalloy producers has been to control the geometry of the liquid pool in solidifying ingots. Accurate pool depth control at appropriate values is expected to result in ingots free of segregation defects. This article describes an industrial VAR experiment in which a 430mm (17 in) diameter Alloy 718 electrode was melted into a 510mm (20 in) ingot. In the experiment, the depth of the liquid pool at the mid-radius was controlled to three different set-points: 137 mm (nominal), 193 mm (deep) and 118 mm (shallow). At each level, the pool depth was marked by a power cutback of several minutes. The ingot was sectioned and longitudinal slices were cut out. Analysis of the photographed ingot revealed that accurate control was obtained for both the nominal and deep pool cases, while the third one was not conclusive.

## 1. Introduction

Alloy 718, a nickel-based superalloy, is the most widely-used superalloy in history. This material can be found extensively in aircraft engines and power-generation turbines. These days, the drive for increased operating efficiency in these applications has resulted in the need for larger, yet structurally ingots for forging stock of Alloy 718 [5]. However, this task has proved challenging due to an increasing tendency for segregation defects at larger diameters.

Engineers have dedicated years to the development of techniques to prevent segregation defects in large ingots of nickel superalloys. Triple melting (TM) by VIM+ESR+VAR was largely successful in that task and enabled the fabrication of ingots up to 510 mm (20 in) in diameter for Alloy 718 and 910 mm (36 in) for Alloy 706 [4]. Moreover, adjustments in chemistry (reducing carbon, nitrogen, and niobium) enabled the production of Alloy 718 ingots up to 910 mm (36 in) in diameter [6].

At the same time, it was observed that the tendency for defect formation in superalloys was determined by the liquid pool profile and ingot solidification patterns during the process. Based on this observation, it was hypothesized that accurate solidification control would result in improved ingots, and potentially larger Alloy 718 ingots free of segregation defects. However, the lack of an appropriate solidification model in VAR prevented the development of a pool profile controller.

The first attempt to control pool shape in VAR was reported by Beaman *et al.*[1].Although promising, the method was not extendable to large ingots due to the model's inability to describe convective heat transfer in the liquid phase. This paper describes the first pool depth control demonstration in large ingots. In the experiment, an alternative form of the controller proposed by Beaman was used to melt a 430 mm (17 in) diameter Alloy 718 electrode into a 510 mm (20 in) ingot.

IOP Conf. Series: Materials Science and Engineering **143** (2016) 012014 doi:10.1088/1757-899X/143/1/012014

The ingot was sectioned and Canada etched to reveal the pool shape profile, which compared favorably to the process set-points.

## 2. Controller design

The layout of the pool depth control system is depicted in Figure 1.Two reference signals are given to the process controller: a reference pool depth $S_{p,ref}$ and a reference electrode gap $G_{ref}$. The controller returns two commands: melting current $I$ and ram drive speed $V_{ram}$, which are obtained by comparing the references to state estimates. Commanded current, drive speed, and several measurements from the furnace (electrode gap $G$, ram position $X$, current $I$, electrode mass $M$ and voltage $V$) are fed to an electrode estimator, that returns estimates of melting efficiency $\mu_e$[1]; electrode gap $G_e$; and bias in measured voltage and current, $U_{b,e}$ and $I_{b,e}$. Estimated melt rate $MR_e$ is calculated from these parameters and fed to BAR[2] along with current $I$, voltage $V$ and cooling gas (helium) pressure $p_{he}$ to obtain predictions of the liquid pool profile $S_{p,BAR}$. These predictions are fed to an ingot estimator that returns estimates of the thermal state of the solidifying ingot. Estimates from both estimators are used by the controller to compute optimal control signals.

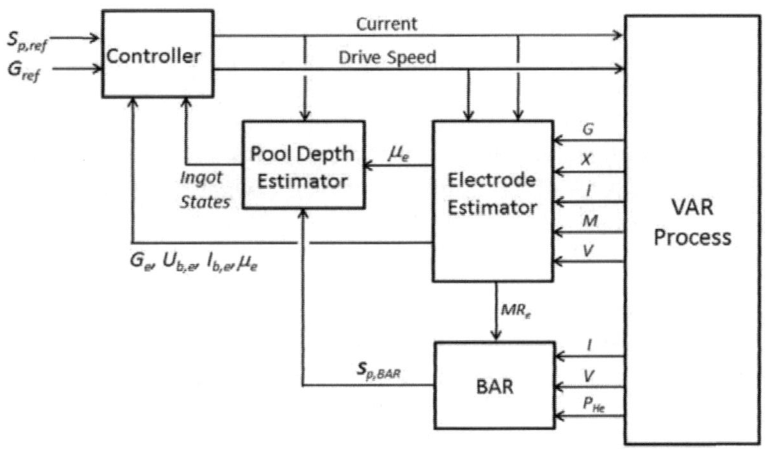


Figure 1. Schematic of the VAR pool depth controller design.

In this application, the process controller consists of two separate sub-systems: an electrode gap controller based on drive speed and a pool depth controller based on melting current. Commanded ram drive speed is set by a proportional controller following the design described in Ref. [2]. Meanwhile, commanded current is given by a Linear-Quadratic-Gaussian (LQG) controller that adds a nominal value to corrections for the estimated state, defined reference, estimated melting efficiency, and helium pressure; similar to the design in Ref. [1].

In this version of the pool depth controller, the ingot solidification model was obtained from a first-order approximation for pool depth dynamics which compared favorably to more accurate models (see Figure 2[3]). This model is able to track pool depth in four locations during the melt: at the center (r = 0.00 R), quarter-radius (r = 0.25 R), mid-radius (r = 0.50 R), and three-quarter-radius (r = 0.75 R). However, for simplicity, only mid-radius pool depth is used for process control.

---

[1]Defined as the instantaneous fraction of supplied power that is used in melting the electrode $\mu = P_{melt} / P_{total}$.

[2] Basic Axisymmetric Remelting (BAR) is the SMPC VAR solidification model. It is used to obtain predictions of liquid pool depth based on electric and thermal parameters of the process [3].

[3]BAR, a finite volume model, was used as benchmark for ingot solidification.

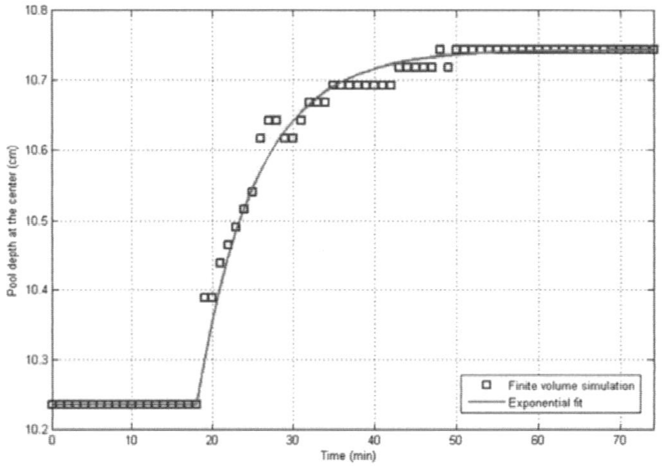


Figure 2. Comparison between BAR (finite volume simulation of ingot solidification) and a first-order exponential response for Alloy 718 in a 152 mm (6 in) to 203 mm (8 in) melt when current is instantaneously perturbed from 2500 A to 2750 A.

## 3. Experiment

The goal of the experiments was to test the proposed control strategy using SMPC's Advanced VAR Controller (AVARC)[4]. AVARC was run from a personal computer which communicated with the furnace control system through Ethernet connection. The experiment was carried out at Special Metals Corp. in New Hartford using a 430 mm (17 in) diameter electrode of Alloy 718, which was melted into a 510 mm (20 in) ingot.

In the experiment, the arc was struck with the host furnace PLC. Current was ramped up to a 6350 A hold. AVARC control was enabled after liquid metal was observed to cover the base plate. The initial control mode of the controller was set to current mode with a 6350 A set-point which was held for 240 minutes. This was followed by a change to pool depth mode with a mode transition time of 30 minutes.

Three mid-radius pool depths were programmed into the controller recipe: 137 mm (5.4 in), 193 mm (7.6 in) and 118 mm (4.6 in). The pool depth holds were separated by 90 minute ramps. Hot-top was programmed to begin with 227 kg of electrode remaining and involved ramping the pool depth down to 520 mm (2.5 in) over 90 minutes, followed by a 2500 A hold.

## 4. Experimental results

Figure 3 shows pool depth and current data for the entire test melt. The vertical dotted red line shows the point at which pool depth control mode was activated. Note the slight increase in current that occurs immediately after the switch, which is necessary to deepen the pool to the mid-radius set-point of 137 mm (5.4 in). The observed mid-radius liquid pool depth, measured with an online simulation of BAR, is plotted in blue and compared to the user-defined reference shown as black dash-dot line. Both lines coincide for nominal conditions and for the shallow pool case, but a mismatch of 7 mm can be observed for the deep pool one.

It is evident from these data that linear pool depth ramps defined in the reference require nonlinear current ramps. The three pool marking events are prominently evident in the current plot, as well as their effect in liquid pool depth.

[4]The Specialty Metals Processing Consortium (SMPC) is a collaborative program between several superalloy producers that conducts research on remelting processes.

IOP Conf. Series: Materials Science and Engineering **143** (2016) 012014 doi:10.1088/1757-899X/143/1/012014

The nonlinear current response that is required to produce agreement between the estimated pool depth and the reference set-point produces concomitant nonlinear responses in melt rate. This is seen in Figure 4 which shows plots of the estimated melt rate and measured pool depth (mid-radius). Note that during the first two pool marking events, melt rate is completely shut off, while it is reduced to about 340 g/min (0.75 lb/min) in the final pool marking event.

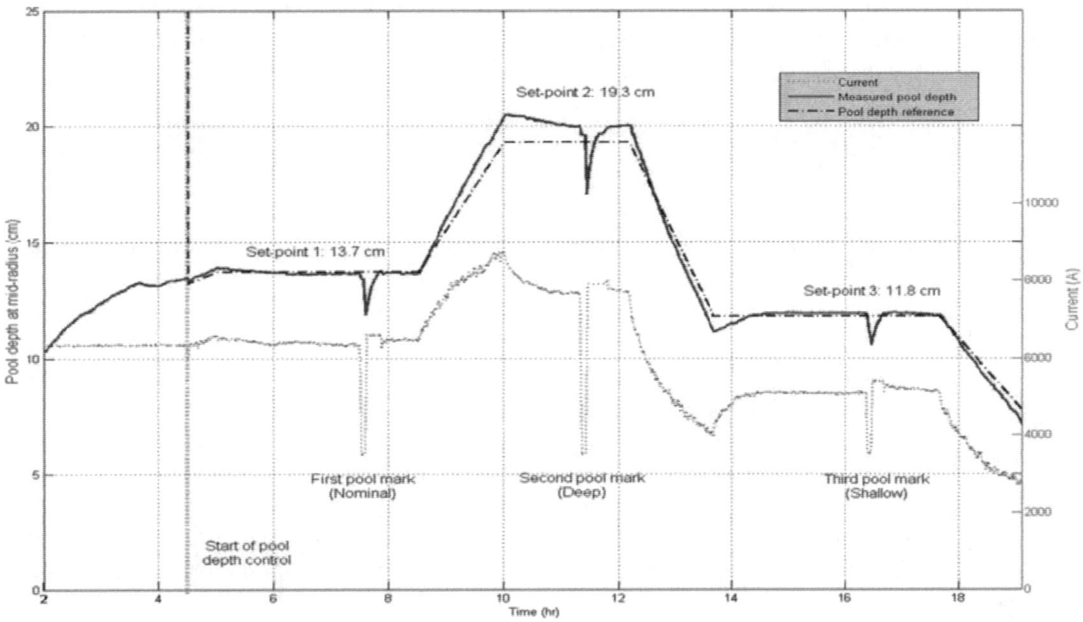


Figure 3. Overview of the test melt showing pool depth and commanded current.

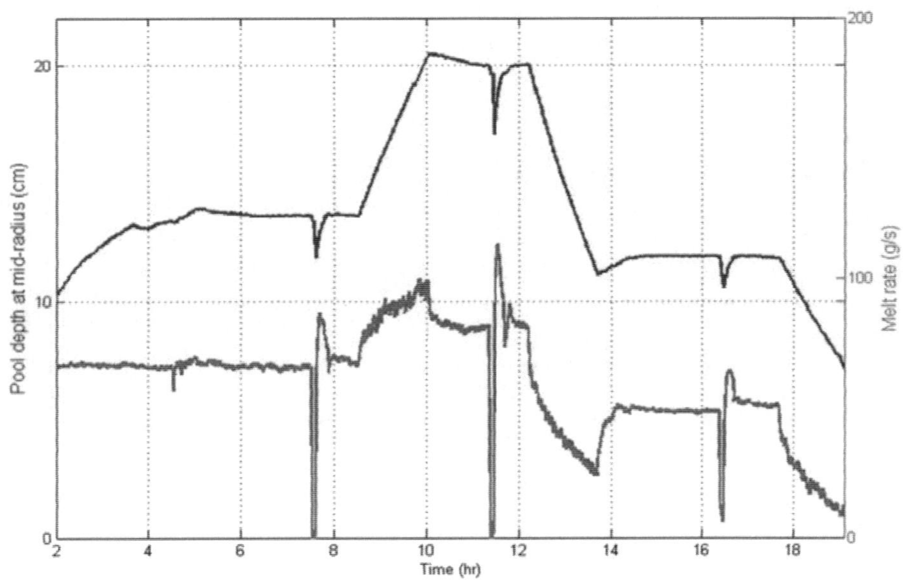


Figure 4. Comparison of pool depth and melt rate in the test melt.

Figure 5 (right) shows an expanded view of the current and electrode mass during the first pool marking event. It is evident from these data that the ingot does not grow significantly during the pool marking event. This is confirmed by the estimated melt rate data shown in Figure 5 (left). The estimated melt rate is zero by the time the minimum current of 3500 A is reached. Near the end of the low current hold it barely rises above zero. Integrating the regions where the current is ramped down and up gives ~ 4.7 kg (10.4 lbs) of electrode melted during this pool marking event, corresponding to ~3 mm (0.1 in) of ingot growth. Because this number is smaller than the error in determining the ingot height at the beginning of a pool marking event, the working assumption during the pool depth analysis is that the ingot does not grow during pool marking.

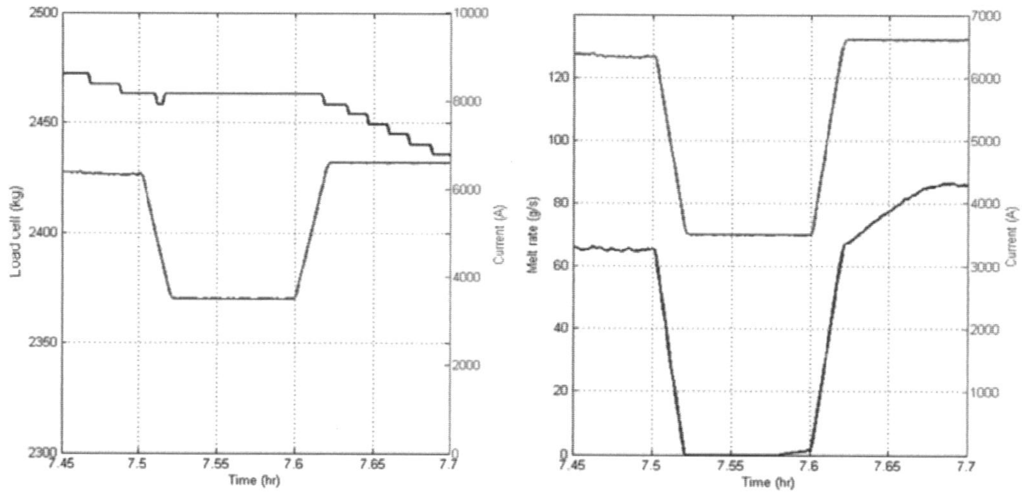


Figure 5. Expanded view of estimated melt rate, electrode mass and current during the first pool marking.

Given the large fluctuations in current and melt rate, as well as the violent perturbations introduced by the pool marking events, the question naturally arises as to how the controller performed with respect to drip-short based electrode gap control. Figure 6 shows plots of drip-short frequency, electrode gap calculated directly from the Zanner function using drip-short frequency and current, and the estimated electrode gap from the electrode estimator [7]. For this material and size, the Zanner function is given by $G=965.0f_{DS}^{-0.595}I^{-0.669}$.

We note that "class zero" drip-shorts were used for gap control and that the ram was not allowed to back up during this trial. What stands out immediately from the plots is that the estimated electrode gap is very smooth and flat relative to the calculated electrode gap through all pool depth holds and ramps. This is in spite of large fluctuations in current and melt rate. Of course, the pool marking events are an exception to this, but even these are well behaved with maximum deviations of only about ±5 mm (±0.2 in). We conclude that gap was well controlled throughout the test melt. The exception to this is a large gap deviation that starts at about 18.1 hours at the end of the test[5].

Data presented in Figures 3-6 were obtained from the process controller. Besides these results, data were also obtained by cutting the test ingot into cylindrical sections, cutting sections in half longitudinally, and then cutting a plate from one of the longitudinal faces of each of the half sections. These sections were then Canada etched to reveal the segregation patterns on the face of each plate for the purpose of detecting the pool profiles generated by the pool marking events.

[5]This event was due to a ram position measurement problem that developed late in the melt and does not reflect controller performance.

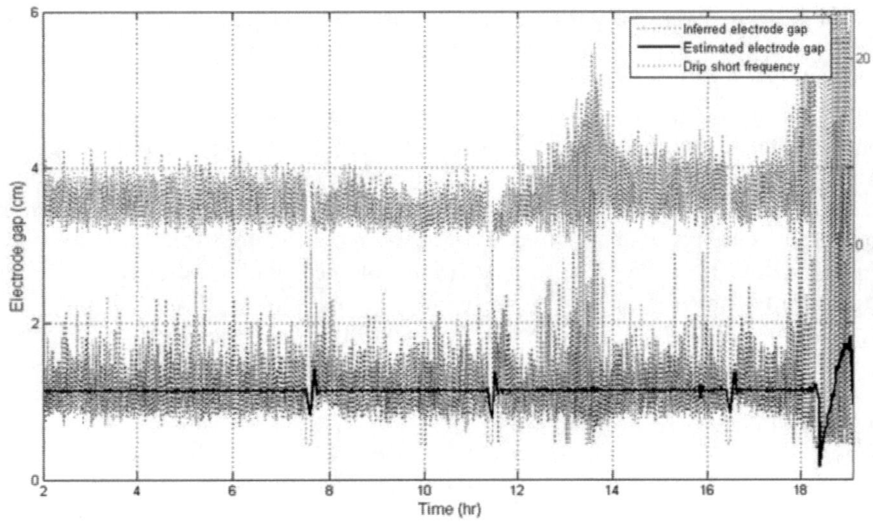


Figure 6. Plots of electrode gap inferred with Zanner's equation, estimated electrode gap, and drip short frequency for the test melt.

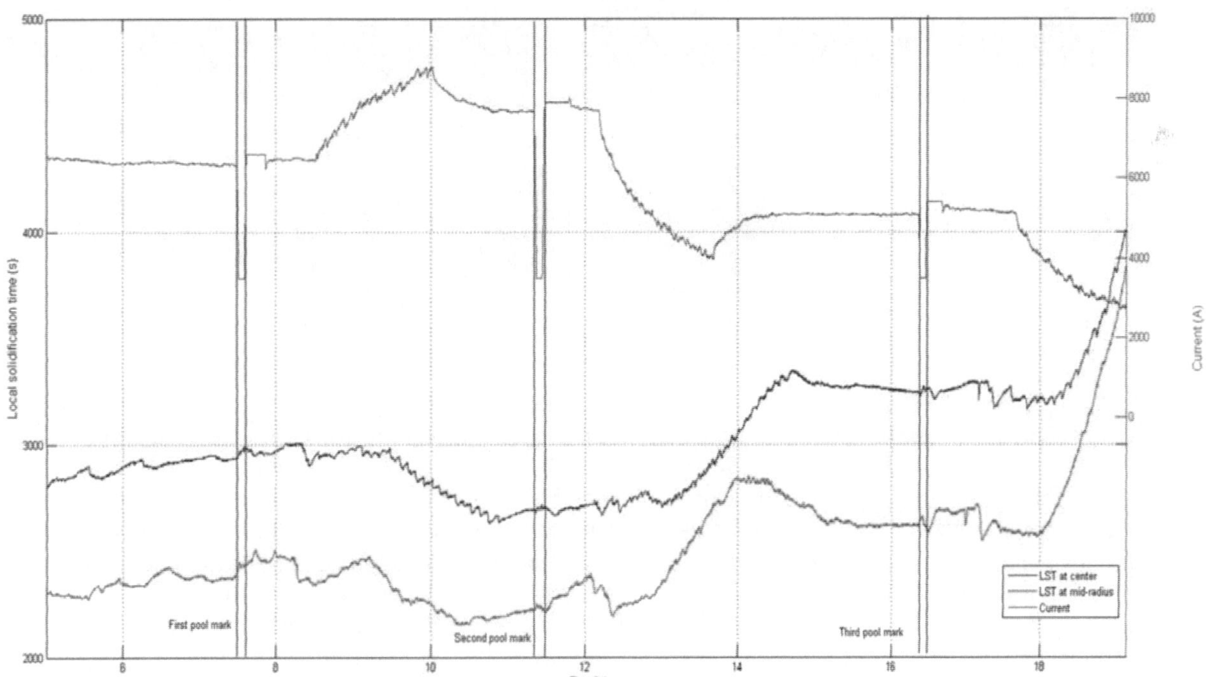


Figure 7. Local solidification time (LST) and current for the test melt.

Figure 7 shows a plot of local solidification time (LST) for the test, along with a plot of current. LSTs at the center and mid –radius were calculated by the controller during melting from BAR, the finite volume model. The trend is as expected: the longest LST is in the center of the ingot and decreases toward the edge. Also the general trend is observed that LST decreases with increasing pool depth and increases with decreasing pool depth, with this trend most pronounced in the center of the ingot. Even though the mushy zone grows with increasing current, the casting rate also increases. It is evident from these data that BAR predicts that the increase in casting rate dominates LST determination under these casting conditions. Note that the perturbations from the pool marking events are felt first near the edge of the ingot and last at the center of the ingot.

Figure 8 (a) shows the results for the first pool marking. The event resulted in a smooth, relatively uniformly grey area with no evident "tree ring" banding. This presumably results from the pool solidifying in during the low power hold. The finite volume solution, shown by the red line in the figure, fits the bottom of this region very well; the mid-radius pool depth of 137 mm (5.4 in) is shown as the yellow line. A similar result is observed for the 193 mm (7.6 in) deep pool marking shown in Figure 8 (b). Both plates show evidence of heavy shelf forming during the low power hold.

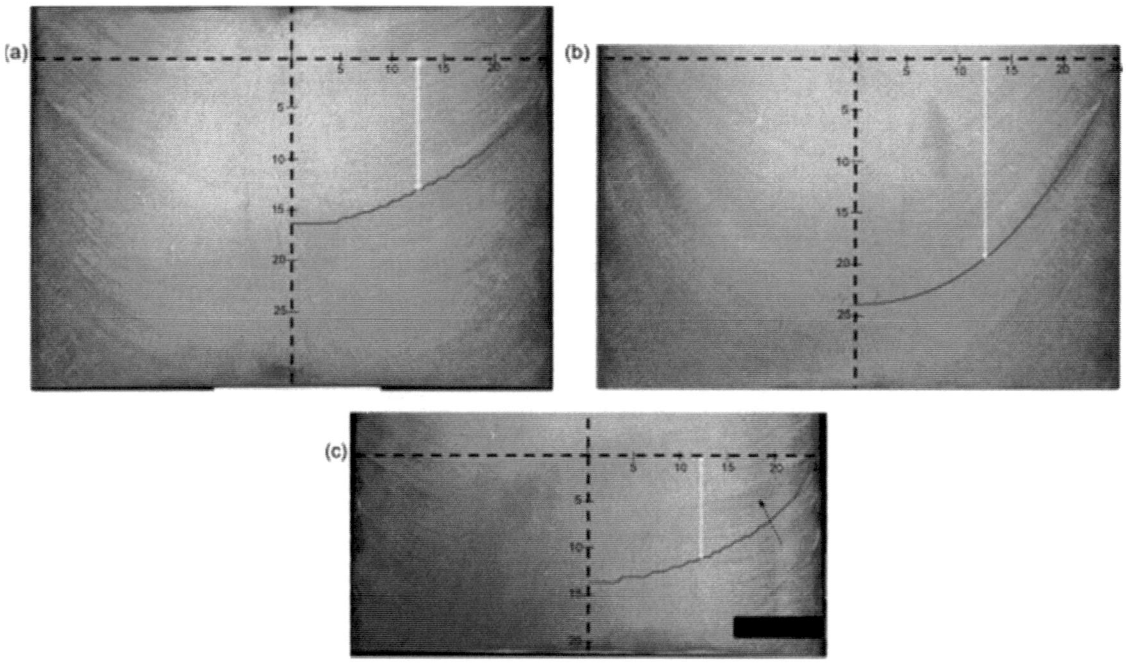


Figure 8.Finite volume pool depth measurement superimposed on an image of segregation (Canada) etched plate for references of: (a) 137 mm, (b) 193mm and (c) 118 mm (Scales in cm).

In spite of this success in controlling to the deeper pool shapes, the lines on the plate shown in Figure 8 (c) seem to indicate a much shallower pool than predicted by BAR or the controller. There is no clear indication of the location of the pool marking event. The black arrow points to light etching region but it is not conclusive. If this is the pool marking event, then the pool looks to be about 69 mm (2.7 in) deep at mid-radius instead of 118 mm (4.6 in). There is evidence of heavy shelf for this entire section of ingot and the photograph does not reveal any pool structure in the central region. More careful metallographic analysis may be required to actually determine the pool shape during the shallow pool depth hold.

## 5. Discussion

In general terms, the controller performed as it was designed to perform. It controlled the estimated mid-radius pool depth to the set-point reference and showed reasonable agreement with the finite volume predictions throughout the test. But, how well is this success reflected in the ingot structure?

The etching results for the first two commanded pool depths are particularly encouraging. Pool shapes revealed by the segregation patterns generally match BAR results and the mid-radius pool depths appear in reasonable agreement with the set-point values. Careful metallographic analysis of the plates will need to be performed if more accurate knowledge of the pool shapes is required, but the results shown in Figures 8 (a) and (b) are certainly very reasonable even if they are not perfectly accurate. Figure 8 (c), on the other hand, indicates a discrepancy between what was theoretically expected and what was actually produced in the ingot.

Going back to Figure 7, which shows calculated LSTs for the test melt, it is clear that a steady-state thermal distribution was not reached in the ingot during any of the pool depth holds. However, BAR data immediately prior to marking the 118 mm pool indicate that the solution is well behaved at this point. Given the evidence of heavy shelf in the segregation etched plate, it is possible that the shelf formed during the power cutback and simply never melted back at the relatively low power required to hold the shallow pool set-point (~5000 A). If this is true, we would expect the centerline pool depth to be closer to the BAR prediction. Confirmation (or refutation) of this hypothesis may be revealed by more careful metallurgical analysis of the plate.

The observation that pool depth may be controlled to a constant set-point under conditions of widely fluctuating LST indicates that this may not be the best solution to controlling the process so as not to form solidification defects. A better approach may be to control LST at one or more radial positions instead.

## 6. Conclusions

A VAR ingot pool depth controller was developed and successfully tested. A single industrial test was performed at Special Metals Corporation in New Hartford, New York. In the experiment, a 430 mm (17 in) diameter Alloy 718 electrode was remelted into a 510 mm (20 in) diameter ingot. Mid-radius ingot pool depth was controlled at three reference set-points during the test melt: 137 mm, 193 mm, and 118 mm. Post mortem analysis of the ingot revealed that control at the nominal and deep reference values was successfully implemented. Ingot analysis from the shallow pool depth was inconclusive. The relevant section of ingot shows signs of heavy shelf possibly left over from the power cutback required to produce the shallow pool depth set-point. What little information that can be gleaned from the ingot analysis indicates that the pool was significantly shallower than the 118 mm set-point, perhaps only 69 mm. Further more careful ingot analysis is warranted at this point.

In spite of the mismatch at the shallow pool depth setting, the controller performed successfully and predictably given the data that were being fed to it. This experiment constitutes the first successful pool depth control test in large ingots of Alloy 718. More testing will be required to determine safe pool depths to be used as references for the production of larger defect-free Alloy 718 ingots.

## 7. Acknowledgements

The author would like to thank the Specialty Metals Processing Consortium for support and Special Metals Corporation for hosting the experiment.

## References

[1] Beaman J, Lopez F and Williamson R 2013 *J. Dyn. Syst.-T. ASME* **136(3)** 031007
[2] Beaman J, Williamson R, Melgaard D and Hamel J 2005 *Proc. IMECE* (Orlando, FL) p 1059
[3] Bertram L, Schunk P, Kempka S, Spadafora F and Minisandram R 1998 *JOM* **50(3)** p 18
[4] Moyer J, Jackman L, Adasczik C, David R and Forbes-Jones R 1994 *Proc. Superalloys 718, 625, 706 and various derivatives* (Warrendale, PA) p 39
[5] Schwant R, Thamboo S, Anderson A, Adasczik C, Bond B, Jackman L and Uginet J 1997 *Proc. Superalloys 718, 625, 706 and various derivatives* (Pittsburgh, PA) p 141
[6] Schwant R, Thamboo S, Yang L and Morra M 2005 *Proc. Int. Symp. Superalloys 718, 625, 706 and derivatives* (Pittsburgh, PA) p 15
[7] Zanner F 1981 *Met. Trans. B* **12(4)** p 721

IOP Conf. Series: Materials Science and Engineering **143** (2016) 012015 doi:10.1088/1757-899X/143/1/012015

# The Development and Application of an Integrated VAR Process Model

**A. Stewart Ballantyne**

MeltMet Technologies

Unit 1205 – 788 Humboldt St, Victoria, BC Canada V8W 4A2

stewball43@yahoo.com

**Abstract.** The VAR ingot has been the focus of several modelling efforts over the years with the result that the thermal regime in the ingot can be simulated quite realistically. Such models provide important insight into solidification of the ingot but present some significant challenges to the casual user such as a process engineer. To provide the process engineer with a tool to assist in the development of a melt practice, a comprehensive model of the complete VAR process has been developed. A radiation heat transfer simulation of the arc has been combined with electrode and ingot models to develop a platform which accepts typical operating variables (voltage, current, and gap) together with process parameters (electrode size, crucible size, orientation, water flow, etc.) as input data. The output consists of heat flow distributions and solidification parameters in the form of text, comma-separated value, and visual toolkit files. The resulting model has been used to examine the relationship between the assumed energy distribution in the arc and the actual energy flux which arrives at the ingot top surface. Utilizing heat balance information generated by the model, the effects of electrode-crucible orientation and arc gap have been explored with regard to the formation of ingot segregation defects.

## 1. Introduction

The development of a VAR melt practice for a new alloy or different sized ingot often utilizes a prior practice for a similar alloy or similar sized ingot as a starting point. Following multiple melt trials and evaluations, a successful practice can usually be found although it may not necessarily be the most optimum practice. The cost of performing multiple optimization trials combined with the urgency of qualifying a successful practice to permit commercialization of the product often preclude an exhaustive investigation of the many possible practices. Moreover, altering a qualified melt practice can be exceedingly costly. To this end, the use of mathematical models to provide additional insight into the solidification conditions during casting can assist tremendously in arriving at an optimum melt practice without an excessive number of trials.

Such models have evolved significantly over the years as improvements in computing power have permitted the use of more advanced numerical algorithms to simulate increasingly detailed physical phenomena. Due to the sophistication of these models, their use generally depends on individuals who have a significant modelling background as opposed to a casual user such as a process engineer. The models also require thermophysical property data and boundary condition data which may not be readily available to a process engineer; and can consume significant computing resources especially for the more sophisticated models. Finally, the models may not accept input data in the form of operating parameters as would be specified by a process engineer. Consequently, from a process engineer's point of view, there is a need which would be filled by a model of the complete VAR process.

## 2. Model Development

The VAR process model (Figure 1) combines three separate models for the electrode, the arc, and the ingot into an integrated model which accepts input parameters that are specified as operating variables – voltage, either current or melt rate, and gap. The electrode model consists of a one-dimensional conduction model to calculate the axial temperature gradient in the electrode. From the axial temperature profile, heat conduction into the electrode and heat losses from the lateral electrode surface can be calculated and incorporated into the arc model.

The arc model is key to linking the electrode and ingot models and is based on an electrical analog developed by Ballantyne[1] for radiative heat exchange between the surfaces which bound the arc – the electrode bottom surface, the ingot top surface, and the crown/crucible wall. The ingot top surface is divided into multiple concentric rings, each representing an isothermal region of the ingot surface, with each in turn constituting a node in the electrical network. The arc itself is comprised of two components,

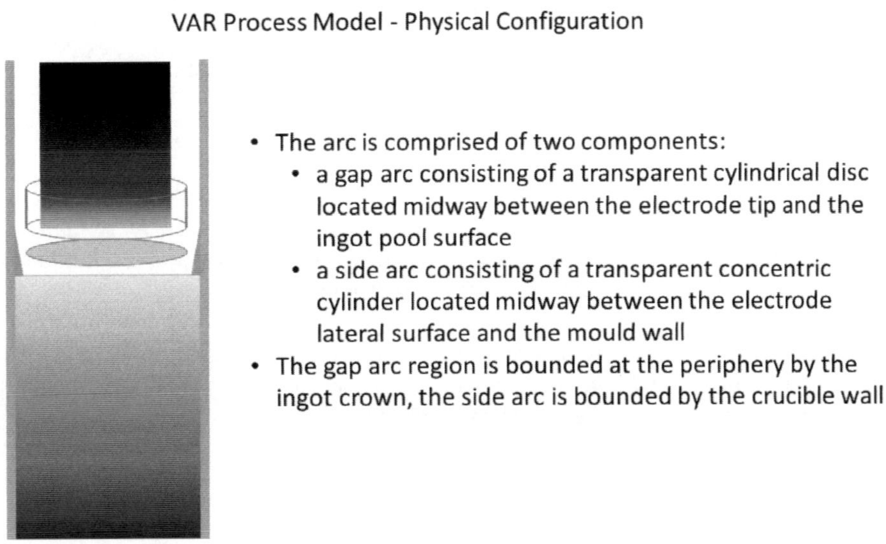


**Figure 1.** Physical Configuration of the VAR Process Model

the first being a circular disc between the electrode tip and the top of the ingot (the gap arc) and the second being a concentric hollow cylinder in the annulus between the lateral electrode wall and the crucible wall (the side arc). The previously described model[1] has been extended to incorporate a gap arc which is divided into multiple concentric rings with each ring constituting a node in the electrical network. The network for this modified model is shown in Figure 2. As before, the side arc constitutes a single node in the network. These arc nodes do not represent physical surfaces but instead represent "heat source nodes" into which "current" is introduced with the amount of current being proportional to the distributed heat input to the system. The input heat distribution in the gap arc may be specified by defining the heat flux distribution as uniform, Gaussian, or linearly increasing from edge to center of the arc. The side arc heat input is defined as a percentage of the total heat input to the system. The former model has also been modified based on the assumption that radiation from the gap arc does not impinge directly on the crucible wall, but rather on the crown that forms above the ingot. Radiation from the side arc impinges on the lateral electrode wall and crucible wall above the crown. The interior surface temperatures of the crown and crucible are estimated iteratively from the radiative heat flux and the thermal resistance from the crown to the crucible wall to the cooling water with the inclusion of helium cooling, forced convection and boiling water heat transfer.

Geometric parameters relating electrode size and weight, arc gap, and crucible dimensions are used to calculate electrode pipe, final ingot size and weight, and the view factors which determine the radiative exchange between the appropriate surfaces. Emissivities for the various surfaces involved in the radiation exchange must be specified but, once defined, should vary very little from case to case. In general, the emissivities of the electrode, ingot pool, and crown are set equal to each other since all have formed in a vacuum.

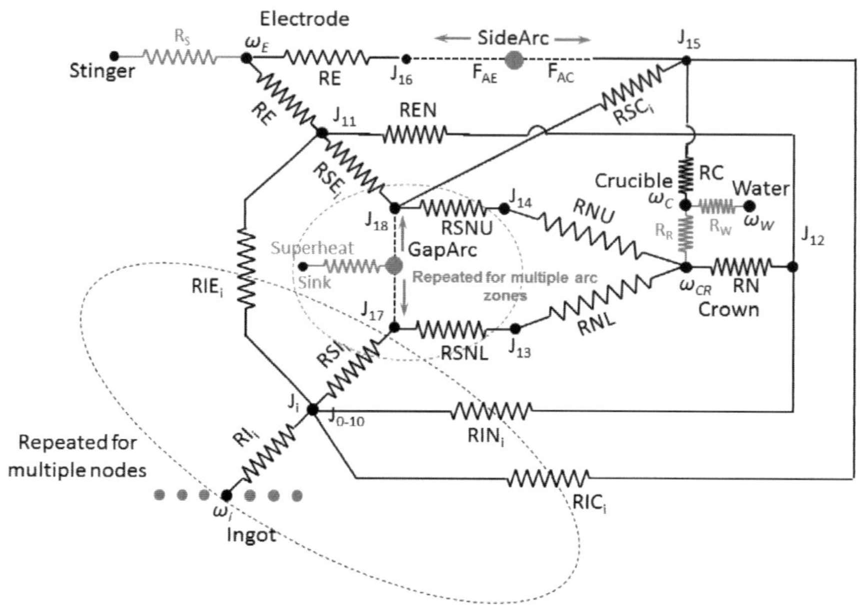


**Figure 2.** Modified network analysis incorporating multiple arc and ingot nodes with the electrode, crown and crucible wall

The application of Kirchoff's law at each node in the circuit leads to a system of simultaneous equations that may be solved for the temperatures in the radiative network. The elements in red (Figure 2) constitute non-radiative terms in the electrical network and comprise heat losses from the electrode side wall, the heat transfer to the droplets as they fall through the gap arc, heat transfer from the crown to the crucible and heat transfer from the crucible wall to the cooling water. The assumption is made that the droplets leave the electrode tip at the liquidus temperature and are superheated to a specified temperature as they pass through the arc.

The ingot model is based on a two-dimensional finite difference model which incorporates temperature-dependent thermophysical properties and boundary conditions. At a time increment commensurate with the melt rate, the number of axial nodes in the ingot is increased and multiple iterations are performed to update the ingot thermal regime. Prior to each iteration, the electrode and arc models are solved to calculate updated heat flows to the top surface of the ingot.

## 3. Model Execution

The model utilizes a Windows-type interface which consists of four panels (Figure 3) for controlling execution, editing input data, reviewing thermophysical properties, and displaying execution results. Results are written in tabular form to an ASCII (TXT) file in addition to being written to comma-separated value (CSV) and visual toolkit (VTK) files for ease of use in further analysis and plotting. Execution times are very fast, typically being 10-15 seconds.

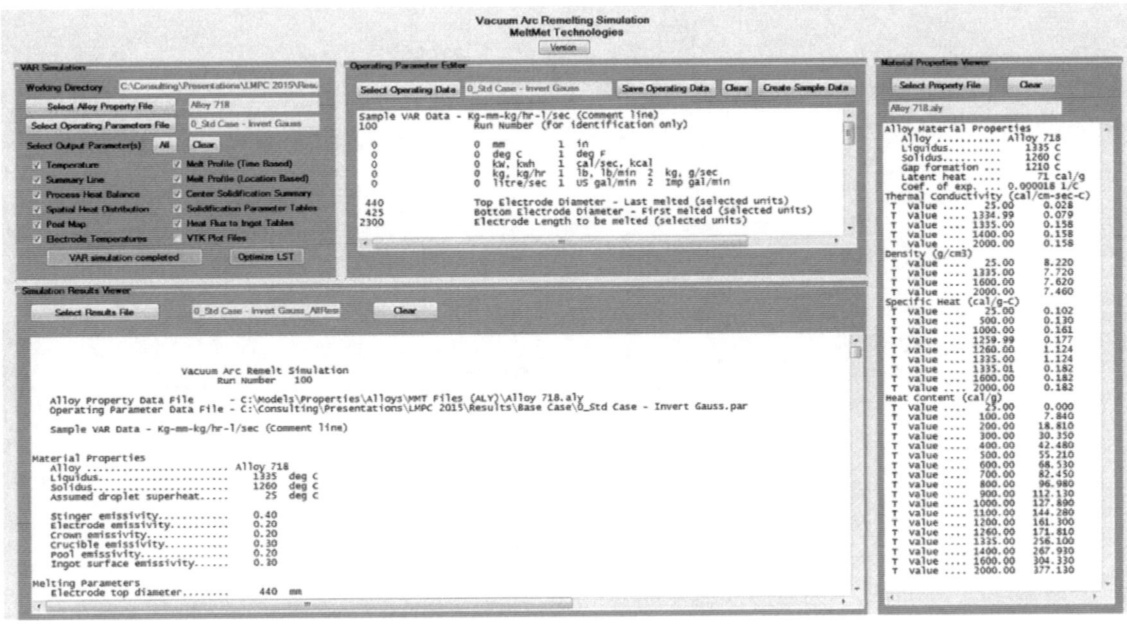


**Figure 3.** User interface for VAR process model

Input parameters to the model include electrode weight, and top and bottom diameters and length for both the electrode and crucible. From these data, the melt configuration (upright, invert, etc.), percent electrode pipe, and the final ingot weight and height are calculated. Depending on the melt configuration, the changing fill ratio at the gap arc interface will affect the view factor calculations in the arc radiation model.

Input parameters require specification of a melt profile either as voltage-current-gap, or voltage-melt rate-gap as a function of either % of the final ingot length or as a function of the remaining electrode weight. Depending on whether the current or melt rate is specified, the model will utilize an iterative procedure to determine the missing factor – melt rate or current.

The evaluation of the "current flows" in the electrical circuit analog permits calculation of the various heat flows in the VAR process from which an overall heat balance and specific power consumption can be calculated.

## 4. Effect of Input Heat Distribution

The input heat distribution is specified as an edge to center heat flux distribution. The choices are uniform, Gaussian, or linearly increasing. Previous modelling studies[2] have made the assumption that the heat flux distribution to the ingot top surface is Gaussian in nature based on the visual observation of an increasing light intensity toward the center of the arc. However an equivalent explanation may be that the increasing light intensity is due to an increasing transverse optical path length through an arc of uniform energy flux distribution. Regardless of the explanation, it is interesting to examine the heat distribution to the ingot which results from uniform and Gaussian heat flux distributions in the arc.

Figures 4 and 5 show two input flux distributions for the same total power input. A Gaussian flux distribution in the arc (Figure 4) is reduced to a much lower, more uniform flux at the ingot surface due to the multiple reflections between the electrode and the ingot surfaces. A uniform flux in the arc (Figure 5) is reduced to a much lower, but still uniform flux at the ingot surface. A comparison of these two ingot fluxes (Figure 6) themselves reveals that the Gaussian flux retains the Gaussian shape but the maximum to minimum ratio at the ingot surface is much lower than it is in the arc distribution.

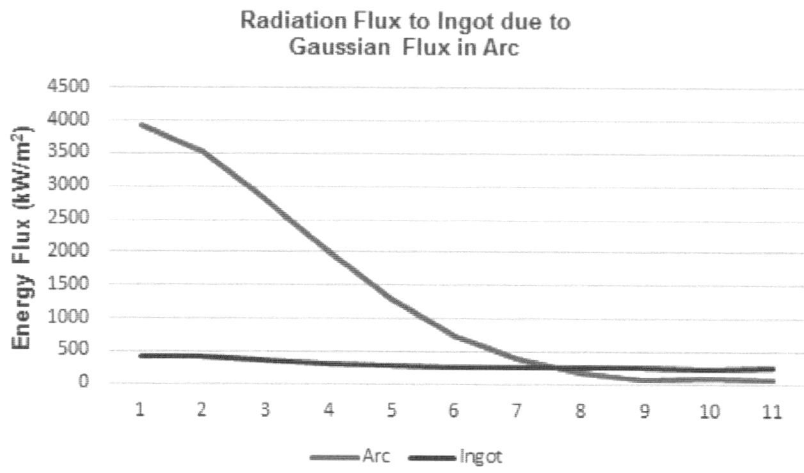


**Figure 4.** Energy flux to the ingot based on a Gaussian flux in arc

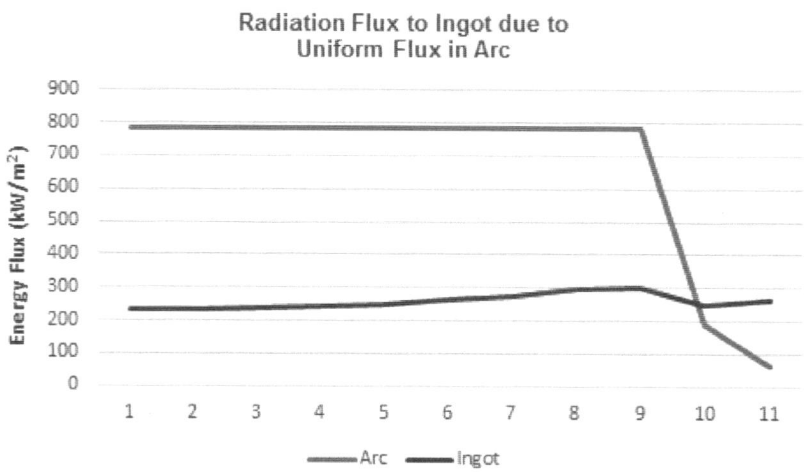


**Figure 5.** Energy flux to the ingot based on a uniform flux in arc

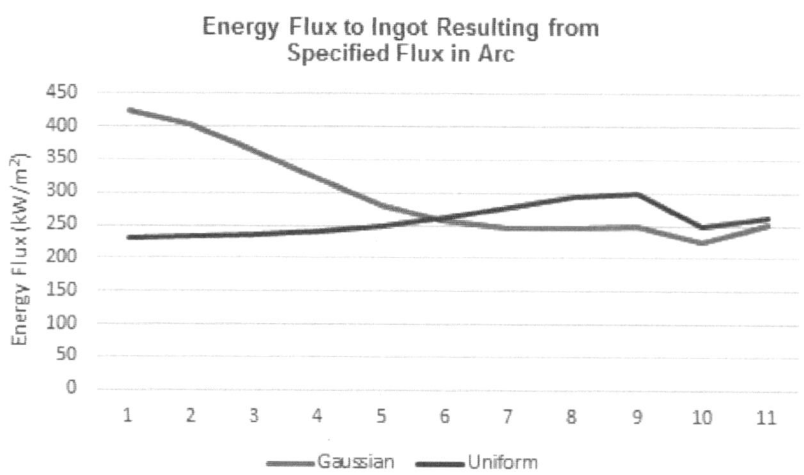


**Figure 6.** Energy fluxes to the ingot based on Gaussian and uniform fluxes in the arc

## 5. Effect of Changing Fill Ratio

One of the outcomes of the process model is the calculation of an overall heat balance of the process at any specific instant, which may then be used to compare different melting scenarios. A typical heat balance in graphical form (Figure 7) at a given instant displays the energy input in the gap and side arcs, the various energy losses, and the internal energy exchanges between various components. A comparison of heat balances for two different melting conditions can yield insight into the effect on the process behaviour.

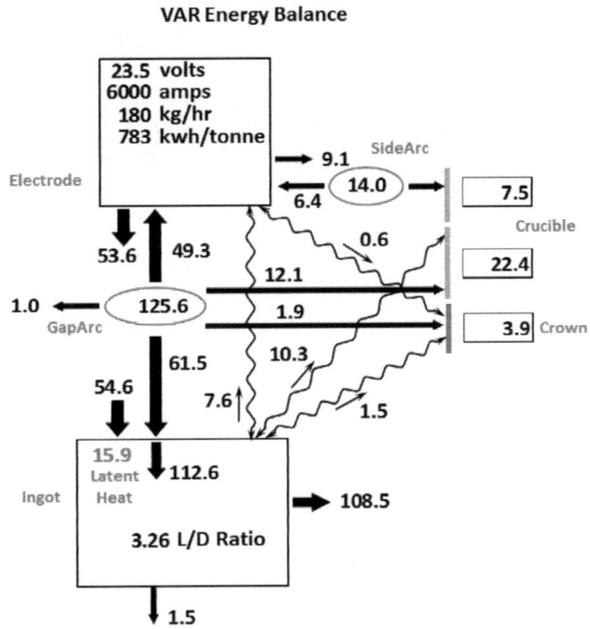


**Figure 7.** Instantaneous energy balance for the VAR process

One such comparison involves the effect of a changing annulus as would arise due to a tapered electrode melting into a tapered crucible.

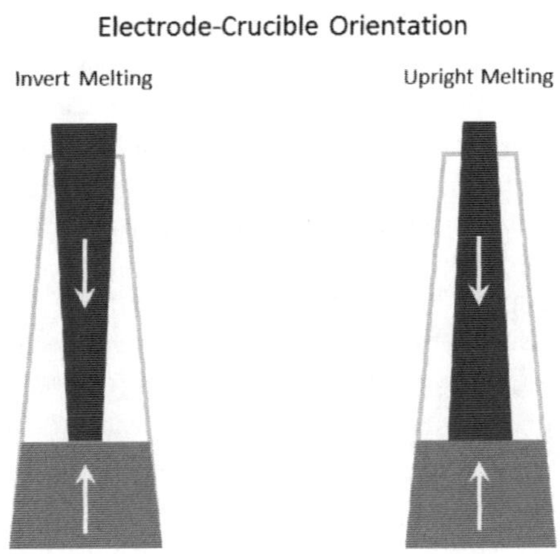


**Figure 8.** Invert and upright melting orientations

Melt geometries for invert and upright melts of a typical 425-440 mm diameter electrode into a 497-515 mm diameter crucible are summarized in Figures 9 and 10. In the case of the invert melt, regardless of the tapers, the annulus always decreases as the ingot height increases. For the upright melt, the annulus may increase or decrease as the ingot is formed, depending on the tapers.

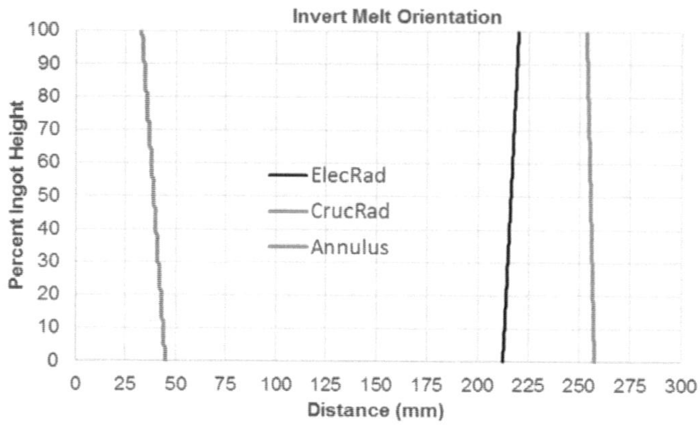


**Figure 9.** Changing melt geometry at the melt interface for an invert melt

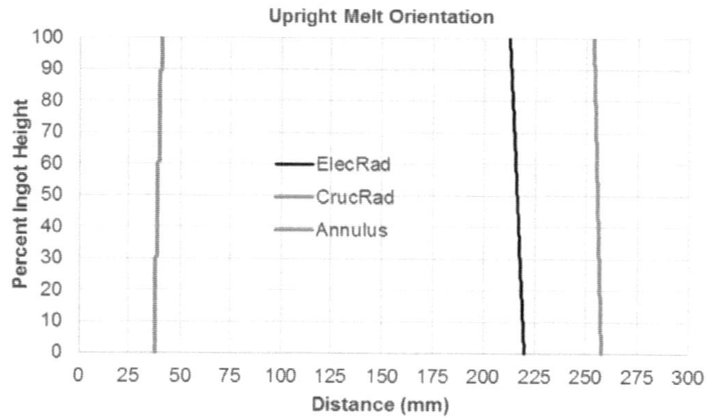


**Figure 10.** Changing melt geometry at the melt interface for an upright melt orientation

Regardless of which configuration is used, the result is most likely to be an annulus which varies throughout the melting operation. The consequences of this changing annulus can be predicted from an examination of the changing heat balance of the VAR process throughout the melt. For the case of a melt with a decreasing annulus at a constant melt rate, a comparison of the heat balance at a 20% ingot length with that at 100% ingot length (Figure 11) shows an increase in melting efficiency (reduced specific power consumption) and a declining input power (current) requirement due to the declining fill ratio.

In contrast, a melt with an increasing annulus at the same melt rate results in a declining melt efficiency (increased specific power consumption) and an ever increasing power (current) requirement (Figure 12). An examination of the ingot top heat flux distribution throughout each melt (Figures 13 and 14) reveals that, in the case of the decreasing annulus, the peak energy flux moves outward from the central region of the ingot in the early stages of melting toward the ingot surface as melting proceeds. In contrast, for the increasing annulus case, the peak energy flux moves inward toward the center of the ingot with the outer periphery receiving a reduced energy input as melting progresses.

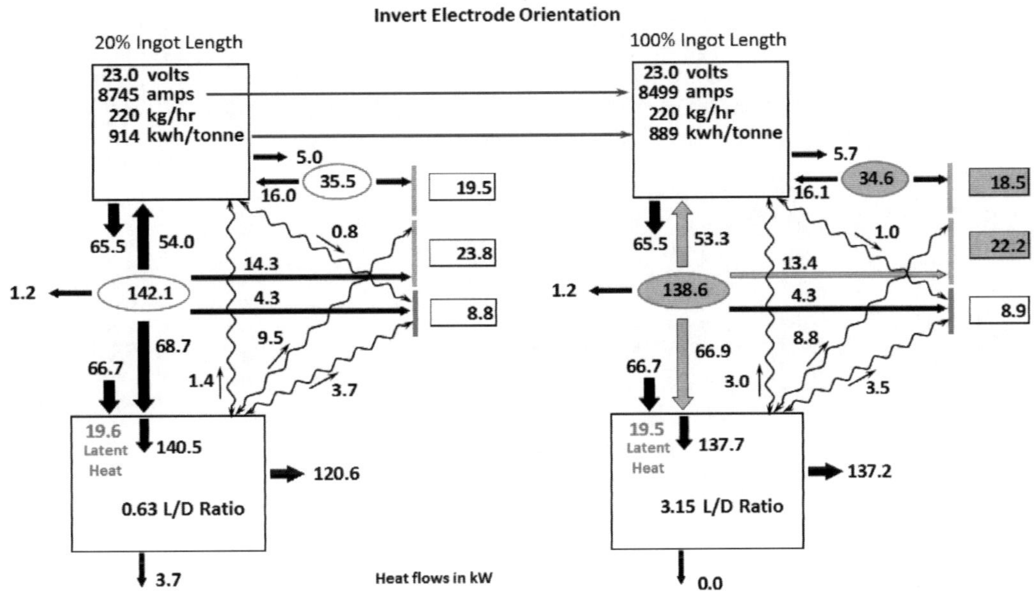


**Figure 11.** Energy balance at 20% and 100% ingot lengths for an invert melt orientation

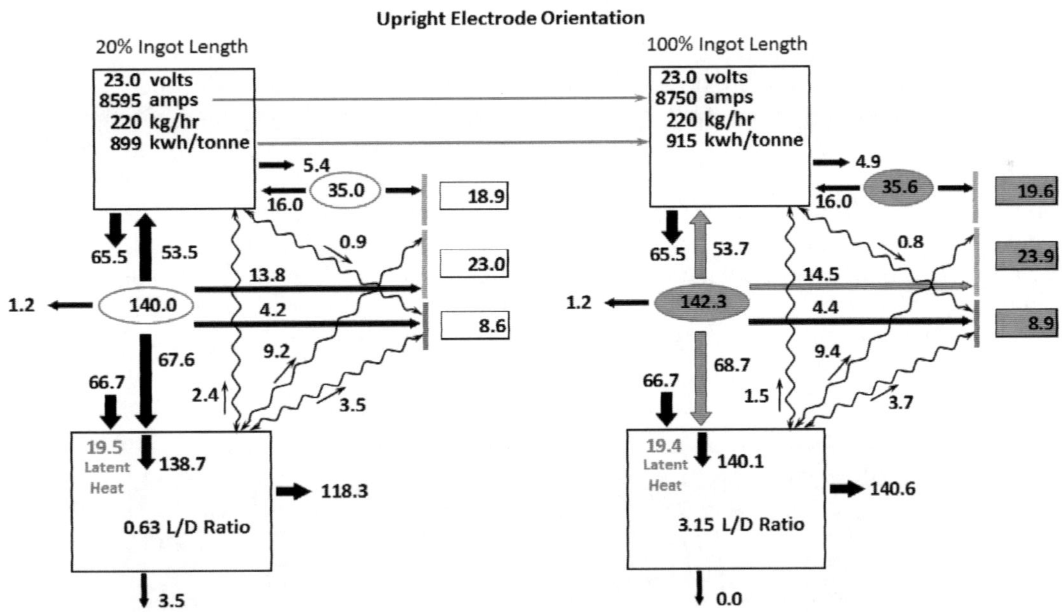


**Figure 12.** Energy balance at 20% and 100% ingot lengths for an upright melt orientation

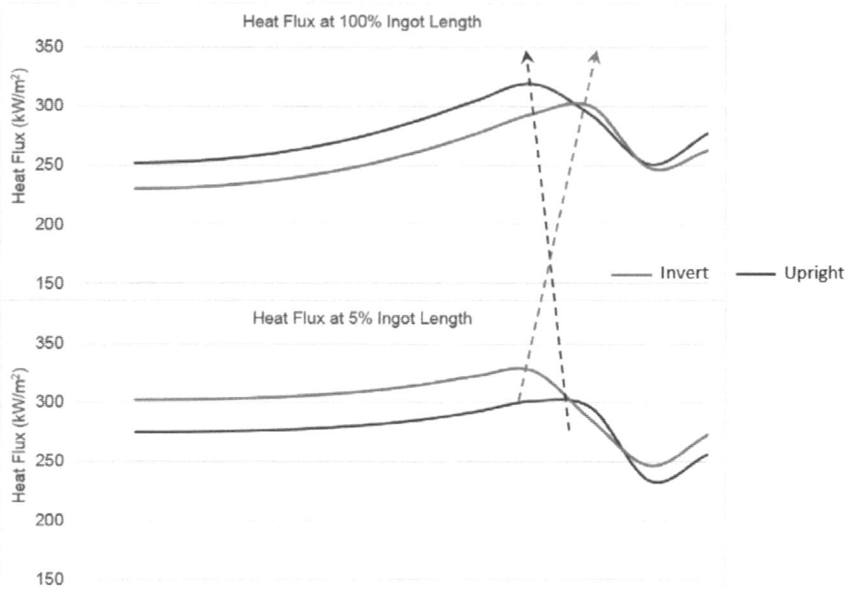


**Figure 13.** Heat flux to ingot top at 5% and 100% ingot lengths for invert and upright melting

Of the two situations, the decreasing annulus case must be considered to be the more robust approach. The declining input power requirement and increasing annular heat flux are both beneficial with regard to reducing the tendencies for positive segregation in the central region and white spot formation in the peripheral region of the ingot. In contrast, the increasing annulus case with its increasing current requirement and more centrally concentrated energy input as melting progresses leads to an increased possibility of segregation defects. While an invert melt practice will always yield a decreasing annulus, the specific electrode and crucible dimensions in an upright practice should be considered carefully in order to avoid an excessive increase in annulus throughout the melt. The evolving trends of a concentrated central heat input and an increasing melt current requirement increase the risk of Lorentz driven flow in the pool thereby raising the tendency toward freckle formation.

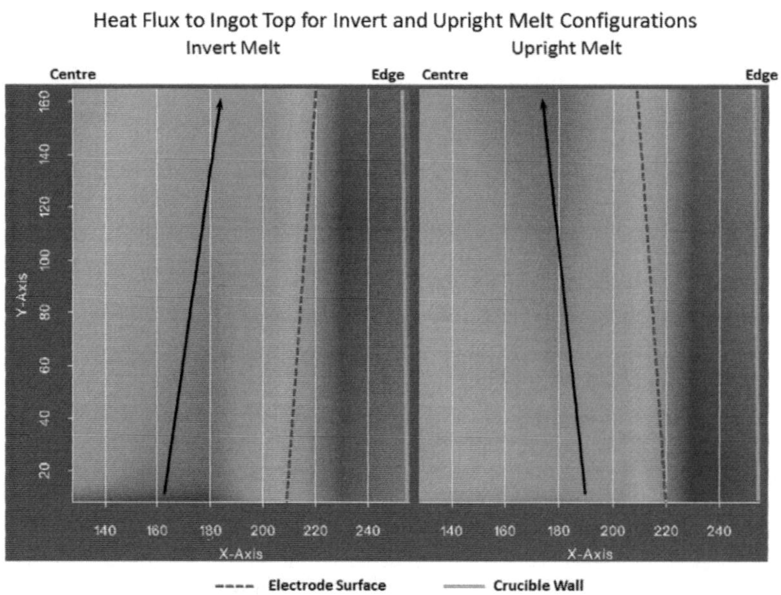


**Figure 14.** Top ingot heat flux in the outer ingot region for invert and upright melt orientations

## 6. Effect of Gap Changes During Melting

Successful melting of segregation prone alloys requires good control of the arc gap between the melting electrode tip and the ingot pool surface. An examination of the heat balance at different electrode gaps illustrates what happens when changes in the gap occur. As seen in Figure 15, under melt rate control, an increase in the arc gap leads to an increased heat loss to the crown/crucible wall which tends toward a drop in melt rate. In response to this, the melt rate controller will increase the current thereby increasing the power input to try to maintain the melt rate set point. This rise in current increases the likelihood of Lorentz driven flow in the liquid pool with the attendant risk of positive segregation in the ingot.

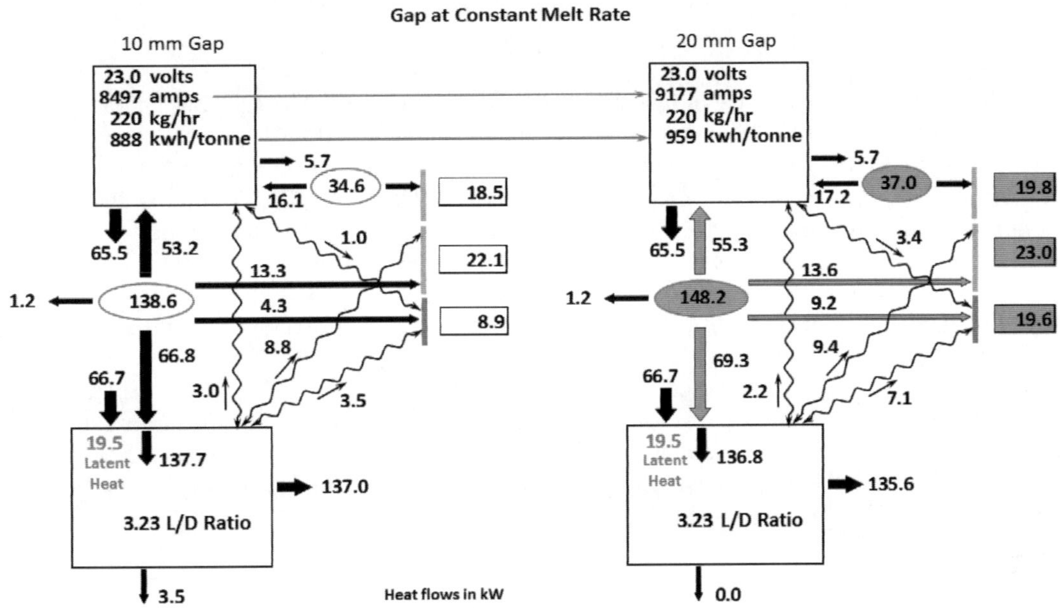


**Figure 15.** Heat balance changes as gap arc widens from 10 mm to 20 mm under melt rate control

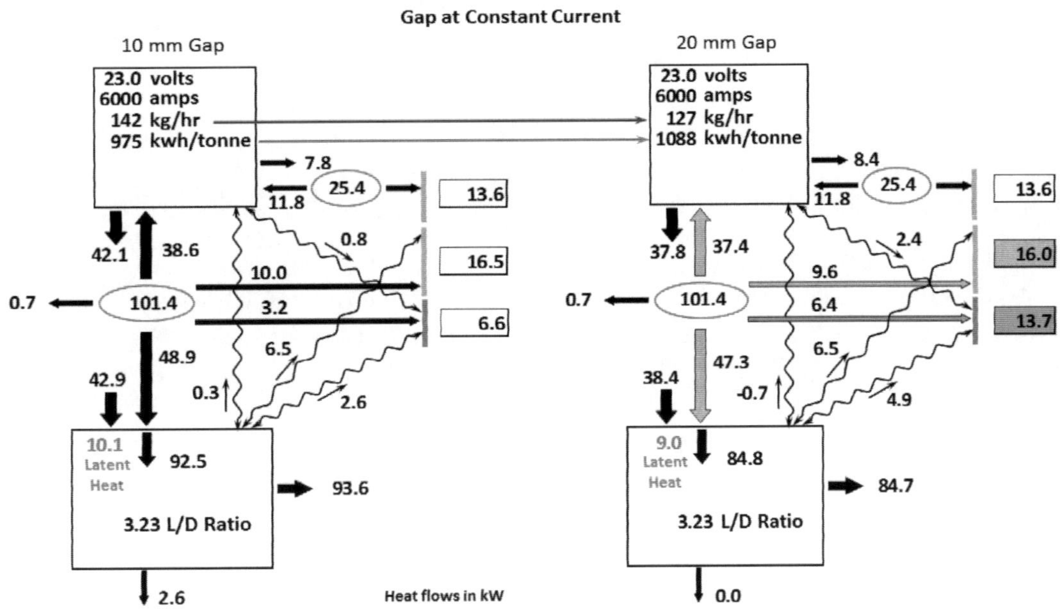


**Figure 16.** Heat balance changes as gap arc widens from 10 mm to 20 mm under current control

The effects of a gap change when melting under constant current are illustrated in Figure 16. Once again, as the gap widens, the heat loss to the crown increases. In this case however, the controller makes no effort to alter the melt current. Assuming the furnace voltage remains constant, the power input to the furnace remains the same but is redistributed. The increased heat loss to the crown leads to a lower heat input to the electrode and ingot which in turn leads to a reduction in the observed melt rate and an increase in the specific power consumption. Thus an increase in gap for both melt rate controlled and current controlled processes results in a drop in efficiency as measured by an increase in specific power consumption. However, at constant melt rate, the increase in gap will move the process toward conditions more conducive to the formation of positive segregation while a gap increase at constant current will move the process toward the formation of white spots.

## 7. Summary

The development of an electrical analog to simulate heat transfer in the VAR arc has enabled the linking of electrode and ingot models leading to development of an integrated VAR process model. The model incorporates the necessary boundary conditions so that a process engineer need only specify operating conditions and alloy properties in order to simulate the VAR process. Simulation results presented as text, comma-separated value, and visual plot files include melt profile conditions, heat flux distributions, solidification conditions, and heat balance results. The model also includes an optimization function which executes multiple simulations over a range of melt rates to show the variation in pool depth, mushy zone depth, and local solidification times with melt rate thereby assisting in the selection of an optimum melt rate.

Model generated results can be used to prepare heat balances of the VAR process which can provide insight into the behavior of the process under different melting conditions. Such insights can provide an indication of whether the process is more or less sensitive to segregation defects as a result of process changes, whether these changes occur from practice to practice or from beginning to end of a specific practice.

## References

[1] Ballantyne, A.S., Proc. of the 2013 Int. Symp. on Liquid Metal Processing and Casting, TMS, 2013.

[2] Bertram, L.A., Adasczik, C.B., et al, Proc. Int. Symp. On Liquid Metal Processing and Castings, 1997, p. 110-132.

[3] Sucec, J., "Heat Transfer", Simon & Schuster, 1975.

IOP Conf. Series: Materials Science and Engineering **143** (2016) 012016 doi:10.1088/1757-899X/143/1/012016

# Examination of an alternative method for the pyrometallurgical production of copper-chromium alloys

**J. Brenk, S. Hassan-Pour, P. Spiess, B. Friedrich**

IME – Metallurgische Prozesstechnik und Metallrecycling, RWTH Aachen, Germany

jbrenk@ime-aachen.de

**Abstract.** In this paper an alternative to the usual route for the production of copper-chromium alloys by Hot Isostatic Pressing (HIP) followed by Vacuum Arc Remelting (VAR) is investigated. Therefore the HIP is replaced by an aluminothermic reduction. As oxidizing agents for this aluminothermic reduction (ATR) chromium oxide ($Cr_2O_3$) and copper oxide (CuO) are used. These oxidants are mixed in a stoichiometric relation with aluminium (Al) to get the following aluminothermic reactions going:

$$Cr_2O_3 + 2\,Al \leftrightarrow 2\,Cr + Al_2O_3 \tag{1}$$
$$3\,CuO + 2\,Al \leftrightarrow 3\,Cu + Al_2O_3 \tag{2}$$

As ATRs always are exothermic reductions, it is important to control the heat output of the reaction. A common simplification for the heat calculation in the field of ATR is the "Shemshushny Factor" (Sh-Factor). This factor determines the rate of energy input per reactant mass:

$$Sh = \frac{\sum H_r}{\sum m}$$

To avoid a secondary melt phase after ATR, inline casting is used to directly obtain vacuum arc remeltable electrodes out of the liquid melt of the ATR.

The second part of this work deals with the remelting of these electrodes via VAR. The first VAR trials aim at finding process parameters for remelting a copper-chromium ingot. As demixing is to avoid, it is necessary to control process parameters within a small range to obtain a shallow melt pool.

## 1. Introduction

Copper-chromium (CuCr) alloys play a very important role in the field of medium voltage switch contacts. The combination of high thermal and electrical conductivity with high strength and a high arc resistance make it the adequate material for these switches. Compared to pure copper the chromium added to the copper matrix increases the resistance against contact breaking sparks. As these sparks are unavoidable and carry an enormous amount of energy at voltages of 1 kV to 145 kV, pure copper would melt which could result in welding the switch contacts together in the worst case. [1]

The challenge coming with CuCr alloys is that there is a miscibility gap in the system of CuCr. [2][3][4] Therefore the process of producing CuCr alloys consists of two complex process steps: consolidation of an electrode and refining of that electrode. The main target of the consolidation step is to achieve a uniform distribution of Cr in the Cu-matrix. This is mostly achieved by hot isostatic pressing. The following remelting step is necessary to receive a homogenous microstructure and to control material-

and solidification defects. In this work the consolidation step of hot isostatic pressing is replaced by an aluminothermic reduction followed by vacuum arc remelting (VAR). The objective is to find a technical usable alloying composition that is feasible regarding its final aluminium content.

## 2. Fundamentals

As mentioned above the production of CuCr alloys will be carried out via an aluminothermic reduction followed by a refining step in the VAR.

### *2.1. Aluminothermic reduction*

The aluminothermic reduction is defined as a reaction between a metal oxide of the target metal(s) and metallic aluminium as reducing agent. During this reaction a metal phase consisting of the target metal(s) and a slag are formed. This slag mainly consists of $Al_2O_3$ but also other slag forming components such as CaO can be added to the system. The principle behind the aluminothermic reduction is the different affinity for oxygen of the target metal(s) compared to aluminium. If this difference is big enough the aluminothermic reduction is distinguished by a strong exothermic behaviour and very high reaction rates. [5]
For the present investigation the two target metals are added as oxides and react as following:

$$Cr_2O_3 + 2\,Al \leftrightarrow 2\,Cr + Al_2O_3 \qquad -\Delta H_{298} = 274 \text{ kJ} \tag{1}$$

$$3\,CuO + 2\,Al \leftrightarrow 3\,Cu + Al_2O_3 \qquad -\Delta H_{298} = 604 \text{ kJ} \tag{2}$$

For all aluminothermic reductions the energy density of the system has to be adjusted properly in order to get a self-propagating reaction front going on the one hand and to not have a dangerously high energy in the system on the other hand. An estimation for the energy density of the system is the so called "Shemshushny Factor" (Sh-Factor). This factor is defined as the amount of energy brought into the system as reaction enthalpy divided by the total mass of the system: [6]

$$Sh = \frac{\sum H_r}{\sum m}$$

Usually to keep an aluminothermic reduction going so called boosters need to be added to the system to raise the energy level of the whole system. Boosters are substances that react with the reducing agent, aluminium, very exothermically and by that add a lot of energy to the system. The products of the reaction between booster and aluminium have to go into the slag or even evaporate to not contaminate the target metal/alloy. But as seen in equation (2) the reaction between CuO and Al is very exothermic. This means that for the system CuCr no boosters are needed. The energy level is even so high that some kind of cooling agent needs to be used to lower the Sh-Factor of the system. In the trials described in this paper lime (CaO) was used to lower the Sh-Factor. Lime on the one hand consumes energy as it is heated up and molten throughout the process and on the other hand helps forming the slag.

#### *2.1.1. Inline casting.*

As the material produced via ATR has to be remelted in the VAR, it needs to have the typical cylindrical shape of VAR electrodes. Usually the metal produced via ATR solidifies in the ATR reactor and therefore has a round, disk like shape. To obtain the cylindrical electrode shape the material would have to be remelted, for example in a vacuum induction furnace (VIM), and cast into a cylindrical mould.
The approach of this work is the so called inline casting. This means that the electrodes are cast directly out of the ATR process. A cylindrical mould is attached to the bottom of the ATR reactor to obtain the desired shape of the electrode. The connection between reactor and mould is closed with a small metal plate which melts as soon as the reaction reaches the bottom of the reactor. In this way the molten metal can flow into the mould but not slag or unreacted material.

*2.2 Vacuum arc remelting*

The vacuum arc remelting is a process where a consumable electrode is being remelted into a water cooled copper mould under a vacuum (~$10^{-3}$ mbar). In general at VAR the energy input is inserted into the metal by an electric arc induced by a direct current between the electrode and the mould or later the ingot. As seen on figure 1 this leads to a dripping of the metal from the electrode to a melt pool inside of the mould.

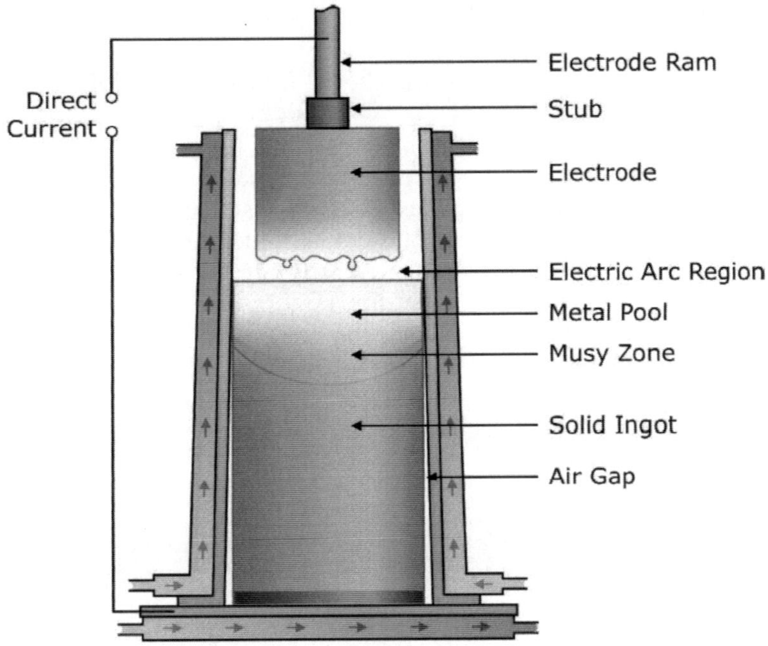


Figure 1: Schematic procedure of vacuum arc remelting

By that it is possible to refine the metal in several ways. First it is possible to remove dissolved gases and volatile elements very easily. This is possible because of the low furnace pressure combined with a very large phase boundary between metal and vacuum. The size of the phase boundary is that big because all the metal from the consumable electrode is dripping to the mould as very small droplets. For this work it means that it is possible to get rid of the unwanted aluminium content up to a certain limit. Second it is possible to get rid of non-metallic inclusions (NMIs) as they will be transferred to the mould walls by a combination of convection and electromagnetic forces caused by the electric arc. So if there are remaining slag particles from the ATR-process they will not survive VAR and therefore not be in the final metal block.

Third there will be a certain solidification structure of the ingot that can be controlled by melt power and cooling rate. [7][8]

## 3. Experimental Procedure

The experimental procedure of this work basically consists of three parts:

1. The ATR with the focus on finding a mixture that is of technical relevance and which has a low aluminium content.

2. The inline casting to obtain an electrode usable in the VAR.

3. The VAR to remove NMIs and to homogenize the ingot.

### *3.1. Aluminothermic reduction*

The aim of the investigations regarding ATR is to find a mixture of copper and chromium that is of technical relevance and that is producible via ATR with reasonable losses and with an aluminium content as low as possible. So firstly there are calculations with a thermodynamic database necessary to select promising mixtures of Cu and Cr. The most promising mixtures then are tried out in first small trials in a 16 kg scale. This size, although seeming quite big, has been chosen because slag separation for smaller ATRs is not satisfactory.

As input mixture oxides of the target metals, CuO and $Cr_2O_3$, as well as aluminium as reducing agent are used. Because the Sh-factor for the studied mixtures for a target of more than 50 wt.-% Cu in the metal phase is very high, lime is added to lower the Sh-factor. The most promising mixtures of the thermodynamic calculations are Cu25Cr and Cu15Cr. Besides of varying the chromium content in the target metal the stoichiometry is also varied. Since one of the targets of this work is to get an aluminium content as low as possible, it is tried to let all the aluminium react and get into the slag phase as $Al_2O_3$. As it is nearly impossible to mix a system good enough to guarantee 100 % reaction, the approach of this work is to add more oxides of the target metals than needed for a stoichiometric reaction. In this way a slight loss of CuO or $Cr_2O_3$ is taken into account to lower the aluminium content of the material. As the affinity of Cr to O is much higher than the affinity of Cu to O it can be assumed that $Cr_2O_3$ needs to be added hyperstoichiometric to reach the target composition in the target metal. The relation of Cu to Cr in the target metal by then is controlled via the Al content in the input mixture, assuming nearly all of the Al to react as there are more reaction partners (excess $Cr_2O_3$) than needed.

Altogether the following mixtures have been tried out:

Table 1: Input mixtures for ATR

| Cu content /wt.-% | Cr content /wt.-% | Stoichiometry $Cr_2O_3$ |
|---|---|---|
| 85 | 15 | stoichiometric |
| 85 | 15 | hyperstoichiometric |
| 75 | 25 | hyperstoichiometric |

The mixture with the best results regarding Al content, accuracy of the target metal and metal yield, Cu15Cr hyperstoichiometric, is scaled up to a larger scale of 115 kg input material. This is the size needed to produce VAR usable electrodes via inline casting later on.

### *3.2. Inline casting*

With the scale up being successful the mould for inline casting is attached to the same reactor used at the scale up. Different types of moulds have been taken into account but finally a mould using the principle of uphill casting is used for casting the electrode. The final design of the reactor with attached mould can be seen on figure 2. On the left hand side of this figure the connection between reactor and mould can be seen.

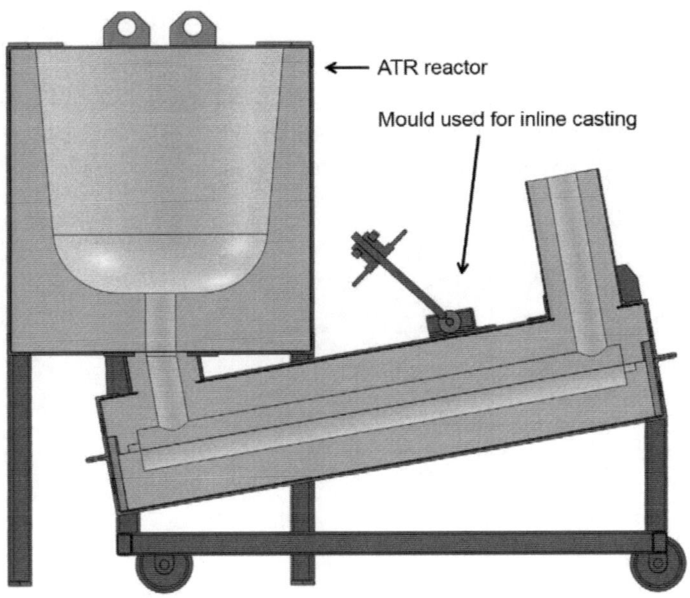


Figure 2: Schematic of the ATR reactor with attached mould

At the bottom of the reactor a small aluminium plate seals the connection to the mould to avoid the mixture of input material falling into the mould before the start of the reaction. After ignition the reaction front migrates through the mixture towards the bottom of the reactor. As soon as the reaction front reaches the aluminium plate the plate is melted by the energy of the reaction and the liquid material flows into the mould. On the other, higher side of the mould there is a hole to let reaction gases and also air stream out of the mould making way for the liquid metal.

### *3.3. Vacuum arc remelting*

To refine the material the electrode produced via ATR and inline casting is remelted in a VAR. The electrode is attached to a stub as can be seen on figure 3 and then inserted into the VAR. The electrode without stub has a length of 520 mm, a diameter of 70 mm and a weight of 19.8 kg. The diameter of the mould is 120 mm.

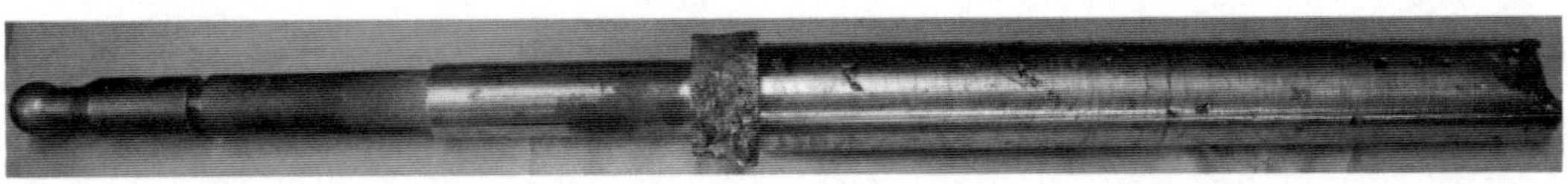

Figure 3: Electrode with attached stub prepared for the VAR process

As there is no history of remelting Cu alloys, especially CuCr, at the IME, process parameters for the VAR such as melt rate/current and drip shorts/voltage can only be assumed and need to be adjusted and enhanced throughout the first remelting.

As copper has a very high thermal conductivity the remelting of CuCr is not as linear as of a standard VAR alloy such as Ni-based superalloys or Ti-based alloys. It takes much longer for the electrode to start moving as not only the bottom of the electrode is heated up but due to the high thermal conductivity the whole electrode is heated up before the electrode tip is hot enough to start melting.

## 4. Results

### *4.1. Aluminothermic reduction*

The results of the preliminary trials demonstrate that higher Cu contents in the target phase and therefore in the system are leading to lower Al concentrations in the metal. Also a hyperstoichiometric addition of $Cr_2O_3$ decreases the Al content in the target metal. This can both be seen on figure 4.

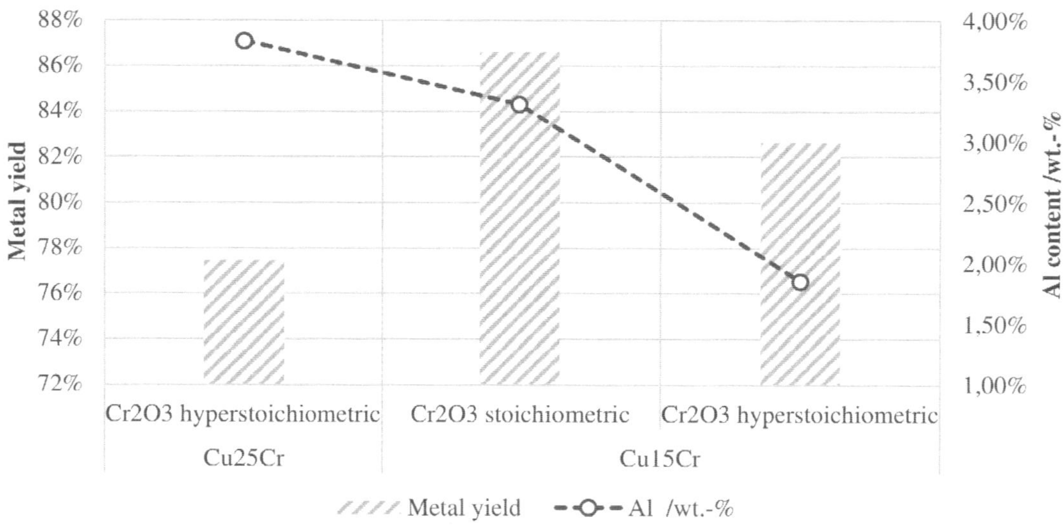


Figure 4: Al content and metal yield of preliminary trials

The metal yield of the trials with the higher Cu content (Cu15Cr) is also significantly higher. This can be explained by the lower affinity of copper to oxygen and therefore a more complete reduction of the target metal. It can also be seen that in order to get rid of aluminium in the target metal a certain metal loss is unavoidable as the trials with the lowest Al content do not have the best, but still a satisfactory, metal yield.

As the most promising mixture (Cu15Cr, hyperstoichiometric) is scaled up to 115 kg the Al content gets down to an even lower level. The final Al content of the electrode obtained by inline casting is 0.35 wt.-%, so it should be possible to get rid of nearly all of this during remelting in the VAR. The metal yield of the scale up trial also is higher than that of the preliminary trials.

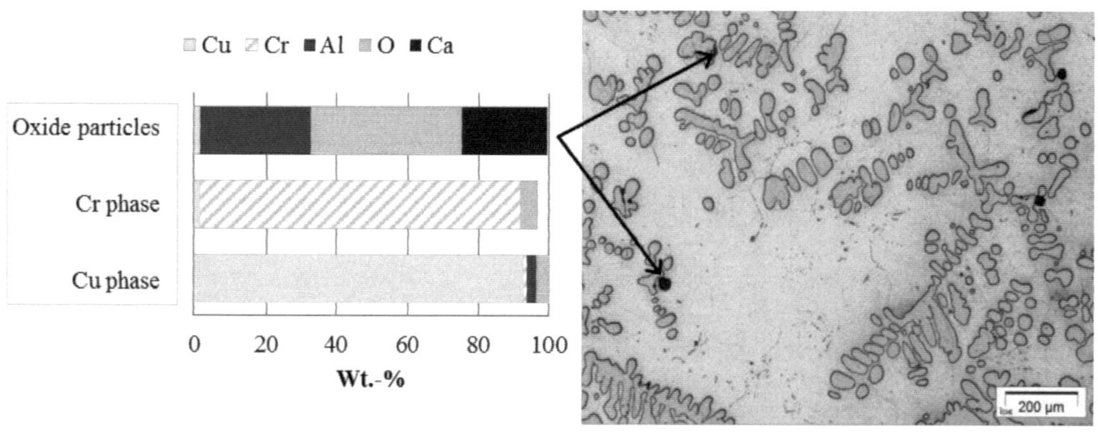


Figure 5: Micro of the microstructure of the scale up Cu15Cr trial with EDS analysis

Figure 5 shows that the Cr in the obtained metal block forms dendrites in the Cu matrix. So Cu and Cr are neither alloyed to one homogeneous alloy nor is the Cr spread in form of small randomly shaped droplets.
On figure 5 there are also black spots visible. EDS phase analysis shows that these spots are oxide particles that did not settle to the slag phase. These oxide particles mainly consist of Al, O and Ca. This reflects the slag composition of CaO and $Al_2O_3$. The oxide particles are always very close to the Cr dendrites. One explanation is that the solid Cr dendrites form before the slag particles can rise to the top of the melt and hold them down.

### *4.2. Inline casting*

As to be seen on figure 6 after casting the mould did not fill completely but therefore the connection between mould and reactor was filled with metal. This was not intended but the electrode after machining is still remeltable in the VAR process. Therefore the ingot casting can be seen as a success although there is room for improvement.

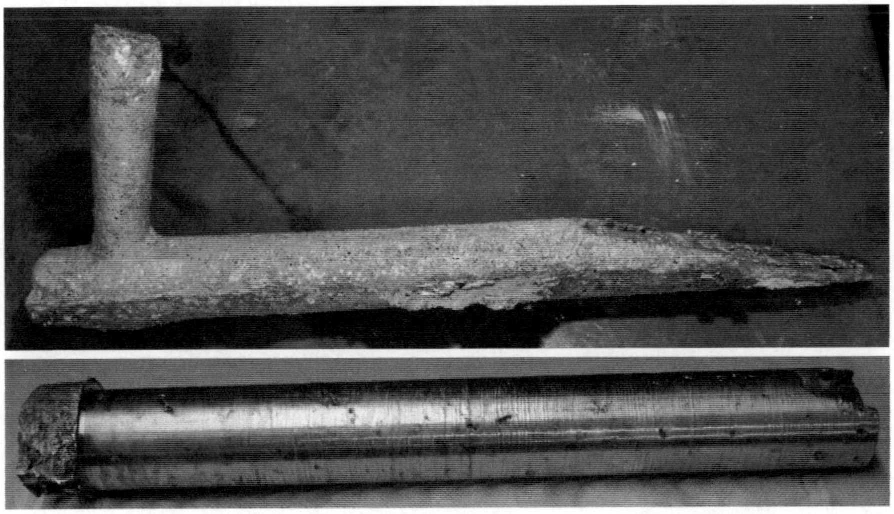

Figure 6: Metal product of ATR inline casting (top) and electrode prepared for VAR (bottom)

### *4.3. Vacuum arc remelting*

The product of the VAR remelting obtains no surface defects and has no shrinkage cavity. After the ingot has been cut in half there are grey spots visible that later on are identified as Cr. So a complete uniform distribution of Cr has not been achieved. But as can be seen at the SEM image on the right side of Figure 7 despite the few grey spots the rest of the Cr is spread in very small dendrites. A comparison of these dendrites with the dendrites formed after ATR shows that VAR leads to a much finer and, apart from the few grey spots, more uniform microstructure than ATR.

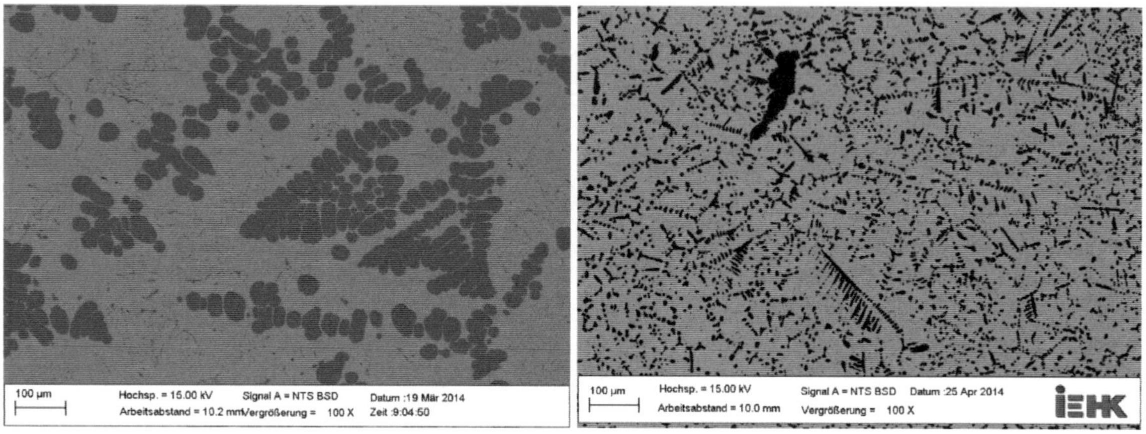


Figure 7: Comparison of the microstructure before (left) and after (right) VAR on SEM images

There is another unwanted effect besides the grey spots of pure Cr after the VAR remelting: There are strong macro segregations of the Cr throughout the Cu. So at the walls of the mould there is hardly any Cr to be found. As figure 8 shows this macro segregations lead to a very discrete separation between the Cu matrix with Cr and a nearly pure Cu phase.

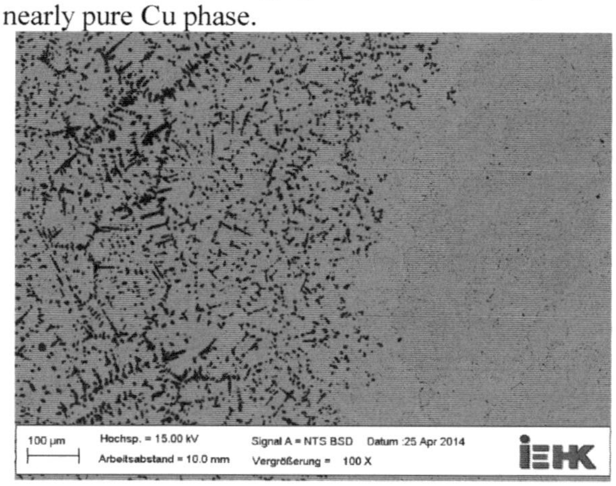


Figure 8: Visible Cr segregations on a SEM image

These segregations can be explained by the very high cooling rate in the VAR. As there is a massive cooling of the copper mould there is an enormous temperature gradient inside of the solidifying ingot. This temperature gradient leads to a segregation of the high melting Cr towards the warmer centre of the ingot.

The Al content of the ingot is below detection limit of a XRF. This means that it is possible to get rid of a certain amount of Al remaining from the ATR.

An approach on eliminating the segregations is using lower melting rates to generate a shallower melt pool and have faster cooling conditions. The next steps will be trying to reduce the melt rate down to the lowest tolerable limit to see what can be done to get rid of the segregations. Also a second VAR step and its effect on segregations will be tested

## 5. Conclusions

This investigation shows that it is possible to produce copper chromium alloys by using a combination of ATR and VAR. To obtain a product usable in medium voltage switches there must be a focus on the aluminium content and the homogeneity of the metal. Inline casting is of great help in obtaining a VAR

IOP Conf. Series: Materials Science and Engineering **143** (2016) 012016 doi:10.1088/1757-899X/143/1/012016

remeltable electrode directly from the ATR process. CuCr alloys are remeltable in a VAR furnace and it is possible to get rid of Al up to a certain limit. But further investigations on how to obtain a homogenous ingot without macro segregations need to be carried out.

**References**


[1] D. Mayer, "Werkstoffe für die Mittelspannungsenergietechnik Kupfer-Chrom-Basis (Cu/Cr), Werkstoffinformation Firma Doduco, 2011

[2] Zhou et al. "On the metastable miscibility gap in liquid Cu-Cr alloys" Journal of Materials Science 44, 2009, pp. 3793-3799 Another reference

[3] K.T. Jacob, S. Priya, Y. Waseda, "A Thermodynamic Study of Liquid Cu-Cr Alloys and Metastable Liquid Immiscibility", Zeitschrift für Metallkunde, Vol. 91, 2000, pp. 594-600

[4] D.J. Chakrabarti, D.E. Laughlin, "The Cr-Cu (Chromium-Copper) System", Bulletin of Alloy Phase Diagrams, Volume 5 No. 1, 1984, pp. 59-68

[5] G. Gottstein, M. Winning, B. Friedrich, "Chemische Technik: Prozesse und Produkte"

[6] W. Dautzenberg, "Untersuchung über die aluminothermische Gewinnung von Manganmetall, Ferrotitan und Chrommetall", RWTH Aachen, 1944

[7] D.K. Melgaard, R.L. Williamson, J.J. Beaman, "Controlling Remelting Processes for Superalloys and Aerospace Ti alloys", Journal of the Minerals, Metals & Materials Society, 1998, pp 13-17

[8] R.L. Boxman, P.J. Martin, D.M. Sanders, "Handbook of Vacuum Arc Science and Technology", Noyes Publications, 1995

IOP Conf. Series: Materials Science and Engineering **143** (2016) 012017 doi:10.1088/1757-899X/143/1/012017

# On the Modeling of Vacuum Arc Remelting Process in Titanium Alloys

**Ashish Patel and Daniel Fiore**

Timet, 900 Hemlock Road, Morgantown Business Park, Morgantown PA 19543

Ashish.patel@timet.com

**Abstract.** Mathematical modeling is routinely used in the process development and production of advanced aerospace alloys to gain greater insight into the effect of process parameters on final properties. This article describes the application of a 2-D mathematical VAR model presented at previous LMPC meetings. The impact of process parameters on melt pool geometry, solidification behavior, fluid-flow and chemistry in a Ti-6Al-4V ingot is discussed. Model predictions are validated against published data from a industrial size ingot, and results of a parametric study on particle dissolution are also discussed.

## 1. Introduction

The basic process of VAR is the continuous melting of a consumable electrode by means of a direct current arc under vacuum on the order of 0.1 to 1 Pa. In titanium, the consumable electrode is manufactured using compacted sponge and master alloys or is the output of a previous melt cycle (either VAR or Cold Hearth melted) /1/. The consumable electrode is suspended in a water cooled copper crucible. A DC electric arc is struck between the bottom of the VAR crucible (stool or baseplate) and the bottom of the electrode. The electrode melts creating an ingot within the crucible /2/. For most titanium and titanium alloys, melting currents are very high, and an external, current carrying coil around the water jacket is used to contain the motion of the arc.

The mathematical model, SOLAR, (SOLidification during Arc Remelting) is routinely used at TIMET for new process development and process improvement efforts. The model was developed at the École des Mines, Nancy, and over the past decade, several papers have been published (including LMPC proceedings) describing the model and highlighting its capabilities. In short, the ingot growth during VAR is simulated using a two-dimensional, axi-symmetric representation. The model computes the current, temperature, turbulent flow, species distribution, and particle dissolution in the ingot by iteratively solving the appropriate governing equations with prescribed boundary conditions using the finite volume method /3/.

The primary focus of this paper is to present additional results of model validation by comparing model predictions to measured data already available in the literature /4/. The predicted pool profile compared well against reported measurements from a 0.914 m (36 inch), Ti-6Al-4V ingot using nominal thermo-physical properties /5/. Here, the effect of processing parameters on pool profile and ingot structure is presented. The likelihood of dissolution of exogenous inclusions is also discussed.

## 2. Results

As reported /6/, a 0.914 m dia. ingot was melted at a constant current of 33 kA and voltage of 44.7 V, which resulted in a melt rate of 0.45 kg/sec. A 2 mT axial stirring field, reversing every 60 seconds, was applied during this melt. These parameters, along with the nominal thermo-physical properties served as input to the model. These VAR conditions produce a very deep pool, reaching a depth of nearly 1.5 m at the center with very steep side walls. The effect of stirring field strength on pool shape is shown in Figure 1. As the field is increased from 0.5 mT to 2 mT, the azimuthal flow resulting from the interaction of the axial stirring field and radial current intensifies. This, in turn, results in higher velocities and a steeper pool in the top half of the ingot. Pool depth also increases as the stirring field is increased.

The flow within the pool is very transient, and fluctuates during the stirring cycle. At the start of the stirring cycle (t=0), the in-plane flow ($V_r$, $V_z$) has one large re-circulating anti-clockwise loop. During the cycle, this flow field changes and can form several loops in different directions. Azimuthal flow, on the other hand, gradually increases throughout the cycle and is highest at the top of the ingot where there is good contact with the crucible wall. The predicted flow field at the beginning of the cycle is shown in Figure 2. The maximum azimuthal velocity is of the order of 0.4 m/sec, whereas the maximum axial velocity is 0.05/sec under these operating conditions.

Material flow and solute partitioning within the mushy zone of the melt pool results in macro-segregation, as shown in Figure 3 for a hypothetical electrode composition and melt profile /7/. During solidification, iron segregates to the liquid ($k_{Fe}<1$), resulting in nearly 15% enrichment of iron towards the center of the ingot. Oxygen, on the other hand, segregates to the solid, ($k_O>1$), and there is nearly 10% depletion of oxygen at the center of the ingot. The overall segregation in the ingot is quantified by calculating a segregation index, $SI = \frac{1}{V}\int\left|\frac{c-c_o}{c_o}\right| dV$ for each solute element /8/. Results indicate that as stirring field intensity increases, the segregation index also increases.

The estimated temperature gradient in the ingot is shown in Figure 4. The gradient is high, on the order of 10 K/mm, close to the lateral surface and drops three orders of magnitude towards the center to nearly 0.01 K/mm. Ingot growth is in the range of 0.05 to 0.01 mm/sec. Referring to the solidification maps available in the literature /9/, would indicate a change in grain structure from columnar at the surface (higher temperature gradients) to equiaxed at center. This is consistent with observation from literature /4/, with the columnar-to-equiaxed transition occurring near 0.1 K/mm.

### 2.1. *Inclusion Dissolution*

The two common exogenous particles found in titanium alloys are low density, high nitrogen inclusions (LDI) and high density refractory metal inclusions (HDI). These inclusions are potentially removed by the diffusion of nitrogen or the refractory element in the molten metal pool. The capability to predict particle dissolution has recently been added to the SOLAR model /10, 11/, and a parametric study is presented in this section.

Figure 5 shows the motion of a 0.5 mm dia. particle with a density nearly 2.5 time that of molten titanium, inserted 1 cm from the centerline of the ingot. The projection of the particle on the r-z plane as it moves in the pool is shown in Fig 5. In this instance, the particle drops straight to the bottom of the pool. Similarly, particles of different initial diameter were inserted into the pool, and the model shows that, for the given operating conditions, particles with initial diameter of 0.5 mm and above survive and remain in the ingot, and only smaller particles, of the order of 0.25 mm diameter, fully dissolve.

Using mass balance at the particle-liquid interface, the dissolution time for a spherical particle can be approximated by the following relationship /12/: $t = \frac{d^2}{D\,(sh\#)(C_i)}\frac{\rho_p}{\rho_{Ti}}$, where d is the initial particle diameter, D is the solute diffusivity, $C_i$ is the initial solute concentration, and $\rho_p$ and $\rho_{Ti}$ are the density of the particle and molten titanium, respectively. The Sherwood number (sh#) represents the ratio of convective to diffusive mass transport. This relationship clearly shows that the dissolution time of an HDI is proportional to the square of the initial particle size and the particle density. The particle residence time within the pool is the ratio of the pool depth to settling velocity, and in order for the particle to fully dissolve, the dissolution time must, of course, be less than the residence time.

The trajectory of 0.5 mm diameter LDI with high nitrogen content and density slightly higher than the molten metal is shown in Figure 6. In this particular case, the LDI follows a tortuous path in the pool and dissolves. Similar to the HDI, the effect of different initial particle size was investigated. Results indicate that the diameter actually increases due to the formation of an outer beta phase at lower nitrogen contents. Under the given conditions, although the residence time is large, it appears that particles with initial diameter greater than 1 mm would survive, and only LDI's with initial diameter of the order of 0.5 mm would dissolve in the pool. Using the mass transfer correlations in the literature with flow velocities between 0.05 and 0.5 m/sec, the nitrogen mass transfer ($h_m$) from the LDI to the bulk is in the range of $2x10^{-4}$ m/sec to $6.5x10^{-4}$ m/sec. Higher flow velocity and higher temperature promote LDI dissolution /13/.

## 3. Conclusions

4. The VAR of a production size ingot of Ti-6Al-4V was simulated using the SOLAR code. Results indicate that the pool profile is significantly affected by the externally applied magnetic field, as the field increases, the flow and overall segregation tendency within the ingot also increases. The model also accurately predicts trends in the temperature gradient within the ingot, and could be used as an indicator of the columnar to equiaxed transition in grain structure. The capability of the model to predict the dissolution of both HDI and LDI appears to be reasonable.

## 5. References

[1] "Vacuum Arc Remelting", ASM Handbook, Vol. 15, 2008, pp.132-138.
[2] K. M. Kelkar et al, "Computational Analysis of the Vacuum Arc Remelting and Electroslag Remelting (ESR) Processes", ASM Handbook, Vol. 22B, 2010, pp. 196-213.
[3] S. Hans et.al, "A Numerical Model for the Prediction of Transient Turbulent Fluid Flow, Heat Transfer and solidification during VAR", Proc. LMPC 1995, A. Mitchell and J. Fernihough Eds, AVS, pp. 143-154
[4] L. A. Bertram, et al., The MacroScale Simulation of Remelting Processes", JOM, 1998, pp 18-21
[5] K. Mills, "Recommended values of thermophysical properties for selected commercial alloys", ASM 2002, pp. 205-217.
[6] A. Patel, D. Tripp, D. Fiore, "Application of a model for simulating the vacuum arc remelting process for titanium alloys", Proc., LMPC, TMS, 2013, pp 241-244.
[7]. S. Fox. et.al, to be presented at Titanium 2015.
[8] D. Zagrebelnyy and M. J. M. Krane, "Segregation development in multiple melt VAR", Met Trans B, vol. 40B, 2009 281-288.
[9] J. A. Dantzig, M. Rappaz, "Solidification", EPFL 2009.
[10] G. Ghazal et. al., "On the Dissolution of nitride Ti defects during VAR of Ti alloys", Met. Trans B., Vol. 41B, 2010, pp 646-659.
[11]. G. Ghazal et. al., "Dissolution of high density inclusions in Titanium Alloys", ISIJ International, Vol. 52, 2012, pp. 1-9.
[12]. A. Yamanaka and H. Ichihashi, "Dissolution of refractory elements to Titanium Ally in VAR", ISIJ International, vol. 32, No. 5, 1992, pp 600-606.
[13] J. P. Bellot, et. al. "Dissolution of hard-alpha inclusions in liquid titanium alloys", Met. Trans B., vol 28B, 1997, pp 1001-1010.

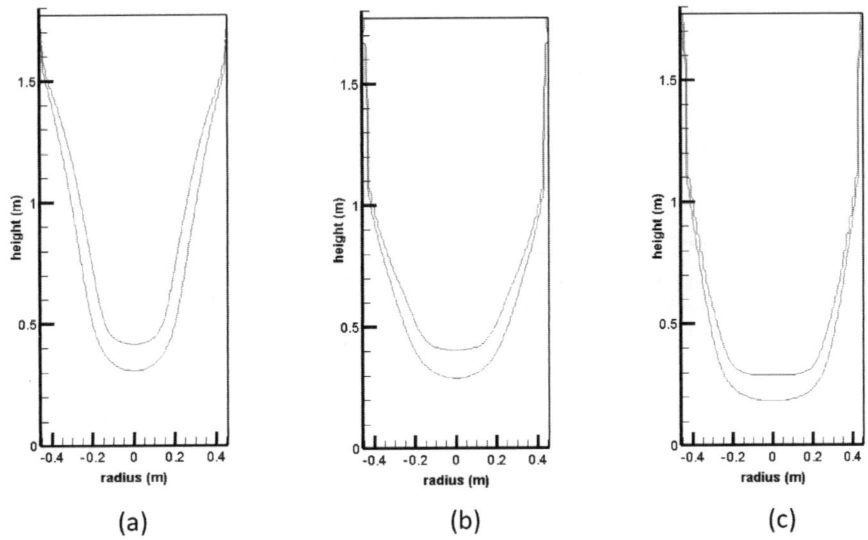


Figure 1. Effect of stirring field strength of pool profile during the VAR of a 0.94 m diameter ingot at full power. (a) 0.5 mT, (b) 1 mT, (c) 2 mT. The stirring field was reversed every 60 sec.

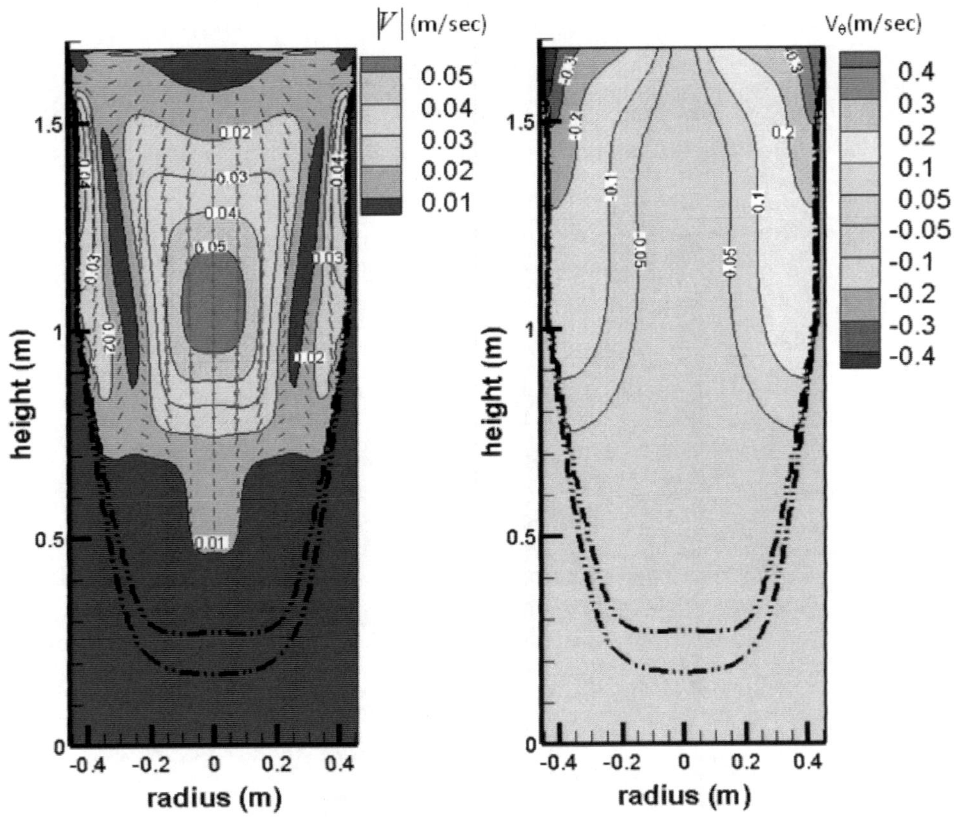


Figure 2. Calculated flow field at the beginning of the stirring cycle during VAR, field of 2 mT . The in-plane velocity is shown on the right ($V_r$, $V_z$) and the azimuthal velocity, $V_\theta$, is shown on right. The dotted line represents the liquidus and solidus isotherms

IOP Conf. Series: Materials Science and Engineering **143** (2016) 012017 doi:10.1088/1757-899X/143/1/012017

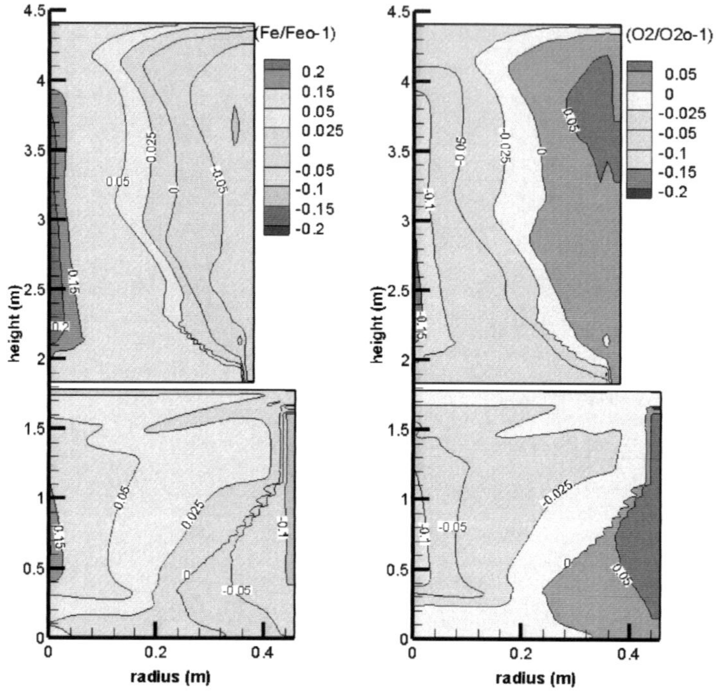


(a) (b)

Figure 3. Computed ingot chemistry (0.94 m dia.) after VAR and starting electrode chemistry ( 0.762 m diameter). (a) iron distribution, (b) oxygen distribution.

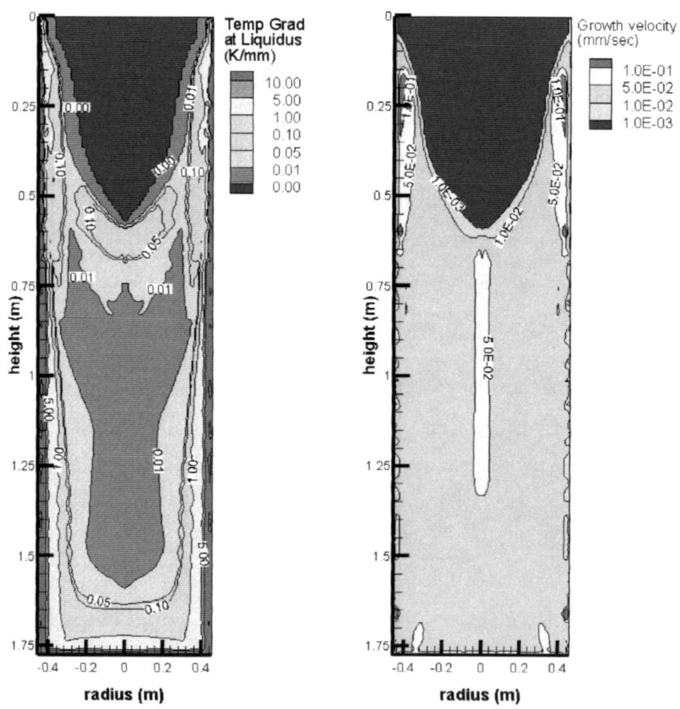


Figure 4. Predicted temperature gradient and growth velocity during solidification of a 0.94 m dia. VAR ingot.

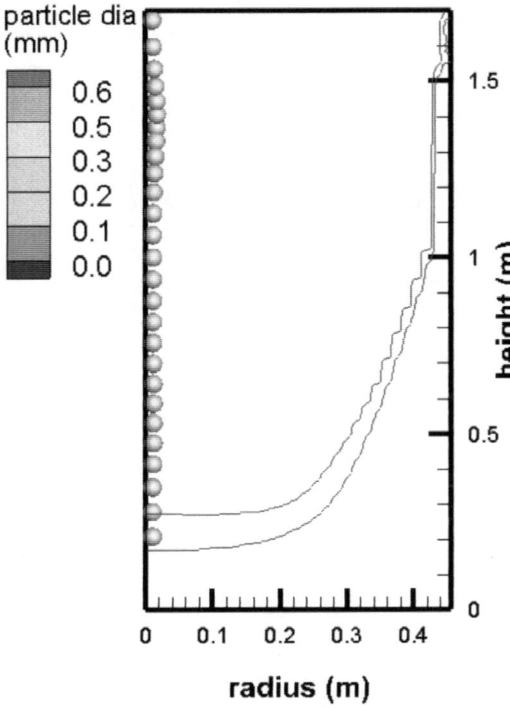


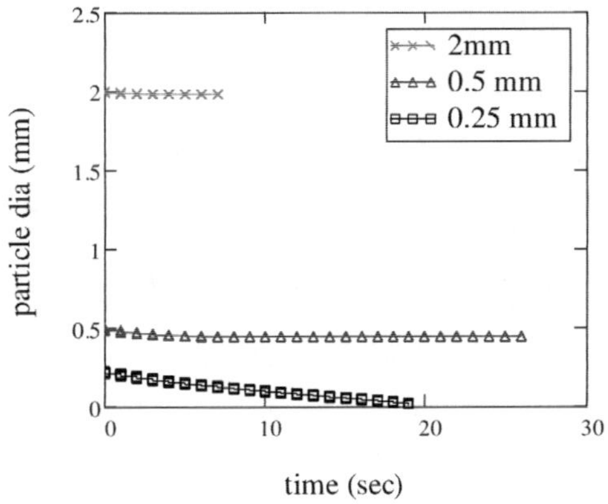


Figure 5: The trajectory of a 0.5 mm dia. HDI inserted at the center of the ingot (left), and HDI dissolution in the VAR pool (above).

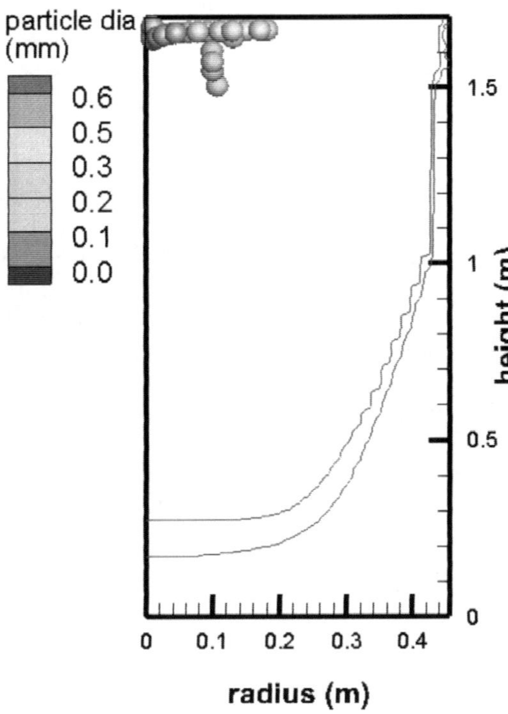


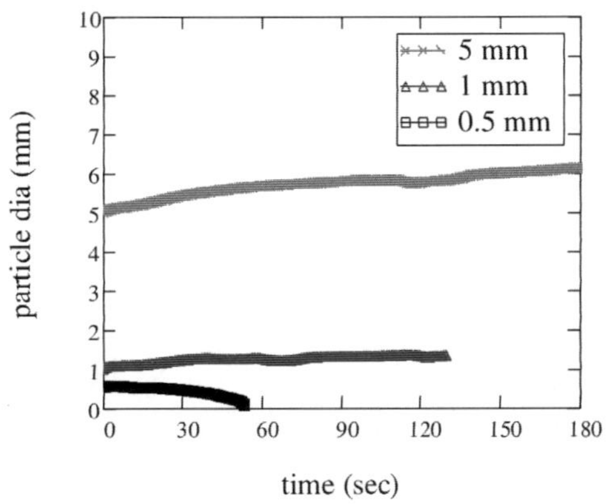


Figure 6: The trajectory of a 0.5 mm dia. LDI inserted at the center of the ingot (left), and LDI dissolution in the VAR pool (above).

# Auxiliary particle filter-model predictive control of the vacuum arc remelting process

**F Lopez, J Beaman, R Williamson**
Department of Mechanical Engineering, University of Texas at Austin, Austin, TX 78705, USA

E-mail: felipelopez@utexas.edu

**Abstract.** Solidification control is required for the suppression of segregation defects in vacuum arc remelting of superalloys. In recent years, process controllers for the VAR process have been proposed based on linear models, which are known to be inaccurate in highly-dynamic conditions, e.g. start-up, hot-top and melt rate perturbations. A novel controller is proposed using auxiliary particle filter-model predictive control based on a nonlinear stochastic model. The auxiliary particle filter approximates the probability of the state, which is fed to a model predictive controller that returns an optimal control signal. For simplicity, the estimation and control problems are solved using Sequential Monte Carlo (SMC) methods. The validity of this approach is verified for a 430 mm (17 in) diameter Alloy 718 electrode melted into a 510 mm (20 in) diameter ingot. Simulation shows a more accurate and smoother performance than the one obtained with an earlier version of the controller.

## 1. Introduction

Vacuum Arc Remelting (VAR) is a melting process used in the production of nickel-based superalloys and various titanium alloys. In this process, a continuous arc is used to melt a metal alloy electrode in a water-cooled copper crucible. Molten metal falls to the bottom of the crucible and solidifies into an ingot. By performing this process in a vacuum, some oxides and other volatile compounds are removed from the material improving its properties [12]. Although ingots obtained with VAR have elevated chemical and mechanical homogeneity, they are prone to segregation defects, such as freckles (solute channels) and white spots.

The study of the mechanisms that describe the formation and growth of solidification defects in remelting processes is still an active area of research, but it is commonly believed that they can be prevented by maintaining appropriate solidification conditions throughout the melt. For example, solute-lean white spots are believed to be caused by material falling into the liquid pool atop the ingot and resulting in solidification defect precursors [7]. For that reason, deep liquid pools - which result in longer local solidification times (LST) - are used to dissolve fall-in material. On the other hand, solute-rich freckles are believed to be caused by convective instabilities and buoyancy-driven flows [5]. It has been reported that the tendency for freckles decreases when melting with shallower liquid pools because LST decreases. Prevention of both freckles and white spots can be thought of as a balancing act that requires having a liquid pool that is deep enough to prevent white spots while being shallow enough to prevent freckles.

IOP Conf. Series: Materials Science and Engineering **143** (2016) 012018 doi:10.1088/1757-899X/143/1/012018

In VAR process control, not only the shape of the solidification front is important for prevention of defects. For example, tree rings - another kind of solidification defect - are found when thermal instabilities occur in the liquid pool [17]. Thus, for accurate prevention of solidification defects in remelted ingots, process controllers must drive the liquid pool profile to desired values (to prevent white spots and freckles) while attenuating process perturbations (to prevent tree rings).

In usual industrial practice, ingot solidification is not monitored nor controlled in the melt. Most process controllers attempt to attain a desired melt rate hoping that it will result in controlled solidification, but this is valid only at steady-state conditions. The first attempt at solidification control was developed by Beaman *et al.*, who designed a model-based controller for pool depth based on melting current [2]. The pool depth controller was designed for quasi-steady conditions - which occur during most of the melt - and worked in cooperation with a separate controller that corrected ram speed to keep a desired electrode gap. The controller was tested experimentally showing close agreement between references and manually-measured pool depths [16]. Although successful in its primary goal, the authors reported that inaccuracies in the ingot solidification model resulted in oscillatory behavior when the reference depth was modified.

This article presents an improved version of the pool depth controller where:

- A nonlinear model is used for improved performance when moving away from the linearization point, e.g. melt rate perturbations, start-up, hot-top, etc.
- A model predictive controller (MPC), an advanced method of process control widely used in chemical applications, is adopted to optimize the predicted system response in the near future.
- The electrode gap dynamics are coupled to the solidifying ingot in a unified system, instead of using two separate controllers as in the previous version of the pool depth controller.
- Both the estimation and control problems were expressed in the form of Sequential Monte Carlo (SMC) problems.

The rest of the paper is organized as follows. Section 2 presents the stochastic model used for estimation and control. In Section 3, an auxiliary particle filter – an SMC method - is used for state estimation. Section 4 describes the MPC problem and presents a solution using SMC methods. The model predictive controller is coupled to the auxiliary particle filter to form a novel method for nonlinear estimation and control of liquid pool depth for the VAR process. The effectiveness of the method is verified via simulation for a melt of Alloy 718 from an electrode of 430 mm (17 in) in diameter to a 510 mm (20 in) ingot in Section 5. Finally, Section 6 contains the concluding remarks.

## 2. Stochastic model

A nonlinear stochastic model is proposed for nonlinear estimation and control of the VAR process. It is a variation of the model proposed by Beaman *et al.*[2] that, although simpler, shows improved accuracy for ingot solidification. The nonlinear ingot solidification model (initially obtained from a numerical approximation of the heat equation) was substituted by a linear model. This is justified by observations where liquid pool depth showed exponential convergence to a new equilibrium when process inputs were perturbed, suggesting a first-order response.

The state at time $n$ is given by $x_n = \begin{bmatrix} \Delta_n & G_n & X_{ram_n} & M_{e_n} & \mu_n & S_{C_n} & S_{M_n} & p_{he_n} & i_n \end{bmatrix}^T$ [1]; where $\Delta$ is the electrode thermal boundary layer, $G$ is the electrode gap, $X_{ram}$ is the ram position, $M_e$ is the electrode mass, $\mu$ is the melting efficiency[2], $S_C$ and $S_M$ are the pool depths at the center of the ingot and mid-radius, $p_{he}$ is the helium pressure in the crucible and $i$ is the current. The process

[1] The superscript $T$ denotes the transpose of a matrix.

[2] Defined as the instantaneous fraction of supplied power that is used in melting the electrode $\mu = P_{melt} / P_{total}$.

is controlled with the control signal $u_n = [i_{c_n} \quad V_{ram_n}]^T$, where $i_{c_n}$ is the commanded current and $V_{ram_n}$ is the commanded ram speed. The model is defined as:

$$\Delta_{n+1} = \Delta_n + \left[\frac{\alpha_r C_{\Delta\Delta}}{\Delta_n} - \frac{C_{\Delta p}}{A_e h_m}\mu_n (V_c + R_i i_n) i_n\right]\Delta t + G_{11} di_n \tag{1}$$

$$G_{n+1} = G_n - \left[\frac{a\alpha_r C_{s\Delta}}{\Delta_n} - \frac{C_{sp} a}{A_e h_m}\mu_n (V_c + R_i i_n) i_n + V_{ram_n}\right]\Delta t + G_{21} di_n - dV_{ram_n} \tag{2}$$

$$X_{ram_{n+1}} = X_{ram_n} + V_{ram_n}\Delta t + dV_{ram_n} \tag{3}$$

$$M_{e_{n+1}} = M_{e_n} + \left[\frac{\rho A_e \alpha_r C_{s\Delta}}{\Delta_n} - \frac{\rho C_{sp}}{h_m}\mu_n (V_c + R_i i_n) i_n\right]\Delta t + G_{41} di_n \tag{4}$$

$$\mu_{n+1} = \mu_n + d\mu_n \tag{5}$$

$$S_{C_{n+1}} = S_{C_n} + [-A_C (S_{C_n} - S_{C_0}) + B_{\Delta C}(\Delta_n - \Delta_0) + B_{iC}(i_n - i_0 + di_n) + B_{\mu C}(\mu_n - \mu_0) + B_{heC}(p_{he_n} - p_{he_0})]\Delta t \tag{6}$$

$$S_{M_{n+1}} = S_{M_n} + [-A_M (S_{M_n} - S_{M_0}) + B_{\Delta M}(\Delta_n - \Delta_0) + B_{iM}(i_n - i_0 + di_n) + B_{\mu M}(\mu_n - \mu_0) + B_{heM}(p_{he_n} - p_{he_0})]\Delta t \tag{7}$$

$$p_{he_{n+1}} = p_{he_n} + dp_{he_n} \tag{8}$$

$$i_{n+1} = i_{c_n} + (i_n - i_{c_n})e^{-\Delta t/\tau} + di_n \tag{9}$$

where $G_{11} = -(V_c + 2R_i i_0)\mu_0 C_{\Delta p} / A_e h_m$, $G_{21} = (V_c + 2R_i i_0) a\mu_0 C_{sp} / A_e h_m$, $G_{41} = -(V_c + 2R_i i_0)\rho\mu_0 C_{sp} / h_m$ [3], and variables $A_{C \text{ or } M}$, $B_{\Delta C \text{ or } \Delta M}$, $B_{iC \text{ or } iM}$, $B_{\mu C \text{ or } \mu M}$ and $B_{heC \text{ or } heM}$ are obtained from a linear model of ingot solidification[4].

The proposed model is stochastic due to the presence of the random inputs $di_n$, $dV_{ram_n}$, $d\mu_n$ and $dp_{he_n}$ which model uncertainty in melting current (caused by unknown current partition and distribution in the arc), ram speed (because of friction and error in the translation mechanism), melting efficiency (due to inhomogeneous electrodes and end effects) and helium pressure in the cooling system, respectively. These random inputs are modeled as $di_n \sim \mathrm{N}(0,\sigma_i^2 \Delta t^2)$, $dV_{ram_n} \sim \mathrm{N}(0,\sigma_{V_{ram}}^2 \Delta t^2)$, $d\mu_n \sim \mathrm{N}(0,\sigma_\mu^2 \Delta t)$ and $dp_{he_n} \sim \mathrm{N}(0,\sigma_{he}^2 \Delta t)$[5].

The stochastic model can be summarized in two equations: 1) the conditional probability of the current state given its previous value, $p(x_n \mid x_{n-1})$, defined by equations (1) to (9); and 2) the

[3] These coefficients come from the electrode melting model proposed in Ref. [3]: $V_c$ is the cathode voltage fall; $a$ is the area ratio given by $a = 1 - A_e / A_i$, where $A_e$ and $A_i$ are the cross sectional areas of the electrode and ingot respectively; and $R_i$ is the electric resistance in the electrode; and the coefficients $C_{\Delta\Delta}$, $C_{\Delta p}$, $C_{s\Delta}$ and $C_{sp}$ depend on thermo-physical properties.

[4] In this application, they were obtained from simulations of Basic Axisymmetric Remelting (BAR) using system identification based on first order systems [4]. The subscript *0* denotes nominal value in steady-state operation.

[5] $x \sim \mathrm{N}(\mu,\sigma^2)$ denotes a random variable $x$ sampled from a normal distribution of mean $\mu$ and variance $\sigma^2$.

probability of measurements given the current state, $p(y_n \mid x_n)$. The second equation is required because, just like in most industrial processes, the state is not observed directly but through an observation sequence $y_n$, described by:

$$p(y_n \mid x_n) \sim \mathrm{N}\left(\begin{bmatrix} G_n & X_{ram_n} & i_n & M_{e_n} & S_{C_n} & S_{M_n} & p_{he_n} \end{bmatrix}^T, R\right) \tag{10}$$

$$R = \mathrm{diag}\left(\sigma_G^2, \sigma_X^2, \sigma_{i_m}^2, \sigma_{M_e}^2, \sigma_C^2, \sigma_M^2, \sigma_{he_m}^2\right)^6. \tag{11}$$

The stochastic model was developed for two purposes: 1) estimation, the determination of the probability of the current state based on all the available observations up to this instant; and 2) control, the computation of the optimal control signal that optimizes the performance in the near future.

## 3. Auxiliary particle filter

State estimation determines the probability of the state $x_n$ by updating the conditional probability $p(x_{n-1} \mid y_{1:n-1})$ when a new set of measurements $y_n$ becomes available. The update is performed using Bayes' rule and it involves stochastic integrals. Unfortunately, these calculations have no analytical solutions in nonlinear non-Gaussian scenarios, such as the one presented in this article. In these cases it is common to construct approximate solutions using particle filters, a family of Monte Carlo methods. These methods use discrete approximations of probability density functions allowing calculations to be performed in the form of finite sums [6].

The proposed VAR stochastic model has more process uncertainty than measurement uncertainty[7]. Studies have shown that the auxiliary particle filter (APF) [6,7] outperforms other particle methods in statistical efficiency in this application [8]. The superior performance was obtained by incorporating knowledge of the next-step measurements in the proposal mechanism, preventing particles from moving to unlikely regions of the state-space[8].

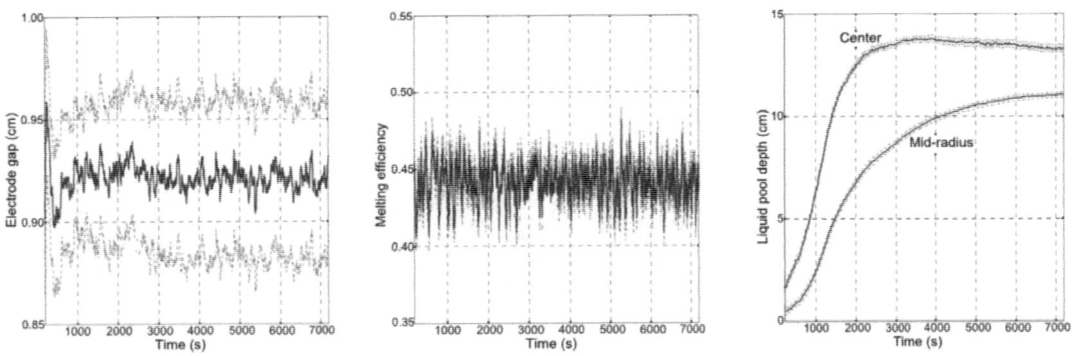


Figure 1. Estimates for electrode gap $(G)$, efficiency $(\mu)$, and pool depth $(S_{C \text{ or } M})$ in a VAR melt of Alloy 718. Maximum likelihood estimates are plotted in blue and 95% confidence intervals in red.

---

[6] $A = \mathrm{diag}(x_1, \ldots, x_n)$ denotes an $n \times n$ diagonal matrix where $A_{ii} = x_i$ for $i = 1, \ldots, n$.

[7] Process is uncertain due to disturbances that may occur in the process that make it difficult to accurately predict how much metal is melted at each sampled time. These uncertain predictions are then corrected with precise measurements of ram position and electrode weight, and noisy electrode gap measurements (inferred from drip-short frequency or voltage).

[8] Incorporation of the measurement reduces the variability of the weights $\{w_n^i\}_{i=1}^N$. And since the variance of the estimates is proportional to the variance of the weights, less variable weights result in more precise predictions.

As in other Monte Carlo methods, precision of the estimates is dependent on the number of samples (particles) used in the approximation. The desired precision for the estimates was defined as 95% confidence intervals of 0.1 cm for electrode gap and 0.04 cm for liquid pool depth[9]. It was observed experimentally that at least 10,000 particles were required to meet the desired precision. As an example, Figure 1 shows state estimates for the APF running with 65,536 particles.

## 4. Model predictive control

The linear pool depth controller proposed by Beaman *et al.* was based on a Linear-Quadratic-Gaussian (LQG) controller, a classical solution in the optimal control literature [1]. In this article, a model predictive controller is proposed as a better alternative. The basic concept in MPC is to use a dynamic model to solve an open-loop optimal control problem at ever time step $n$ for a prediction horizon of length *P*. At a fixed time $n$, the cost function is defined as $J(x_n, \bar{u}_{n:n+P})$

$$\min_{\bar{u}_{n:n+P}} J(x_n, \bar{u}_{n:n+P}) = \min_{\bar{u}_{n:n+P}} \sum_{j=n}^{n+P} \left\| \bar{u}_j - \bar{u}_{j-1} \right\|_Q^2 + \sum_{j=n+1}^{n+P} \left\| s_j - \bar{y}_{j-1} \right\|_R^2 \quad (12)$$

[10]

which penalizes variations in control signals and deviations from desired references. In this expression, $\bar{u}_{n:n+P}$ is the tentative control input, $\bar{x}_{n:n+P}$ the predicted states, $\bar{y}_{n:n+P}$ the predicted observations and $s_j$ are the process set points (desired trajectory) [13][11].

Monte Carlo methods, previously described as estimators, have also been explored as tools to solve the MPC problem. The solution described in this article is based on Stahl and Hauth's PF-MPC (Particle filter-Model predictive control), where the MPC optimization problem is expressed in the form of an estimation problem and two particle filters are used in the same application: one for estimation and one for control [14]. Because estimation is performed in the usual way, only the control problem is discussed in this section.

In this approach, the proposed control sequence $\bar{u}_j$ and the desired reference $s_j$ are treated as random variables. The closer the prediction and reference are, the more optimal the proposed control sequence will be. Predictions are made following $\bar{x}_j = f(\bar{x}_{j-1}, \bar{u}_{j-1})$ for the state and $\bar{y}_j = g(\bar{x}_j)$ for the observations[12]. In these predictions, every entry of the control sequence has a normal distribution around the previous value of the sequence, i.e. $\bar{u}_j \sim \mathrm{N}(\bar{u}_{j-1}, Q^{-1})$. The likelihood of each control sequence is evaluated with a normal distribution that penalizes the error between the predicted output and the reference, following $p(s_j \mid \bar{y}_j) = \mathrm{N}(\bar{y}_j, R^{-1})$. In this approach, the solution of the optimization problem is found at the mode of the conditional distribution $p(\bar{u}_{n:n+P} \mid s_{n+1:n+P})$.

The method is illustrated in Figure 2, where measurements $y_n$ are taken from the plant and fed to the first particle filter, which acts as an estimator. The estimator takes the previous state estimate, in the form of samples $x_{n-1}^i$ and their respective weights $w_{n-1}^i$, inputs $u_{n-1}$ and measurements $y_n$; and

[9]Electrode gap and liquid pool depth are the most significant control variables in liquid pool depth control. Error bounds were defined based on usual practice in other liquid pool depth control experiments.

[10] $\|x\|_M^2$ denotes the 2-norm of a vector $x$ evaluated with a weighing matrix $M$.

[11] A bar as a superscript means that the variable is a prediction of the future performance of the system based on the model.

[12] The functions $f$ and $g$ are deterministic versions of the two conditional probabilities used in the stochastic model. Function $f$ returns the most likely new state given its previous value and the control inputs, and function $g$ returns the most likely observations for a given state.

returns updated samples $x_n^i$ and weights $w_n^i$. The samples are fed to the second particle filter, which evaluates control sequences and their likelihood in resulting in the desired reference. In this stage, the samples and weights from the particle filter are used to predict states $\overline{x}_j^i$ and observations $\overline{y}_j^i$, for a proposed control sequence $\overline{u}_j^i$. A control sequence is proposed for each sample from the particle filter, and a weight $\overline{w}_j^i$ is assigned to each control sequence dependent on how close the predictions are to the reference $s_j$. The control sequences and their weights are propagated to a point estimator that returns the optimal signal $u_n$, which is sent back to the plant, closing the loop.

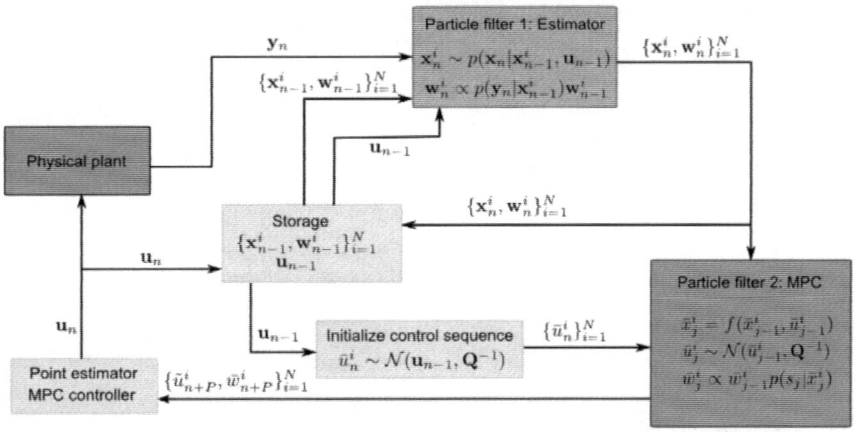


Figure 2. Schematic of a PF-MPC controller.

The original method, as proposed by Stahl and Hauth, was modified for this application to use an auxiliary particle filter (hence, APF-MPC) in the estimation stage while still having a basic particle filter in the control one. The approaches are interchangeable as they both return discrete approximations of the same probability density function.

## 5. Numerical study

The performance of the proposed APF-MPC controller was tested via simulation for a VAR melt of Alloy 718, from an electrode that was 430 mm (17 in) in diameter to a 510 mm (20 in) ingot[13][14][15]. In

---

[13] The thermo-physical properties used for Alloy 718 are: thermal diffusivities at 300 K and 1623 K ($\alpha_r$ and $\alpha_m$) equal to $2.4\times10^{-2}$cm$^2$/s and $6.0\times10^{-2}$cm$^2$/s, while enthalpies at 1623 K and 1673 K ($h_m$ and $h_s$) are equal to $5.4\times10^3$J/cm$^3$ and $8.1\times10^3$J/cm$^3$, respectively. The electrical parameters are: electric resistance ($R_i$) of $4.37\times10^{-4}\Omega$ and cathode voltage fall ($V_C$) of 21.2 V. The nominal operating conditions are: centerline and mid-radius pool depths ($S_{C,0}$ and $S_{M,0}$) of 15.7 cm and 13.2 cm, current ($i_0$) of 6000 A, melt rate ($m_0$) of 60 g/s, helium pressure ($p_{he,0}$) of 3.0 Torr and melting efficiency ($\mu_0$) of 0.44. Other model parameters include: $C_{\Delta\Delta}$ equal to 40, $C_{\Delta p}$ is 3.8, $C_{s\Delta}$ is 6.7, $C_{sp}$ is 1.3, $A_C$ is $1.9\times10^{-3}$1/s, $A_M$ is $1.4\times10^{-3}$1/s, $B_{\Delta C}$ is $2.6\times10^{-5}$ 1/s, $B_{\Delta M}$ is $-1.3\times10^{-4}$ 1/s, $B_{iC}$ is $6.6\times10^{-6}$ cm/s-A, $B_{iM}$ is $3.2\times10^{-6}$ cm/s-A, $B_{\mu C}$ is $6.9\times10^{-4}$ cm/s, $B_{\mu M}$ is $2.7\times10^{-4}$ cm/s, $B_{heC}$ is $-8.1\times10^{-4}$ cm/s-Torr, $B_{heM}$ is $-6.5\times10^{-4}$ cm/s-Torr, $G_{11}$ is $-5.6\times10^{-6}$ cm/A, $G_{21}$ is $5.4\times10^{-7}$ cm/A, $G_{41}$ is $-2.2\times10^{-2}$ g/A, $\Delta t$ is 6s and $\tau$ is 1s.

[14] Standard deviations of process uncertainties are defined as: 20 A for current ($\sigma_i$), $5.0\times10^{-4}$ cm/s for ram speed ($\sigma_{Vram}$), $0.01\mu_0$ for melting efficiency ($\sigma_\mu$), and $0.001p_{he0}$ for helium pressure ($\sigma_{he}$). Meanwhile, standard deviations of measurement uncertainties are defined as: 0.2 cm for electrode gap ($\sigma_G$), 0.005 cm for ram position ($\sigma_X$), 15 A for current ($\sigma_{im}$), 200 g for electrode mass ($\sigma_{Me}$), 1.0 cm for both centerline and mid-radius pool depths ($\sigma_C$ and $\sigma_M$), and 0.01 Torr for helium pressure ($\sigma_{hem}$).

[15] In the optimization step, the length of the prediction horizon ($P$) was set to 15. The optimization weights were set as: $10^4/i_0^2$ for current variations ($Q_{1,1}$), $10^4/V^2_{ram,0}$ for ram speed variations ($Q_{2,2}$), $50^2/G^2_0$ for electrode gap and $50^2/S^2_{C,0}$ for pool depth.

this simulation, a constant reference was defined for electrode gap while a time-varying one is used for liquid pool depth at the center of the ingot. Initially, the reference starts at nominal conditions and is driven 5% deeper in a 20 minutes ramp, where it stays for 60 minutes. Then, it is ramped down to 95% of the nominal value in 20 minutes, where it stays for 60 minutes more. Finally, it is brought back to nominal conditions in a 20 minutes ramp, where it stays until the end of the simulation.

The closed-loop system was simulated for 10,000 particles, and nominal helium pressure and melting efficiency, resulting in Figure 3. Precise control is achieved for both electrode gap and liquid pool depth. The prediction horizon chosen for this simulation was 1.5 minutes, but longer prediction horizons (close to 5 minutes) show a smoother behavior, preventing oscillations in input signals at the expense of increasing the computational cost of the algorithm.

One of the main advantages of the proposed APF-MPC controller as compared to the LQG controller is that future variations in pool depth reference are taken into account when computing optimal control signals, resulting in a superior response as the commanded current no longer shows jumps. The controller proposed by Beaman *et al.* was able to drive the ingot to desired pool depths at the expense of having big current perturbations, which resulted in tree rings [16]. This problem seems to have been solved. Additionally, this controller is expected to result in improved performance in highly-nonlinear operation, e.g. start-up and hot-top.

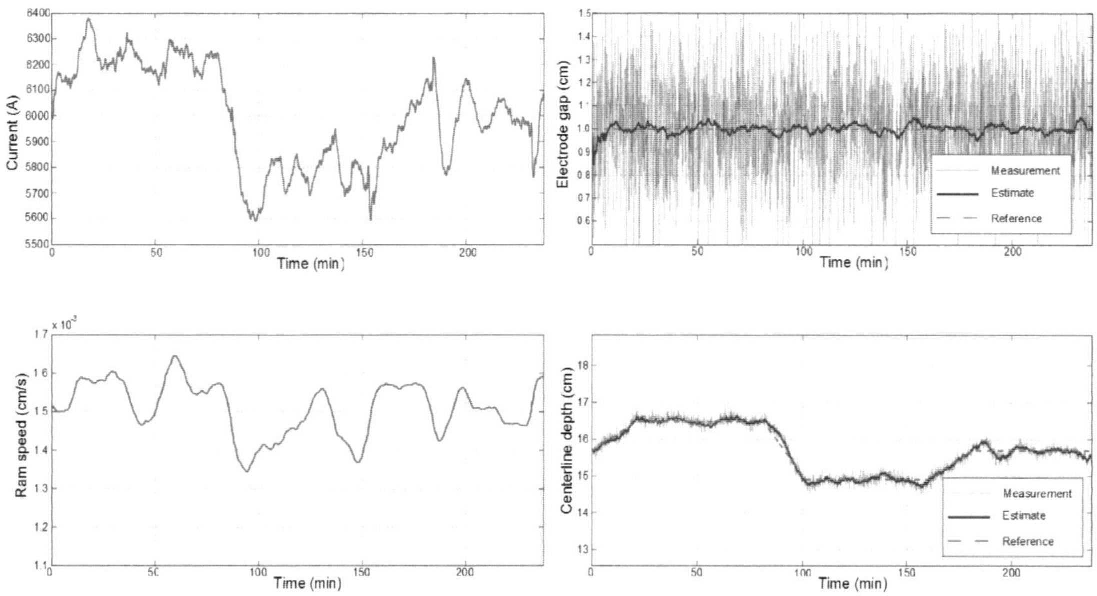


Figure 3. Process inputs and outputs in the closed-loop stochastic simulation of the VAR system using APF-MPC.

Although promising, the APF-MPC controller is too slow to be implemented in closed-loop operation using a personal computer. The controller, when simulated with MATLAB on an Intel Core i7 CPU 860 at 2.8 GHz x 8 processor with Windows 7, was 14 times slower than the sampling time of the process. The inability to perform in real time prevents its implementation in process control applications. However, the method can be accelerated if implemented on appropriate computer architectures. Both estimation and control problems were formulated as SMC problems, which are parallel by nature. Studies suggest that the implementation of these algorithms in parallel computer architectures accelerates their performance by several orders of magnitude. Graphics processing units (GPUs), common in the gaming industry, allow massive parallelization by distributing independent tasks over different computing units, speeding up the total execution time [9].

The application of GPU architectures to accelerate estimation in VAR has been reported by Lopez *et al.* [8]. A similar study was performed for the APF-MPC controller using NVIDIA GTX TITAN. The simulation took 5.8 ms for 16,384 particles, 9.6 ms for 32,768 particles, 14 ms for 65,536 particles, 23 ms for 131,072 particles, 40 ms for 262,144 particles, 75 ms for 524,288 particles and 144 ms for 1,048,576 particles. The obtained results suggest that when implemented on GPUs, the APF-MPC controller is much faster than typical sampling times for VAR processes (around 5 seconds for large Nickel ingots).

## 6. Conclusions

In this paper it was shown that model predictive control, widely used in chemical applications, should be explored as an alternative to the LQG approach followed in the pool depth controller developed by Beaman *et al.* The most substantial advantage of the proposed approach is that nonlinearities – important to describe the thermal dynamics of the melting electrode – are included in the dynamic model, ensuring a more accurate behavior during highly-dynamic conditions, e.g. melt rate perturbations, start-up, hot-top, etc. Also, in the proposed approach a centralized controller monitors both electrode gap and liquid pool depth control, instead of having two separate controllers that ignored the interaction between the two sub-systems. And finally, this approach takes into account the desired near future when choosing an optimal input signal, resulting in smoother current variations. The major drawback is that melt recipes will have to be pre-defined and not changed during operation.

SMC methods, commonly used as estimators, have been proposed to solve the MPC problem in a semi-stochastic form following the formulation of Stahl and Hauth. The computational cost of the algorithm would be prohibitive if implemented on personal computers. However, simulation time is well under typical sampling times when parallel computer architectures (such as GPUs) are used.

## 7. Acknowledgements

This work was supported by NSF Grant No. 1239343. The authors wish to thank Lixun Zhang and Aloysius Mok from the Department of Computer Science at the University of Texas at Austin.

**References**

[1] Athans M 1971 *IEEE Trans. Autom. Control* **16(6)** p 529
[2] Beaman J, Lopez F and Williamson R 2013 *J. Dyn. Syst.-T. ASME* **136(3)** 031007
[3] Beaman J, Williamson R, Melgaard D and Hamel J 2005 *Proc. IMECE* (Orlando, FL) p 1059
[4] Bertram L, Schunk P, Kempka S, Spadafora F and Minisandram R 1998 *JOM* **50(3)** p 18
[5] Copley S, Giamei A, Johnson S and Hornbecker M 1970 *Metall Trans B* **1(8)** p 2193
[6] Doucet A, de Freitas N and Gordon N 2001 *Sequential Monte Carlo methods in practice* (Series: Information Science and Statistics) (Springer-Verlag New York)
[7] Jackman L, Maurer G and Widge S 1994 *Proc. Superalloys 718, 625, 706 and various derivatives* (Warrendale, PA) p 153
[8] Lopez F, Zhang L, Mok A and Beaman J 2015 *Comput. Ind.* **71** p 116
[9] NVIDIA Corporation 2014 *GPU accelerated applications* (NVIDIA Corporation)
[10] Pitt M and Shephard N 1999 *J. Amer. Statist. Assoc.* **94(446)** 590
[11] Pitt M and Shephard N 2001 Auxiliary variable based particle filters *Sequential Monte Carlo methods in practice* (Series: Information Science and Statistics) ed Doucet A, de Freitas N and Gordon N (Springer-Verlag New York) p 271-293
[12] Pollock T and Tin S 2006 *J. Propul. Power* **22(2)** p 361
[13] Rawlings J and Mayne D 2009 *Model predictive control: Theory and design* (Nob Hill Publishing) chapter 1 pp 1-87
[14] Stahl D and Hauth J 2011 *Syst. Control Lett.* **60(8)** p 632
[15] Van Den Avyle J, Brooks J and Powell A 1998 *JOM* **50(3)** p 22
[16] Watt T, Taleff E, Lopez F, Beaman J and Williamson R 2013 *Proc. LMPC* (Austin, TX) p 261
[17] Xu X, Ward R, Jacobs M, Lee P and McLean M 2002 *Metall. Mater. Trans. A* **33A** p 1795

IOP Conf. Series: Materials Science and Engineering **143** (2016) 012019 doi:10.1088/1757-899X/143/1/012019

# A full 3D model of fluid flow and heat transfer in an E.B. heated liquid metal bath

**A. Matveichev, A. Jardy, J.P. Bellot**

Institut Jean Lamour - UMR 7198 CNRS/Université de Lorraine,
Laboratory of Excellence DAMAS, CS 50840, 54011 Nancy Cedex, France

E-mail: alain.jardy@univ-lorraine.fr

**Abstract**. In order to study the dissolution of exogeneous inclusions in the liquid metal during processing of titanium alloys, a series of dipping experiments has been performed in an Electron Beam Melting laboratory furnace. Precise determination of the dissolution kinetics requires knowing and mastering the exact thermohydrodynamic behavior of the melt pool, which implies full 3D modeling of the process. To achieve this goal, one needs to describe momentum and heat transfer, phase change, as well as the development of flow turbulence in the liquid. EB power input, thermal radiation, heat loss through the cooling circuit, surface tension effects (i.e. Marangoni-induced flow) must also be addressed in the model. Therefore a new solver dealing with all these phenomena was implemented within OpenFOAM platform. Numerical results were compared with experimental data from actual Ti melting, showing a pretty good agreement. In the second stage, the immersion of a refractory sample rod in the liquid pool was simulated. Results of the simulations showed that the introduction of the sample slightly disturbs the flow field inside the bath. The amount of such disturbance depends on the exact location of the dipping.

## 1. Introduction

The aim of this paper is to present recent development in IJL Nancy of a full 3D modeling of coupled fluid flow and heat transfer during the processing of liquid metal, and its implementation within the open source platform OpenFOAM. The model has been applied to the numerical modelling of Electron Beam heating and melting of a titanium alloy. A Ti button, placed in a water-cooled hemispherical copper crucible, is heated by an electron beam which moves very rapidly to repeat a given pattern of energy input. The flux density created by the beam leads to the melting of the metal and subsequent formation of a liquid bath.

Previously, an experimental study was performed to investigate the kinetics of dissolution of refractory metals such as W or Mo in liquid titanium alloys [1], in order to assess the possibility of HDI removal during Ti processing. This had been done by dissolving cylindrical rods dipped in molten titanium for controlled periods of time. A schematic representation of such experiment is given on Fig. 1.

Kinetics of the dissolution strongly depends on the liquid bath agitation and temperature, hence the necessity of a precise knowledge of the local temperature and velocity fields at the rod dipping position. Since experimental measurement of these quantities is very difficult, it appeared that numerical simulation of the thermohydrodynamic behavior of the liquid bath was the best approach.

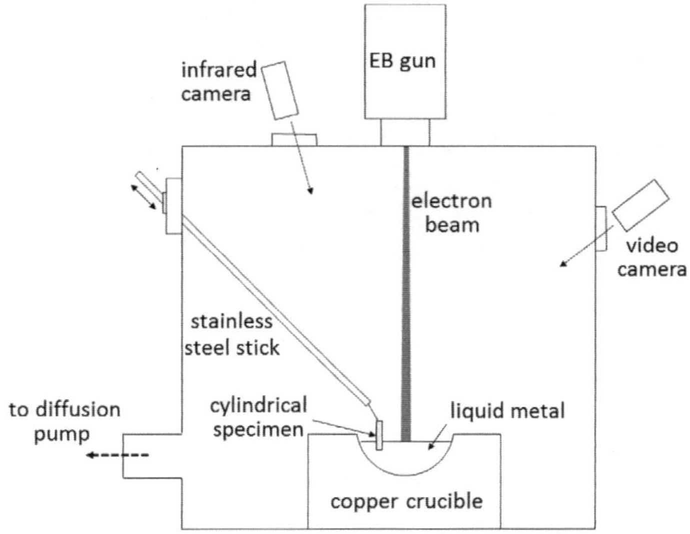


**Figure 1.** Scheme of a typical dipping experiment in the laboratory-scale 100 kW EB furnace at IJL Nancy.

## 2. Principle of the numerical model

The mathematical model accounts for various features that are responsible for the behavior of the metallic bath. Intense surface heating creates a Marangoni driven flow that, along with buoyancy effects, influences heat transfer and the steady-state liquid pool shape. Also, high surface velocities lead to the development of flow turbulence. Due to significant surface temperature, the radiative heat transfer is high and must be accounted for, as well as the heat loss through the crucible cooling circuit. The latter depends on the quality of the contact between the button and crucible. Therefore, the major physical phenomena which are simulated are listed in Table 1.

**Table 1.** The major physical phenomena considered in the model.

| |
|---|
| Heat transfer<br>• Heat transfer within the Ti sample by convection and conduction,<br>• Heat input by the electron beam,<br>• Radiation at the top surface,<br>• Phase change,<br>• Heat extraction by the cooling water circuit.<br><br>Momentum transfer<br>• Buoyancy caused by volumetric temperature gradients, and surface tension effects (Marangoni flow) related to surface temperature gradient,<br>• Turbulence development,<br>• Interaction between the solid and liquid within the mushy zone. |

Space limitations do not allow us to give details of the constitutive equations used in the mathematical model, nor of their boundary conditions. Basically, they are similar to the ones used in classical CFD axisymmetrical models developed at IJL Nancy, such as the SOLAR and SOLECS codes used for simulating the Vacuum Arc Remelting and ElectroSlag Refining operations [2–5]. Certain specific features of the model are reported below, along with simplifying assumptions.

Typical timescale of the process is several minutes, much longer than the beam dwell time (a few milliseconds), so the energy input can be considered as uniformly distributed over the pattern, neglecting the actual motion of the beam. Assuming a constant back-scattering coefficient $\xi$, the incoming flux density $\varphi_{EB}$ takes the following form:

$$\varphi_{EB} = (1-\xi)\,\frac{W_{EB}}{A_{pat.}}$$

where $W_{EB}$ is the total beam power and $A_{pat.}$ is the area covered by the beam pattern. Also, the deformation of the top free surface is neglected. Therefore, the shear stress $\tau$ caused by the tangential variation in surface tension with the temperature is expressed as:

$$\tau = \frac{\partial \gamma}{\partial T}\left(\vec{\nabla} T - \vec{n}\,(\vec{\nabla} T\,.\,\vec{n})\right)$$

where $\gamma$ is the surface tension of liquid Ti alloy.

Velocity is set to zero at the metal/crucible interface (no slip). At this interface, heat transfer is calculated using the following equation:

$$\varphi_{cool} = \zeta\,(z)\,h_{cool}\,(T - T_w)$$

in which $\varphi_{cool}$ is the heat loss, $h_{cool}$ is a heat transfer coefficient that accounts for conduction in the crucible and copper-water interaction, and $T_w$ is the temperature of the cooling water. Parameter $\zeta(z)$ was introduced to describe the quality of the contact between the metal and crucible. Basically, a better contact is obtained near the top of the button, while the cooling of the metal is liable to cause shrinkage near the bottom.

The set of coupled partial differential equations, which constitute the model, were solved numerically with a hybrid PISO-SIMPLE [6,7] algorithm. Time discretization was based on an implicit Euler scheme, while either the first-order upwind or the second-order van Leer scheme [8] was used for the discretization of convective terms. Laplacian (i.e. diffusive) terms were treated with a linear 2nd order scheme. The finite volume matrix was solved using preconditioned conjugate/bi-conjugate gradient methods [9,10].

Before advancing to the new time step, one needs to ensure full convergence of the solution. Following [11], residuals of the temperature, velocity, and pressure were checked, and the whole system was iterated until values between $10^{-2}$ and $10^{-6}$ were reached.

Asymmetric EB pattern requires a full 3D modeling. The solver was implemented within the OpenFOAM [12] framework so it is fully 3D, parallel and easily extensible with new physical models. The total number of cells was typically 500,000, while the time step for the calculations was initially set equal to $10^{-3}$ s and then controlled by a CFL condition. A weak parallelization (between 32 and 48 calculation nodes) was employed. The computational time required to obtain the results presented in next section was around 10 hrs.

## 3. Computed thermohydrodynamic behavior of the Ti button

In this section, typical results of the 3D model are presented, when applied to the case of Ti melting in the EB furnace. In order to validate the model, some simulation results are compared with experimental data on actual melting of a titanium button in our facility. Total beam power was 16 kW and the back-scattering coefficient was estimated at 0.3.

In the experiment, the 1.25 kg button was preheated with a circular beam pattern for the first 10 minutes, after which the beam was switched to a more complex 3D pattern and the button heated during another 20 minutes. The 3D beam pattern design consisted of one horizontal rectangle, two vertical rectangles and an arc, as shown on Fig. 2 (left). Such a horseshoe pattern was used as it allows to dip a refractory sample in a deep liquid pool at a location which does not directly interacts with the beam.

The actual pattern was reconstructed from successive frames extracted from a high-speed video of the process. The horseshoe shape is clearly visible on Fig. 2 (right), which shows that the center of the top arc was displaced 7 mm to the east and 3 mm to the north from the center of the button top surface. Vertical rectangles of the pattern were 2.1 × 3.5 cm in size, while the horizontal rectangle was 6.2 × 1.4 cm.

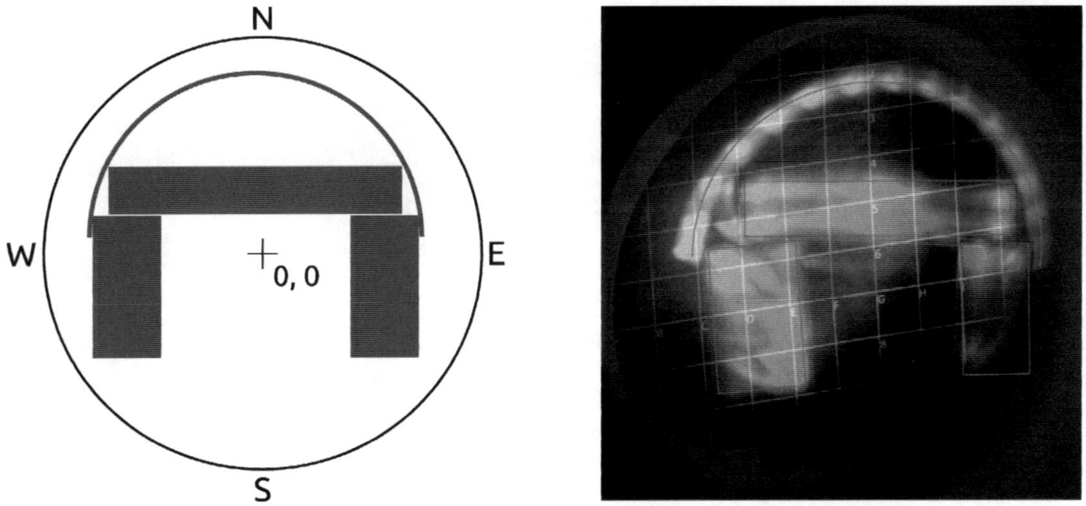


**Figure 2.** (left) design of the horseshoe beam pattern used during the last 20 minutes, (right) reconstruction of the actual pattern from high-speed films.

At t = 0, the solid metal is supposed to be at the room temperature. Figs 3 and 4 report the computed evolution of the liquid volume and the heating (by the beam) and cooling (radiation and contact) powers. Discontinuities at t = 600 s correspond to the abrupt change in the beam pattern, from circular to horseshoe. From these figures, it appears clearly that a steady state was fully reached after 30 minutes.

During the experiment, copper was introduced after 30 minutes into the liquid pool to mark its shape. The EB power was switched off and the button cooled rapidly. After full solidification and cooling, the button was cut every 5 mm in N-S and E-W directions (see Fig. 2 left) for reconstruction of the 3D liquid pool profile. The cutting scheme is shown on Fig. 5.

IOP Conf. Series: Materials Science and Engineering **143** (2016) 012019 doi:10.1088/1757-899X/143/1/012019

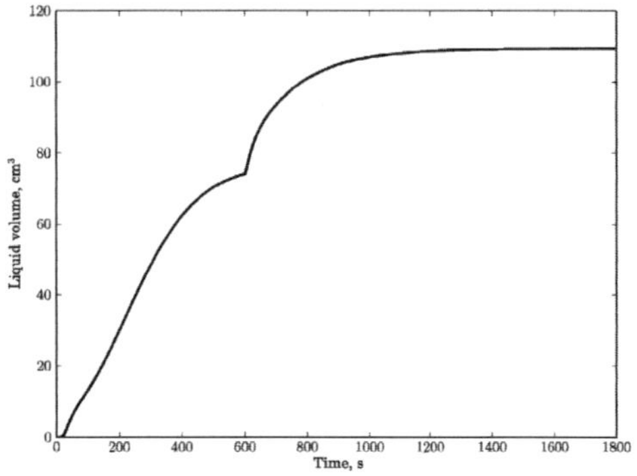


**Figure 3.** Computed evolution of the liquid volume

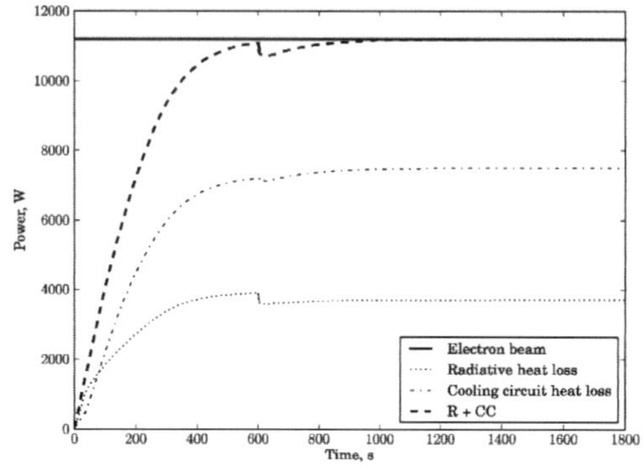


**Figure 4.** Computed evolution of the heating and cooling powers

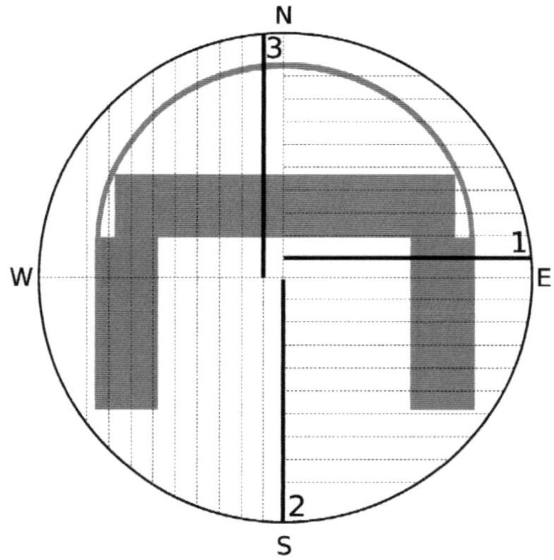


**Figure 5.** Cutting scheme of the solidified button. Bold numbered lines correspond to the the macrographs used to compare the observed profile shape with the simulation results

As shown on Fig. 5, three macrographs have been extracted for the sake of comparison between the computed and observed pool shape. These comparisons are the object of Figs 6 to 8.

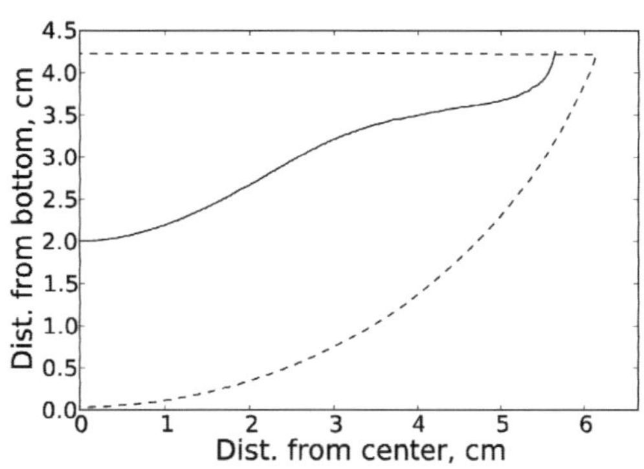


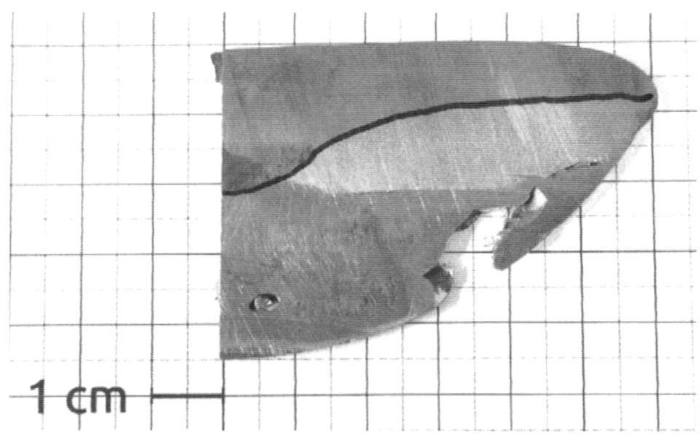


**Figure 6.** Melt pool profile intersection with plane 1 – (left) computed (right) observed

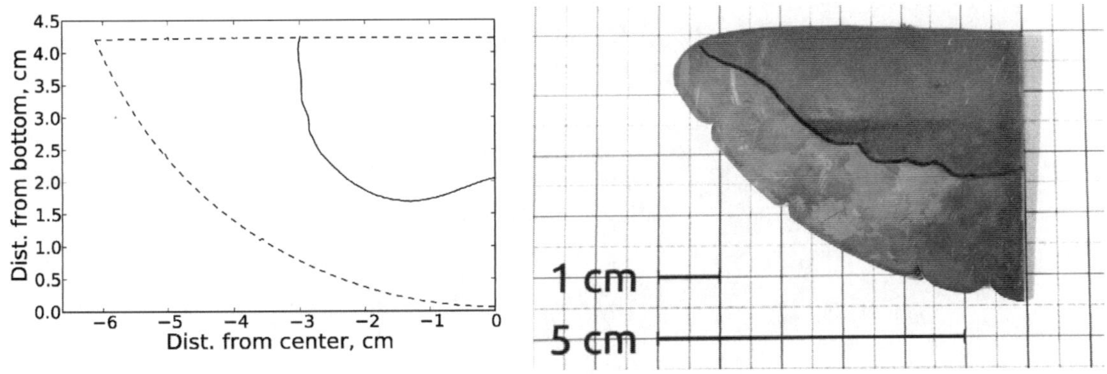


**Figure 7.** Melt pool profile intersection with plane 2 – (left) computed (right) observed

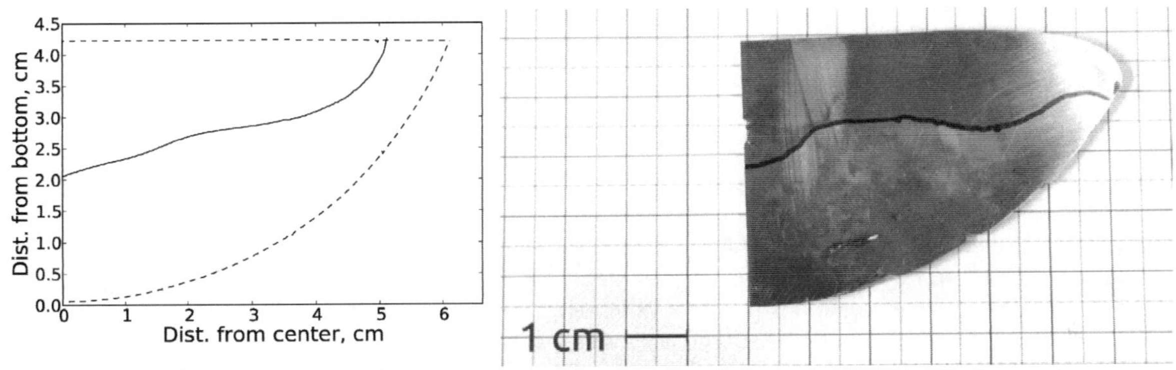


**Figure 8.** Melt pool profile intersection with plane 3 – (left) computed (right) observed

Qualitatively, the comparison is correct and the calculation reveals the main features of the observed pool: it is significantly larger than the beam pattern and its deepest part is located between the two vertical rectangles of the pattern. These two characteristics of the liquid pool behavior are directly related to the importance of convective heat transfer caused by the Marangoni driven flow. Figs 9 and 10 show the computed velocity field in the N-S center-plane and on top surface.

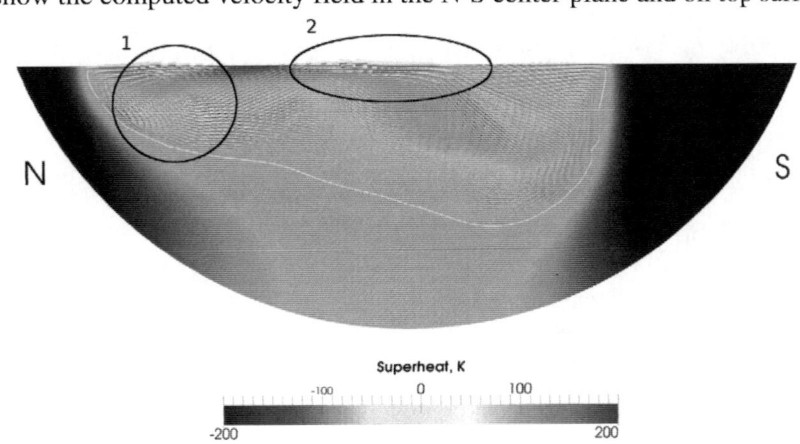


**Figure 9.** Computed velocity field in the N-S center-plane of the button. 1 – area of strong recirculation in the northern part of the button; 2 – intense flow caused by two jets formed by cumulating Marangoni flow from horizontal and vertical EB pattern rectangles

IOP Conf. Series: Materials Science and Engineering **143** (2016) 012019 doi:10.1088/1757-899X/143/1/012019

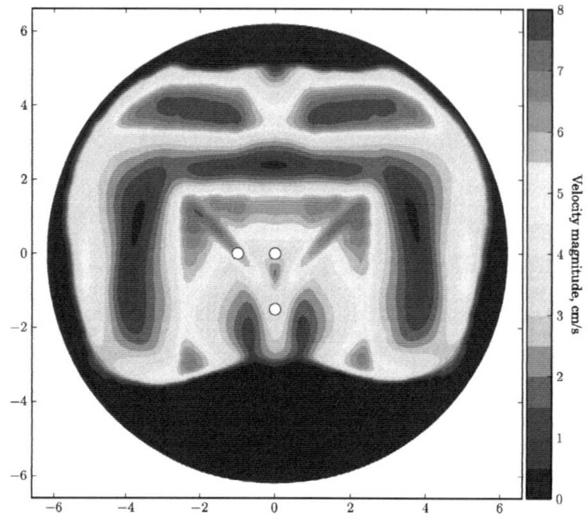


**Figure 10.** Computed velocity field at the top surface of the button. Rod dipping positions (see next section) are marked with white dots

Indeed, the flow, only weakly turbulent, is fully 3D. It appears to be mainly driven by Marangoni convection: a strong recirculation occurs at the North of the horizontal rectangle, while Marangoni driven jets are created between the two vertical rectangles. The maximum superheat reached at the surface is 210 K, while the maximum velocity is around 8 cm/s. It can be deduced from Figs 6 to 8 that the validity of the prediction is poorer near the edge because of the strong assumption of a flat top surface. A more detailed simulation should account for the deformation of the free surface.

## 4. Simulation of the influence of rod dipping

Let us recall that, during dissolution experiments, solid material was immersed into the liquid titanium bath [1]. The simple dissolution model requires local values of the velocity, turbulence and temperature in the liquid metal. As it was discussed in the previous section, the data can be obtained from the 3D model described in the paper. However, this approach assumes that presence of the inserted rod itself does not influence significantly the local fluid flow and heat transfer mechanisms.

To check the validity of this assumption, a new set of simulations were performed, where a solid rod was inserted in the steady-state liquid pool formed by the horseshoe electron beam pattern. The solid cylinder had a 4 mm diameter, while the full inserted length was 1.5 cm, which is a bit lower than the maximum pool depth. In order to simplify the calculation and maximize the effect of the presence of the rod, the latter was assumed to remain solid, i.e. no dissolution was accounted for. At the rod-liquid interface, the new version of the model used simple no-slip boundary conditions for fluid flow, wall-functions for turbulence and assumed no heat transfer.

Three dipping positions were simulated, as represented on Fig. 10. The first one is the location of the predicted maximum pool depth, where two jets formed by the Marangoni-driven flow meet, the second one is the geometrical center of the top surface, while the third immersion is localized in one of the Marangoni jets.

The initial liquid pool state corresponds to the steady state computed in the previous section. During the development of the new thermohydrodynamic behavior, first-order discretization schemes were used for the first 5 seconds, after which a switch to second-order schemes was made till the new steady state was reached.

In the case of central dipping position, after an initial perturbation caused by the sudden immersion, a new steady state was reached, which did not differ much from the previous one. Therefore, it can be concluded that the immersion of the rod had only a slight influence on the liquid metal state. In the case of dipping at the location of the deepest pool, the flow field near the free surface was not affected much. However, some perturbations occurred inside the pool. Indeed, more titanium was solidified, so that the solidification front actually reached the tip of the cylinder (no real dissolution of the rod was allowed). In the bulk, the temperature field was not heavily changed.

From the above discussion, it appears that in most cases, results of the 3D model, without actually accounting for the immersion of the refractory sample, can be used in a first approximation as a data set for a simple model of dissolution kinetics. Even though a precise quantitative prediction would require accounting for the effect of the presence of the rod, we must not forget that the dissolution process itself would certainly limit the influence of the sample on the pool behavior.

## 5. Conclusions

A numerical model of the coupled fluid flow and heat transfer in the liquid bath created by the EB melting of titanium in a hemispherical crucible has been set up. The corresponding 3D solver code was implemented within the OpenFOAM software. Validation was achieved, at least qualitatively, by comparing the predicted 3D pool profile with the observation of an EB melted Ti button in our experimental facility. The model was applied to the calculation of velocities and temperatures during dissolution experiments based on the immersion of refractory samples into the bath. While the 3D model enables us to reach data necessary for dissolution calculations, such an approach does not account for the influence of the presence of a dissolving rod on the behavior of the bath. Therefore, the immersion of a solid cylindrical body which represents a non-dissolving refractory sample was also simulated. Results revealed that the foreign body may slightly disturb the flow in its vicinity. The amount of the disturbance depends on the exact dipping location.

The main features required for improving the model are a better representation of the free surface, assumed here to be flat, and an actual modelling of the dissolution of the immersed rod, not accounted for yet. Anyway, these first results are promising enough so that a 3D modelling of the continuous growth and solidification of a remelted ingot, either an EB or a VAR one, has started at IJL Nancy.

## References

[1] G. Ghazal, P. Chapelle, A. Jardy et al., *ISIJ International* **52** (2012) 1-9.
[2] T. Quatravaux, S. Ryberon, S. Hans et al., *Journal of Materials Science* **39** 2004) 7183-7191.
[3] A. Jardy, D. Ablitzer, *Materials Science and Technology* **25** (2009) 163-169.
[4] V. Weber, A. Jardy, B. Dussoubs et al., *Met. and Mat. Transactions B,* **40** (2009) 271-280.
[5] M. Revil-Baudard, A. Jardy, H. Combeau et al., *Met. and Mat. Transactions B,* **45** (2014) 51-57.
[6] S.V. Patankar, D.B. Spalding, *Int. Journal Heat and Mass Transfer* **15** (1972) 1787-1806.
[7] R.I. Issa, *Solution of the implicit discretized fluid flow equations by operator splitting,* Mechanical Engineering Rep. FS-82-15, Imperial College London, 1982.
[8] B. van Leer, *Journal of Computational Physics* **14** (1974) 361-370.
[9] M.R. Hestenes, *Conjugate directions methods in optimization,* Springer, 1980.
[10] H.A. van der Vorst, *SIAM J. Sci. and Stat. Comput.,* **13** (1992) 631-644.
[11] H. Jasak, *Error analysis and estimation for the Finite Volume method with applications to fluid flows,* PhD thesis, Imperial College London, 1996.
[12] OpenFOAM Foundation, http://openfoam.org.
[13] O. Bertrand, B. Binet, H. Combeau et al., *Int. Journal of Thermal Sciences,* **38** (1999) 5-26.
[14] N. Hannoun, V. Alexiades, T.Z. Mai, *Numerical Heat Transfer B,* **44** (2003) 253-276.
[15] G. Vidalain, L. Gosselin, M. Lacroix, *Int. Journal Heat and Mass Transfer,* **52** (2009) 1753-1760.

# Dynamic melting and impurity particle tracking in continuously adjustable AC magnetic field

**V. Bojarevics, K. Pericleous**

University of Greenwich, London SE10 9LS, UK

v.bojarevics@gre.ac.uk

**Abstract.** The analysis of semi-levitation melting is extended to account for the presence of particles (impurities, broken metal dendrite agglomerates, bubbles) during the full melting cycle simulated numerically using the pseudo-spectral schemes. The AC coil is dynamically moving with the melt front progress, while the generated Joule heat serves to enhance the melting rate. The electromagnetic force is decomposed into the time average and the oscillating parts. The time average effects on the particle transport are investigated previously using approximations derived for a locally uniform magnetic field. This paper presents expressions for the skin-layer type of the AC force containing also the pulsating part which contributes to the particle drag by the 'history' and 'added mass' contributions. The intense turbulence in the bulk of molten metal additionally contributes to the particle dispersion. The paper attempts to demonstrate the importance of each of the mentioned effects onto the particle transport during the melting until the final pouring stage. The method could be extended to similar AC field controlled melting/solidification processes.

## 1. Introduction

Aluminium, magnesium and titanium alloys are the light structural metals attractive for aerospace and automotive applications that offer mechanical properties at a significant weight saving. Research efforts are aimed at developing new casting techniques and practices to ease inherent processing difficulties of aluminium and magnesium alloys to achieve more desirable mechanical properties for critical automotive structural and kinetic components. One major barrier to casting these components is melting and subsequently casting homogeneous alloys free of contamination and severe segregation, which may lead to undesirable properties. These factors can affect the melt microstructural features leading to the poor grain structure, solute segregation and porosity, and eventually the reduced mechanical properties of cast product.

A finer microstructure usually improves the mechanical properties of the solidified alloy. Finer microstructure can be achieved by increasing the cooling rate, adding dispersed micro to nano size particles of other materials serving as grain refiners. The latter technique is preferable if it can be controlled by additional stirring, and it eases the restrictions on the cooling rate during the casting process. It is well known that forced stirring of the melt by electromagnetic forces can significantly promote the detachment of dendrite arms and transporting the crystal fragments in a disperse fashion due to the presence of high turbulence [1]. However, a unidirectional stirring in laminar conditions could induce undesirable channel segregation in the solidifying material [2]. Therefore the electromagnetic stirring must be applied in an optimized way, specifically designed for a particular solidification process.

The semi-levitation melting (SM) is a containerless melting process where the electromagnetic confinement force field is generated using induction coils [3,4]. The electromagnetic field induces a strong turbulent flow in the molten pool, which is beneficial for producing a homogeneous melt with uniform temperature. The process combines induction melting and casting into a single self-controlling operation. The charge is gradually melted from top to bottom through movement of the induction coil as illustrated in Figure 1. Pouring of the charge through the hole in the chill block into the mold below occurs when the liquid melting front reaches the base of the metal charge, Figure 2.

The AC coil is dynamically moving with the melt front progress, while the generated Joule heat serves to enhance the melting rate when the coil proximity is adjusted. The electromagnetic force in the liquid melt can be decomposed into the time average over the AC period part, which serves to confine the relatively slower moving melt from leaking and produces intense large scale mixing. The oscillating nature of the total magnetic force means the presence of additional pulsating force component which exerts a significant effect on the microscopic particles in the melt. The particles travel due to the hydrodynamic drag, buoyancy [5], turbulent fluctuations [6] and the electromagnetic force action via the local pressure distribution around the particle [7]. The time average effects are investigated previously [8-11], but an incomplete approximation was used which was derived for a locally uniform electric and magnetic fields. The latter contradicts the skin-layer type of the exponential force distribution in the AC field [12,13]. Moreover, the total force in AC field contains also the pulsating part which contributes to the oscillating particle drag by affecting the 'history' and 'added mass' contributions [14]. The high mixing rates ensure the intense turbulence in the bulk of molten metal, additionally contributing to the particle dispersion.

This paper attempts to demonstrate the importance of each of the mentioned effects to the organized particle 'swarm' like transport and the eventual distribution when pouring melt in the bottom hole after the complete melting is reached. The semi-levitation dynamic melting principles could be extended to the similar AC field controlled melting processes, like the electrode melting for gas atomization, the contactless top coil 'immersion' mixing [15] and possibly other arrangements involving the dynamic AC field adjustment for high proximity electromagnetic interaction. The Lagrangian particle tracking in the presence of strong electromagnetic force field brings new perspectives to the understanding of the melting and casting processes.

## 2. Mathematical modelling

The mathematical modelling of the complex time dependent problem describing initially a solid body heating due to the AC electromagnetic field, then the melting process and subsequent liquid metal shape adjustment to the imposed force balance. The free surface confinement, internal fluid flow and the temperature evolution is accomplished using the SPHINX code developed at the University of Greenwich [16-18]. The time dependent Joule heating and the gradual melting requires the model to account for the phase transition and the following liquid material shape change by the action of the EM force time variable distribution.

The fluid flow is solved using the Reynolds averaged turbulent flow model for an incompressible fluid with the effective viscosity $v_e$ and the turbulent diffusivity $\alpha_e$ modified heat transfer equations:

$$\partial_t \mathbf{v} + (\mathbf{v}\cdot\nabla)\mathbf{v} = -\rho^{-1}\nabla p + \nabla\cdot(\nu_e(\nabla\mathbf{v} + \nabla\mathbf{v}^T)) + \rho^{-1}\mathbf{f}_{AC+DC} + \mathbf{g}\ , \qquad (1)$$

$$\nabla\cdot\mathbf{v} = 0\ , \qquad (2)$$

$$C_p^*(\partial_t T + \mathbf{v}\cdot\nabla T) = \nabla\cdot(C_p\alpha_e\nabla T) + \rho^{-1}\,|\,\mathbf{J}\,|^2/\sigma\ , \qquad (3)$$

where the usual notation is used, given in detail elsewhere [17]. The boundary conditions are prescribed dynamically on the liquid interface defined in spherical coordinates with the transformation function:

$$\mathbf{R} = \mathbf{e_R} R(\theta, \varphi, t) \ . \tag{4}$$

The surface dynamics are resolved following the velocity on the interface by the kinematic condition:

$$\mathbf{e_n} \cdot \mathbf{v} = \mathbf{e_n} \cdot \partial_t \mathbf{R} \ . \tag{5}$$

The free surface boundary conditions are determined by the tangential and normal stress components accounting for the temperature dependent surface tension $\gamma(T) = \gamma(T_m) + \gamma_T \cdot (T - T_m)$ :

$$\Pi_{n\tau} = \gamma_T \mathbf{e}_\tau \nabla T ; \quad \Pi_{nn} = \gamma K; \quad K = \nabla \cdot \mathbf{e}_n \ . \tag{6}$$

Note, that to preserve the momentum conservation the temperature dependent surface tension is used both for the normal and the tangential stress (Marangoni effect).

The solution method is based on the continuous co-ordinate transformation adapting to the free surface and the containing vessel shape. The present implementation adds the Lagrangian passive particle tracking algorithm. The position $\mathbf{R_p}$(x,y,z,t) of an individual particle can be determined following its path and the variable total force $\mathbf{F_p}$(x,y,z,t) acting on the particle by solving the set of two equations [5,6]:

$$\partial_t \mathbf{R} = \mathbf{u}_p, \quad m_p \partial_t \mathbf{u}_p = \mathbf{F}, \tag{7}$$

where $\mathbf{u}_p$ is the particle velocity and $m_p$ its mass. The force $\mathbf{F_p}$ acting locally on a spherical particle can be decomposed into the fluid drag force $\mathbf{F}_d$, the buoyancy force $\mathbf{F}_g$ and the effective electromagnetic force $\mathbf{F}_e$ . The drag force depends on the Reynolds number $\mathrm{Re}_p = r_p |\Delta \mathbf{u}| / \nu$, where $r_p$ is the particle radius, $\Delta\mathbf{u} = \mathbf{u} - \mathbf{u}_p$ is the (slip) velocity relative to the fluid velocity $\mathbf{u}$, ν – the kinematic viscosity. For small particles (1nm -100 μm) the Reynolds number $\mathrm{Re}_p$ is rather small, of the order 0.1 - 10, therefore the drag force can be approximated as instantaneous modified Stokes formula [5,14]

$$\mathbf{F}_d = \frac{3C_d^* \rho \pi d_p^2}{Re_p} |\Delta\mathbf{u}| \Delta\mathbf{u} = 3C_d^* \nu \rho \pi d_p \Delta\mathbf{u}, \qquad C_d^* = 1 + 0.15 Re_p^{0.687}, \quad (Re_p < 1000). \tag{8}$$

The buoyancy force due to the gravity $\mathbf{g}$ action on the particle and the surrounding fluid is

$$\mathbf{F}_g = (\rho_p - \rho) V_p \mathbf{g}, \tag{9}$$

$\rho$ is the fluid density, $\rho_p$ – the particle density, $V_p$ – the particle volume.

The electromagnetic force acts on the induced electric current in the particle and due to the fluid pressure redistribution on the surface of the particle [7]. For the case of electrically non-conducting particles (oxides, carbides, dendrite fragments, bubbles etc.) the electromagnetic force in the AC skin-layer can be derived as

$$\mathbf{F}_e = -V_p (\tfrac{3}{2} \bar{\mathbf{f}}_e + \mathbf{f}'(t)), \tag{10}$$

the sum of the time average and the oscillating components of the electromagnetic force computed in the fluid at the location of the particle. Note, the Leenov, Kolin [7] expression derived for DC constant fields,

$$\mathbf{F}_{eDC} = -V_p \tfrac{3}{4}\mathbf{f}_e \tag{11}$$

gives two times smaller force magnitude acting on a non-conducting particle.

The presence of the oscillating force component requires the modified drag expression for an oscillating spherical particle [14]:

$$\mathbf{F}_{\mathbf{d}} = 6\pi\nu\rho r_p[(1+\varepsilon)\Delta\mathbf{u} + \frac{\varepsilon}{\omega}(1+\frac{2}{9}\varepsilon)\frac{d\Delta\mathbf{u}}{dt}], \qquad \varepsilon = \sqrt{\frac{\omega r_p^2}{2\nu}}, \tag{12}$$

where the Stokes number $\varepsilon$ can reach an order 1 in an AC field oscillating at typical frequencies $\omega$ $\sim 10^3$ - $10^4$ Hz. The equation (12) contains the instantaneous Stokes drag, the memory term (Basset force) and the added mass force.

The typical mixing flow is turbulent ($Re \sim 10^3$ - $10^4$), which further requires a modification due to the stochastic part of velocity in accordance to the resolved turbulent kinetic energy $k$ and the local eddy life time (or the particle transit time in that eddy, whichever is the shortest) [6]. The use of $k$- $\omega$ turbulence model in the SPHINX code facilitates to obtain these quantities, which are locally interpolated to the particle position at each time step. The numerical integration of equations (7-10, 12) is performed for each individual particle of various properties depending on the initial seeding locations. The time dependent forces $\mathbf{F_p}$ are sensitive to the location and the instantaneous update of local slip velocity $\Delta\mathbf{u}$ (including the stochastic contribution). Therefore, the stability of long term integration along the particle tracks can be adversely affected by the choice of numerical integration scheme. The classical explicit integration schemes are limited to extremely small time steps, which could make the numerical solution difficult for the dynamic flow conditions. The exponential numerical scheme, proposed in [13], permits stable time integration of the particle tracks with the dynamically adjustable time steps of the order 0.1 – 1 ms used for the unsteady fluid flow solution in the examples considered below.

## 3. Numerical results

The electromagnetic field induces a strong turbulent flow in the molten pool, which is beneficial for producing a homogeneous melt with uniform temperature. The process combines induction melting and casting into a single self-controlling operation. The charge is melted from top to bottom through movement of the induction coil as illustrated in the Figures 1 and 2. The electromagnetic force needs continuous adjustment while the free surface moves in the molten state. Figure 1 demonstrates the difference in the distribution of the current and the force at the initial stage and the stage when almost full volume of the melt is liquid just before pouring starts. Pouring of the charge through the hole in the chill block into the mold below occurs when the liquid melting front reaches the base of the metal charge.

In this study the main interest is about the behavior of the particles in the melt. The Figure 3 compares the particle instantaneous positions after the random seeding in the initial solid cylinder, melting and mixing for 340 s, then pouring while the magnetic confinement and mixing is maintained. If the effects of the turbulent fluctuations and the oscillating force component are neglected, the mixing patterns are organized in particle 'loops', while the larger size particle loops collapse to concentrated localizations on the free surface. This may indicate a coagulation process for the larger particles. The electromagnetic force effect on the particles of sizes 50 -100 μm is comparable to the drag force, leaving the particles at positions where the gravity (buoyancy) force is balanced by the electromagnetic expulsion and the drag. This corresponds to observations for the small metallic samples (about 2-3 cm in diameter) where, after fast solidification, the particles are localized at certain surface positions depending on their size [8,13].

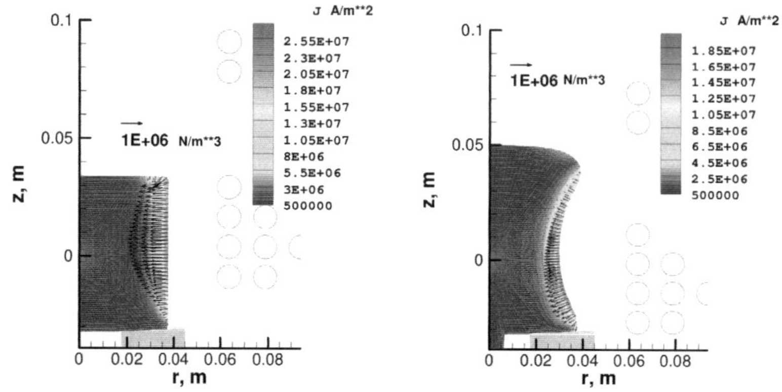


**Figure 1.** The induced electric current density and the time average electromagnetic force distribution in the melting ingot (d=7.5 cm) at the initial time (left) and at the final stage when liquid pouring starts (right) for the effective current magnitude in the coil 1094 A at the frequency 3717 Hz.

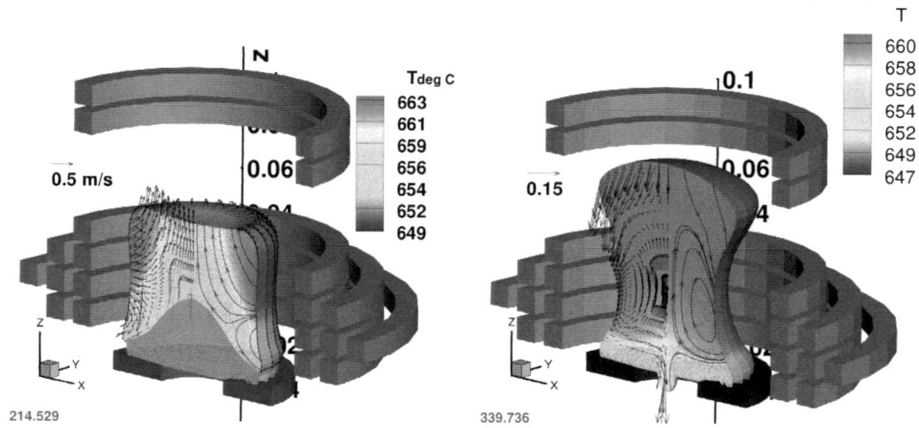


**Figure 2.** Melting front progress shown as two snapshots of the simulated results: the flow and the temperature at the intermediate stage 214 s and at the final stage when liquid pouring starts 339 s.

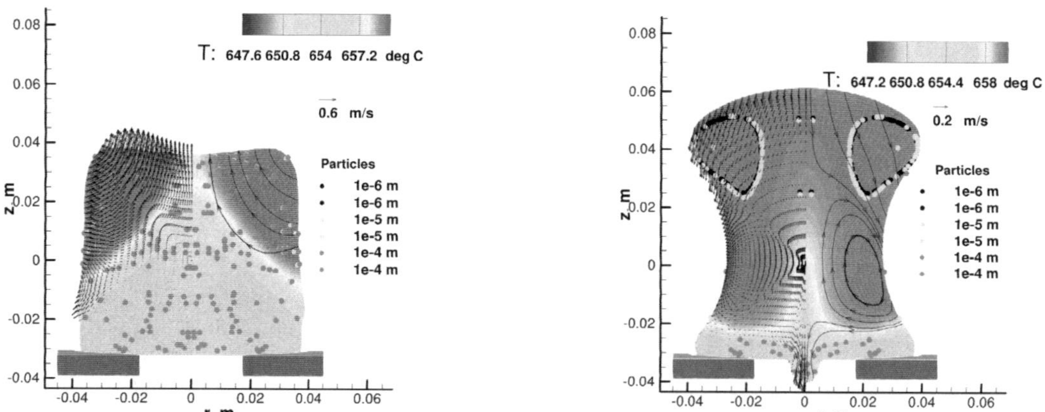


**Figure 3**. The particle positions at two snapshots of the simulated results: at the initial stage (214 s) and at the intermediate stage when liquid pouring starts (339 s). The particle tracking does not include the oscillating force and the stochastic turbulent fluctuation effects.

The effect of turbulent fluctuations is more pronounced in the semi-levitation case, having a larger size metallic volume (the initial ingot is 7.5 cm in diameter). In addition, the full AC force including the oscillating part contribution creates more realistic particle distribution shown in the sequence of snapshots in the Figure 4. The larger particles are still trapped on the surface, however their position is more dynamic, eventually shifting to lower height closer to the solid front. The larger particles are the last to exit the liquid zone when poured to the hole at the bottom.

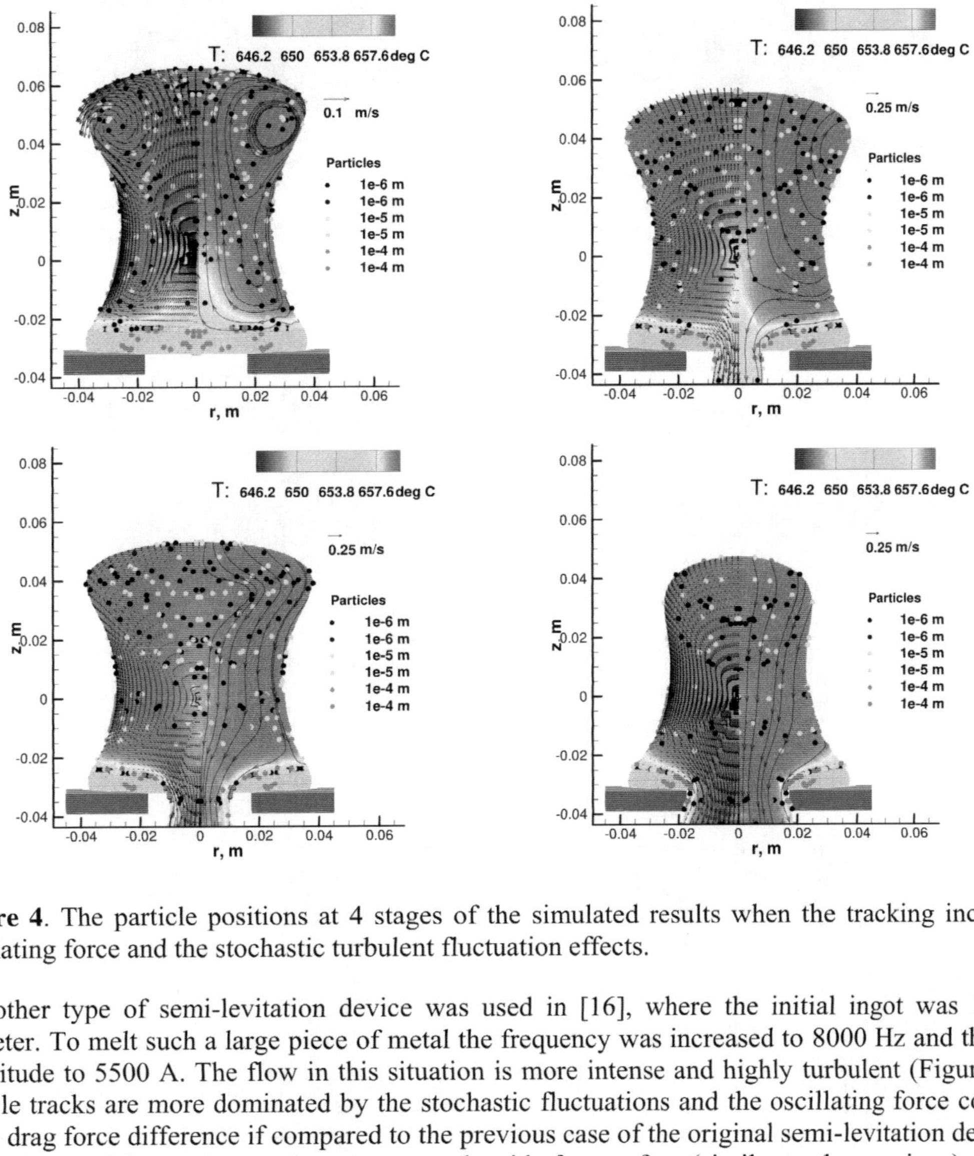


**Figure 4**. The particle positions at 4 stages of the simulated results when the tracking includes the oscillating force and the stochastic turbulent fluctuation effects.

The other type of semi-levitation device was used in [16], where the initial ingot was 12 cm of diameter. To melt such a large piece of metal the frequency was increased to 8000 Hz and the current magnitude to 5500 A. The flow in this situation is more intense and highly turbulent (Figure 5). The particle tracks are more dominated by the stochastic fluctuations and the oscillating force contributes to the drag force difference if compared to the previous case of the original semi-levitation device. The larger size particles tend to agglomerate near the side free surface (similar to the previous) and to the top of the liquid metal column sustained by the magnetic forces. The smaller size particles appear to be removed first from the liquid column to the hole at the bottom when pouring starts, while the large particles are retained until the end stage and reach the exit at the very end of the pouring process (Figure 6).

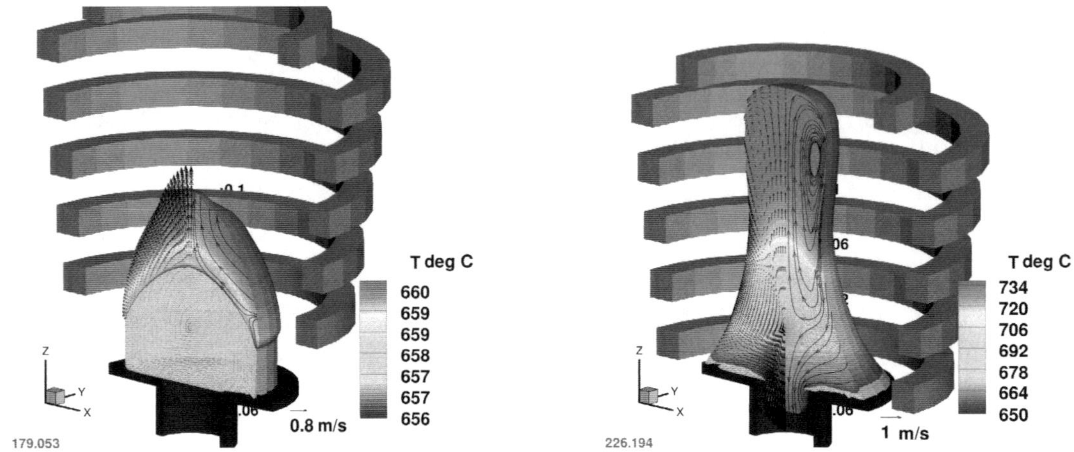


**Figure 5**. Melting of the large ingot (d=12 cm) shown as two snapshots of the simulated results: the flow and the temperature at the intermediate stage 179 s and at the final stage when liquid pouring starts 226 s. (For the effective current magnitude in the coil 5500 A at the frequency 8000 Hz).

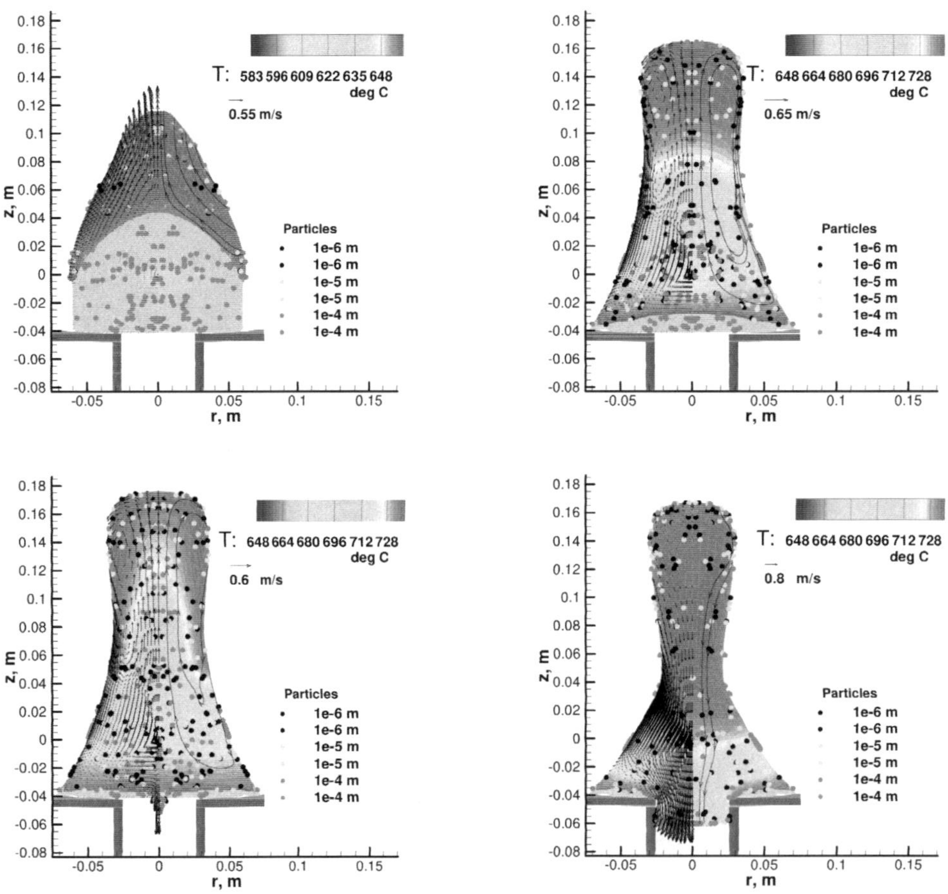


**Figure 6**. The particle positions at 4 stages of the simulated results for the large ingot when the tracking includes the oscillating force and the stochastic turbulent fluctuation effects.

IOP Conf. Series: Materials Science and Engineering **143** (2016) 012020 doi:10.1088/1757-899X/143/1/012020

## Conclusion

The results of modelling show the importance of the electromagnetic effects on the particle distribution when pouring the molten metal in the semi-levitation process, indicating the potentially beneficial feature to supply well dispersed small particles at the initial and mid stages of the mold filling, separating the agglomerates until the last stage when the mold could be already full. This would result in a fine grained casting structure.

## Acknowledgment

The authors acknowledge financial support from the ExoMet Project (co-funded by the European Commission (contract FP7-NMP3-LA-2012-280421), by the European Space Agency and by the individual partner organizations), and from the project SIKELOR of the European Union's Seventh Framework Programme for research, technological development and demonstration under grant agreement No. 603-718.

## References

[1] Campanella T, Charbon C and Rappaz M 2004 Grain refinement induced by electromagnetic stirring: a dendrite fragmentation criterion *Metallurgical and Materials Transactions* **35A** 3201-3210

[2] Medina M, Du Terrail Y, Durand F and Fautrelle Y 2004 Channel segregation during solidification and the effects of an alternating travelling magnetic field *Metallurgical and Materials Transactions* **35B** 743-754

[3] El-Kaddah N, Piwonka T S and Berry J T 1991 *US Patent* 5033948

[4] Adams J and El-Kaddah N 1993 Containerless processing and characterization of high purity aluminum alloys *Proceedings of the 1st International Conference on Processing Materials for Properties* (TMS Warrendale OH) 905-908

[5] Clift R, Grace J R, Weber M E 2005 *Bubbles, Drops, and Particles* (Dover Publications Mineola New York)

[6] Tucker P G 2001 *Journ. Fluids Eng* **123** 372-381

[7] Leenov D and Kolin A 1954 *Journ. Chem. Phys* **22** 4 683-688

[8] Toh T, Yamamura H, Kondo H, Wakoh M, Shimasaki Sh, Taniguchi S 2007 *ISIJ International* **47** 11 1625-1632

[9] Haverkort J W, Peeters T W J 2010 *Metallurgical and Materials Transactions* **41B** 1240-1246

[10] Taniguchi S, Kikuchi A 2000 *Proc. 3rd Int. Symp. Electromagn. Proc. Materials* (ISIJ, Nagoya) 315-320

[11] Zhang L, Wang Sh, Dong A, Gao J, Damoah L N W 2014 *Metall. Materials Trans* **45B** 2153-2185

[12] Etay J 2009 private communication

[13] Bojarevics V, Pericleous K and Brooks R 2009 *Metall.Materials Trans* **40B** 3 328 – 336

[14] Landau L D and Lifshitz E M 1987 *Fluid Mechanics* (Pergamon Press)

[15] Bojarevics V, Djambazov G, Pericleous K 2015 *Metall Materials Trans* **46A** 2884 – 2892

[16] Bojarevics V, Pericleous K and Cross M 2000 Modeling the Dynamics of Magnetic Semi-levitation Melting *Metall. Materials Trans* **31B** 179-189

[17] Bojarevics V, Harding R A, Pericleous K and Wickins M 2004 The development and validation of a numerical model of an Induction Skull Melting Furnace *Metallurgical and Materials Transactions* **35B** 785-803

[18] Bojarevics A, Bojarevics V, Gelfgat J and Pericleous K 1999 Liquid Metal Turbulent Flow Dynamics in a Cylindrical Container with Free Surface: Experiment and Numerical Analysis *Magnetohydrodynamics* **35** 3 258-277

IOP Conf. Series: Materials Science and Engineering **143** (2016) 012021 doi:10.1088/1757-899X/143/1/012021

# Experimental study on directional solidification of Al-Si alloys under the influence of electric currents

**D Räbiger, Y Zhang, V Galindo, S Franke, B Willers and S Eckert**[1]

Helmholz-Zentrum Dresden-Rossendorf, 01314, Germany

E-mail: s.eckert@hzdr.de

**Abstract.** The application of electric currents during solidification can cause grain refinement in metallic alloys. However, the knowledge about the mechanisms underlying the decrease in grain size remains fragmentary. This study considers the solidification of Al-Si alloys under the influence of electric currents for the configuration of two parallel electrodes at the free surface. Solidification experiments were performed under the influence of both direct currents (DC) and rectangular electric current pulses (ECP). The interaction between the applied current and its own induced magnetic field causes a Lorentz force which produces an electro-vortex flow. Numerical simulations were conducted to calculate the Lorentz force, the Joule heating and the induced melt flow. The numerical predictions were confirmed by isothermal flow measurements in eutectic GaInSn. The results demonstrate that the grain refining effect observed in our experiments can be ascribed solely to the forced melt flow driven by the Lorentz force.

## 1. Introduction

The adjustment of fine grain morphologies is a crucial issue for improving characteristics and properties of cast and wrought aluminium alloys. Several methods are known to achieve grain refinement in solidification processes: add-on of grain refiners [1], rapid cooling conditions [2], mechanical or electromagnetic stirring [3], or ultrasonic treatment [4].

Grain refinement by inoculants is well reported as a very effective and well-known method which is widely used in foundry industry. It can be realized easily by addition of Al-Ti or Al-Ti-B master alloys and provides beneficial effects, for instance better feeding, reducing of hot tearing and improvement of strength [5,6]. However, the inoculation of the melt by grain refining particles may produce undesired side effects appearing as impurities or local defects in the microstructure. The electromagnetic treatment of solidifying alloys in form of electromagnetic stirring or the so-called Electric Current Pulse technique (ECP) unlocks a remarkable potential to influence the solidified macrostructure of metals during the solidification process. Many studies have shown that beneficial effects like a distinct grain refinement or the promotion of the transition from a columnar to an equiaxed dendritic growth (CET) can be achieved (see for instance [7-9]). However, the physical mechanism of the grain refinement effect caused by ECP has not been understood so far. Various effects are under discussion, such as the fragmentation of dendrite induced by the electric current [7], the reduction of the nucleation activation energy [8], or the break out and the transport of little grains from the boundary

[1] To whom any correspondence should be addressed.

by the periodic Lorentz force [9]. However, the possibility that intense Lorentz forces resulting from the interaction between the strong electrical current and the self-induced magnetic field can create significant melt flows was only insufficiently considered by previous studies. This situation was the motivation for us to perform an experimental study for investigating fluid flow effects during solidification of AlSi alloys under the impact of electric currents.

## 2. Experimental setup

Fig. 1(a) shows a schematic view of the experimental setup. A cylindrical vessel made of Perspex with an aspect ratio $A = H_0/2 \cdot R_0 = 1.2$ was used. The size of the inner diameter $D = 2 \cdot R_0$ and the filling height $H_0$ was chosen to be 50 mm and 60 mm, respectively. For fluid flow measurements the cylinder was filled with the eutectic alloy GaInSn. Two pencil electrodes (diameter 8 mm) are dipped through the free surface into the melt up to an immersion depth of 10 mm at opposite radial positions of $r_E = 18$ mm. The lateral surface of the electrodes is coated by an electrically insulating material in order to ensure that the electric current enters the liquid metal only through the front surface of the electrodes. The ultrasound Doppler velocimetry (UDV) was applied to measure the fluid flow inside the cylinder. More details about the measuring principle can be found in [10]. To achieve the acoustic coupling between sensor and fluid we attached the ultrasonic transducer to the bottom wall of the fluid container. A new measuring system of linear sensor arrays was used to realize a flow mapping of the meridional flow. Details of the measuring system are given in [11]. The spatial resolution in lateral direction varied from 5 mm at the sensor to approximately 7.5 mm at the lid of the fluid vessel. In axial direction a spatial resolution of about 1.5 mm was achieved. The velocity data were acquired with sampling frequencies between 0.5 and 6 Hz.

Solidification experiments were performed using an Al-7wt%Si alloy prepared from 99.99% Al and 99.999% Si according to the designated composition. The raw alloy was modified with 200ppm Sr. We give only a brief description here concerning the solidification experiment. Further details can be found in [12]. The Al-7wt%Si samples were solidified directionally from the bottom in a double-walled, cylindrical stainless steel mold. The mold has an internal diameter of 50 mm and a height of 100 mm. The schematic drawing of the experimental set-up can be found in Fig. 1(b). Preparatory specimens were made with a predetermined weight of 310 g corresponding to a filling height of 63 mm of the liquid in the mold. After melting the mold was positioned at a water-cooled copper chill which was kept at a constant temperature of about 20°C. In the experiments presented here the cooling of the sample and the influence of the electromagnetic field starts nearly at the same time.

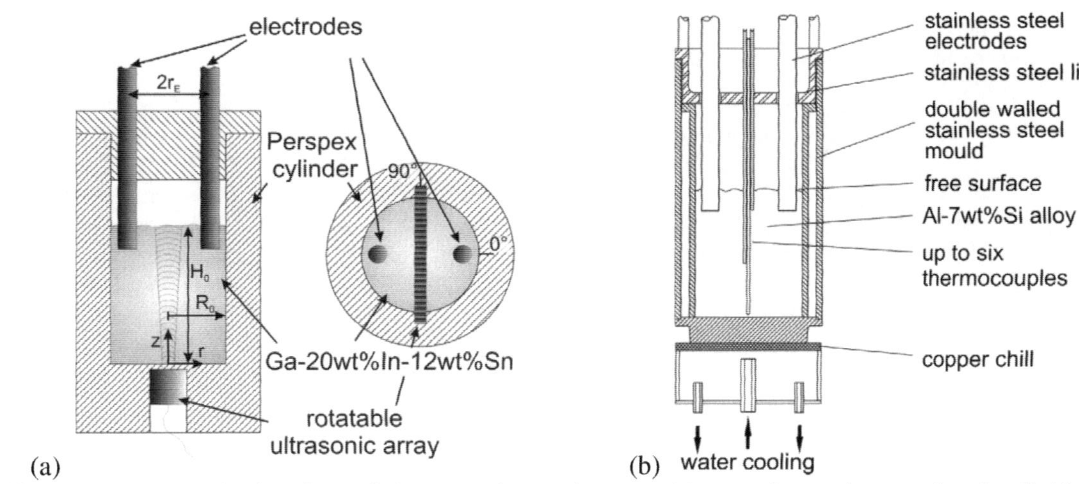


**Figure 1.** Schematic drawing of the experimental setup: (a) experimental setup for the fluid flow measurements, (b) experimental setup of the solidification experiments

To avoid an immediate formation of a solid shell the electrodes were preheated until achieving the same temperature as the melt before cooling from the bottom was initiated. The cables between the electrodes and the power supply were fixed at a considerable distance to the mold in order to minimize the interaction between the induced magnet field around the cable and the electric current flowing in the solidifying melt. The DC electric current and the current pulses are generated by the pulse reverse power supply pe86CWD (plating electronic). This device gives us the possibility to generate rectangular current pulses of variable pulse length at adjustable frequencies up to 200 Hz. Temperature measurements were made within the mold cavity during solidification using 6 type-K thermocouples arranged vertically along the axis of the mold. Temperature measurements were conducted at vertical positions of 1.5, 11, 22, 33, 44 and 55 mm above the bottom of the mold with an overall sampling rate of 5 Hz.

## 3. Numerical scheme

Numerical computations have been performed to calculate the actual distributions of electric current and magnetic field as well as the resulting Lorentz force. Furthermore, these simulations provide predictions of the flow field in the liquid metal column.

For the numerical simulations of the electric current $\boldsymbol{J}$ and the magnetic induction $\boldsymbol{B}$ in the melt and all electrically conducting parts of the facility the finite element code OPERA (Cobham plc. [13]) was used. In the case of a direct current imposed on the electrodes we solve the Laplace equation for the electric potential $\nabla^2 \varphi = 0$ taking into account the charge conservation and the continuity of the current density $J$ at the interface between two regions with different electric conductivities: $J_n = -\sigma_1 \boldsymbol{n} \cdot \nabla \varphi_1 = -\sigma_2 \boldsymbol{n} \cdot \nabla \varphi_2$. The magnetic induction $\boldsymbol{B}$ was calculated from the current density $\boldsymbol{J} = \sigma\,(-\nabla \varphi)$ using the Biot-Savart law:

$$\boldsymbol{B}(\boldsymbol{r}) = \frac{\mu_0}{4\,\pi} \int d\,\boldsymbol{r}'^3 \, \frac{\boldsymbol{J}\,(\boldsymbol{r}') \times (\boldsymbol{r} - \boldsymbol{r}')}{|\,\boldsymbol{r} - \boldsymbol{r}'\,|^3} \tag{1}$$

The boundary conditions for the electric potential are defined using the amplitude of the imposed electric current at the end of the electrodes: $I_0 = \int_A d\,\boldsymbol{s} \cdot \boldsymbol{J}$, where $A$ is the area of the electrode's cross section.

The flow in the volume containing the melt was simulated numerically by means of the open source library OpenFOAM [14] solving the Navier-Stokes equation together with the incompressibility condition and including an electromagnetic force density term:

$$\rho \left( \frac{\partial\,\boldsymbol{u}}{\partial\,t} + (\boldsymbol{u} \cdot \nabla)\,\boldsymbol{u} \right) = -\nabla\,p + \eta\,\nabla^2\,\boldsymbol{u} + \boldsymbol{J} \times \boldsymbol{B} \tag{2}$$

The boundary conditions for the flow field are the no-slip condition $\boldsymbol{u} = \boldsymbol{0}$ at the solid container walls. For the melt surface either $\boldsymbol{u} = \boldsymbol{0}$ or the conditions for a stress-free, non-deformable surface $u_n = 0$ and $\partial\,u_t \,/\, \partial\,z = 0$ are applied depending on whether the melt flow is evaluated in an open or an enclosed container. A computational grid with 650000 volume elements was used.

In case of an applied alternating current the induced electromagnetic fields have to be taken into account. For that reason we used the force density term $\boldsymbol{f}_L = <\boldsymbol{J} \times \boldsymbol{B}>_T$ averaged over one period $T = 2\,\pi \,/\, \omega$.

The correct determination of the electromagnetic fields and the resulting Lorentz force density $\boldsymbol{f}_L = <\boldsymbol{J} \times \boldsymbol{B}>_T$ is an obvious prerequisite for the numerical simulation of the flow field.

## 4. Results

Parameter variations in case of a treatment with pulsed electric current concern the current amplitude $I_P$, the frequency $f$ and the pulse length $t_p$. The pulse scheme is shown in figure 2(a). A first series of flow measurements were performed to investigate the impact of the current parameter on the fluid flow. Our flow measurements revealed that the global structure of the flow does not depend remarkably on the choice of parameter sets [15]. A typical flow field as predicted by numerical simulations is shown in figure 2(b). The melt flow is driven by a Lorentz force which is generated by the interaction between the radially diverging electric current at the tips of the electrodes and the almost azimuthally induced magnetic field. A strong downwards Lorentz force exists just below the electrodes, as shown in figure 2(c). This Lorentz force drives a flow in form of two descending jets. A recirculating flow occurs close to the cylinder walls within the lateral interspace on both sides of the electrodes. The numerical predictions are confirmed by corresponding flow measurements [15].

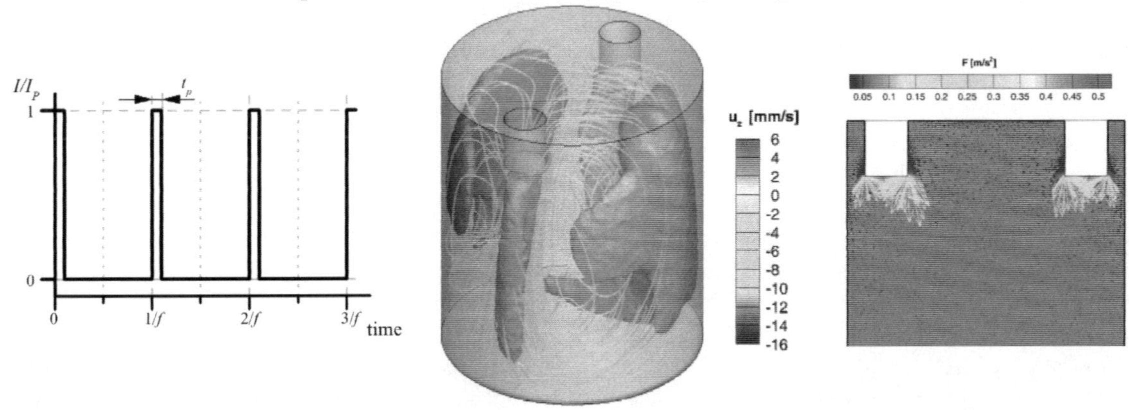


(a) (b) (c)

**Figure 2.** Schematic drawing of: (a) the pulse scheme, (b) numerical results showing the three-dimensional flow structure caused by the electric current and the interaction with the self-induced magnetic field (the iso-surfaces illustrate the vertical velocity, where negative and positive values represent an downward and upward flow, respectively) and (c) calculated Lorentz-force in the plane containing the electrodes

While the flow pattern remains similar for all parameter variations made here, the flow intensity is dependent on the parameters $I_P$, $f$ and $t_P$. For comparison we introduce the effective value of the applied current amplitude $I_{eff} = I_P{\cdot}(t_P{\cdot}f)^{0.5}$. This value is equivalent with the energy input into the system. In the case of a direct current $I_{eff}$ is equal to the applied $I_{DC}$.

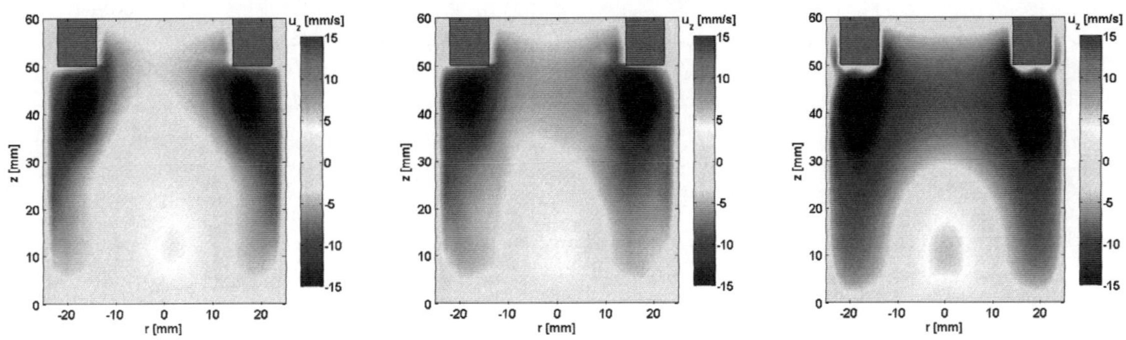


(a) (b) (c)

**Figure 3.** Two-dimensional distribution of the mean vertical velocity in the plane containing the electrodes measured at a) $I_{DC} = 48$ A, b) pulsed current with $I_P = 152$ A, $f = 200$ Hz, $t_P = 0.5$ ms ($I_{eff} = 48$ A) and c) numerical simulation with $I_{DC} = 48$ A

Figure 3(a) and (b) presents measurements of the averaged vertical velocity ($u_z$) recorded for a DC value of 48 A and a pulse current with $I_{eff}$ = 48 A. The measurements show the flow structure in the plane containing the electrodes. However, the mean flow intensities and flow structure are almost identical. Figure 3(c) depicts the calculated results from the numeric with a DC current of 48 A. The general agreement of the numerical results is obviously. A more detailed picture of the flow structure and dynamic can be obtained in [16].

It is well-known from previous investigations, where the influence of melt stirring on solidification was studied, that the melt flow determines the temperature distribution in the sample. Figure 4(a) shows the evolution of the temperature along the sample axis at different vertical positions for different kinds of melt treatment. Figure 4(b) depicts the mean vertical temperature gradient averaged over the six measuring points along the axis. In case without any melt agitation (black line) we observe the development of an axial temperature gradient after the initiation of the cooling process. This temperature gradient is only slightly affected by applying electric current pulses at low effective currents ($I_{eff}$ = 53.7 A – green line). An almost complete homogenization of the temperature field in the liquid phase is achieved by a further increase of the effective current ($I_{eff}$ = 152 A – blue line). Distinct oscillations of the temperature signal indicate the occurrence of turbulent fluctuations in the liquid. Dendrite fragmentation by partial remelting can be promoted by the occurrence of such temperature and coupled solute fluctuations at the solidification front. This finding of a coupled temperature and solute fluctuation would be indirect affirmed by experiments in pure aluminium [17, 18]. The energy input caused by joule heating has no effect on the solidification time (achievement of solidus at the upper thermocouple (dashed line)). Figure 4 also contains a corresponding measurement obtained from a solidification experiment under the influence of electromagnetic stirring by a travelling magnetic field (TMF – red line). The field strength of the TMF ($B_{TMF}$ = 12 mT) was chosen in such a way that the intensities of the generated melt flows (the mean intensity of the vertical velocities along the axis were taken as reference) can be expected to be almost the same.

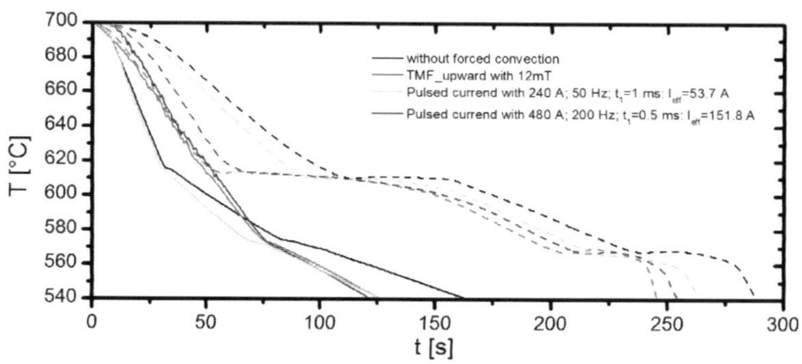


a)

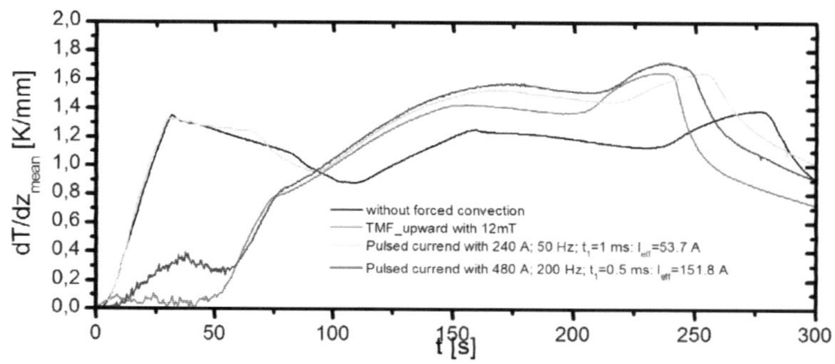


b)

**Figure 4.** Measurements of the temperature along the sample axis during the solidification for varying kinds of melt forcing (a) for the vertical position z = 1.5 mm (solid line) and z = 55 mm (dashed line) and (b) the mean vertical temperature gradient averaged over six thermocouples

The TMF generates a flow pattern in form of a torus, where the melt flows downwards at the axis and upwards along the side walls of the cylinder [19, 20]. Thus, we can assume that the global flow structures driven by the TMF or the electric current in the experimental configuration considered here are not identical, but, rather similar.
The macrostructure of solidified samples in the longitudinal section is shown in figure 5. Coarse, columnar grains grow parallel to temperature gradient in the reference experiment without applied electric current (fig. 5(a)). A transition to an equiaxed structure of large grains occurs only just below the surface. The application of electric current pulses or a TMF changes the grain morphology completely. The section is covered by equiaxed grains which are significantly smaller as found for the untreated case. A few columnar grains still exist in the lower part of the sample which was processed under the influence of the electric current pulses (fig. 5(b)). It becomes obvious that these columnar dendrites are not perfectly aligned in vertical direction. It is known that the deflection of the columnar dendrites can be caused by a lateral incident flow [21, 22] or by an asymmetrical flow structure as shown in [23] for a TMF-driven flow where a small shift of the symmetry lines between magnetic field and mould axis caused significant changes of the flow structure. For our experiment we have to expect a slight asymmetry in the current-driven flow due to a non-symmetric configuration of the power lines around the experimental facility. This assumption is supported by the fact, that the measured flow field shows a deflection too, which change its direction if the electric current is applied with reversed polarity [16].

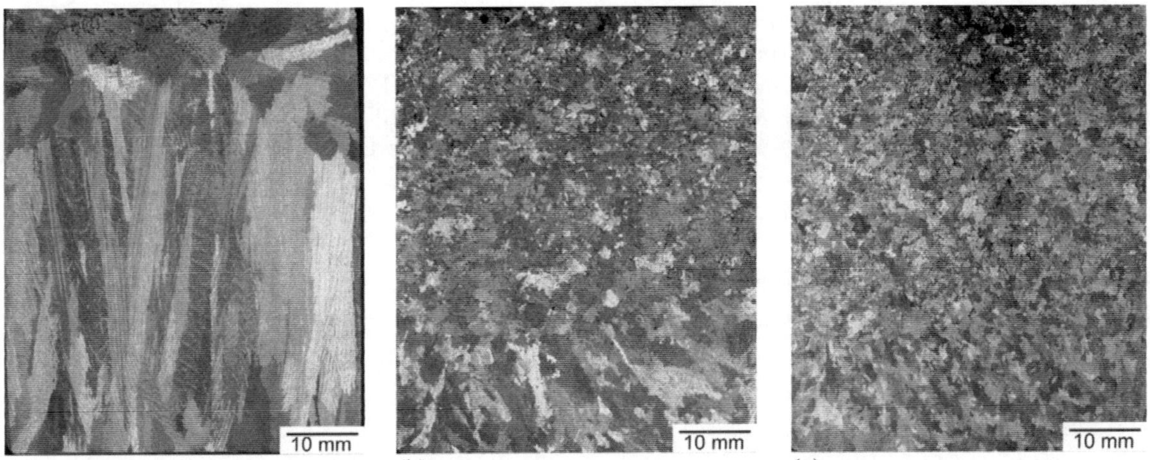


(a) (b) (c)

**Figure 5.** Macrostructure on the longitudinal section perpendicular to the plane containing the electrodes of the solidified samples: (a) without forced convection, (b) with $I_{DC}$ = 152 A and (c) with downward TMF ($B_{TMF}$ = 12 mT)

## 5. Conclusions

The present paper considers the impact of electric current on grain refinement in solidifying Al-Si alloys and the role of melt flow in this process. This so-called electro-vortex flow is driven by a Lorentz force caused by the interaction between the applied electric current and its own induced magnetic field. The flow intensity is determined by the time-averaged input of electric power which can be expressed by the effective current $I_{eff}$. A recent study revealed a reproducible correlation between $I_{eff}$ and the grain size in the solidified samples, whereas the solidified structure becomes finer with increasing effective current. There are no remarkable differences with respect to the influence of a direct current or a treatment using electric current pulses on the mean grain size and the area of equiaxed zone in the solidified samples. The results demonstrate that the grain refining effect observed in our experiments can be ascribed solely to the forced melt flow driven by the Lorentz force. A subsequent, systematic study is planned to investigate in detail the flow structure for other electrode configurations and a wide range of parameters. The results from this work will provide the basis for further research activities focusing on an optimization of the electric current treatment during solidification.

## 6. Acknowledgments

The research is supported by the German Helmholtz Association in form of the Helmholtz-Alliance "LIMTECH".

## 7. References

[1] McCartney DG 1989 Int. Mater. Rev. **34** 247
[2] Trivedi RK 1994 Mater. Sci. Eng. A **178** 129
[3] Flemings MC 1991 Metall. Trans. A **22** 957
[4] Eskin GJ 1998: Ultrasound treatment of Light Alloy Melt: Gordon and Breach, Amsterdam
[5] Murty BS, Kori SA and Chakraborty M 2002 Int. Mat. Rev. **47**(1) 3
[6] Quested TE 2004 Mater. Sci. Techn. **20** 1357
[7] Nakada M, Shiohara Y and Flemings MC 1990 ISIJ Int **30** 27
[8] Dolinsky Y and Elperin T 1993 Phys Rev B **47** 14778
[9] Li J, Ma JH, Gao YL and Zhai QJ 2008 Mater Sci Eng A **490** 452
[10] Takeda Y 1991 Nucl. Eng. Design **126** 277
[11] Franke S, Lieske H, Fischer A, Büttner L, Czarske J, Räbiger D, Eckert S 2013 Ultrasonics **53** Issue 3 691
[12] Willers B, Eckert S, Nikrityuk PA, Räbiger D, Dong J, Eckert K and Gerbeth G 2008 Metall. Mater. Trans. **B 39** 304
[13] http://www.comham.com/emdesign
[14] http://www.openfoam.org
[15] Räbiger D, Zhang Y, Galindo V, Franke S, Willers B and Eckert S 2014 Acta Mater. **79** 327
[16] Franke S, Räbiger D, Galindo V, Zhang Y and Eckert S 2015 Flow Meas Inst in press
[17] Edry I, Erukhimovitch V, Shoihet A, Mordekovitz Y, Frage N and Hayun S 2013 J Mater Sci **48** 8438
[18] Zhang Y, Räbiger D, Eckert S 2016 J Mater Sci **51** 2153
[19] Cramer A, Zhang C, Eckert S 2004 Flow Meas Instrum 2004 **15** 145
[20] Metan V, Eigenfeld K, Räbiger D, Eckert S 2009 J Alloy Compnds **487** 163
[21] Eckert S, Willers B, Nikrityuk P A, Eckert K, Michel U and Zouhar G 2005 Mater Sci. Eng. **A413-414** 211
[22] Eckert S, Räbiger D, Matthes M, Zimmermann G, Schaberger-Zimmermann E 2012 IOP Conference Series: Materials Science and Engineering **27** 012051
[23] Cramer A, Pal J, Koal K, Tschisgale S, Stiller J, Gerbeth G 2011 J of Crystal Growth **321** 142

IOP Conf. Series: Materials Science and Engineering **143** (2016) 012022 doi:10.1088/1757-899X/143/1/012022

# Velocity and flow rate measurement of liquid metal by contactless electromagnetic Lorentz force technique

**N Dubovikova, C Karcher and Y Kolesnikov**
Technische Universität Ilmenau
Institute of Thermodynamics and Fluid Mechanics
P.O.Box 100565, D-98684 Ilmenau, Germany

E-mail: nataliia.dubovikova@tu-ilmenau.de

**Abstract.** Providing flow analysis in case of aggressive and hot liquids is a complicated task, especially when liquid's composition and, hence, its physical properties, are unknown. Contactless techniques are the most promising methods for liquid metal flow rate control and some of these methods are based on electromagnetic induction of breaking force acting on an electrically conductive fluid which is moving through a magnetic field. One of the techniques is time-of-flight Lorentz force velocimetry (LFV). By using the method one can estimate volumetric flow rate without knowing of electrical conductivity, magnitude of magnetic field or characteristic dimension. The most important and crucial challenge within the technique is detection of small fluctuations of Lorentz force value. In this article we will focus on application and investigation of time-of-flight LFV.

## 1. Introduction

Electromagnetic flow meters find practical use during several decades for laboratory and industrial flows implying on the principle known since Faraday time [1]. Further, the theory of electromagnetic flow meters has been developed and comprehensively summarized by Shercliff [2]. From the magnetohydrodynamics it is well known that the relative motion of an electrically conducting material and magnetic field lines induces electric currents in the conductor. This principle forms the basis of electromagnetic flow investigation and flow measurements. According to main principle of electromagnetic velocimetry, electromotive force is generated within conductive liquid that moves in magnetic field; the force acts in orthogonal direction to movement and magnetic field lines and brakes relative motion of conductive body and magnetic field. One of the fields, where the electromagnetic principle can be successfully applied, is the investigation of conductive fluids flow rate. Providing of velocimetry or flow rate measurements in case of aggressive and hot fluids like liquid metals – is a complicated task. Liquid metals are not transparent to use any optical methods, and conditions inside them make it not possible to employ any mechanical probes. Therefore contactless techniques are the most promising methods for liquid metal's flow rate control. The method of electromagnetic flow control we present here - Time-of-flight Lorentz force velocimetry [3]. The technique (figure 1) is based on difference between two force signals that are measured by Lorentz force flow meters. The difference occurs by a vortex, that moving with a flow through magnetic field of both flow meters one by one. It is well known principle to use vortices as signal particles, but usually contact elements like electrodes in conductive fluids [4] or piezo sensors in commercial vortex shedding flow meters [5] are applied for vortex detection. These methods have significant disadvantage for liquid metal flow measurements – they need mechanical contact between detector and a liquid.

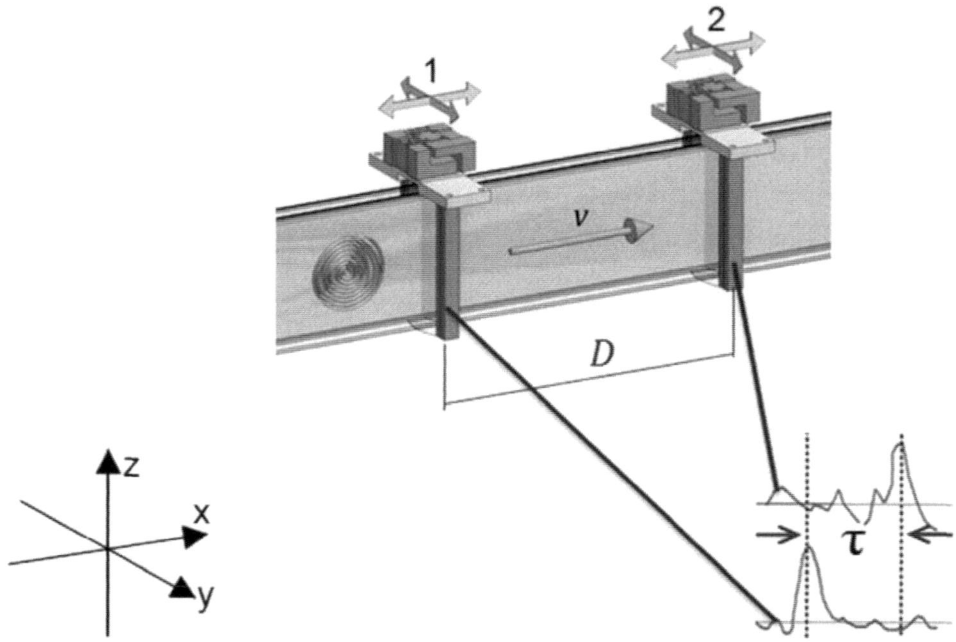


1, 2 – flow meters, $D$ – distance between flow meters, τ – time delay between disturbances
**Figure 1.** The principle of time-of-flight Lorentz force velocimetry

The time-of-flight Lorentz force flow meter is a system of contactless sensors and consist of two permanent magnets, rigidly connected to precise force sensor, so when naturally or artificially created vortex is moving through their magnetic field, it disturbs velocity and, hence, the measured Lorentz force. The distance between measurement systems $D$ is known, time shift $\tau$ between disturbances (peaks of two signals) can be obtained from cross-correlation function of force signals. With this data, flow rate of liquid metal can be calculated by velocity definition: distance per time multiplying duct cross-section area $A$:

$$\boldsymbol{Q = Av = Ak\frac{D}{\tau}}$$

where $k$ –coefficient, obtained experimentally for every specific liquid and duct.

## 2. Measurement results

Our measurements were performed under next conditions and materials: the GaInSn (gallium, indium and tin alloy) is used as experimental fluid, the electromagnetic pump generates flow with Reynolds numbers from 4000 till 60000. The duct has cross-section 80 mm height per 10 mm width.

3D force sensors have measurement range 2 N, so magnet holding system should be as light as possible to not overload strain gauge sensor component in gravity direction, and the goal was reached by applying standard aluminium profiles for construction. Besides this, the magnet holding system allows mount permanent magnets with 1 mm distance to the duct, which increased sensitivity of force measurement.

The current way of signal processing contains three main steps: segmentation, filtration and correlation. Six force signals are obtained from two 3D strain gauge sensors – in x-direction (along the flow), in y-direction (orthogonal to the flow) and in z-direction (gravity line). Each experiment is repeating three times and least one minute, that is sufficient to get time delay in ms range. The

obtained signal divided into three parts 20 s each. It is made to avoid the influence of random outer noise. The received signal segments undergo band-pass filtering in low frequency range (3 5 Hz, 4 6 Hz, 5 7 Hz and 6 8 Hz) and the cross-correlation functions of alignment pairs of filtered signals (x1x2, y1y2 and z1z2) gives the desired value of time delay.
The time delay values, calculated by cross-correlation of x- and y-directions, don't represent change of velocity with increase of pump rotation frequency and, thus, cannot be used in velocity estimation. In x-direction constant bulk flow force is in evidence and it is a complicated task to distinguish useful signal from strong turbulent disturbances. In y-direction case flow is parallel to magnetic field lines and if this occurs there is no Lorentz force generation possible. The only component, that shows visible change of time delay during experiment, is z-component of force, so the mean velocity was estimated by this component.
Experiments were performed in four positions of measurement system:

- normal position – magnetic field lines penetrate all height of the duct;
- half position – magnetic field influence half duct height;
- with artificially created disturbances;
- with naturally appeared disturbances.

The half position measurements were performed to check influence of different flow meters mounting on non-flow direction components of force. Flow meters were mounted for the test in such way that ends of magnets were situated over the middle of the duct height.
As reference value of mean velocity the electro-potential probe (Vives-probe [6]) readings were used. The Vives probe is a contact local velocity measurement method, so the inserted in liquid metal part of the probe (diameter 5 mm) was also used for experiment as artificial disturbances generator to create Karman Street and was situated before both flow meters according to flow direction.
Under naturally appeared disturbances I mean the Vives-probe positioned between two flow meters, so it does not create disturbances that can be used as tracking particles for both and, moreover, implies additional noise effect. But even in this case I have measured effective signals and can estimate velocity. The detected disturbances are created due to duct geometry – sharp duct turns, including heat exchanger plate, electromagnetic fluctuations caused by rotation of the pump, etc.
Measurement system and signal processing improvement enabled to estimate mean velocity of liquid metal in the duct (figure 2). Electromagnetic pump creates flow within the duct, so its rotation frequency is used as horizontal coordinate to refer velocity.
The measurement results show high error value, as one can see by error bars on all four diagrams. This is observed due to low signal-to-noise ratio. More specifically, signal is getting lost against the background noise, what is easy to see on comparison between cross-correlation function of two signals and their autocorrelation function: if peaks of cross- and auto-correlations coincide we can detect only noise.
Another observation, that can be made due to presented diagrams – the time-of-flight velocity value mostly lays lower than measurement results of Vives-probe. This can be explained by the velocity difference between mean flow in the duct centre and moving vortices or other disturbances that are slowed down according to their size and direction by boundary layers or internal friction in liquid.

## 3. Conclusions

Experimental research of time-of-flight Lorentz force velocimetry is very challenging and interesting task. And the method itself is a prospective technique for mean velocity and volumetric flow rate measurements especially in relation to its industry applications. This method is successfully used in laboratory conditions for flow control of turbulent liquid metal and gained results show sufficient correlation between time-of-flight and reference velocity estimation without applying any additional experimental coefficient. The time-of-flight Lorentz force velocimetry doesn't directly depend on

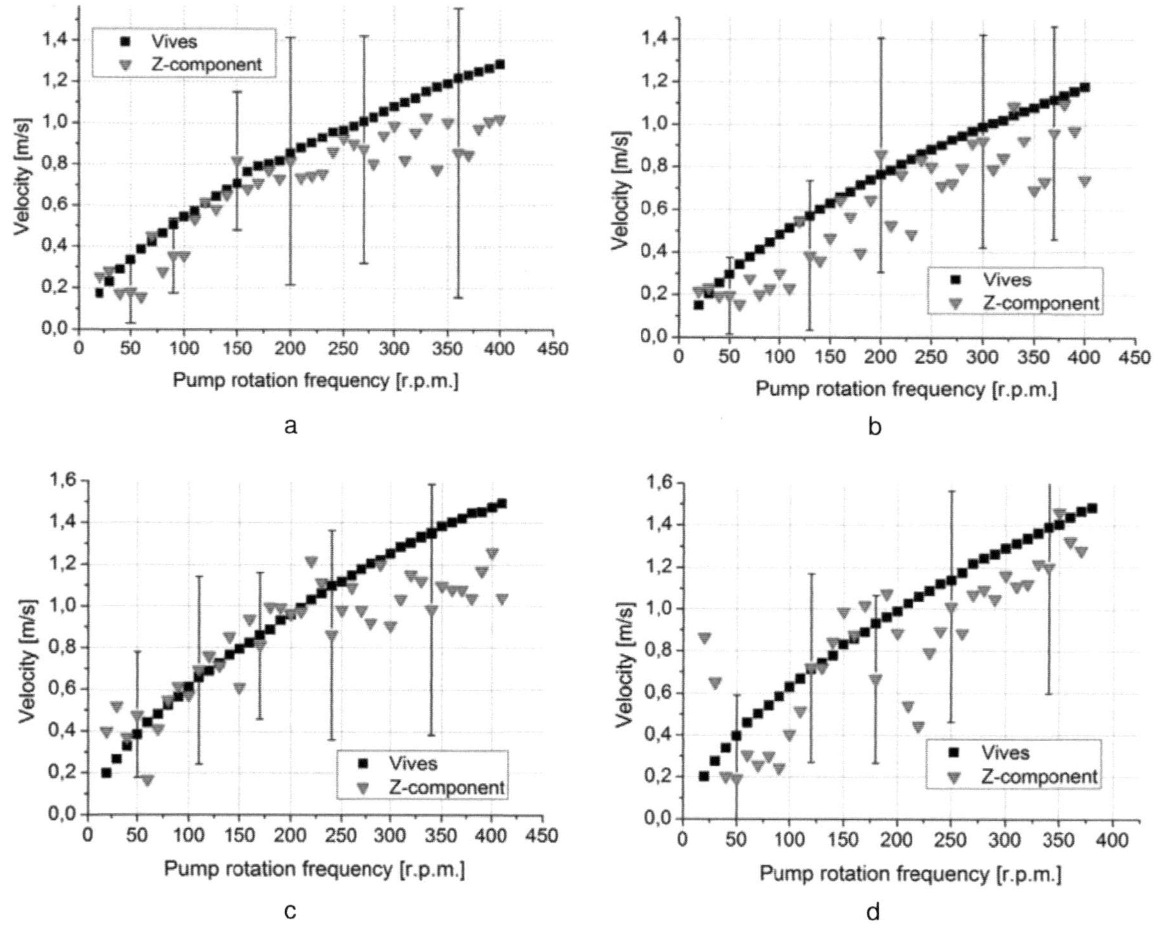


**Figure 2.** Time-of-flight measurements results in comparison with reference method – Vives-probe. Velocity on (a) and (b) are presented for normal position of sensors with and without artificial vortex generation correspondingly; (c) and (d) present half position sensors with and without artificial vortex generation

physical properties of liquid metal and gives wide opportunities to investigate turbulent duct flow under moderate, high Reynolds number in the sense of behaviour of vortices of different nature.

## 4. References


[1] Davidson, P.A. An introduction to Magnetohydrodynamics, 2001. Cambridge: Cambridge University Press, 432.

[2] Shercliff, J.A. The theory of electromagnetic flow-measurement, 1962. Cambridge: Cambridge University Press, 146.

[3] Jian, D., Karcher, Ch., Electromagnetic flow measurements in liquid metals using time-of-flight Lorentz force velocimetry. 2012 *Meas. Sci. Technol.* **23** (7).

[4] H. Yu, D. Li, R. C. Roberts, K. Xu and N. C. Tien A time-of-flight flow sensor for the volume measurement of trace amount of interstitial fluid. 2012 *J. Micromech. Microeng.* **22** 055009.

[5] Burgess Thomas H., patent US 3874234 A Vortex shedding flowmeter Fischer & Porter C0., Warminster 1975.

[6] Ricou R, Vives C Local velocity and mass transfer measurements in molten metals using an incorporated probe. 1982 *Int J Heat Mass Transfer* **25**: 1579–1588.

# Contactless inductive flow tomography: basic principles and first applications in the experimental modelling of continuous casting

**F Stefani, S Eckert, M Ratajczak, K Timmel and T Wondrak**
Helmholtz-Zentrum Dresden – Rossendorf, Bautzner Landstraße 400, 01328 Dresden, Germany

E-mail: F.Stefani@hzdr.de

**Abstract.** Contactless inductive flow tomography (CIFT) aims at reconstructing the flow structure of a liquid metal from the magnetic fields measured at various positions outside the fluid body which are induced by the flow under the influence of one or multiple applied magnetic fields. We recap the basic mathematical principles of CIFT and the results of an experiment in which the propeller-driven three-dimensional flow in a cylindrical had been reconstructed. We also summarize the recent activities to utilize CIFT in various problems connected with the experimental simulation of the continuous casting process. These include flow reconstructions in single-phase and two-phase flow problems in the Mini-LIMMCAST model of slab-casting, studies of the specific effects of an electromagnetic stirrer attached to the Submerged Entry Nozzle (SEN), as well as first successful applications of CIFT on the background of a strong electromagnetic brake field. We conclude by discussing some remaining obstacles for the deployment of CIFT in a real caster.

## 1. Introduction

In many metallurgical technologies it is desirable to monitor the velocity field of the liquid metal. Its opaqueness usually prevents the deployment of optical measurement techniques, such as laser Doppler anemometry (LDA) or particle image velocimetry (PIV). Non-optical flow measurement techniques, using ultrasonic beams or intrusive pressure or potential probes, face serious problems if the liquid metal is hot and/or chemically aggressive. Any sort of contactless flow measurement would be most helpful in these cases.

The contactless inductive flow tomography (CIFT) [1] is able to provide a rough picture of the flow by applying primary magnetic fields to the metallic melt and by measuring the flow induced (secondary) magnetic field perturbation outside the fluid volume. The primary magnetic field can be applied in various directions. This allows collecting more information about (basically) the same velocity field, as long as the typical time for velocity changes is larger than the switching times between different applied fields. The remaining non-uniqueness of the inverse problem can be overcome by using standard regularization methods, such as Tikhonov regularization and the L-curve technique [2].

We start by delineating the basic mathematics for the forward and the inverse problem as it had been developed in previous publications [3-5]. In Section 3, we recap the results of a first test experiment which had shown the applicability of CIFT for the reconstruction of three-dimensional velocity fields [1]. With view on the application to the continuous casting of steel, we focus then on the reduction of CIFT to the (essentially) two-dimensional (2D) geometry of slab-casting for which some experimental results are illustrated. We also present some new results evidencing the feasibility of CIFT in the presence of a strong magnetic brake field. The paper concludes with an outlook on future methodological developments and possible industrial implementations of CIFT.

## 2. Basic mathematics

In this section we give a short outline of the basic mathematics underlying CIFT. We consider an electrically conductive fluid with a velocity field **u**, that is assumed to be steady or at least slowly varying on the magnetic diffusion time scale. Imagine this moving fluid to be exposed to a stationary magnetic field **B**. Then, according to Ohm's law in moving conductors, the current density

$$\mathbf{j} = \sigma(\mathbf{u} \times \mathbf{B} - \nabla\varphi) \tag{1}$$

is induced, where $\sigma$ is the electrical conductivity of the fluid and $\varphi$ denotes the electric scalar potential. According to Biot-Savart's law, this current density **j** induces now a secondary magnetic field **b** at a position **r**:

$$\mathbf{b}(\mathbf{r}) = \frac{\sigma\mu_0}{4\pi}\int_D \frac{(\mathbf{u}(\mathbf{r}') \times \mathbf{B}(\mathbf{r}')) \times (\mathbf{r} - \mathbf{r}')}{|\mathbf{r} - \mathbf{r}'|^3} dV' + \frac{\sigma\mu_0}{4\pi}\oiint_S \varphi(\mathbf{s}') \frac{\mathbf{r} - \mathbf{s}'}{|\mathbf{r} - \mathbf{s}'|^3} \times \mathbf{n}(\mathbf{s}') dS' \tag{2}$$

In the volume integral, $dV'$ denotes the volume element and $\mathbf{r}'$ the position vector in the volume. In the surface integral, $dS'$ denotes a surface element and $\mathbf{n}(\mathbf{s}')$ denotes the normal vector of the surface at the position $\mathbf{s}'$. It is essential to take into account the second term on the r.h.s. of (2). Only in very special cases, for example, if a simple poloidal flow is exposed to an axial magnetic field so that the induced current flows only in azimuthal direction, this second term is zero. If, however, a purely azimuthal flow is exposed to an axial field, the second term completely cancels out the first term. Utilizing the fact that **j** is divergence-free, one can derive from (1) a Poisson equation for the electric potential whose solution also fulfills the boundary integral equation

$$\varphi(\mathbf{s}) = \frac{1}{2\pi}\int_D \frac{(\mathbf{u}(\mathbf{r}') \times \mathbf{B}(\mathbf{r}')) \cdot (\mathbf{s} - \mathbf{r}')}{|\mathbf{s} - \mathbf{r}'|^3} dV' - \frac{1}{2\pi}\oiint_S \varphi(\mathbf{s}') \frac{\mathbf{s} - \mathbf{s}'}{|\mathbf{s} - \mathbf{s}'|^3} \cdot \mathbf{n}(\mathbf{s}') dS' \tag{3}$$

if insulating boundaries are assumed.

In general, the total magnetic field **B** under the integrals of (2) and (3) is the sum of an externally applied primary magnetic field $\mathbf{B}_0$ and the induced secondary magnetic field **b** itself. In magnetohydrodynamics, the ratio between **b** and $\mathbf{B}_0$ is known to be proportional to the magnetic Reynolds number $Rm$, defined as

$$Rm = \mu_0 \sigma L U \tag{4}$$

with $L$ and $U$ denoting characteristic length and velocity scales of the fluid, respectively. For large values of $Rm$ and suitable flow topologies, it is possible to achieve self-excitation of a magnetic field. Such homogeneous dynamos have indeed been studied [6,7] by solving the integral equation system (2) and (3). However, in most industrial applications $Rm$ is smaller than 1. To take continuous casting of steel as a typical example, with a mould width of 1 m, a typical flow velocity of 0.1 m/s, and a conductivity of the liquid steel of $7 \cdot 10^5$ S/m, we arrive approximately at $Rm$= 0.1. In such cases, **B** can be replaced by $\mathbf{B}_0$ under the integrals in (2) and (3). In this approximation, we get a linear inverse problem for the determination of the velocity field **u** from the induced magnetic field **b** measured in the exterior of the fluid. Of course, the unknown electric potential at the fluid boundary must not be neglected and has to be treated in due manner.

Assume now $K$ different external magnetic fields $\mathbf{B}_0^k$ to be applied to the fluid. Suppose, for each $\mathbf{B}_0^k$, all measured induced magnetic field components to be collected into an $N_b$-dimensional vector with the entries $b_i^k$ . Accordingly, the discretized electric potential at the surface $S$ is collected into an $N_\varphi$-dimensional vector with the entries $\varphi_m^k$, and the desired velocity **u** in the volume $V$ is discretized as an $N_V$-dimensional vector with the entries $u_l$. Then, (2) and (3) can formally be written in the form:

$$b_i^k = K_{il}^k u_l + L_{im} \varphi_m^k \tag{5}$$

$$\varphi_m^k = M_{ml}^k u_l + N_{mn} \varphi_n^k \tag{6}$$

Note that only the matrices $\mathbf{K}^k$ and $\mathbf{M}^k$ depend on the applied magnetic field $\mathbf{B}_{0,k}$, whereas the matrices **L** and **N** depend on the geometry only. A peculiarity of equation (6) is well known from magnetoencephalography (MEG) [8]. Actually, the matrix $(\mathbf{I} - \mathbf{N})$ is singular which indicates the fact that the electric potential is defined only up to an additional constant. This problem can be circumvented by using the so-called deflation method [8], replacing $(\mathbf{I} - \mathbf{N})$ by the well-conditioned matrix $(\mathbf{I} - \mathbf{N})^{\text{defl}}$. Inserting the solution of (deflated) equation (6) into (5) we end up with a single linear relationship between the desired velocity field and the measured magnetic field in the form

$$b_i^k = K_{il}^k u_l + L_{im} (I - N)_{mn}^{-1,defl} M_{nl}^k u_l \tag{7}$$

For further details of the numerical solution of this equation, in particular concerning the accurate computation of the boundary integrals, see [9].

We can now derive the velocity by solving the normal equations arising from the minimization of the following total functional of the velocity field:

$$F[\mathbf{u}] = \sum_{k=1}^{K} F_k[\mathbf{u}] + F_{\text{div}}[\mathbf{u}] + F_{\text{pen}}[\mathbf{u}] \tag{8}$$

with

$$F_k[\mathbf{u}] = \sum_{i=1}^{N_b} \frac{1}{\sigma_{i,k}^2} \left( b_{i,\text{meas}}^k - b_{i,\text{model}}^k[\mathbf{u}] \right)^2 \tag{9}$$

$$F_{\text{div}}[\mathbf{u}] = \frac{1}{\sigma_{\text{div}}^2} \sum_{l=1}^{N_V} (\nabla \cdot \mathbf{u})_l^2 \, \Delta V_l \tag{10}$$

$$F_{\text{pen}}[\mathbf{u}] = \frac{1}{\sigma_{\text{pen}}^2} \sum_{l=1}^{N_V} \mathbf{u}_l^2 \, \Delta V_l \tag{11}$$

where $\Delta V_l$ denotes the volume of the element at the volume grid point $l$. The first sum over the functionals $F_k[\mathbf{u}]$ is the most essential one as it represents the mean squared residual deviation of the measured magnetic fields $b_{i,}{}^k{}_{\text{meas}}$ from the fields $b_{i,}{}^k{}_{\text{model}}$ modeled according to (7).

The functional $F_{\text{div}}[\mathbf{u}]$ enforces the velocity field to be solenoidal, and the last functional $F_{\text{pen}}[\mathbf{u}]$ is the regularization (penalty) functional which tries to minimize the kinetic energy. The parameters $\sigma_{i,k}$ are the a priori errors for the measured induced magnetic fields. $\sigma_{\text{div}}$ is chosen very small in order to ensure the divergence-free condition. The parameter to play with in the regularization procedure is $\sigma_{\text{pen}}$. Here we employ the so-called Tikhonov regularization [3]: by increasing the regularization parameter $\sigma_{\text{pen}}$, one obtains solutions of the inverse problem with increasing kinetic energy of the flow. The optimal solution is then found at the point of strongest curvature of Tikhonov's L-curve which is done by an automatic search described in [9].

The above equation system (7-11) is valid for the case of different applied magnetic fields and a fully three-dimensional (3D) flow. It can, however, also be adapted to the essentially two-dimensional (2D) case of slab-casting, where only one magnetic field $B_0$ is applied. In this case, the dimensional reduction of the velocity structure from 3D to 2D can be realized by adding an additional functional which enforces the velocity to be 2D. Optionally, any available prior information, e.g. on the velocity at the inlet which is available from the casting speed, can also be considered by including a further appropriate quadratic functional into (8).

It should be noted that regularization based on some reasonable norm of the velocity is actually more than only a trick since in most cases the assumption of a rather smooth velocity field is sensible from a physical point of view. For the same reason it is clear, however, that the method is not suited to detect very small vortices in a strongly turbulent flow (see [4,5] for the corresponding uniqueness problem).

## 3. A three-dimensional flow reconstruction problem

The aim of a first demonstration experiment [1] for CIFT was to reconstruct a propeller-driven three-dimensional flow of a liquid metal in a compact cylindrical vessel (see Figure 1(a)). The flow of the eutectic alloy GaInSn was stirred by a motor driven propeller with a diameter of 6 cm inside a cylindrical polypropylene vessel with radius $R = 9$ cm and height 17.2 cm.

For determining both the poloidal flow (in radial and axial direction) and the toroidal flow (in azimuthal direction), we applied subsequently a vertical $B_{0,z}$ and a horizontal $B_{0,x}$ magnetic field, both produced by Helmholtz-like coil pairs. The switching between the two fields occured every 3 seconds, so that after 6 seconds all the magnetic field information was available for the velocity reconstruction. This time resolution could be significantly enhanced, with a physical limitation given by the magnetic decay time $\mu_0\sigma R^2$, which is in the order of 0.05 s for the demonstration experiment.

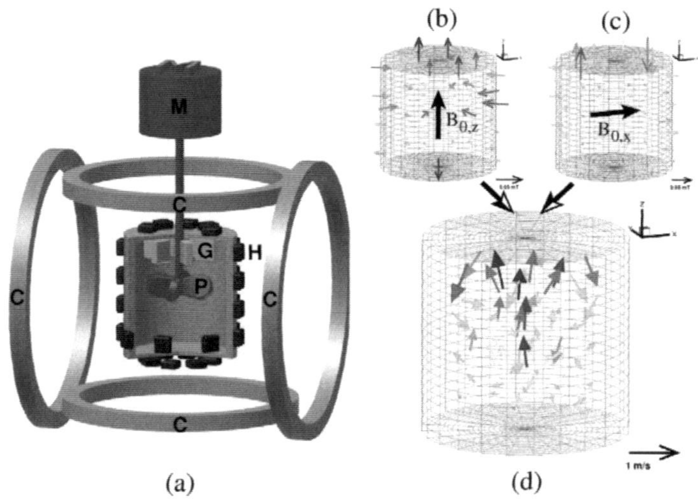


**Figure 1.** (a) Schematic sketch of the experiment for the validation of CIFT at a propeller driven 3D flow using GaInSn. M—motor, C—coils, P—propeller, G—guiding blades, and H—Hall sensors. (b) Measured induced field at 48 positions for upward pumping propeller for applied $B_{0,z}$. The arrows indicate the field component measured by the Hall sensors. (c) Same as (b), but for applied $B_{0,x}$. (d) Velocity field as reconstructed by CIFT from (b) and (c).

For each of the two applied fields $B_{0,z}$ and $B_{0,x}$ the components of the induced fields normal to the surface were measured at 48 positions, which are rather homogeneously distributed all over the surface of the cylindrical vessel. For the measurement KSY 44 Hall sensors (Infineon) were used. The small ratio of around $10^{-3}...10^{-2}$ between the induced and the applied fields requires a stable current source of the Helmholtz-like coils, a very stiff relative position of Hall sensors and coils, and significant effort to compensate drift and sensitivity changes of the Hall sensors (e.g., due to varying temperature).

By inverting the measured induced magnetic fields, it was possible to clearly distinguish between upward and downward pumping of the propeller, with the rotational component being deliberately reduced by guiding blades in the case of upward pumping. For this particular case, Figures 1(b) and 1(c) show the measured induced fields for applied $B_{0,z}$ and $B_{0,x}$, respectively, and Figure 1(d) depicts the velocity field as reconstructed from these two sets of information. The comparison with UDV measurements had shown a good coincidence of the resulting velocity fields [1]. Later, an independent confirmation of the flow structure was obtained by Lorentz force velocimetry [10].

## 4. Application to slab casting problems

We now move from the general 3D flow reconstruction problem to a problem which can be treated as a 2D problem. In the continuous casting of steel slabs, the liquid metal flows from a tundish through a submerged entry nozzle (SEN), with typically two sideward directed ports at the lower end, into the mould where it starts to solidify at the water cooled copper walls. The flow structure in the mould is known to play a key role for the quality of the produced steel [11]. A stable double-roll flow structure, in which the two jets emanating from the SEN ports first reach the narrow-faces of the mould and then split into upward and downward directed branches, is considered as most desirable. The so-called single-roll structure, however, with the jets being sharply bent upward after leaving the SEN ports, represents a danger since it could lead to entrainment of casting powder into the steel. Any sort of online-monitoring of the detailed flow-structure in the mould would be helpful for an active control of the casting process, with the prospect to increase the possible casting speed significantly. As a contactless method, CIFT suggests itself for such monitoring, although various problems connected with copper-mould oscillations and the use of coatings with ferromagnetic Nickel makes its implementation in the industry still a formidable task.

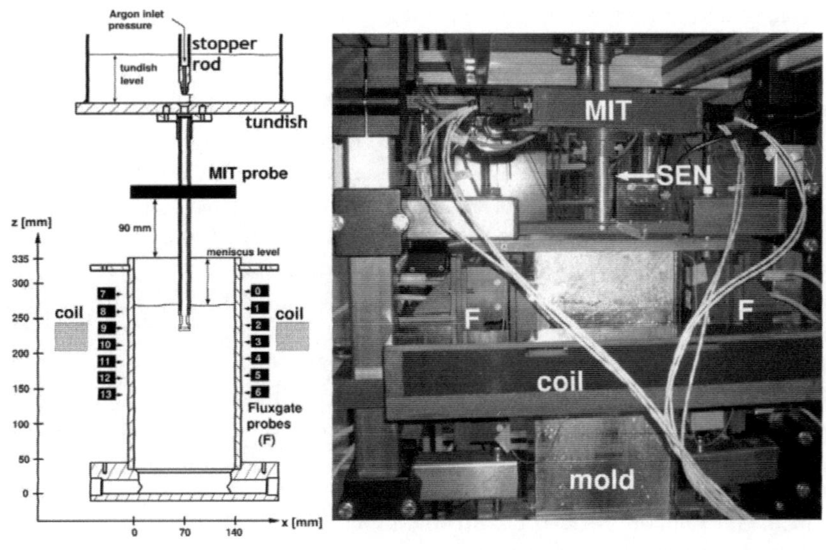


**Figure 2.** Schematic sketch (left) and photography (right) of the central part of the Mini-LIMMCAST facility.

As a very first step in this direction, we have installed a simplified CIFT system at the small continuous casting model Mini-LIMMCAST, working also with the GaInSn alloy (Figure 2). The simplification concerns the restriction of the CIFT system to a single magnetic field coil which produces a mainly vertical magnetic field. This configuration is sufficient for the determination of the velocity component parallel to the wide faces of the mould [12], which is indeed the dominant one for the particular case of slab-casting. The induced fields are measured by fluxgate sensors positioned at the narrow faces of the mould, typically at 7 positions on either side. In addition to CIFT, we have utilized a mutual inductance tomography (MIT) system for determining the conductivity distribution in the SEN, which allows visualizing the details of the two-phase GaInSn/Argon flow [13,14]. The simultaneous utilization of CIFT and MIT leads to a detailed understanding of the two-phase flow in the SEN and of the resulting flow in the mould [15]. Figure 3 illustrates two typical flow structures that had been detected during one run of the experiment, i.e. double-roll and single-roll structures occurring on different sides of the mould.

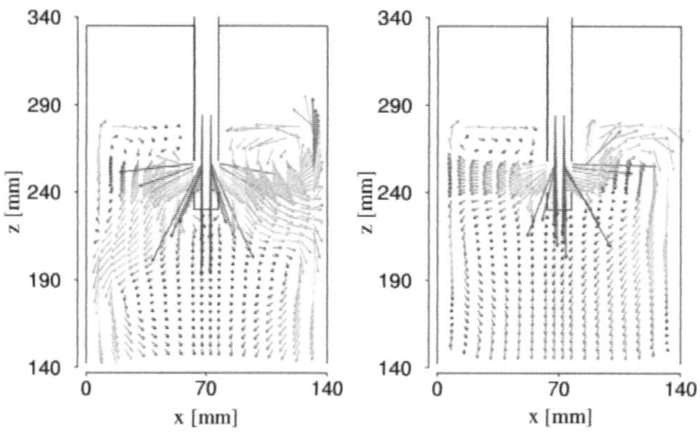


**Figure 3.** CIFT-reconstructed velocity obtained at the Mini-LIMMCAST facility, for the case that Argon is being injected at the top of the SEN. At one instant, the desired double role structure (left) appears, somewhat later, and the undesired single role structure dominates (right).

## 5. CIFT in presence of a strong electromagnetic brake

For decades, electromagnetic brakes (EMBrs) have been used with the goal to stabilize the flow in the mould and to dampen its oscillations, while recent experiments have challenged this simple view [16]. One might ask whether CIFT is able to work even under the extreme condition that a strong static magnetic field is applied in addition to the measuring field. Some related experiments were conducted at the Mini-LIMMCAST slab caster model. A ruler-type brake was used to generate a DC magnetic field perpendicular to the wide faces of the mould and hence to the main flow direction (Figure 4). Under these circumstances it was necessary to enhance the emitting system by a second coil. The induced magnetic fields were measured with $2 \times 7$ cylindrical induction coils positioned at the narrow faces of the caster model (the presence of the strong brake field prohibits the use of fluxgate sensors which would saturate approximately at 2 mT). Each induction coil has 340,000 windings with a conductor diameter of 25 $\mu$m. The signals from the coils were digitized by a 24-bit ADC from LTT Tasler. In the post-processing step the amplitude and the phase of the signal in comparison to the current through the excitation coil is calculated using the quadrature demodulation.

Figure 5 shows the time-dependence of the measured magnetic fields at the $2 \times 7$ positions for three different settings. Figure 5a corresponds to the case that the brake field is switched off. Figure 5b corresponds to the case that a brake field of 300 mT is switched on, and the walls of the "mould" are insulating. Apart from an overall change of the structure of the induced field (which reflects an upward shift of the jet), we also observe a slow oscillation (with a typical period of 5 seconds). This confirms the observations of [16] that the magnetic brake, contrary to previous expectations, can indeed lead to stronger fluctuations. Finally, figure 5c shows the signal for switched-on brake field, this time however, with conducting walls. Evidently the long period oscillations have disappeared, again in agreement with [16].

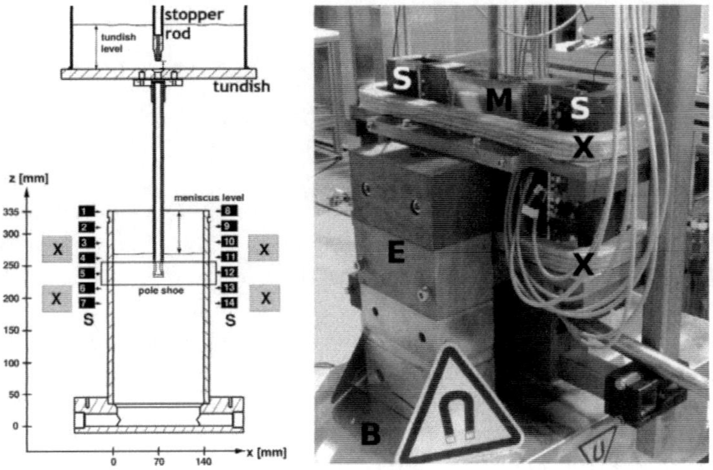


(a) (b)

**Figure 4.** Schematic sketch (left) and photography (right) of the central part of the Mini-LIMMCAST facility with a ruler-type electromagnetic brake applied, and an enhanced emitter coil system. X—two emitter coils, S—sensor system, E— EMBr, B—brake coils.

While more detailed investigations of these effects are still necessary, it is quite remarkable that on the background of a strong brake field of 300 mT it is possible to measure safely the induced magnetic fields which are 6-7 orders of magnitude smaller and to reconstruct from them the velocity.

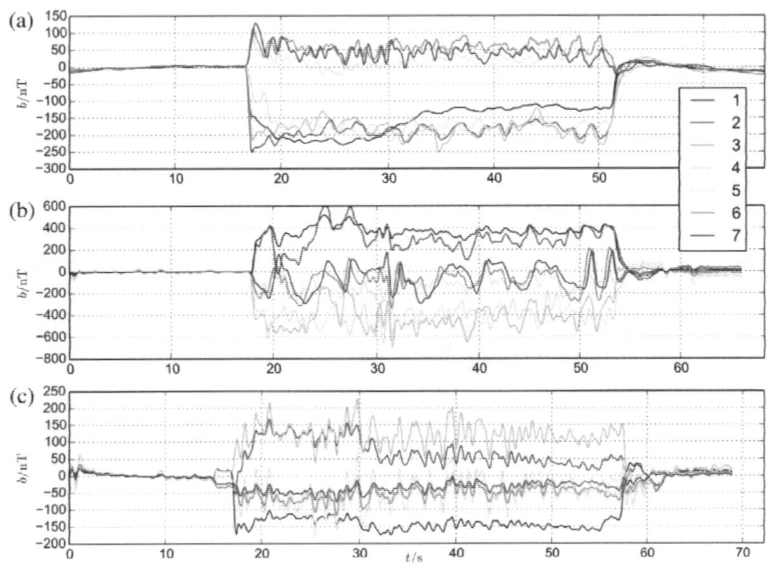


**Figure 4.** Measurements of the induced magnetic field for the magnetic brake switched off (a), switched on with insulating walls (b), and with conducting walls (c).

## 6. Outlook

Future work will be devoted to adapting CIFT for its deployment under the harsh mechanical and electromagnetic conditions in a real casting plant. A main challenge is certainly the up-and-down oscillation of the copper mould. Even if the measuring coil, together with the sensors, could be stably installed at the mould, one has to be aware of disturbing induction effects due to the relative motion of the mould to the rest of the facility. A further problem is posed by the widely used Nickel coating of the copper could. At present it is not clear whether or not the magnetic field deformations due to the magnetic properties of Nickel can be mastered.

Another possible application of CIFT, beyond steel casting, may be the Czochralski crystal growth of silicon. First tests have already been made at various pullers. On the laboratory scale, CIFT-like flow inference methods have also been applied for the identification of current-driven instabilities [17] with relevance for liquid metal batteries, and for the flow reconstruction in a Rayleigh-Bénard cell with liquid metal.

Another project for the future is to combine into one single scheme the CIFT method with the mutual inductance tomography. A simultaneous but yet spatially separated application of both methods at the Mini-LIMMCAST facility was already documented in [13,14]. The question is whether both conductivity variation and the velocity field could be obtained by one single measurement system. A first step in this direction was the reformulation of CIFT in terms of the sensitivity matrix [18].

## 7. Acknowledgments

This work was supported by Helmholtz-Gemeinschaft Deutscher Forschungszentren (HGF) in frame of the Helmholtz Alliance LIMTECH.

## 8. References

[1] Stefani F, Gundrum T and Gerbeth G 2004 *Phys. Rev. E* **70** 056306
[2] Hansen PC 1992 *SIAM Rev.* **34** 561
[3] Stefani F and Gerbeth G 1999 *Inverse Probl.* **15** 771
[4] Stefani F and Gerbeth 2000 *Inverse Probl.* **16** 1
[5] Stefani F and Gerbeth G 2000 *Meas. Sci. Techn.* **11** 758
[6] Stefani F, Gerbeth G and Rädler KH 2000 *Astron. Nachr.* **321** 65
[7] Xu M, Stefani F and Gerbeth G 2004 *J. Comp. Phys.* **196** 102
[8] Hämäläinen R et al. 1993 *Rev. Mod. Phys.* **65** 413
[9] Wondrak T, Stefani F, Gundrum T and Gerbeth G 2009 *Int. J. Appl. Electromagn. Mech.* **30** 255
[10] Heinicke C and Wondrak T 2014 *Meas. Sci. Techn.* **25** 055302
[11] Zhang L, Yang S, Cai K, Li J and Thomas BG 2007 *Met. Mater. Trans. B* **38** 63
[12] Wondrak T et al. 2010 *Meas. Sci. Tech.* **21** 045402
[13] Terzija N et al. 2011 *Flow Meas. Instr.* **22** 10
[14] Terzija N et al. 2011 *Meas. Sci. Techn.* **22** 015501
[15] Wondrak T et al. 2011 *Met. Mater. Trans. B* **42** 1201
[16] Timmel K, Eckert S and Gerbeth G 2011 *Met. Mater. Trans. B* **42** 68
[17] Seilmayer M et al. 2012 *Phys. Rev. Lett.* **108** 244501
[18] Yin W, Peyton AJ, Stefani F and Gerbeth G 2009 *Meas. Sci. Tech.* **20** 105503

IOP Conf. Series: Materials Science and Engineering **143** (2016) 012024 doi:10.1088/1757-899X/143/1/012024

# Magnetohydrodynamic effects in liquid metal batteries

**F Stefani, V Galindo, C Kasprzyk, S Landgraf, M Seilmayer, M Starace, N Weber and T Weier**
Helmholtz-Zentrum Dresden – Rossendorf, Bautzner Landstraße 400, 01328 Dresden, Germany

E-mail: F.Stefani@hzdr.de

**Abstract.** Liquid metal batteries (LMBs) consist of two liquid metal electrodes and a molten salt ionic conductor sandwiched between them. The density ratios allow for a stable stratification of the three layers. LMBs were already considered as part of energy conversion systems in the 1960s and have recently received renewed interest for economical large-scale energy storage. In this paper, we concentrate on the magnetohydrodynamic aspects of this cell type with special focus on electro-vortex flows and possible effects of the Tayler instability.

## 1. Introduction

The increasing deployment of strongly fluctuating renewable energies requires a corresponding large-scale expansion of electricity storage. While electrolytically generated hydrogen, partly processed to synthesized hydrocarbons (power-to-gas, power-to-liquids), is widely considered the only viable way for long-term storage on the TWh scale, electrochemical storage is an attractive candidate for short-term and mid-term storage. Stationary Sodium-Sulfur (NaS) batteries, which are already sold as an industrial product on the MW scale, may have a promising future, in particular if the mass storage of Na and S can be spatially separated from the actual energy converter.

Liquid metal batteries (LMBs) are presently discussed as an alternative route of economical grid-scale energy storage. LMBs had been intensively studied in the 1960s [1], and experience a remarkable renaissance nowadays [2]. They consist of two liquid metal electrodes and a molten salt ionic conductor sandwiched between them. The respective density ratios allow for a stable stratification of the three layers. On discharge, the upper alkaline (Li, Na...) or earth-alkaline (Mg, Ca...) metal is oxidized, and its cations cross the fused salts into the lower layer where they are reduced and alloy with a heavy metal (Pb, Bi...) or metalloid (Sb).

The main advantage of this setup is the ultrafast charge transfer kinetics due to the liquid-liquid electrode-electrolyte interface, and the fast mass transport in liquids, allowing for charging and discharging current densities of about 4-100 kA $m^{-2}$ (depending on the choice of the anode, cathode, and electrolyte material). The second advantage is the potentially low cost which results from the abundance of electrode materials such as Na and Pb. Further to this, the absence of any degradation effects due to the use of liquid metals and salts promises unprecedented cycle life.

Disadvantages of LMBs are the comparably low cell voltages (typically < 1 V), high operating temperatures to keep the metals and the salt liquid, several corrosion problems, and high self-discharge rates for some of the used metals (including, unfortunately, Na) due to their significant solubility in the molten salt electrolyte.

In this paper, we will be concerned with another issue. While fluid dynamics does not play any role in most traditional battery systems, it cannot be neglected in LMBs (see Figure 1). It goes without saying that LMBs are not suitable for mobile applications since any acceleration would lead to uncontrollable movements of the three layers and very likely to a rupture of the electrolyte, with possibly drastic consequences due to the resulting heat. But even for stationary applications there are flow-driving energy sources which should be carefully considered. Small batteries, with sufficiently thick electrolytes, have been safely operated for months. The ultimate goal is, however, to construct LMBs as large as possible with electrolyte layers as thin as possible. The latter derives from the demand to minimize the unavoidable voltage loss $V_{\rm loss}=d\ j/\sigma$ within the electrolyte of thickness $d$ and conductivity $\sigma$, for a charging/discharging current with density $j$. The former request is a consequence of the (rough) estimate that the active materials costs are (presently) only one fourth (or so) of the total battery costs, and that any increase of the size would lead to a significant decrease of the costs, both per kW and per kWh.

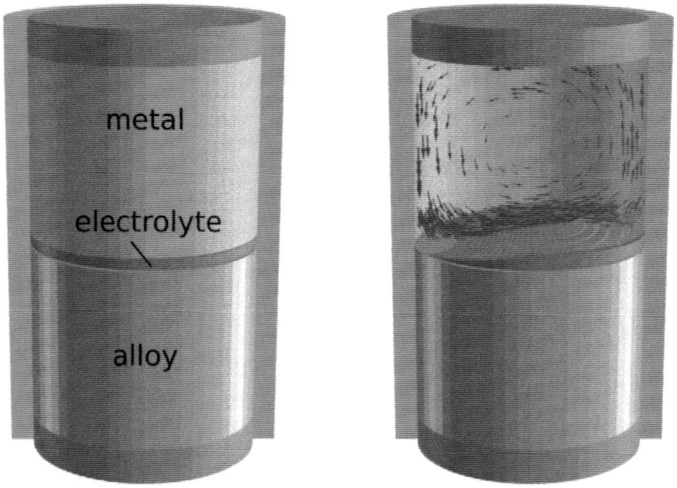


**Figure 1.** Sketch of a liquid metal battery with typical inventory. The electrolyte, which separates the two liquids, acts as an ion conductor. A strong enough flow might wipe the electrolyte and lead to a short circuit.

This paper deals, therefore, with the following question: what kind of limitations are set by (magneto)-hydrodynamic effects on the maximum size of LMBs and/or on the minimum thickness of the electrolyte layer, and what countermeasures could be taken to overcome those limitations?

## 2. Magnetohydrodynamics of Liquid Metal Batteries

The full hydrodynamics of LMBs is extremely involved. It includes Rayleigh-Bénard convection due to resistive heating of the electrolyte layer, surface-tension driven flows, capillary effects at the walls etc. Here, we focus on the magnetohydrodynamic (MHD) aspects, including electro-vortex flows and various current-driven instabilities. Even this restricted problem gets rather complex, and includes, e.g. the mutual interaction of current-driven flow instabilities and long and short wave interface instabilities, whose numerical study has just begun.

In the following, we will start to tackle the issue from two sides. We will first discuss the so-called Tayler instability (TI), a kink type current-driven instability, and second the electro-vortex flow (EVF) which appears when current distributions are inhomogeneous. While the latter might indeed be more relevant for real-world LMBs with their complicated wiring system and current-collectors of finite thickness, the TI still marks a paradigmatic reference point. Actually, for large total currents the TI would lead to vigorous flows, and the ultimate disruption of the electrolyte layer, even in the idealized case that the charging/discharging current could be made perfectly homogeneous. We will also show some first results on the interaction of TI and EVF, and their influence on the metal-salt interface.

### *2.1. Tayler instability*

The kink-type Tayler instability had first been discussed in the context of fusion related and astrophysical plasma physics. The self-interaction of a homogeneous axial current in a cylinder with its own azimuthal magnetic field produces an inwardly directed pinch force that becomes unstable either to an axisymmetric ("sausage") or to a non-axisymmetric ("kink") instability. In the idealized case without any viscosity or resistivity of the fluid, the kink instability can be suppressed either by an additional axial magnetic field (leading to the celebrated Kruskal-Shafranov condition of fusion research [3]), by stable stratification [4], or by rotation [5]. Comparably late, the stabilizing effect of viscosity and resistivity was investigated in detail in [6] where the key role of the Hartmann number for the onset of the TI was also identified. This Hartmann number $Ha=B_\varphi(R)\,R\,(\sigma/\rho\nu)^{1/2}$ (with $B_\varphi(R)$ denoting the azimuthal field at cylinder radius $R$, and $\sigma$, $\rho$, $\nu$ denoting conductivity, density, and viscosity of the fluid), which measures the ratio of magnetic forces to viscous forces, must be above 20 for the TI to set in [7]. This critical number, together with typical growth rates of the TI, has recently been confirmed in a liquid metal experiment using electrical currents up to 8 kA [8].

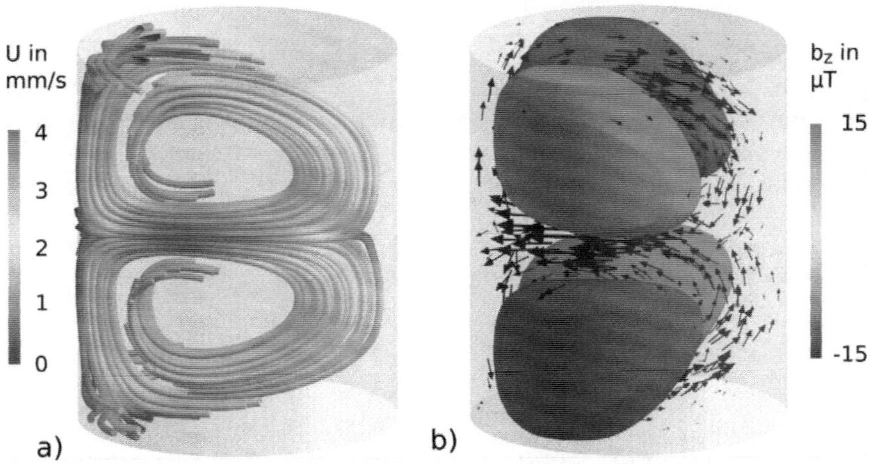


**Figure 2.** Simulated Tayler instability for a cylinder filled with liquid Li, at a current of 4 kA. (a) Flow structure, (b) induced axial magnetic field component.

In order to simulate the effects of the TI as well as of EVFs on LMBs, we have set up a numerical model [9] on the basis of the OpenFOAM library which is extended by a Poisson solver for the electric potential and an implementation of Biot-Savart's law for the induced magnetic field. Figure 2 gives a first impression of the TI for the case of a liquid lithium column with a total current of 4 kA. We can easily recognize the asymmetric, kink-type structure of the flow which is, however, superposed by an axisymmetric component arising in the non-linear saturation regime of the TI.

In a follow-up paper [10] we had discussed several means for preventing the TI. The simplest way is to avoid high columns of liquid metals so that the TI, which appears with a typical wavelength, cannot develop within the limited height. However, this height constraint represents a limitation of the storable electric energy. As a rough estimate, valid for Na and a current density of 10 kA m$^{-2}$, one hour corresponds to 1 cm. If one plans to store energy for a few days, one easily ends up with heights in the order of 1 m.

Three alternative methods to suppress the TI are sketched in figure 3. A return current through a central bore changes the radial structure of the azimuthal field in such a way that the TI disappears [12]. An axial magnetic field $B_z$ suppresses the TI according to the Kruskal-Shafranov condition [3]. Note, however, that this additional $B_z$ can give rise to a sloshing motion of the interface, as it is known from aluminium reduction cells. A slight stabilization could also be achieved by applying a horizontal magnetic field $B_x$.

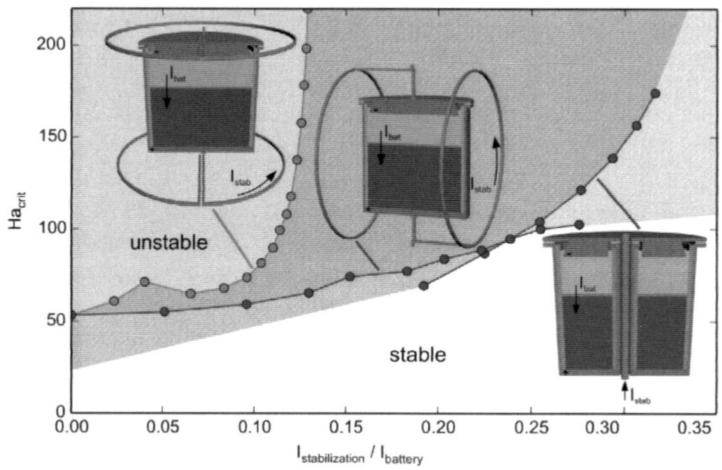


**Figure 3.** Three methods of suppressing the TI in liquid metal batteries. A return current through a central bore changes the radial structure of azimuthal field in such a way that the TI disappears. An additional axial magnetic field $B_z$ suppresses the TI according to the Kruskal-Shafranov condition. A similar effect is achieved by a horizontal field $B_x$.

*2.2. Electro-vortex flow*

We consider now the case that the electrical current is not completely homogeneous. This is already relevant if the current collectors above and below the battery have a finite length (which is always the case). Imagine that this finite-length battery is connected to wires, the current will acquire radial components which are still present in the parts of the fluid close to the collectors. Despite their smallness compared to the total current, these radial components together with the azimuthal field produce a Lorentz force that points away from the current collectors. The resulting EVF will compete with the TI in a non-trivial manner [11].

For a simplified geometry, this is illustrated in Figure 4 where we distinguish three cases. In the first case, "fixed potential over D", the effect of the current collectors is simply neglected and it is assumed that the applied current is completely homogeneous in the fluid. Then, the TI develops as usual. If we now consider a current collector of height $h=3D$, with a constant potential applied over a circular diameter $d=D/2$, we see that initially an EVF is developing, with its typical (more or less axisymmetric) double-vortex structure driven by the Lorentz forces close to the current collectors. After some time, however, the TI comes into play and drives the systems to velocities which are slightly larger than the usual ones. If we decrease the height of the current collectors to $h=2D$, the EVF develops faster, and reaches such velocities which apparently lead to suppression of the TI. Obviously the interaction of TI and EVF is quite intricate.

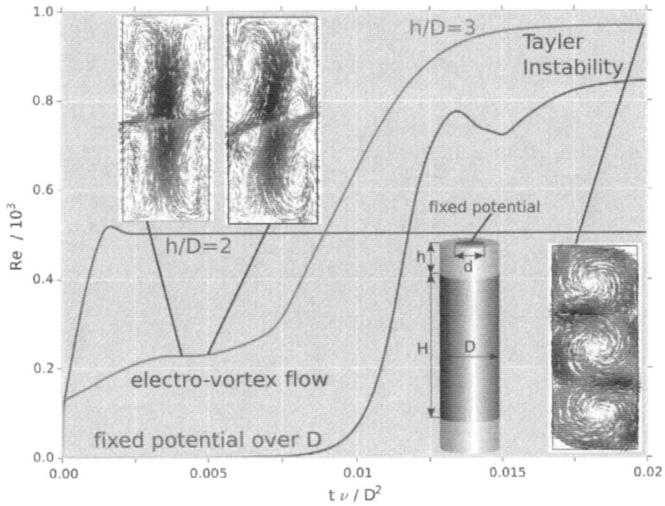


**Figure 4.** Interaction of TI and EVF. Mean velocity in a cylindrical liquid metal column ($D$ = 1 m, $H$ = 2.4 m) for an applied current corresponding to $Ha$ = 100. Simulation is carried out without current collector and a current collector of aspect ratio 2 or 3 with an electrical conductivity 5 times larger than that of the fluid.

*2.3. EVF and the integrity of the electrolyte layer*

Let us focus now on the case of a pure EVF, and consider its effect on the multiphase stratification in a real battery. For this purpose, we consider a Na-Bi system with a molten salt as electrolyte. The current collectors are assumed as rather thin so that a very vigorous EVF appears. The shape of the electrolyte layer is computed by a volume-of-fluids (VOF) method. Figure 5 shows the time evolution of the EVF and the salt layer. From the thickness of the salt layer we can infer at which current the layer is disrupted.

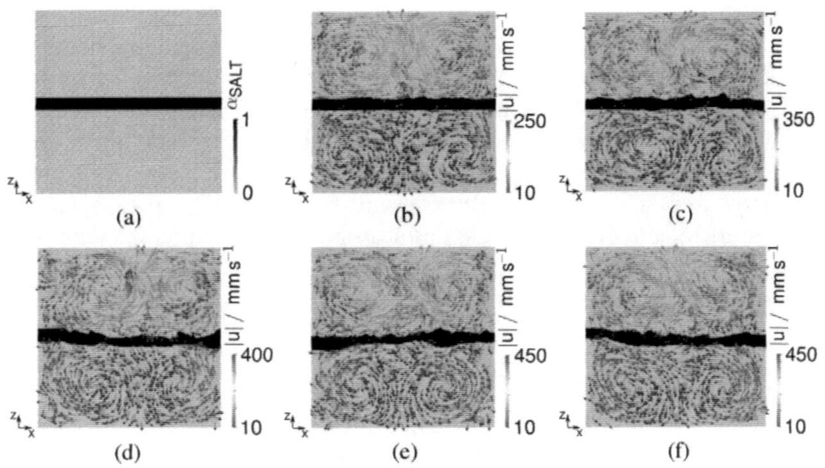


**Figure 5.** Time evolution of the flow and the electrolyte layer for an EVF in a cell with aspect ratio 1. The snapshots (a-f) correspond to the instants 0, 4 s,... , 20 s, respectively. Initially, the thickness of the salt layer is 0.09 of the total height. The current is 1 kA.

Taking into account possible real current densities in LMBs, we then change the electrical current and the initial thickness $h_0$ of the salt layer to obtain a variety of combinations of relevant parameters. Motivated by [13], we expect that the governing parameter for the integrity of the layer is the Richardson number $Ri_{Na}=(\rho_{salt}/\rho_{Na}-1)\, g\, h_0/\langle v_{Na}\rangle^2$ which is the ratio of the potential energy to the kinetic energy ($g$ is the gravitational acceleration, and $\langle v_{Na}\rangle$ is the averaged velocity magnitude in the Na layer). Figure 6 confirms that this is indeed the case: approximately at $Ri_{Na}=1$, the minimum layer thickness bends sharply to zero and the risk of a potential battery failure increases.

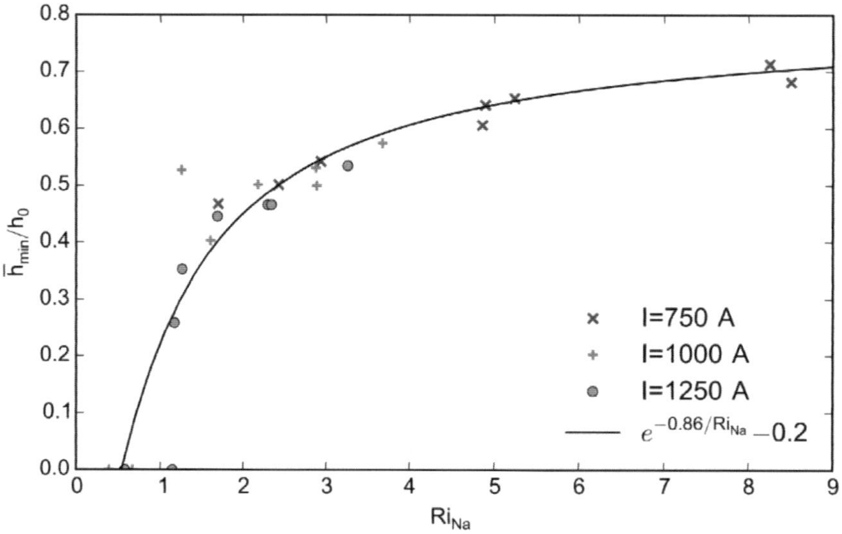


**Figure 6.** Dependence of the normalized minimal thickness of the salt layer on the Richardson number for Na. The solid line corresponds to a fit with an exponential function.

## 3. Towards a real battery

In this section, we show the very first results of a small LMB experiment carried out at HZDR. We have chosen the traditional Na-Bi system, but have replaced the pure Na-salts by the mixture LiCl-KCl-NaCl, mainly for reasons of its lower melting temperature. Figure 7 shows the experimental set-up, including a copper vessel filled with Bi at the bottom, and the salt on top of it. The central position of the Na drop at the top is guaranteed by holding it with a Ta wire. Figure 7 also shows the voltage versus time upon charge and discharge. The thickness of the electrolyte is approximately 4 mm. The interface area between the electrolyte and the sodium drop is not very precisely known. A reasonable estimate is one half of the surface of the sphere. In any case, with measured currents of 300 mA, we obtain a remarkable current density in the order of 10 kA $m^{-2}$.

## 4. Conclusions and prospects

In this paper, we have discussed some particular aspects of the complex (magneto)-hydrodynamics of LMBs. We have started with the idealized setting of a perfectly homogeneous current, which becomes prone to the TI if the total current increases to a value that corresponds to a Hartmann number in the order of 20. In this context, we have discussed various means for the suppression of the TI. The consideration of EVF has shown that even minor deviations from the idealized homogenous current assumption can lead to vigorous flows. A first simulation of a simplified three-layer system confirmed the relevance of the Richardson number for integrity of the stratification, which was already discussed in [14]. Some very preliminary experimental results with a Na-Bi battery system, using a LiCl-KCl-NaCl electrolyte, have reassured that current densities in the order of 10 kA $m^{-2}$ are indeed realistic.

The way to a competitive, large-scale LMB is still long, and will need a close collaboration of chemistry, numerics, and advanced engineering. Hopefully, large-scale LMBs will soon be set up and tested in the framework of the DRESDYN project at HZDR [15].

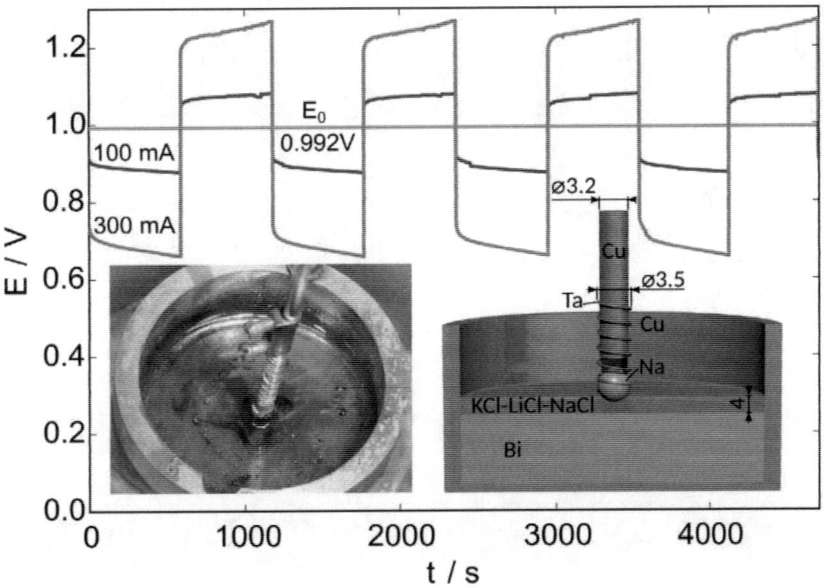


**Figure 7.** A real battery with Na-Bi and KCl-LiCl-NaCl as electrolyte, and the voltage versus time upon charging/discharging.

## 5. Acknowledgments

This work was supported by Helmholtz-Gemeinschaft Deutscher Forschungszentren (HGF) in frame of the Helmholtz Alliance LIMTECH, as well as by Deutsche Forschungsgemeinschaft in frame of the SPP 1488 ("PlanetMag"). Fruitful discussions with Wietze Herreman, Caroline Nore, and Janis Priede are gratefully acknowledged.

## 6. References

[1] Cairns EJ and Shimotake H 1969 *Science* **164** 1347
[2] Kim H et al. 2013 *Chem. Rev.* **113** 2075
[3] Kruskal M and Tuck J 1958 *Proc R. Soc.* A **245** 222
[4] Tayler RJ 1973 *Mon. Not. R. Astron. Soc.* **161** 365
[5] Pitts E and Tayler RJ 1985 *Mon. Not. R. Astron. Soc.* **216** 139
[6] Spies GO 1988 *Plasma Phys. Control. Fusion* **30** 1025
[7] Rüdiger G, Schultz M and Gellert M 2011 *Astron. Nachr.* **10** 1
[8] Seilmayer M et al 2012 *Phys. Rev. Lett.* **108** 244501
[9] Weber N, Galindo V, Stefani F, Weier T and Wondrak T 2013 *New J. Phys.* **15** 043034
[10] Weber N, Galindo V, Stefani F and Weier T 2014 *J. Power Sources* **265** 166
[11] Weber N, Galindo V, Priede J, Stefani F and Weier T 2015 *Phys. Fluids* **27** 014103

[12] Stefani F, Weier T, Gundrum T and Gerbeth G 2011 *Energy Convers. Manag.* **52** 2982
[13] Banks RB and Bhavamai A 1965 *J. Fluid Mech.* **23** 229
[14] Herreman W, Nore C, Cappanera L and Guermond J-L 2015 Tayler instability in liquid metal columns and liquid metal batteries 2015 J. Fluid Mech. **771** 79
[15] Stefani F et al. 2012 *Magnetohydrodynamics* **48** 103

IOP Conf. Series: Materials Science and Engineering **143** (2016) 012025 doi:10.1088/1757-899X/143/1/012025

# Comparison of different Methods to model Transient Turbulent Magnetohydrodynamic Flow in Continuous Casting Molds

**C Kratzsch, A Asad, R Schwarze**

TU Bergakademie Freiberg,
Institute of Mechanics and Fluid Dynamics,
Lampadiusstraße 4, 09599 Freiberg

christoph.kratzsch@imfd.tu-freiberg.de

**Abstract**. Modeling of the processes in the continuous casting mold engaged many scientists once the computer-technology was able to accomplish that task. Despite that, CFD modeling of the fluid flow is still challenging. The methods allow deeper and deeper inside views into transient flow processes. Mostly two kinds of methods are applied for this purpose. URANS simulations are used for a coarse overview of the transient behavior on scales determined by the big rollers inside the mold. Besides, LES were done to study the processes on smaller scales. Unfortunately, the effort to set up a LES is orders of magnitude higher in time and space compared to URANS. Often, the flow determining processes take place in small areas inside the flow domain. Hence, scale resolving methods (SRS) came up, which resolve the turbulence in some amount in these regions, whereas they go back to URANS in the regions of less importance. It becomes more complex when dealing with magnetic fields in terms of EMBr devices. The impact of electromagnetically forces changes the flow structure remarkably. Many important effects occur, e.g. MHD turbulence, which are attributable to processes on large turbulent scales. To understand the underlying phenomena in detail, SRS allows a good inside view by resolving these processes partially. This study compares two of these methods, namely the Scale Adaptive Simulation (SAS) and the Delayed Detached Eddy Simulation (DDES), with respect to rendition of the results known from experiments and URANS simulation. The results show, that the SAS as well as the DDES are able to deliver good results with higher mesh resolutions in important regions in the flow domain

## 1. Introduction

Magnetohydrodynamic flows play an important role especially in casting processes where desired flow situations are created by means of so called electromagnetic braking (EMBr). The control of the jet flow inside a continuous casting mold can be seen as a good example. In the process, liquid metal is released from a tundish into the water-cooled copper mold through a submerged entry nozzle (SEN). By virtue of the two outlets positioned at the circumference of the SEN, usually two submerged jets build up. They impinge on the narrow mold walls, forcing the flow to go up to the free surface and down to the solidification region as well. As a result, typical two mirror-symmetrical recirculation roll pairs are created. In consequence of a whole range of different mechanisms (fluid flow, solidification, multiphase flow, clogging, etc.), strong variations in the structure of these recirculation rolls can be observed. This can be the source of detrimental effects for the final product, e.g. through inclusion transport in the solidification zone. A good overview of plenty of these mechanisms gives Lee et al. [1].

Magnetic fields give the possibility to handle or curb instabilities in the fluid flow. This is done by reduction of jet velocity through Lorentz forces pointing against the flow velocity. Therefore, the name "electromagnetic braking" is given to this principle. However, the reduction of jet velocity can lead to acceleration at other flow regions because of mass and momentum conservation. Also the turbulence is influenced by the magnetic fields resulting in elongated vortex structure and forced anisotropy in every scale (magnetohydrodynamic turbulence). Unfortunately all these complex effects can lead to flows, which are initially not desired and can change the flow situation considerably.

This study compares different methods to model the turbulent flow, starting from two equation unsteady RANS (URANS) models. They are mostly used to describe turbulent flows in technical applications. Unfortunately, when it comes to complex flow physics like magnetohydrodynamic turbulence and relaminarization, the assumptions made for using classical RANS models are not fulfilled anymore. The next step can be the use of Reynolds stress models (RSM), but often a Large Eddy Simulation (LES) is used, which resolves the big turbulent eddies and models the physics on smaller scales through so called subgrid-scale models. The difference in the computational amount is huge. Hence, for a long time every engineer or physician had to make a compromise between accuracy and calculation speed. The problem comes also up if time resolved turbulence data is needed. As a result, hybrid methods are formulated which try to fill the gap between URANS and LES. Two of them are the Scale Adaptive Simulation (SAS) and the Delayed Detached Eddy Simulation (DDES). These are studied here in respect of the rendition of time-averaged and time-resolved data. The comparison is made by using experimental data from the liquid metal mold model mini-LIMMCAST from the Helmholtz-Center Dresden-Rossendorf [2].

## 2. Governing Equations and Methods

The CFD model of the isothermal, incompressible, turbulent mass- and the momentum transport in the mold flow is based on the Reynolds-averaged or filtered Navier-Stokes equations, respectively.

$$\nabla \cdot \overline{\underline{u}} = 0 \tag{1}$$

$$\rho \frac{\partial \overline{\underline{u}}}{\partial t} + \rho \left( \overline{\underline{u}} \cdot \nabla \right) \overline{\underline{u}} = -\nabla \overline{p} + \eta \Delta \overline{\underline{u}} - \nabla \cdot \left( \rho \overline{\underline{u}'\underline{u}'} \right) + \overline{\underline{f}_L} \tag{2}$$

Here, $\overline{\underline{u}}$ denotes the flow velocity, $\overline{p}$ denotes the pressure, $\eta$ is the dynamic viscosity and $\rho$ is the density. The term containing the fluctuation velocities $\nabla \cdot \left( \rho \overline{\underline{u}'\underline{u}'} \right)$ is called Reynolds stresses. It accounts for effects attributable to turbulence. RANS-based models assume only momentum diffusion due to turbulence and relate the effect to turbulent properties, e.g. turbulent kinetic energy $k$. RSM model this term directly through a conservation equation for the Reynolds stresses, while a LES resolves the term in some amount. The rest is solved with a subgrid scale model. In between both prior methods the scale-resolving (SRS) methods try to mix both approaches with the purpose to reduce the computational amount compared to an LES while resolving the turbulence partially. The first investigated method in this study is the SAS. Its implementation in OpenFOAM can be found in Kratzsch et al. [3]. In short, the SAS based on a SST model [4] where turbulent quantities are adaptively changed by means of another turbulent length scale, the von Kàrmàn length scale. Eventually, this results in a reduction of the eddy viscosity in these regions. Hence, LES-like unsteadiness inside a URANS simulation is created. For this method a strong disturbance is necessary to get the LES-like behavior. Contrarily, a DDES is stronger connected to an LES because it solves the core flow by LES whereas the near wall flow is solved by a favourable RANS turbulence model. The DDES method used in OpenFOAM is based on the work of Spalart et al. [5], where a Spalart-Allmaras one equation URANS turbulence model is used for the near wall flow.

As seen in equation (2), the Lorentz forces are added to the Navier-Stokes equations by $\overline{\underline{f}}_L$. Assuming that neither the length scale $l$ nor the characteristic flow velocity $U$ of the melt flow is high, the magnetic Reynolds number $R_m$ is low.

$$R_m = \frac{Ul}{\lambda} = \mu\sigma Ul \leq 1 \tag{3}$$

The corresponding properties are the magnetic permeability $\mu$ and the electrical conductivity σ, respectively. This low magnetic Reynold-number suffices the inductionless approximation, where induced magnetic fields can be neglected [6]. It is a loose coupling of velocity field and magnetic field. The Lorentz force is calculated from:

$$\overline{\underline{f}}_L = \overline{\underline{j}} \times \underline{B}_0 \tag{4}$$

$$\overline{\underline{j}} = \sigma\left(-\nabla\overline{\psi} + \overline{\underline{u}} \times \underline{B}_0\right) \tag{5}$$

$$\nabla^2\overline{\psi} = \nabla \cdot \left(\overline{\underline{u}} \times \underline{B}_0\right). \tag{6}$$

Here, the electric field is replaced by an electric potential $\overline{\psi}$ due to vector identity in Faraday's law for static magnetic fields. In order to estimate the strength of the influence of the magnetic field onto the flow, the Hartmann number can be constructed:

$$Ha = lB_0\sqrt{\frac{\sigma}{\eta}}. \tag{7}$$

The Hartmann number measures the ratio of electromagnetic forces to the friction forces. Hence, the Hartmann number is strongly connected to the generation of so called Hartmann-layers, which can be seen as electromagnetically influenced boundary layers. In the case studied here, the Hartmann number is $Ha_{mold} = 417$. Thus, the Hartmann-layers are very thin (approx. tenth of millimetres). Resolving these layers is not engaged in this study, which may result in an accuracy loss.

## 3. Governing Equations and Methods

In this study, we use the open-source CFD software OpenFOAM in version 2.3.x. The common flow solvers are extended to consider magnetohydrodynamic (MHD) flows with the inductionless approximation. The solver distributes the eddy current density to the cell faces. Then, the electric potential can be evaluated at the faces. Thereby, the solver includes balance of fluxes through cell faces to ensure charge conservation and a proper interpolation of $\overline{\underline{j}}$ from cell face to cell centre [7]. This is of great importance due to the amplification of errors in calculation of $\overline{\underline{j}}$. Also, correction terms needed for arbitrary non orthogonal collocated meshes were included. The validation of the solver is realised by setting up a Shercliff-type laminar MHD channel flow and can be found in [8]. The flow domain of the CFD model mold is deduced from the mini-LIMMCAST facility located at Helmholtz-Zentrum Dresden-Rossendorf, which is a laboratory scale model of a continuous casting mold working with the eutectic liquid metal GaInSn as model melt. The case of a magnetic field in shape of a ruler is studied. The indication value of the magnetic field reads 0.31T. The distribution is shown in figure 2. Experimental flow data were captured by Ultrasound-Doppler velocimetry (UDV) [2].

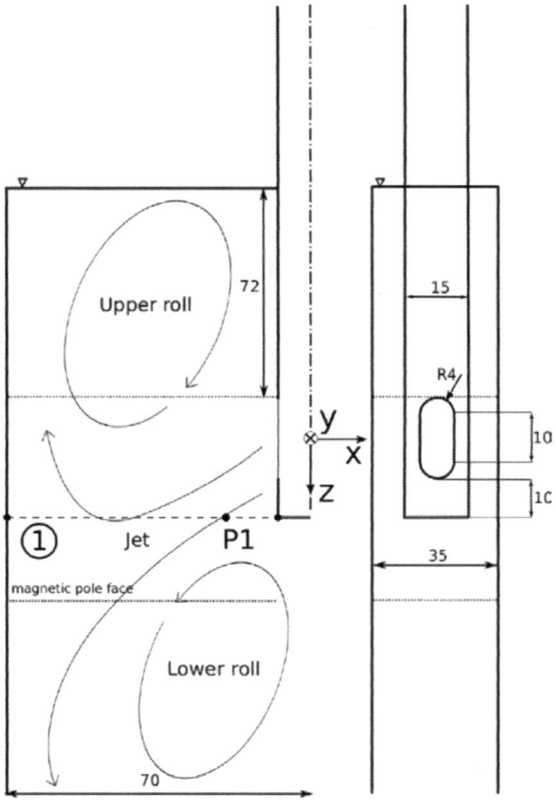


**Figure 1.** Overview of the sampled line and point in the midplane of the mold geometry

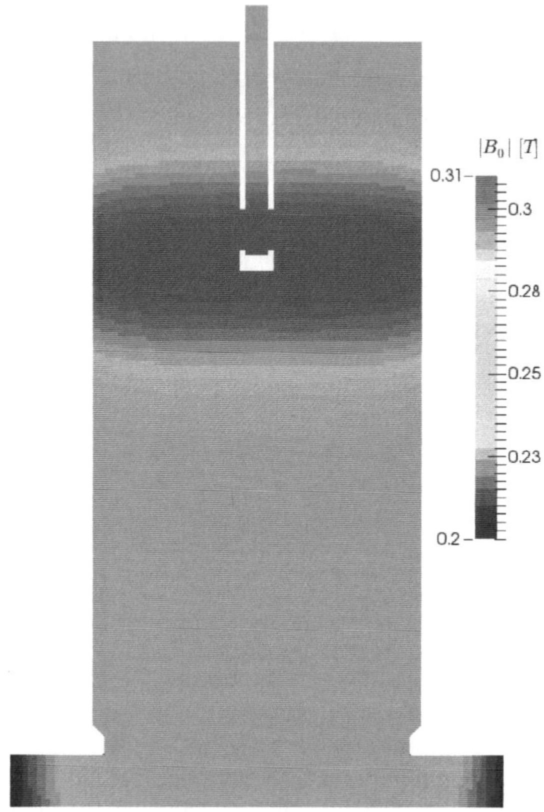


**Figure 2.** Magnetic field distribution in the midplane of the mold domain.

For the CFD simulations, all the meshes are generated using the OpenFOAM meshing tool snappyHexMesh. The mesh contains about 1.5 million cells. Earlier studies reveal that the k-ω-SST model [4] by Menter yields best results for pure hydrodynamic flow [9]. Therefore, it is chosen for the URANS simulations. The scale resolving approaches (SAS and DDES) get their initial field from a SAS simulation on a coarse mesh which runs for $t = \frac{V}{\dot{V}} = 13s$, the characteristic flow time. Here, $V$ denotes the volume of the mold and $\dot{V}$ denotes the flow rate at the inlet. Data were written out every 0.05s for a time of 20s. Point-probes are recorded with the simulation timestep ($\Delta t \approx 10^{-4} s$). Effects by MHD turbulence, like forced vortex anisotropy due to fluctuating Lorentz forces are not considered in the URANS turbulence model. However, the scale resolving methods should capture this effect in some amount. Regarding the discretization, higher order accuracy schemes are used preferentially in order to minimize the effects of numerical diffusion. For all the simulations, the Courant number is held below unity.

## 4. Results

In order to evaluate the simulations, mean and instantaneous velocities are plotted in various figures. Figure 1 gives an overview over the sampled lines, points and slices in this section. First, the mean velocity field in the midplane of the mold geometry is depicted in figure 3. It is obvious, that only the DDES simulation yields a symmetrical mean in the upper mold between the left and the right side, while the SAS- and the URANS-velocity fields are asymmetrical. Despite that, every simulation yields two recirculation rolls in the upper mold, whereas the lower mold does not have a distinct twin roll structure. This behavior is pointing towards the already known jet oscillation enforced by the magnetic field.

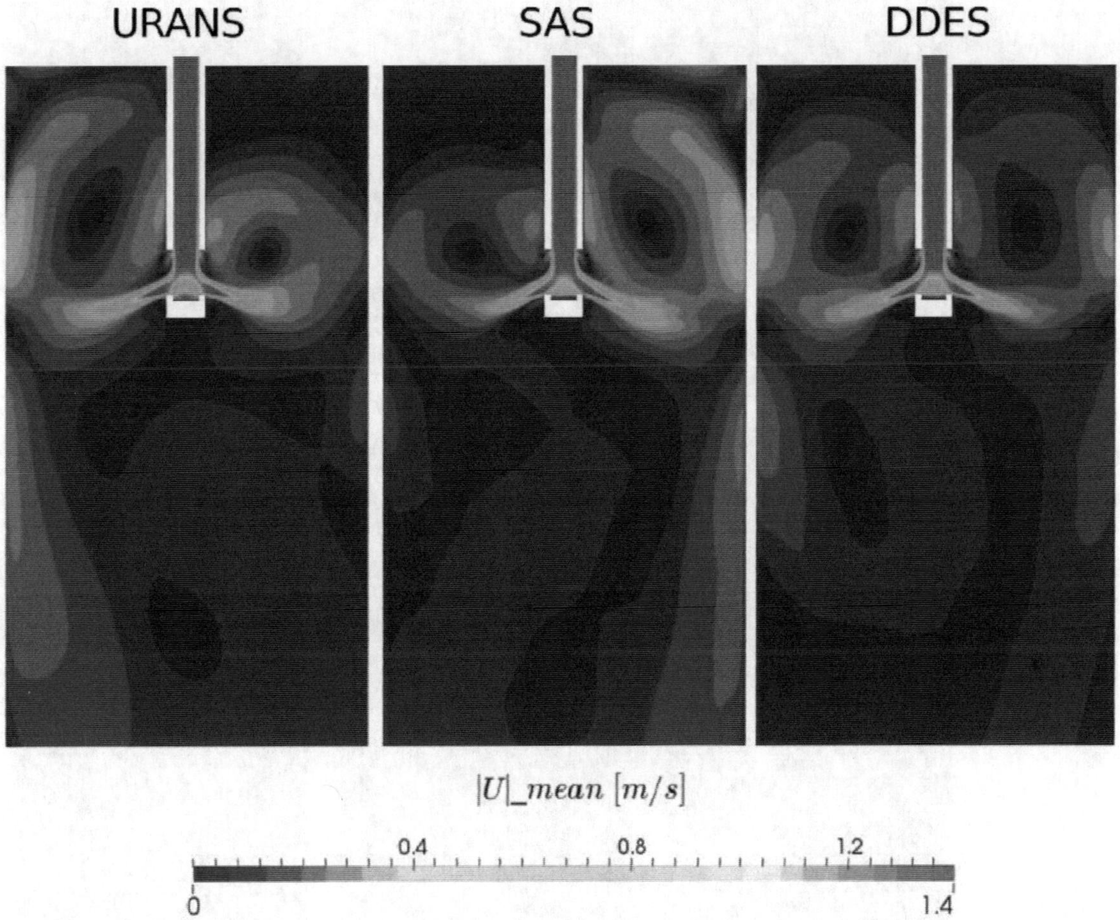


**Figure 3.** Mean velocity magnitude in the midplane of the mold geometry for the URANS, SAS, and DDES.

The oscillation is elucidated by figure 4 for the approaches SAS and DDES. Three timesteps are shown there. It can be seen, that both methods show a more or less periodical structure with downbending of one jet, a central position where both jets have equal angles and afterwards the downbending of the other jet. They are able to resolve a small amount of turbulent scales, like it can be seen at the jet breakup. Apparently, the magnetic field is strong enough at the SEN outlets to damp disturbances through the impingement at the SEN well. Hence, the flow in the SEN right before the release in the mold cavity is fixed compared to a pure hydrodynamic flow. Another reason for the lack of scales in the early jet region can be an unsufficient mesh resolution.

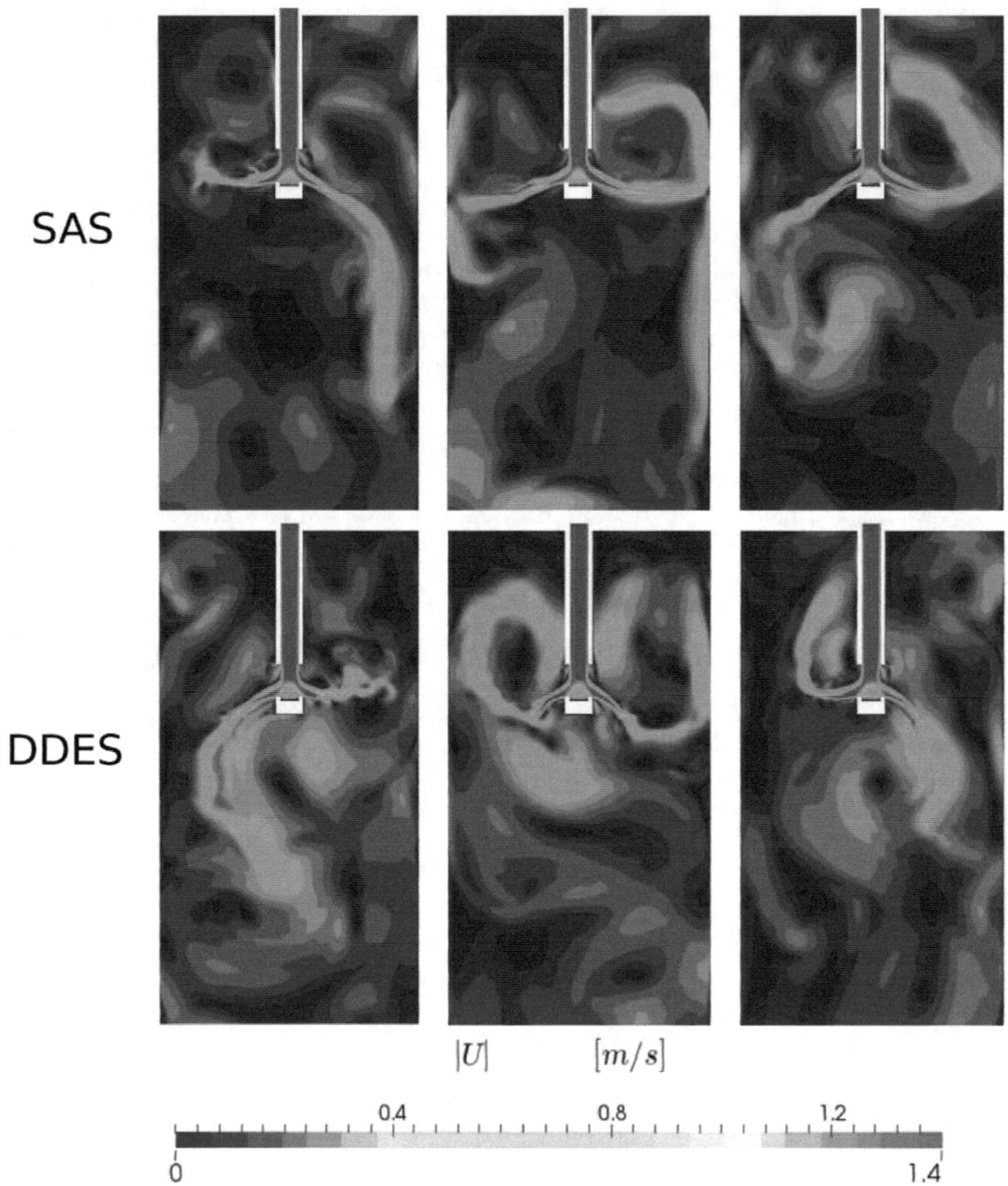


**Figure 4.** Time-dependent velocity field of the magnitude velocity at three snapshots. SAS snapshots – 18.6s, 19.3s, 19.9s; DDES snapshots – 8.0s, 9.0s, 10.0s.

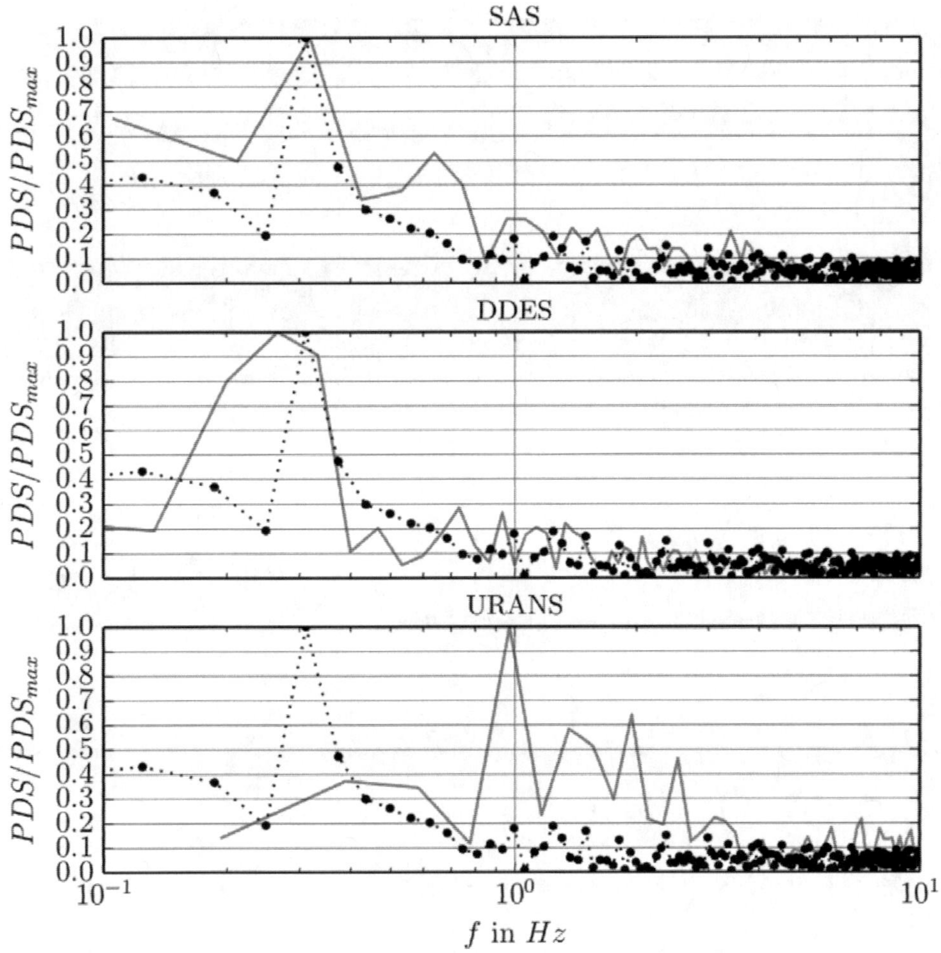


**Figure 5.** FFT of the horizontal velocity at point P1. SAS (top, continuous line), DDES (middle, continuous line), URANS (bottom, continuous line) vs. UDV results (dotted line).

In order to evaluate the results quantitatively, the time-dependent lateral velocity at point P1 is analyzed with a Fast Fourier Transformation (FFT) depicted by figure 5. As figure 1 shows, the point is in close vicinity to the jet outlet and is located at the lower side of the jet. Thus, the point can contain characteristics from the flow of the lower mold because of the infrequent upward directed flow in the lower mold. The plot of the experimentally captured velocity shows one dominant single peak at 0.3 Hz. Comparing this results with the three approaches studied here, clear distinctions can be found. By taking a look on the SAS results, the best accordance in matching the experimental frequency can be observed. Besides, a sub-peak of about 0.6 Hz exists. The DDES graph shows a good agreement with the experimental data. It matches well the experimentally observed frequency without distinct sub-peaks. With the URANS simulations a main-peak can be observed at about 1Hz. Besides, multiple sub-peaks exist, which points to a less pronounced oscillatory behavior.

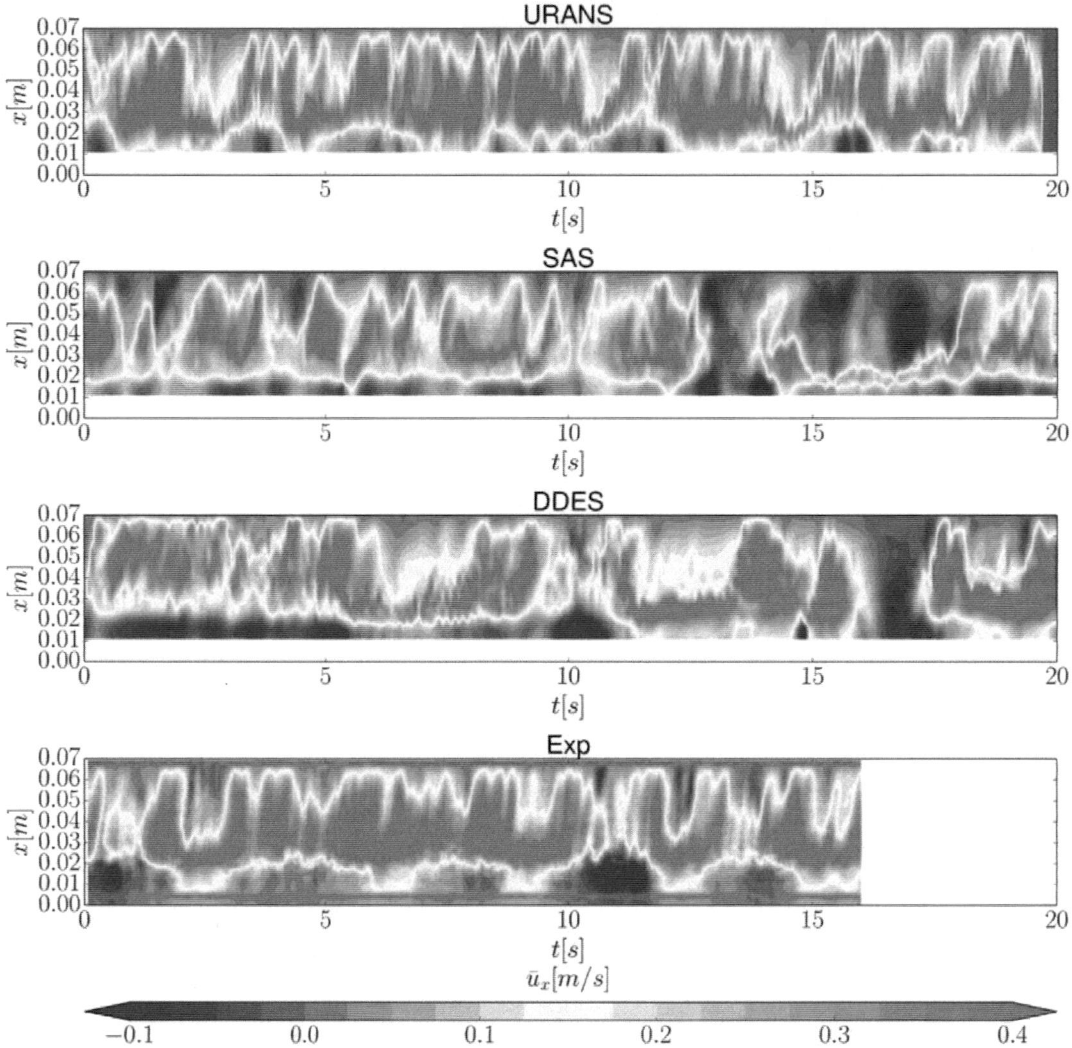


**Figure 6.** Spatio-temporal plot of the horizontal velocity $\bar{u}_x$ over Line 1.

It is clear that this peak can be correlated with the jet oscillation and the lower mold flow structure created by the strong magnetic field. High frequencies as a source of the turbulence are damped by the magnetic field and vanish due to their low energetic level. The flow is left with large scale oscillations which have enough energy to preserve. The jets experience these oscillations and therefore the whole flow situation changes. Regarding the deviations between the URANS and the SRS methods, it seems that the attempt to resolve the regions where the strongest interaction of the magnetic field and the turbulence exist seem to work.

A more elaborate depiction is a spatio-temporal plot of a line, which crosses the jet (in this case line ① sketched in figure 1). Therein, the jet oscillations can be seen over the half width of the mold. An upbended jet is denoted by high velocities in lateral direction while a downbended jet is denoted by low velocities. Low values in x denoting close vicinity to the SEN. High values in x denoting close vicinity to the narrow wall. It can be seen that the SAS and DDES results resemble the experimental results the best. However, the jet fluctuations towards the SEN (lower values in x) match better with the experiments. The disruption patterns of the jet are hard to compare as well as the fluctuations towards the narrow walls.

## 5. Conclusion

The study has shown, that SRS methods can resolve turbulence in a limited amount with mesh resolutions compared to typical URANS simulations. This can give a brighter inside view in the flow behavior and the mechanisms of MHD turbulence and represents a justifiable step away from URANS simulations towards LES. It is confirmed that SRS approaches render a magnetic field enforced oscillation in good accordance with experimental data. The URANS approach lacks in reproducing this behavior.

Apparently, the URANS model have to be augmented by adding sink terms for the effect of the MHD turbulence, e.g. by means of [10]. The evaluation of the SRS approaches will be pursued by comparison with LES using revised subgrid scale modeling regarding anisotropy and relaminarization.

## References

[1] Lee, Peter D., et al., Review: The "butterfly effect" in continuous casting, Ironmaking & Steelmaking (2012), p. 244

[2] Timmel, K., et al., Experimental Investigation of the Flow in a Continuous-Casting Mold under the Influence of a Transverse, Direct Current Magnetic Field, Metallurgical and Materials Transactions B (2011), p.68 – 80.

[3] Kratzsch, Christoph, Asad, Amjad and Rüdiger Schwarze. "CFD of the MHD Mold Flow by Means of Hybrid LES/RANS Turbulence Modeling." Journal for Manufacturing Science and Production.

[4] Menter, F.R., Two-equation eddy-viscosity turbulence models for engineering applications, AIAA Journal (1994)

[5] Spalart, Philippe R., et al. "A new version of detached-eddy simulation, resistant to ambiguous grid densities." Theoretical and computational fluid dynamics 20.3 (2006): 181-195.

[6] Davidson, Peter Alan. *An introduction to magnetohydrodynamics*. Vol. 25. Cambridge university press, 2001.

[7] Ni, M.-J., et al., A current density conservative scheme for incompressible MHD flows at a low magnetic Reynolds number. Part II: On an arbitrary collocated mesh, Journal of Computational Physics (2007), p. 205 – 228.

[8] Kratzsch, Christoph, Schwarze, Rüdiger. "Modeling of the continuous casting mold flow under influence of magnetic fields." *8th European Continuous Casting Conference*. 23 – 26 June 2014, Graz, Austria. pp. 239 – 245.

[9] Kratzsch, Christoph, et al. "URANS simulation of continuous casting mold flow: Assessment of revised turbulence models." *steel research international* 86.4 (2015), pp. 400 – 410.

[10] Widlund, Ola, Said Zahrai, and Fritz H. Bark. "Development of a Reynolds stress closure for modeling of homogeneous MHD turbulence." *Physics of Fluids (1994-present)* 10.8 (1998): 1987-1996.

IOP Conf. Series: Materials Science and Engineering **143** (2016) 012026 doi:10.1088/1757-899X/143/1/012026

# Assessment of Different Turbulence Models for the Motion of Non-metallic Inclusion in Induction Crucible Furnace

**H. Barati, M. Wu, A. Kharicha, A. Ludwig**
Chair for Modeling and Simulation of Metallurgical Processes, Department of Metallurgy, University of Leoben, Franz-Josef Street 18, 8700 Leoben, Austria

E-mail: menghuai.wu@unileoben.ac.at

**Abstract.** Turbulent fluid flow due to the electromagnetic forces in induction crucible furnace (ICF) is modeled using k-ε, k-ω SST and Large Eddy Simulation (LES) turbulence models. Fluid flow patterns calculated by different turbulence models and their effects on the motion of non-metallic inclusions (NMI) in the bulk melt have been investigated. Results show that the conventional k-ε model cannot solve the transient flow in ICF properly. With k-ω model transient flow and oscillation behavior of the flow pattern can be solved, and the motion of NMI can be tracked fairly well. LES model delivers the best modeling result on both details of the transient flow pattern and motion trajectories of NMI without the limitation of NMI size. The drawback of LES model is the long calculation time. Therefore, for general purpose to estimate the dynamic behavior of NMI in ICF both k-ω SST and LES are recommended. For the precise calculation of the motion of NMI smaller than 10 μm only LES model is appropriate.

## 1. Introduction

Understanding the turbulent flow and other transport phenomena, like dissolution of alloying elements and removal of non-metallic inclusions in induction crucible furnace (ICF) is one of the important subjects in steel metallurgy. The flow pattern in the melt is mainly governed Lorentz force. The interaction between the induced electrical current and magnetic field results in Lorentz force acting on the bulk melt in the direction towards the center of crucible. Typically two or more eddies with turbulent pulsations between them might form [1]. This phenomenon has been subject of various researches [2-5].

Jardy et al. [6] carried out experimental and theoretical study to characterize the hydrodynamic behavior and the transport of dissolved elements in the molten steel in vacuum induction furnace. 2D simulation using k-ε turbulence model, with a standard set of constants, showed adequate results [7, 8]. Nevertheless, in spite of the k-ε results demonstrating highest values of turbulent kinetic energy in the eddy centers and the lowest between the eddies, the experimental results showed that the maximum of the turbulent energy is between the vortices of the averaged flow and close to the wall of the crucible [5, 9]. Umbrashko et al. [5] simulated turbulent flow in an induction furnace for a low melting temperature metal and compared averaged with non-averaged Navier Stokes models. Their study revealed that the low-frequency velocity oscillations play the major role in the convective heat and mass transfer when flow structure contains two or more large vortexes of the mean flow. Therefore, they concluded that the Large Eddy Simulation (LES) model is more reliable in the melt agitation in the induction furnace.

LES model has also been applied for induction channel furnace [10]. Turbulent heat and mass transfer in the melt using a 3D transient LES model were simulated. On this base, distributions of alloying additions into the melt and disjointed impurities due to channel erosion were investigated.

Scepanskis et al. [1] focused on particle-laden recirculated turbulent melt flow driven by electromagnetic forces, and compared numerical and experimental results for particle distribution on the melt surface. In another work [11] a methodology is presented that consists of the regression model for the identification of the time that is necessary to homogenize alloying inclusions in liquid metal inside an ICF. Such calculation can be carried out in the rational time and, therefore, it is possible to perform the industrial optimization of the mixing time as the function of size and density of the alloying inclusions.

Single-phase flow in induction furnace is well researched. However, the particle motion is still not sufficiently investigated and knowledge about non-metallic inclusion inside the melt especially when it approaches crucible wall boundary is weak. Therefore, this paper presents a primary study of non-metallic inclusion behaviors in molten steel of laboratory ICF. We should firstly verify the melt flow and particle motion. So, velocity and turbulence inside melt are studied using different models in 2D and 3D. Single particle motion with different properties under the effects of different forces such as drag, lift, virtual mass and pressure gradient is investigated.

## 2. Model Description

A laboratory induction furnace is considered, as schematically shown in Figure 1(a). Six water cooled coils are located around an alumina-silica crucible. Dimensions, parameters, and physical properties used in simulations are listed in Table 1.

Numerical studies of the electromagnetically driven flow have shown that the influence of the pulsating magnetic field can be neglected if its frequency is higher than 5 Hz [12]. Thus in one way coupling manner, electromagnetic forces are calculated using COMSOL Multiphysics software and the time-averaged values are imported as momentum source terms in ANSYS FLUENT for the turbulent flow calculation. Figure 1(b) shows that most forces are toward centerline and localized in a 3 cm thick layer adjacent to the crucible wall. For the turbulent flow three models, k-ε, k-ω SST and Large Eddy Simulation (LES) are used in 2D and 3D cases.

Alumina particles have been considered as typical inclusions in steelmaking processes with different sizes and effect of diverse forces on their motion in the bulk melt are studied. Details of the model for the motion of non-metallic inclusions in the melt can be found in the previous works [13, 14]. Among all forces acting on a particle, buoyancy and drag are main factors of the particle motion and are not negligible. Due to the deformed electrical current path around the non-conductive particle in the conductive melt an electromagnetic force is also applied on the particle, and the maximum value of the electromagnetic force acting on an alumina particle near the crucible wall is about 1% of buoyancy force. Therefore it can be neglected in this case. Other forces are lift, virtual mass and pressure gradient. Since main agent of creation of both virtual mass and pressure gradient force is velocity acceleration (dU/dt) [15], effects of these two forces are taken in account together. To consider dispersion of particles due to the turbulence a stochastic tracking model can be implemented in the momentum calculations, especially in Reynolds Average Navier Stokes (RANS) models like k-ε and k-ω. Fluctuating velocity components defined by Gaussian probability distribution include an artificial normally distributed random number. In order to compare particle trajectory caused by different forces, this stochastic tracking called random walk is here disabled. Moreover, random walk model might lead to unrealistic result when particles enter laminar sublayers close to the crucible wall.

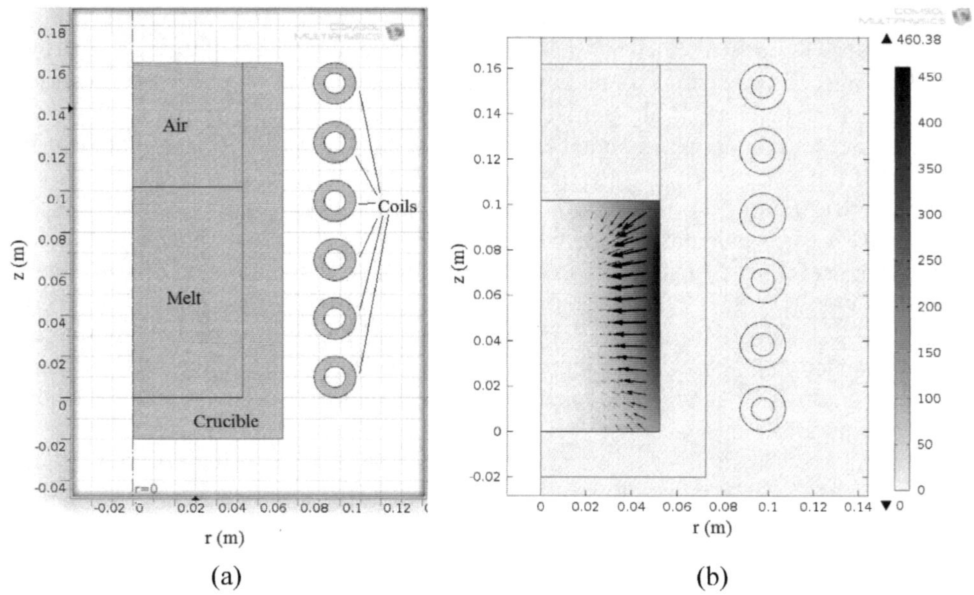


**Figure 1.** (left) Geometry, (right) Direction and magnitude of Lorentz force (N/m$^3$).

**Table 1.** Dimensions, parameters, and physical properties.

| Property | Unit | Value |
|---|---|---|
| Inner radius of crucible ($R_{crucible}$) | mm | 52.5 |
| Radius of coil ($R_{coil}$) | mm | 97.5 |
| Height of melt ($H_{melt}$) | mm | 102 |
| Length of coil ($L_{coil}$) | mm | 162 |
| Inner radius of coil cross section ($r_{in}$) | mm | 5 |
| Outer radius of coil cross section ($r_{out}$) | mm | 10 |
| Thickness of crucible ($Th_{crucible}$) | mm | 20 |
| Electrical conductivity of melt ($\sigma$) | S/m | $2.86\times10^5$ |
| Imposed electrical current ($J_e$) | A | 4000 |
| Frequency ($f$) | Hz | 50 |
| Density of melt ($\rho$) | kg/m$^3$ | 7020 |
| Viscosity of melt ($\mu$) | kg/(m.s) | 0.006 |
| Density of alumina ($\rho_p$) | kg/m$^3$ | 3700 |

## 3. Results and discussion

To validate the calculated Lorentz forces a comparison has been done with experimental results [16]. Figure 2 indicates that the simulation results match those gained using Gauss meter in both cases with and without an aluminum core inside the coil. $B_\infty$ is theoretical value of magnetic flux density at the axis of the coil.

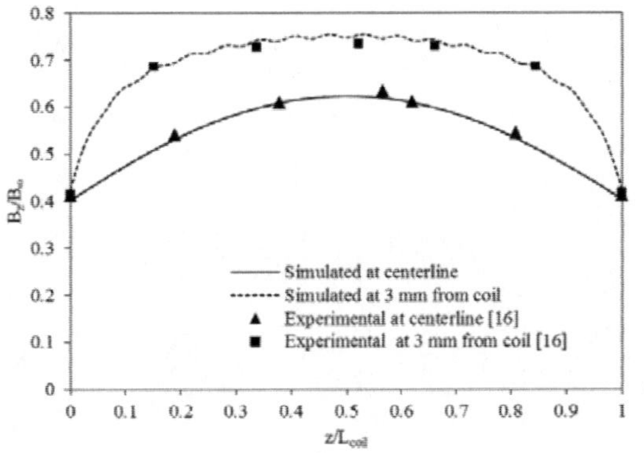


(a)

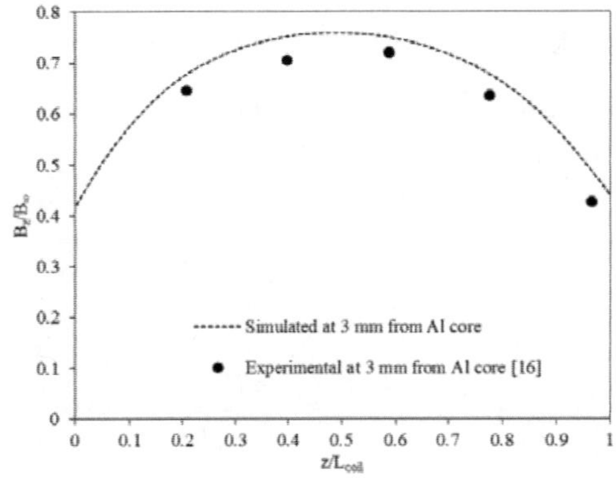


(b)

**Figure 2.** Dimensionless magnetic flux density in z direction inside a coil for the simulation and experiments [16]: (a) coreless coil, (b) coil with an aluminum core.

Velocity distributions for two turbulence models, k-ω SST and LES, are shown in Figure 3 that 2D-axisymmetric and 3D simulations are compared. For better presentation of 3D flow pattern velocity vectors are displayed on two main coordinate planes. Despite of the k-ε model which is claimed to be proper for completely turbulent flow and gives a steady flow pattern for the melt [5], the k-ω model delivered a transient result of the flow pattern in both 2D-axisymmetric and 3D calculations. Like typical time averaged simulation results [5, 11] two main eddies form in upper and lower zones. Size, strength, and location of main eddies change frequently. Moreover some local weak vortices are created and vanished between main eddies.

Because of mathematical filtering in the calculation of variables in LES, instantaneous velocity results are more time dependent. Therefore, there is no stabilized flow pattern. Several erratic vortices form whose locations and sizes vary frequently. It causes oscillation of velocity at different points. The same phenomena were observed by experimental investigations [5, 9].

Instantaneous variation of velocity components at the center of crucible for k-ε, k-ω and LES are plotted in Figure 4. It can be observed that k-ω model can solve the oscillation of velocity like LES while k-ε results are generally constant for all components. Hence, in the cases where accurate local and instantaneous velocity is not of interest, k-ω SST is a reasonable model for the transient simulation of transport phenomena in induction furnace.

After study of fluid flow in the ICF, particle motions under diverse circumstances are calculated. Figure 5 shows trajectories of a particle with 5 μm diameter started from the center of the crucible. In Figure 5(a) k-ω results depict that consideration of lift force results in no change in particle path comparing with drag only condition. However, applying virtual mass and pressure gradient forces makes little distinction. Hence, it can be concluded that the velocity gradient across the particle, i. e. lift force, is almost zero, but the velocity acceleration, i. e. virtual mass and pressure gradient force, is effective to some extent. Nevertheless, all three calculations by k-ω model indicate that the drag force is dominant.

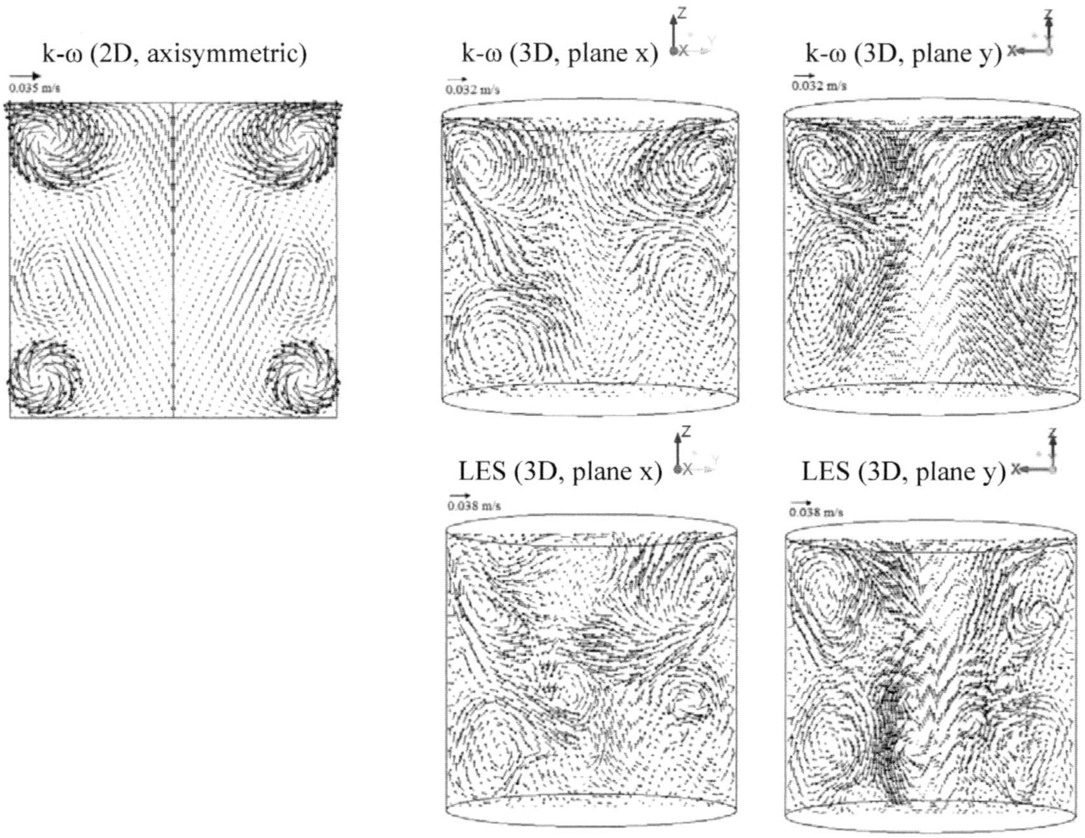


**Figure 3.** Snapshots of velocity field calculated by k-ω and LES models.

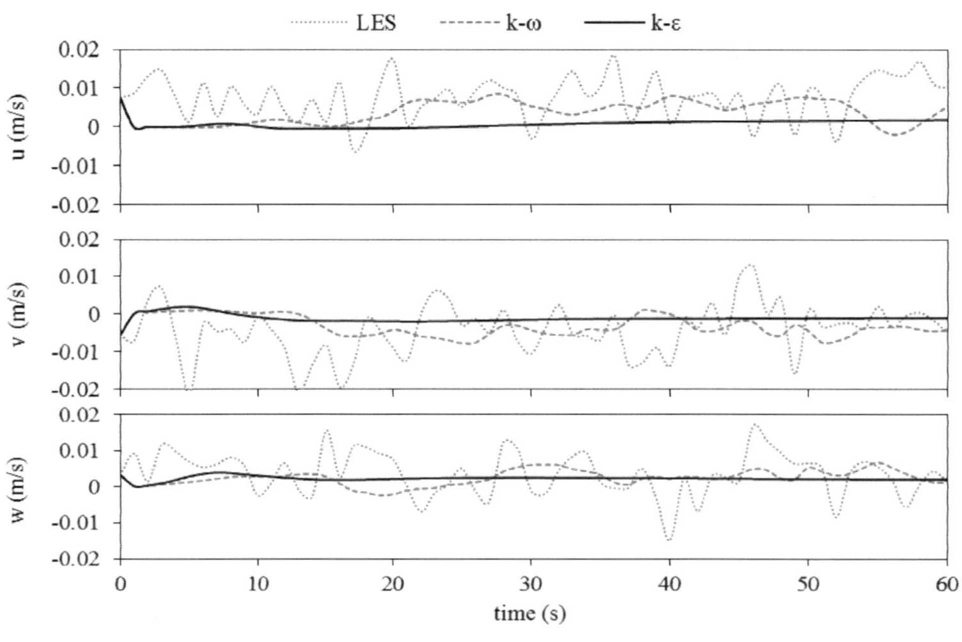


**Figure 4.** Transient velocity components of k-ε, k-ω and LES at center point of crucible.

IOP Conf. Series: Materials Science and Engineering **143** (2016) 012026 doi:10.1088/1757-899X/143/1/012026

LES result, Figure 5(b), indicate that in first few seconds particle under different forces moves in the same paths and then chooses completely different paths. These results are observed repeatedly. Regarding disabled random walk model the only reason resulting in particle movement through diverse paths is due to the different forces. Intense oscillations of velocity in LES (Figure 4) make more locally shear flow regions so lift force can effectively change particle path and moves it into one of numerous local eddies. Also it is obvious from Figure 4 that dU/dt in LES model, i. e. virtual mass and pressure gradient forces, is higher with respect to k-ω. Therefore, these forces in LES model are more effective. On the other hand, local eddies in LES are many and variable, so a short deviation in the particle motion may cause a totally different path. Hence, effective particle motion is fulfilled precisely using LES model.

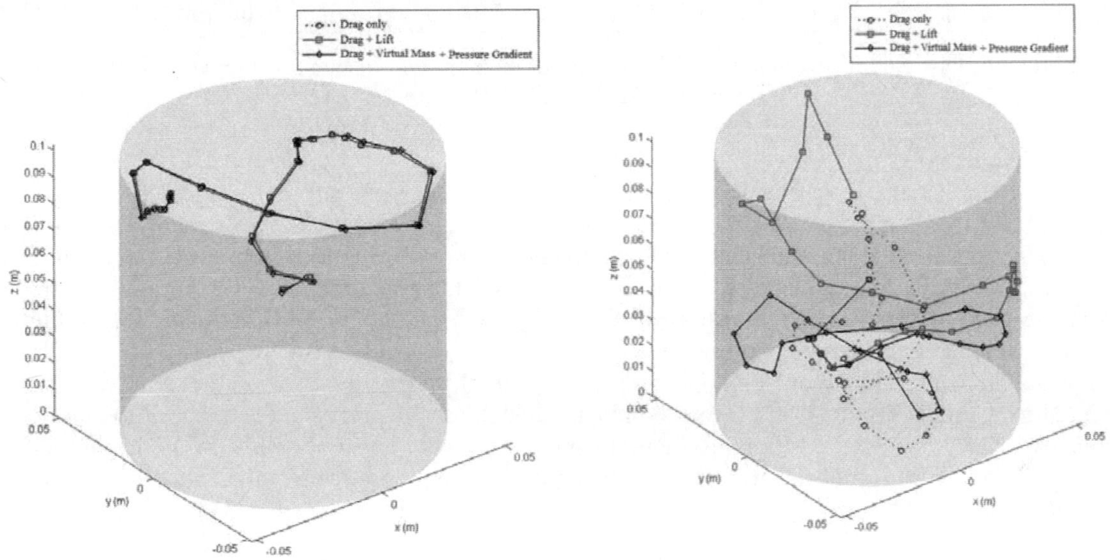


(a) k-ω SST (b) LES

**Figure 5.** Particle trajectories calculated by considering different forces in k-ω and LES turbulence models. (Time interval between sequential points is 1 s.)

Most of non-metallic inclusions in primary steelmaking process such as $Al_2O_3$, $SiO_2$ and MgO are lighter than molten steel and their density vary in a small range, 3970, 2320 and 3650 kg/m$^3$ respectively. So, for comparison of different inclusions, density is not impressing factor. To perform a parameter study of particle properties trajectories of the particle with different size from the same beginning point and at different injection times are studied with k-ω model, Figure 6. Difference between injection times is 0.05 s and all forces mentioned before have been applied. It can be observed in Figure 6 that particles with small diameter (5 and 10 μm) have almost the same trajectories. However, particles with diameter of 20 μm go through paths like small ones in almost 10 s then select different paths and the largest particles (50 μm) are separated in few beginning seconds. Results of different particle sizes for LES model are too chaotic, as shown in Figure 5, therefore the trajectories are not comparable with each other.

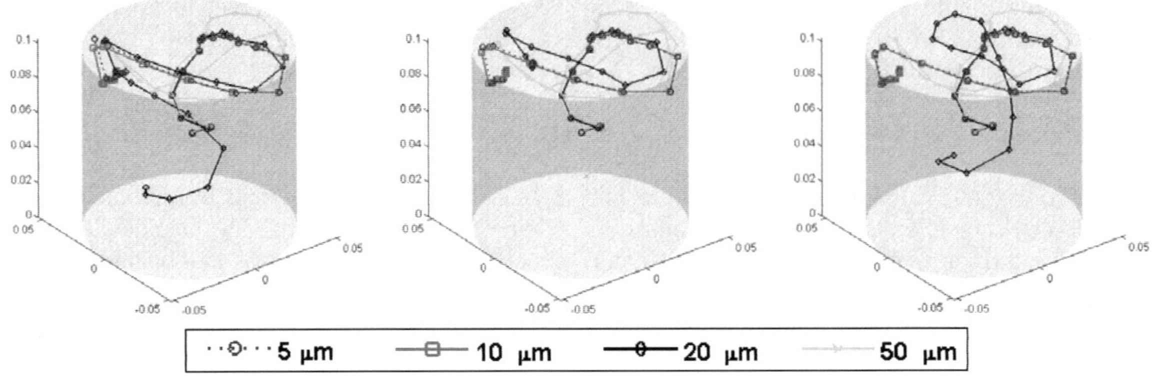


**Figure 6.** Particle trajectories with different particle diameters in k-ω turbulence model. (Time interval between sequential points is 1 s.)

## 4. Conclusions

Numerical simulations have been done to model the turbulent flow and the motion of non-metallic inclusion (NMI) in induction crucible furnace (ICF). Studies focused on k-ω SST and LES turbulence models. Results showed that transient flow pattern and velocity oscillations in the ICF can be modeled using k-ω, while the conventional k-ε is unable to model this phenomenon. The motion of NMI can also be tracked by the k-ω model fairly well. When precise trajectory of NMI in the melt is of interest, LES model is the most suitable option. Of course, LES model requires fine mesh and small time step size which increase calculation costs. Therefore, for general purpose to estimate the dynamic behavior of NMI in ICF both k-ω SST and LES are recommended. For the precise calculation of the motion of NMI, especially for those NMIs with diameter smaller than 10 μm, only LES model is appropriate. However, to confirm above statements further modeling and experimental evaluation efforts are desired.

## 5. References

[1] Scepanskis M, Jakovics A, Baake E and Nacke B 2014 *Int. J. Multiphase Flow* **64** 19–27
[2] Bojarevics V, Djambazov G, Harding R A, Pericleous K and Wickins M 2003 *Magnetohydrodynamics* **4** 395-402
[3] Schwarze R and Obermeier F 2004 *Modell. Simul. Mater. Sci. Eng.* **12** 985-93
[4] Baake E, Nacke B, Jakovics A and Umbrashko A 2001 *Magnetohydrodynamics* **37** 13-22
[5] Umbrashko A, Baake E, Nacke B and Jakovics A 2006 *Met. Mater. Trans. B* **37** 831-838
[6] Jardy A, Gosset D, Ablitzer D and Witzke S 1993 *J. Phys. IV France* **3** C7-1151-C7-1156
[7] El-Kaddah N and Szekely J 1983 *J. Fluid. Mech.* **133** 37–46
[8] Baake E, Muhlbauer A, Jakowitsch A and Andree W 1995 *Met. Mater. Trans. B* **26B** 529–536
[9] Kirpo M, Jakovics A, Baake E and Nacke B 2007 *Magnetohydrodynamics* **43** 161–172
[10] Pavlovs S, Jakovics A, Baake E, Nacke B and Kirpo M 2011 *Magnetohydrodynamics* **47** 399–412
[11] Scepanskis M, Jakovics A, Baake E and Nacke B 2015 *Steel Research Int.* **86** 169–174
[12] Felten F, Fautrelle Y, Du Terrail Y and Metais O 2004 *Appl. Math. Model.* **28** 15-27
[13] Pfeiler C, Wu M and Ludwig A 2005 *Mater. Sci. Eng., A* **413–414** 115–120
[14] Pfeiler C, Thomas B G, Wu M, Ludwig A and Kharicha A 2008 *Steel Research Int.* 599-607
[15] Crowe C, Sommerfeld M and Tsuji Y 1998 *Multiphase Flows with Droplets and Particles* (Boca Raton: CRC Press LLC)
[16] Kennedy M W, Akhtar S, Bakken J A and Aune R E 2011 *COMSOL Users Conference* (Stuttgart)

IOP Conf. Series: Materials Science and Engineering **143** (2016) 012027 doi:10.1088/1757-899X/143/1/012027

# Numeric simulations of a liquid metal model of a bloom caster under the effect of rotary electromagnetic stirring

**M. Barna[1], M. Javurek[1], B. Willers[2], S. Eckert[2] and J. Reiter[3]**

[1] Johannes Kepler University (JKU) Linz, Altenbergerstraße 69, 4040 Linz, Austria

[2] Helmholtz-Zentrum Dresden-Rossendorf (HZDR), P.O. Box 510119, 01314 Dresden, Germany

[3] voestalpine Stahl Donawitz GmbH, Kerpelystraße 199, 8700 Leoben, Austria

martin.barna@jku.at

**Abstract**. At the voestalpine Stahl Donawitz GmbH the continuous casting of round steel blooms is commonly supported by electromagnetically induced stirring of the liquid steel flow. A number of beneficial effects are attributed to electromagnetic stirring in the mould region (M-EMS), e.g. the enhanced transition from columnar to equiaxed solidification, the homogenization of the liquid steel flow or the reduction of surface and subsurface defects. Although the positive effects of M-EMS can be seen on the blooms (e.g. in etchings), the link between electromagnetic stirring of the steel melt and the quality of the solidified bloom is not sufficiently understood. Theoretical considerations are often limited to general cases and their results are therefore not directly applicable to real continuous casting geometries. On the other hand, plant measurements can only be performed to a limited extent due to the harsh conditions and other restrictions (e.g. safety regulations). In this work an alternative approach is used to investigate the steel flow in a round bloom caster under the influence of M-EMS. In a 1:3 scale Perspex model of a round bloom strand, measurements of the flow under the influence of a rotating magnetic field can be conducted. These measurements provide a validation benchmark for the numeric simulations. A numeric model of the before mentioned 1:3 scale model is implemented, encompassing the strand, the submerged entry nozzle as well as the M-EMS device. In the modelling approach, the bidirectional coupling between liquid steel flow and the electromagnetic field/forces has to be considered because otherwise the resulting tangential velocities will be overestimated. With the validated modelling approach, simulations of real casting machines can then be conducted, stirring parameter influences can be shown and conclusions for the real casting process can be drawn.

## Introduction

In the continuous steel casting, electromagnetic stirring for (round) bloom casting is a quite widespread technique. It shall help to increase the productivity while maintaining/enhancing the quality of the product. Electromagnetic stirrers are used at various positions along the strand, with mould-electromagnetic stirrers (M-EMS) being the most established ones. These stirrers produce a magnetic field, which is nearly homogeneous over the cross section of the strand and rotates with a given frequency. Lorentz forces are induced, which excite a rotary motion in the liquid core of the strand. This imposed azimuthal flow of the liquid steel provides numerous benefits. Most important

IOP Conf. Series: Materials Science and Engineering **143** (2016) 012027 doi:10.1088/1757-899X/143/1/012027

for the product quality are the enhanced transition from globular to equiaxed solidification, the reduction of surface and subsurface defects and the homogenization of the liquid steel's temperature and composition. For the steel production, a correlation between the stirring parameters and a quality index (e.g. the percentage of equiaxed zone) is crucial to ensure a good product quality. These correlations can be found by employing plant measurements, experiments and/or numerical simulations

In this work the continuous casting of round blooms with mould-electromagnetic stirring is investigated primarily through numerical simulations validated by measurements made at a 1:3 strand model. A description of the experimental built-up and the considerations behind it are also included, while a detailed description of the experiment can be found in the paper of Willers et al. [1].

**Mathematical and numerical modelling**

The liquid steel flow inside the cast strand is governed by the Navier-Stokes equations. The high velocities of the liquid steel exiting the straight-through submerged entry nozzle (SEN) lead to a turbulent flow in the core, which is mathematically described by the Reynolds averaged Navier-Stokes equations (RANS), using the realizable k-ε model [2] for closure. The validation cases are simulated with the eutectic alloy $Ga_{68}In_{20}Sn_{12}$, which is treated as a Newtonian, incompressible and isothermal fluid. In the strand geometry, no symmetry is used in order to allow for asymmetric flow fluctuations. The strand walls and the mould level are modelled as rigid walls with the no-slip condition.

If electromagnetic fields are employed to modify the liquid steel flow, Maxwell's equations and Ohm's law are used to describe the magnetic field and its interaction with the liquid steel flow. Either a formulation for the magnetic vector potential $A$ or the magnetic flux density $B$ is used. The interaction between the liquid steel flow and the electromagnetic field is given on one hand by the velocity field and on the other hand by the electromagnetic Lorentz forces. So in general, a bi-directional coupling is used to correctly capture the interaction. In the present case, an inverse problem utilizing the potential theory is solved for the magnetic field of the stirrer. The resulting field distribution can then be used for e.g. the Spitzer method [3], [4].

$$\begin{aligned} \overline{F}_r &= -\frac{1}{8}B_0^2\left(\omega - \frac{v_t}{r}\right)^2 \sigma^2 \mu_M r^3 \\ \overline{F}_t &= \frac{1}{2}B_0^2\left(\omega - \frac{v_t}{r}\right)\sigma r \end{aligned} \tag{1}$$

Equation 1 describes the radial and tangential force densities in dependence of the magnetic flux density $B_0$, the angular frequency of the magnetic field $\omega$, the radial position $r$ and the fluid's tangential velocity $v_t$. Also the material parameter, the electrical conductivity $\sigma$ and the magnetic permeability $\mu_M$, are needed. Alternatively, instead of using the solution of the inverse problem to directly compute the forces, it can be used in ANSYS Emag as a boundary condition on a spatially reduced electromagnetic model, which only encompasses the strand.

**Physical modelling**

A 1:3 scale model of the round bloom caster at voestalpine Stahl Donawitz GmbH (Figure 1) is built up at HZDR, made of acrylic glass. The model strand has an inner diameter of d=80 mm and a length of about L=800 mm. The inner diameter of the straight-through SEN is $d_i$=10 mm and its submersion depth is $h_{SEN}$=40 mm. The mould level is kept approximately 25 mm below the upper edge of the model strand. The strand curvature and the solidified shell are neglected.

The stirrer is located about 250 mm from the upper edge of the model strand, with an inner diameter of 200 mm and a height of 310 mm. The stirrer has a rather complex layout which makes it unfeasible for numerical simulation (Figure 2). So instead of building a model of the stirrer, an inverse problem, detailed below, is solved to approximate the magnetic field.

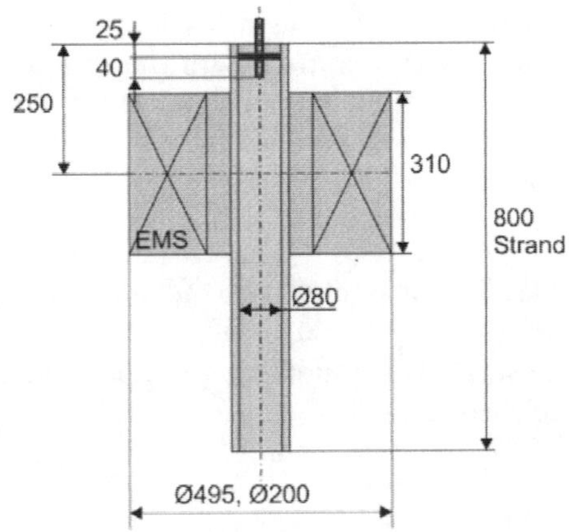


**Figure 1: Draft of the 1:3 round bloom model with mould-electromagnetic stirring; grey: mould and SEN; blue: electromagnetic stirrer; the horizontal, purple line indicates the free surface**

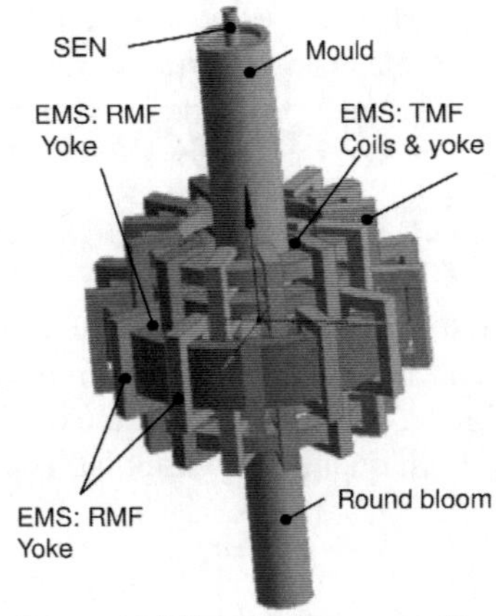


**Figure 2: Layout of the electromagnetic stirrer from HZDR with mould, strand and submerged entry nozzle (SEN); RMF: rotating magnetic field, TMF: traveling magnetic field**

For the validation, the time-averaged tangential velocities are measured at two distances from the stand axis, d=15 mm and d=30 mm. Additional details regarding the experimental setup, flow measurement and the UDV sensors and can be found in Willers et al. [1].

## Results and Discussion

*Dimensioning of the validation model:*

In order to investigate the electromagnetic stirring for the continuous round bloom casting of steel, one must ensure that all phenomena, which are relevant for the real casting process, can be reproduced by the experiment. This can be easily achieved by consulting various dimensionless numbers. In the literature (e.g. [5]–[7]) two distinct numbers are used, the shielding parameter *S* (equation 4) and the interaction number *N* (also called Stuart number, equation 5).

$$S = \mu_{\mathrm{M}} \sigma \omega R^2 = \frac{\text{diffusion time}}{\text{rotation period}} \tag{4}$$

$$N = \frac{\sigma B_0^2}{\rho \varpi} = \frac{\text{electromagnetic forces}}{\text{inertia forces}} \tag{5}$$

As the real casting slightly deviates from the above mentioned analytical case, additional parameters have to be considered. Turbulence plays an important role for the liquid steel flow of continuous casting. So the Taylor number *Ta*, which is the equivalent of the Reynolds number for rotating flows, is added to the spectrum of key figures.

$$\mathrm{Re}_{EMS} \propto Ta = \frac{\varpi^2 R^4}{\nu^2} = \frac{\text{inertia forces}}{\text{viscous forces}} \tag{6}$$

Electromagnetic stirring bears analogy to mechanical mixing/stirring, which is commonly used in various industries and extensively investigated. This knowledge was also used for dimensioning of the experiment and to ensure that it will properly represent the real continuous casting process. In Uhl and Gray [8] a figure can be found where the power number, essentially a flow drag coefficient, is plotted in dependence of the Reynolds number[1]. For values higher than $\mathrm{Re}_{\mathrm{EMS}}>10^4$ the flow is turbulent and the power number is approximately constant. Depending on the geometry of the mechanic stirrer, this drag coefficient lies between 0.2 and 0.8.

[1] In mechanical mixing/stirring, the Reynolds number is usually build with the square of the mixer's diameter, the rotation speed in RPM and the viscosity of the fluid.

To characterize the flow induced by electromagnetic stirring, one can use this knowledge and equilibrate the electromagnetic power, imposed by the electromagnetic stirrer, with the drag of the fluid flow. The electromagnetic stirring power is given by equation 7. Here the force density formulation, derived by Spitzer [3], is used with the angular frequency of the fluid $\omega_{Fl}$.

$$P_{mag} = \int_V \frac{1}{2} B_0^2(r,z)\left[\omega - \omega_{Fl}(r,z)\right]\sigma r^2 \omega_{Fl}(r,z)dV \tag{7}$$

For a first result, some simplifications have to be made: The magnitude of the magnetic flux density $B_0$ and the fluid rotational frequency $\omega_{Fl}$ are assumed to be independent of the radial and the axial coordinate, which will not be exactly true. In reality the stirrer has a finite length; hence an axial variation of both quantities will exist. Equation 7 can be simplified to

$$P_{mag} = \frac{1}{2}\overline{B_0}^2\left[\omega - \overline{\omega_{Fl}}\right]\overline{\omega_{Fl}}\sigma \frac{r^4}{4}\pi h, \tag{8}$$

where $h=C_{EMS}\cdot r$ represents the height of the stirrer and the quantities with the bar, $\overline{B_0}, \overline{\omega_{Fl}}$, are averaged values. The drag of the fluid flow is deduced from Uhl and Gray [8], by transforming the power number. As mentioned before, it is constant for $Re > 10^4$

$$P_{drag} = \zeta_{Fl}\rho \overline{\omega_{Fl}}^3 (2r)^5 \tag{9}$$

Equation 8 and 9 can now be put in equilibrium. The resulting polynomial equation of second order can be solved for the angular frequency of the fluid to estimate the average angular fluid flow, if the drag and stirring coefficient are known (see e.g. [9]). However, under the assumption that the fluid rotates on average at half the angular speed of the magnetic field, one can further simplify the equation to

$$\underbrace{\frac{1}{2^8}C_{EMS}\pi}_{\zeta_{mag}}\overline{B_0}^2\left[\omega - \overline{\omega_{Fl}}\right]\overline{\omega_{Fl}}\sigma r^5 = \zeta_{Fl}\rho\overline{\omega_{Fl}}^3 r^5. \tag{10}$$

In equation 10, the drag coefficient and the stirring coefficient are moved to the left side and the right side has the same structure as the interaction number $N$ (equation 11). It is desirable that the experiment has the same ratio of electromagnetic forces to inertia forces as the real caster. This can be easily achieved with equation 13. For convenience the fluid's angular frequency of both setups shall be the same.

$$\frac{\zeta_{Fl}}{\zeta_{mag}} = \frac{\sigma \overline{B_0}^2}{\rho\overline{\omega_{Fl}}} = N_{calc} \rightarrow \text{const.} \tag{11}$$

$$\overline{B_{0,\text{experiment}}}^2 = \frac{\sigma_{\text{caster}}}{\sigma_{\text{experiment}}}\frac{\rho_{\text{experiment}}}{\rho_{\text{caster}}}\underbrace{\frac{\overline{\omega_{\text{Fl,experiment}}}}{\overline{\omega_{\text{Fl,caster}}}}}_{\overset{!}{=}1}\overline{B_{0,\text{caster}}}^2 \tag{13}$$

If the ratio of stirrer height to strand radius differs between experiment and real caster, it can also be considered in equation 13. This approximation can be used to correctly set the stirring parameters of the experiment similar to the real casting conditions.

Figure 3 depicts the range of all relevant key figures for experiment and real casting geometry next to each other. In addition, the critical values for the key figures are also presented in the figure (black horizontal lines). Arrows indicate whether it's favourable to stay above or below the critical value. It can be seen that the experiment covers much broader ranges than the real caster, which means that all important phenomena occurring at the caster are reproducible by the experiment. Only the range of the Reynolds number for the SEN is smaller in the experimental setup, but this should not pose a problem as the investigation is concentrated mainly on the electromagnetic stirring and less on the submerged jet. For both setups the range of the shielding parameter $S$ is near the critical value (of 1). So some distortion of the magnetic field may be observed, but the effect is accounted for in the numeric model.

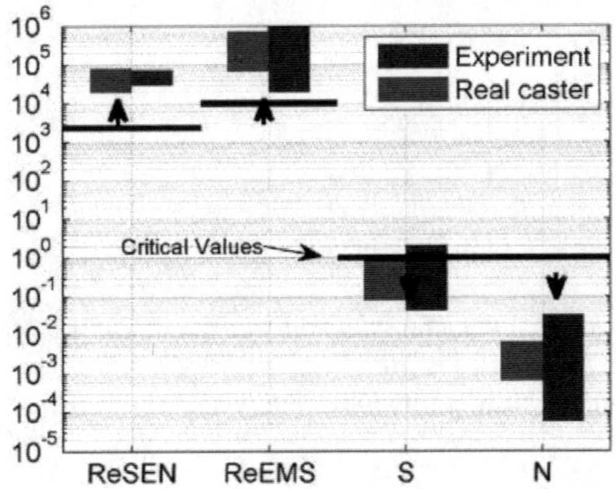


**Figure 3: Graphical representation of the range of the key figures (Reynolds number of the SEN flow $Re_{SEN}$ and the stirred flow $Re_{EMS}$, shielding parameter S and interaction number N) for the real caster (red) and the experiment (blue); critical values are designated by the black lines and arrows indicate the favourable region**

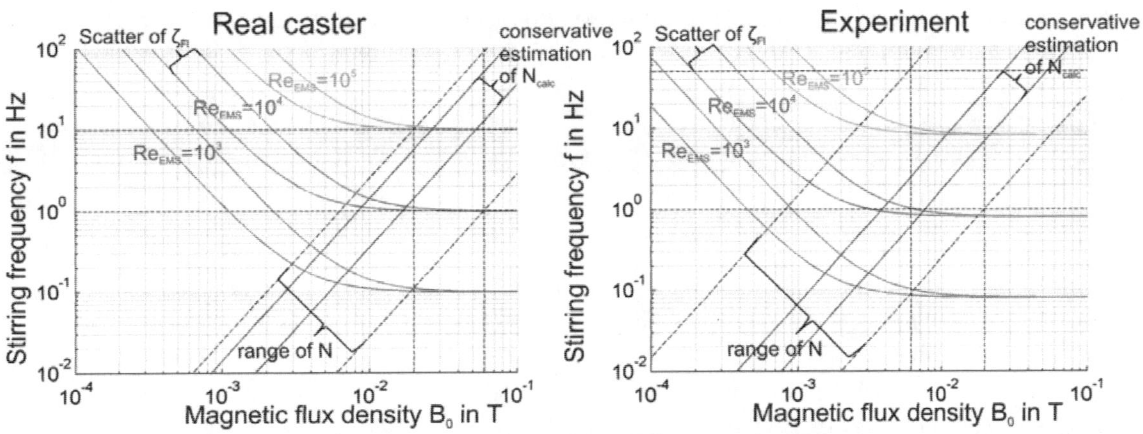


**Figure 4: Range of magnetic flux density $B_0$ and stirring frequency f for the real caster and the experiment; interaction number N and Reynolds number $Re_{EMS}$ for the strand are also plotted into the graph; due to the scatter of the drag coefficient (see [8]), there is some uncertainty in the Reynolds number; the blue lines show a conservative estimation of the calculated interaction number $N_{calc}$ (equation 11) for both setups; as long as the Reynolds number is higher than $10^4$ (turbulent regime) the flow regime in both operation windows will be similar and the experiment is able to represent the real casting situation.**

The operating windows for the caster and the experiment are shown in Figure 4. The ranges of magnetic flux density and the stirring frequency give the limits of operation, which also provide the minimum and maximum of the interaction number. In addition, the Reynolds number of the rotating flow $Re_{EMS}$ is plotted (with regard to the scattering of the drag coefficient, see [8]). Both cases, real caster and experiment, are operated in the turbulent flow regime (above $Re_{EMS}>10^4$) and within the operation limits. So the flow regime will be the same for the real caster and the experiment and the latter can correctly represent the flow in continuous steel casters with electromagnetic stirring.

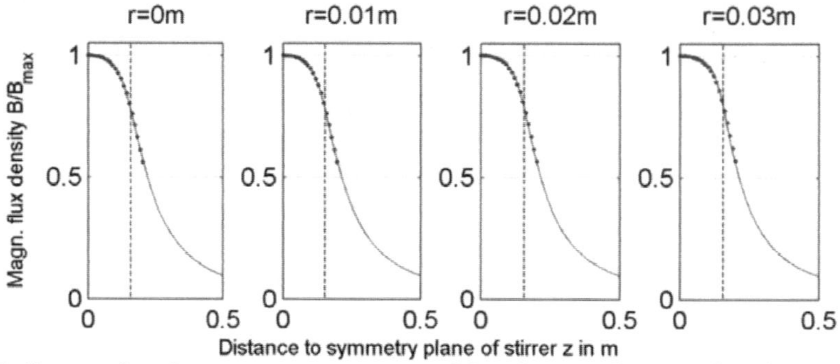


**Figure 5: Comparison between measurements (blue dots) and approximation (green line) of electromagnetic stirring field for different radii plotted over the distance to the stirrer's middle plane, the black dashed line marks the edge of the stirrer**

The electromagnetic field is approximated with the help of the potential theory. This guarantees that the solution will be solenoidal and without rotation, meaning per se no eddy currents will be induced. The complexity of the problem is reduced by assuming a radially and axially symmetric 2D arrangement of sinks and sources, which rotates around the strand with the stirring frequency. The resulting (magnetic) vector potential can then be employed either as a boundary condition for the electromagnetic solver or the magnetic flux density field can be calculated for the CFD model. Figure 5 compares the available measurements and the solution of the inverse problem. The curves show an overall good agreement, only at bigger radii (r≥0.04 m) outside the strand slight differences arise.

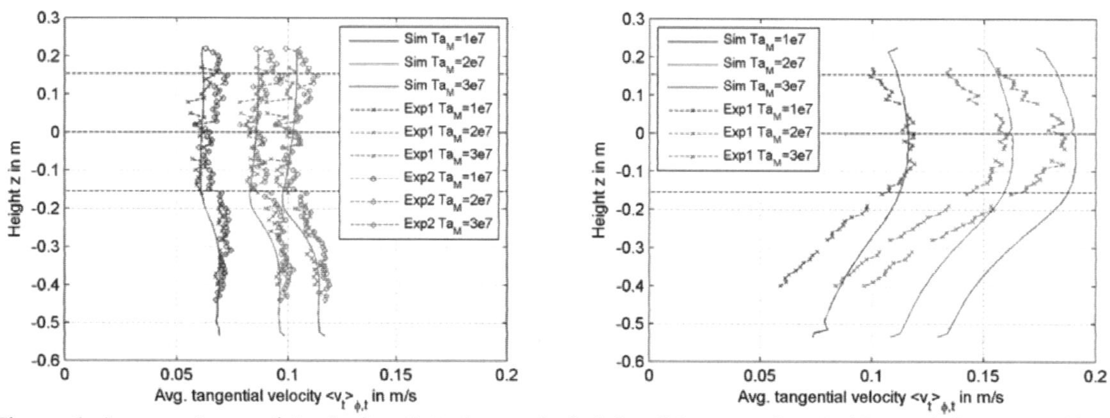


**Figure 6: Average tangential velocity plotted over the height of the strand at d=15 mm (left) and d=30 mm (right) for the magnetic Taylor numbers $Ta_M$=1e7 (blue), $Ta_M$ =2e7 (green) and $Ta_M$ =3e7 (red); solid lines: simulation; markers (X, O) signify two separate measurements; horizontal lines mark the upper and lower edge of the stirrer as well as the symmetry plane**

As stated before, analytical solutions exist for the electromagnetic stirring of an infinitely long cylinder (e.g. [3]–[6]) – even under the influence of axially non-uniform magnetic fields (e.g. [10], [11]). In real casters, contrasting to most analytical cases, the flow is turbulent and the stirring field often has a quite complex distribution. These facts make experimental and/or numerical approaches the more favorable means of investigation.
The first validation is done for three different values of the magnetic flux densities $B_0$={10.58; 14.96; 18.32} mT, which equal magnetic Taylor numbers $Ta_M$={1e7; 2e7; 3e7}[2]. For these three magnetic field strengths, cases solely with electromagnetic stirring are investigated. Afterwards the interaction of stirring with a mass flow through a SEN are investigated for $Ta_M$={1e7; 2e7}. The frequency of the

[2] The magnetic Taylor number is the ratio of the electromagnetic forces to the viscous forces.

stirrer is kept fixed at the stirring frequency used at the steel plant. The Spitzer method is used in conjunction with the approximation of the stirrer field for these validation cases.

Figure 6 depicts the average tangential velocity for a distance of d=15 mm and d=30 mm from the strand axis respectively for the simulation and the experiment. While the measurements are only averaged over time, the simulation results are in addition azimuthally averaged. For the distance d=30 mm the maximum tangential velocity is similar to the measurement results, but the tangential velocity decreases axially far slower (see Figure 6 right diagram). A possible reason could be an overestimation of the turbulent viscosity by the employed turbulence models. The left diagram in Figure 6 shows that, at the smaller distance d=15 mm, the measurements agree well with the simulation results. The increase of tangential velocity in the lower part can be correctly reproduced. The increase of the tangential velocity in the lower part of the strand is probably caused by the transport of angular momentum due to the secondary flow.

An axial offset between simulation results and measurements is observed in all investigated cases, those without and those with an imposed axial mass flow. A possible explanation is that the stirrer is not perfectly aligned inside its casing, which is used as reference for the positioning of strand and stirrer in the experiment.

*Electromagnetic stirring and imposed axial mass flow:*

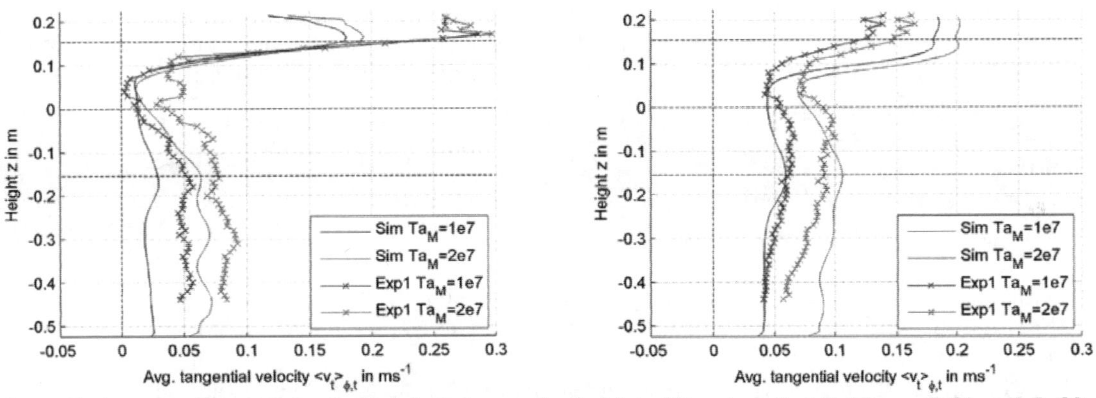


**Figure 7: Average tangential velocity plotted over the height of the strand at d=15 mm (left) and d=30 mm (right) for the magnetic Taylor numbers $Ta_M$=1e7 (blue) and $Ta_M$=2e7 (green); inlet velocity $v_{SEN}$ =1.3 $ms^{-1}$; solid lines: simulation; marker X: measurements; the horizontal lines mark the upper and lower edge of the stirrer as well as the symmetry plane**

With the stirring validated, the next step is to investigate the influence of the submerged jet on to the flow in the strand. The velocity in the SEN is chosen to correspond to real casting situations ($v_{SEN}$=1.3 $ms^{-1}$ equals a casting speed of ~1.0 $mmin^{-1}$). The frequency and stirring field strength of the electromagnetic stirring are kept fixed.

In the simulation a rigid plane wall is assumed, but in the experiment a thin, deformable oxide skin covers the free surface. Also in the vicinity of the SEN, a vortex (randomly) forms and rotates around the SEN. Both phenomena impact the velocity field in the vicinity of the free surface and through convection also the distribution of the tangential velocity further down the strand. Nonetheless, a comparison between the experimental and the numeric simulation is made. Qualitatively the results correspond well. Figure 7 depicts the measurements and simulation results at both radii. A good overall agreement is found (if the axial offset is compensated). But at the distance d=15 mm the simulation results underestimate the effect of the electromagnetic stirring and the transport of angular momentum from the mould region down into the strand (Figure 7 left diagram). The average tangential velocities for both cases are smaller than those in the measurements with the effect being more pronounced for the case $Ta_M$=1e7.

## Conclusion

For the investigation of electromagnetic stirring in continuous casting of round blooms, a 1:3 scale model of the round strand and the mould-electromagnetic stirrer is designed. Analytical considerations show that this model is able to reproduce the flow in the round bloom caster of voestalpine Stahl Donawitz GmbH under the influence of the mould-stirrer quite well. An inverse problem is solved to approximate the magnetic stirring field.

Three different magnetic flux densities are investigated for sole electromagnetic stirring. The average tangential velocities are in good agreement for d=15 mm. For the distance d=30 mm, the simulations can capture the maximum quite well, but the measurements show a stronger axial decrease of the tangential velocity. Throughout the validation an axial offset between measurements and simulation could be observed. For the second part of the validation, an axial mass flow is added. The tangential velocity at d=30 mm is in overall good agreement. For the distance d=15 mm, the simulated tangential velocities are smaller than the measured ones. This deviation is attributed to the deformation of the free surface and a vortex at the SEN, which occurred (randomly) during the experiment and are not considered in the numeric model. The numeric model was validated with measurements and a detailed insight into an electromagnetically stirred flow could be gained. Based on this work, the numeric model can be used to further investigate electromagnetic stirring in continuous casting and investigate the impact of different parameter sets prior to real plant trials.

## Acknowledgments

This work has been financially supported within the Austrian competence centre programme COMET by the BMVIT; by the BMWFJ; by the provinces of Upper Austria, Styria and Tyrol; by the SFG and by the Tiroler Zukunftsstiftung.

## References

[1] B. Willers, M. Barna, J. Reiter, and S. Eckert, "Experimental investigation of electromagnetic stirring affecting the mould flow in a liquid metal model," in *Proceedings of 2nd ESTAD 2015*, 2015.

[2] T.-H. Shih, W. W. Liou, A. Shabbir, Z. Yang, and J. Zhu, "A New k-e Eddy Viscosity Model for High Reynolds Number Turbulent Flows Model Development and Validation," Cleveland, 1994.

[3] K.-H. Spitzer, "Berechnung von Strömungen beim elektromagnetischen Rotationsrühren von Rundsträngen," Technische Universität Clausthal, 1985.

[4] K.-H. Spitzer, M. Dubke, and K. Schwerdtfeger, "Rotational Electromagnetic Stirring in Continuous Casting of Round Strands," *Metall. Mater. Trans. B*, vol. 17, pp. 119–131, 1986.

[5] R. Moreau, *Magnetohydrodynamics*. Dordrecht: Kluwer Academic Publishers, 1990.

[6] A. Alemany and R. Moreau, "Ècoulement d'un fluide conducteur de l'électricité en présence d'un champ magnétique tournant," *J. Mécanique*, vol. 16, no. 4, pp. 625–646, 1977.

[7] H. K. Moffatt, "On fluid flow induced by a rotating magnetic field," *J. Fluid Mech.*, vol. 22, no. 3, pp. 521–528, 1965.

[8] V. Uhl and J. B. Gray, *Mixing – Theory and Practice*. New York, NY: Academic Press, 1966.

[9] M. Barna, "Untersuchungen der Flüssigstahlströmung beim Rundstranggießen mit elektromagnetischem Kokillenrührer," Johannes Kepler Universität Linz, 2013.

[10] P. A. Davidson, "Analytical models of rotary electromagnetic stirring," *IMA J. Manag. Math.*, vol. 7, no. 1, pp. 89–108, Jan. 1996.

[11] P. A. Davidson and J. C. R. Hunt, "Swirling recirculating flow an a liquid-metal column generated by a rotating magnetic field," *J. Fluid Mech.*, vol. 185, pp. 67–106, Apr. 1987.

# Uncertainty Quantification of Modelling of Equiaxed Solidification

**K Fezi and M J M Krane**

Purdue Center for Metal Casting Research, School of Materials Engineering, Purdue University

kfezi@purdue.edu

**Abstract.** Numerical simulations of metal alloy solidification are used to gain insight into physical phenomena that cannot be observed experimentally. Often validation of such models has been done through comparison to sparse experimental data, to which agreement can be misinterpreted due to both model and experimental uncertainty. Uncertainty quantification (UQ) and sensitivity analysis are performed on a transient model of solidification of Al-4.5 wt.% Cu in a rectangular cavity, with equiaxed (grain refined) solidification morphology. This model solves equations for momentum, temperature, and species conservation; UQ and sensitivity analysis are performed for the degree of macrosegregation. A Smolyak sparse grid algorithm is used to select input values to construct a response surface fit to model outputs. The response surface is then used as a surrogate for the solidification model to determine the sensitivities and probability density functions of the model outputs. Uncertain model inputs of interest include the secondary dendrite arm spacing, equiaxed particle size, and fraction solid at which the rigid mushy zone forms. Similar analysis was also performed on a transient model of direct chill casting of the same alloy.

## 1. Introduction

Sophisticated numerical models of solidification processes have been developed in which confidence in the results is dependent on how well the input parameters are known and comparison with limited experimental data. Some of the well-established physics in these models include diffusive and convective transport of heat, mass, and momentum, electromagnetics, and solidification thermodynamics. When validating a model with experimental data, the level of agreement is typically discussed qualitatively and frequently the assumption is made that more sophisticated physics would improve agreement. Quantification of model uncertainty will help quantify comparison to experiments and to identify which inputs it would be most profitable to improve.

Monte Carlo uncertainty quantification (UQ) methods consist of repetitive sampling random values of the model inputs or system parameters and evaluating the solution [1]. This process generates probability distribution functions (PDF) for output quantities of interest as a function of input uncertainties. Although Monte Carlo sampling methods are very effective, they quickly become very computationally expensive for complex numerical models. Use of polynomial chaos expansion (PCE) or global polynomial chaos (gPC), which create surrogate models, has significantly reduced computational expense without sacrificing accuracy [2]. The present study aims to quantify the sensitivities and uncertainties in predicting macrosegregation in grain refined, equiaxed solidification in a rectangular domain. Insight into uncertainty propagation in that system can be applied to models of more sophisticated processes models, like direct chill casting.

## 2. Model Description

### *2.1 Description of Deterministic 2D Solidification Model*

Static casting of a simple binary eutectic alloy (Al-4.5wt%Cu) is modeled. The rectangular cavity (Fig.1) is cooled from the left and an inlet on top allows for feeding due to solidification shrinkage.

The momentum equations are from Vreeman et al. [3], who included the effect of free-floating solid grains. This solid forms on grain refiners initially and has increasing interactions among themselves as they solidify. Once a critical volume fraction solid is reached, $g_{s,pack}$, the free-floating particles coalesce to form a rigid, stationary ($u_s = v_s = 0$), permeable solid structure. This packing fraction is generally treated (as it is here) as a constant and uniform quantity, although it is recognized to be related to the size and morphology of the equiaxed dendrites and the direction and magnitude of the liquid and solid motion relative to the rigid solid [4]. The value of $g_{s,pack}$ is taken here to be in the range up to 0.3 [3,5]. The solid and liquid velocities are found from Stokes law,

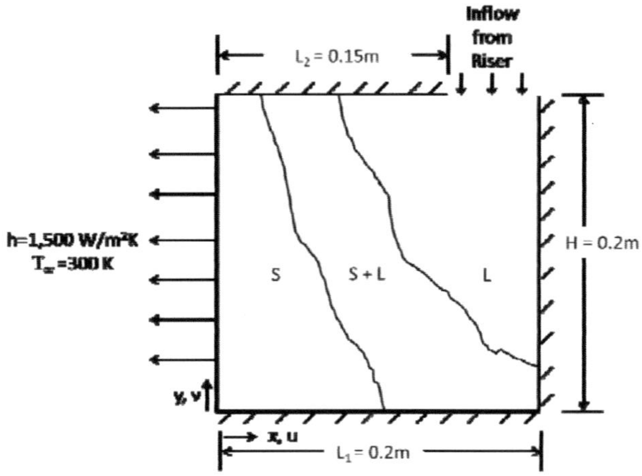


**Figure 1.** Schematic of simulation domain with thermal boundary conditions.

$$\vec{V}_s - \vec{V}_l = \frac{(1-g_s)}{18\mu_m}(\rho_s - \rho_l)d^2\vec{g} \quad , \tag{1}$$

in which $d$ is the average particle diameter, $\mu_m$ is the mixture viscosity, and $g_s$ is the volume fraction solid. This equation treats the equiaxed dendrites as spheres of a uniform, specified diameter. While the unattached solid does not impede the fluid motion, the packed, rigid array of solid dendrites slows the flow in the mushy zone. The mushy zone permeability model used is the Blake-Kozeny formulation: $K = \lambda^2 g_l^3 / 180 g_s^2$. The species and energy conservation equations are shown in equations (2) and (3) in which $f_s$ is the mass fraction solid. The conservation equations are discretized using the finite volume method and the flow field is solved with the SIMPLER algorithm.

$$\frac{\partial}{\partial t}\left(\rho C^i\right) + \nabla \bullet \left(\rho f_l \vec{V} C^i\right) = \nabla \bullet \left(\rho f_l D \nabla C^i\right) + \nabla \bullet \left(\rho D \nabla\left(C_l^i - C^i\right)\right) - \nabla \bullet \left(\rho f_s \vec{V}\left(C_l^i - C^i\right)\right) \tag{2}$$

$$\begin{aligned} &\frac{\partial}{\partial t}\left(\rho c_p T\right) + \nabla \bullet \left(\rho c_p T \vec{V}\right) = \nabla \bullet (k \nabla T) - \frac{\partial}{\partial t}\left(\rho f_l L_f\right) \\ &- \nabla \bullet \left(\rho f_l L_f \vec{V}\right) - \nabla \bullet \left\{ \rho f_s \left[ \left(c_{pl} - c_{ps}\right) T + L_f \right] \left[\vec{V} - V_s\right] \right\} \end{aligned} \tag{3}$$

The simulations begin with quiescent liquid with a superheat of 21.8K (940K). The inlet to the domain feeds the solidification shrinkage with liquid at the nominal composition and superheated temperature. Alloy properties are in (Table 1) and further details of the model are in [3].

IOP Conf. Series: Materials Science and Engineering **143** (2016) 012028 doi:10.1088/1757-899X/143/1/012028

**Table 1.** Thermophysical properties of Al-4.5 wt.% Cu.

| | | | |
|---|---|---|---|
| Liquid Density [kg/m3] | 2460 | Eutectic Temperature [K] | 821 |
| Solid Density [kg/m3] | 2750 | Liquid Thermal Expansion [1/K] | $1.17 \times 10^{-4}$ |
| Specific Heat [J/kg K] | 1006 | Solid Thermal Expansion [1/K] | $2.25 \times 10^{-5}$ |
| Latent Heat [J/kg] | $3.9 \times 10^{5}$ | Liquid Solutal Expansion [1/K] | -0.73 |
| Thermal Conductivity [W/m K] | 137.5 | Solid Solutal Expansion [1/K] | -0.87 |
| Liquid Viscosity [kg/m s] | 0.0023 | Partition Coefficient | 0.17 |

### *2.2 Description of Uncertainty Quantification Model*

This work uses existing UQ techniques, so only a brief description of the methodology will be given here. The UQ model calculates a PDF for each of the output quantities of interest given the known uncertainty of select inputs. Here, the outputs of interest are measures of the levels of macrosegregation, the first of which is the macrosegregation number, $M = \sqrt{(1/V)\iiint (C/C_o - 1)^2\, dV}$, which is a normalized standard deviation and has been used to quantify the overall level of segregation for castings starting with Prescott [6] and Schneider and Beckermann [7]. Another metric, similar to one recently developed by Voller and Vušanović for quantifying positive segregation [8], is the standard deviation of the Weibull distribution, $W$. The last macrosegregation metric is the volume fraction of ingot outside the compositional specification, $V_{spec}$. For this study, the Cu compositional specification of Al alloy 2014 (ASTM B209) is used, which has a range of 3.9 – 5.0 wt.% Cu. These three measures of macrosegregation can be seen in Figure 2, in which the shaded region is the specification range. Uncertainty is assigned to input parameters in the form of normal or uniform distributions. Outputs are then sampled from the full numerical model to generate a response surface, which then acts as a surrogate to running the full solidification model. In this work, the computationally efficient Smolyak sparse grid algorithm [2,9–11] is used to sample the numerical model. Different polynomial functions can be fitted to model outputs of interest to form response surfaces based on the level of the Smolyak algorithm. The fit of the response surface to the sampled points is quantified by using the root mean square error (RMSE). Each input parameter is randomly sampled 10,000 times and the response surface is used to calculate the output PDFs, which are characterized by their normal standard deviation and their actual shapes are shown below. The resulting output PDFs can aid in evaluating the validity of numerical models by comparing the uncertainty of the experimental measurement to that of the numerical model.

The relative sensitivities of the outputs to changes in the input parameters are calculated using the sampled points from the model and the Elementary Effects Method (EEM) [12,13]. The EEM calculates two sensitivity measures to determine whether the inputs have negligible, linear, or nonlinear effects on the outputs. A distribution of elementary effects is calculated for each input parameter by varying them independently with a constant step change across all levels of the other parameters, and calculating the result with the response surface for the array of inputs. The mean sensitivity, $\mu^*$, is the estimate of the mean of the distribution of the absolute value of the elementary effects. Unlike the analytical sensitivity, $\mu^*$ is always positive. For $\mu^*$ values near zero, the input has a small or negligible effect on the output, values much greater indicate the input plays a significant role in determining the output. The second parameter, $\sigma^*$, is a measure of the standard deviation of the output sensitivity distribution. A high value of $\sigma^*$ (a wide distribution), indicates the effects are strongly influenced by other inputs. A low value of $\sigma^*$ indicates a linear dependence on the input. The algorithms and methods described in this section have been compiled into a single publically available framework along with a graphical user interface, the PRISM Uncertainty Quantification (PUQ) tool [14,15]. This tool assumes the chosen numerical model incorporates the basic physics to adequately predict the output of interest.

## 3. Results and Discussion

Macrosegregation begins with solute partitioning during solidification and is affected by subsequent liquid motion relative to the solid. In grain refined alloys, there are two different mechanisms by which this occurs, settling of free-floating solid grains and flow in the rigid mushy zone. For Al- 4.5 wt.% Cu the flow in the rigid mushy zone is driven primarily by shrinkage induced flow. The settling of particles is dependent on their size, the flow in the mushy zone is dependent on the dendrite arm spacing, and both phenomena depend on the formation of the rigid mushy zone and relative size of the slurry.

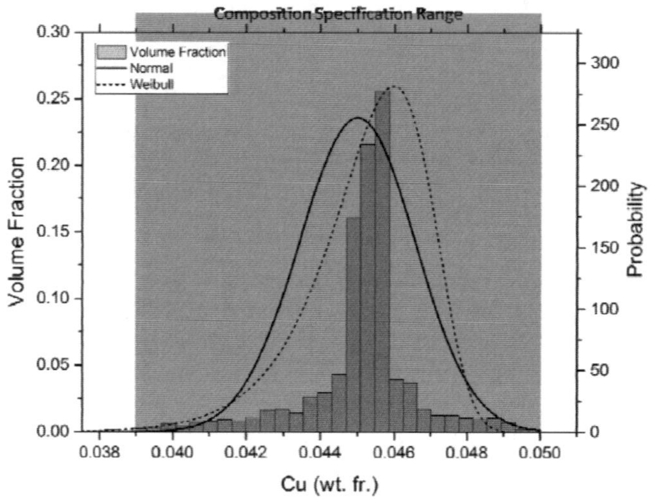


**Figure 2.** Composition distribution of a grain refined Al-4.5 wt% Cu alloy showing the ingot volume fraction occupied at each composition, normal distribution of the data, and the Weibull distribution of the data. The shaded area is the specified range of alloy 2014.

*3.1 2D Model of Solidification in a Rectangular Domain*

The uncertain model parameters used for this study that influence macrosegregation are shown in Table 2. The inputs are given either as a normal distribution with a mean, $\mu$, and deviation, $\sigma$, or a uniform distribution with a minimum and maximum. The packing fraction, $g_{s,pack}$, uncertainty distribution is treated as uniform because no one value in the range is more probable than the rest.

**Table 2.** Uncertainty of input parameters

| Input Parameter | Case A | Case B | Case C |
|---|---|---|---|
| Particle size ($d$) [m] | $\mu = 30x10^{-6}$<br>$\sigma = 7.5x10^{-6}$ | $3x10^{-6}$ | $\mu = 30x10^{-}$<br>$\sigma = 7.5x10^{-6}$ |
| Dendrite arm spacing ($\lambda$) [m] | $91x10^{-6}$ | $\mu = 91x10^{-6}$<br>$\sigma = 9.5x10^{-6}$ | $\mu = 91x10^{-6}$<br>$\sigma = 9.5x10^{-6}$ |
| $g_{s,pack}$ [-] | 0.15 | 0.15 | Min = 0.0<br>Max = 0.3 |

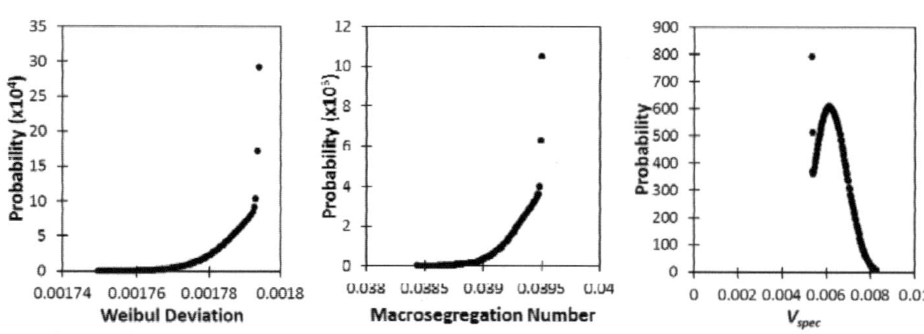


**Figure 3.** Output PDFs for Case A, where $d$ has a standard deviation of 25%.

IOP Conf. Series: Materials Science and Engineering **143** (2016) 012028 doi:10.1088/1757-899X/143/1/012028

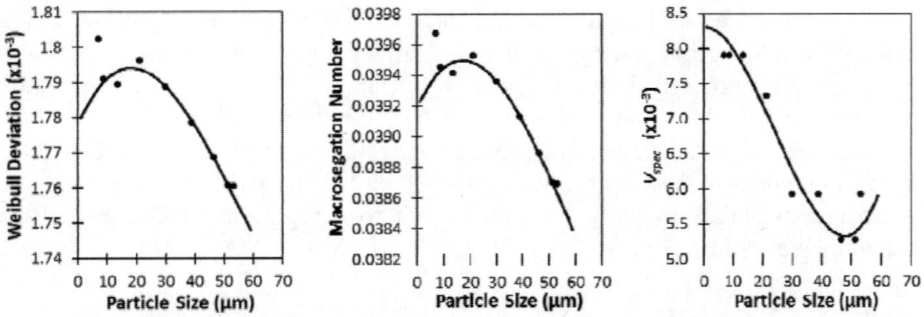


**Figure 4.** Response surfaces of $W$, $M$, and $V_{spec}$ for Case A.

The output PDFs from Case A are shown in Figure 3, which were produced with a level 3 Smolyak run resulting in a cubic polynomial response surface. The response surface for $W$ and $V_{spec}$ are shown in Figure 4 along with the model outputs, in which $W$ and $M$ have similar response surfaces. The settling velocity increases with $d$, causing the free-floating solid to form a rigid mushy more easily and changing the shape of the rigid mushy zone as more particles settle along the bottom of the cavity. This increased settling rate leads to less segregation because the rigid zone grows more rapidly from the bottom of the domain. There is also less horizontal shrinkage flow leading to segregation and more vertical shrinkage flow. The macrosegregation metrics all decrease with increasing $d$, but the function for $V_{spec}$ behaves differently at extreme values of $d$ leading to a different PDF. The resulting PDFs for $W$, $M$, and $V_{spec}$ are not very sensitive to $d$, as their PDFs vary over small ranges.

The dendrite arm spacing, λ, affects the permeability of the rigid mushy zone as dictated by the Blake-Kozeny model and is assigned a standard deviation of 10% for its uncertainty distribution. The output PDFs from Case B are shown in Figure 5, where the response surfaces were produced with a level 3 Smolyak. The response surfaces for $W$, $M$, and $V_{spec}$ are in Figure 6. The relationship between $\lambda$ and $W$ and $M$ is almost linear, so the resulting PDFs are normal distributions. However, the relationship between $\lambda$ and $V_{spec}$ is highly nonlinear and, for small $\lambda$, the casting is entirely within the specification limits. These different λ relationships demonstrate differences between the macrosegregation metrics. $W$ and $M$ show a more or less normal distribution, while the $V_{spec}$ metric shows a high probability (in this range of λ) that the input will be in spec and a very low probability that any more than 3% of the ingot volume will be out of the composition range.

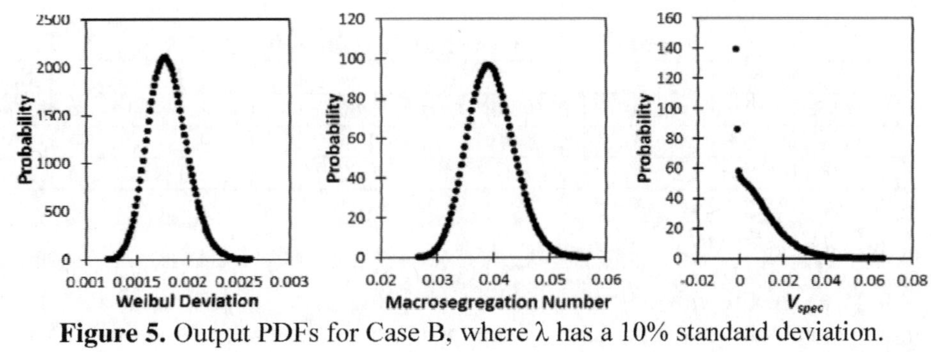


**Figure 5.** Output PDFs for Case B, where λ has a 10% standard deviation.

The critical packing fraction, $g_{s,pack}$, controls the location of the rigid mushy zone formation and indirectly controls the pattern of particle settling. Because $g_{s,pack}$ is the lowest $g_s$ value in the rigid mushy zone, it also determines the highest permeability and so the strongest flow in that region. Figure 7 shows the PDFs for the Case C macrosegregation, based on the same uncertainties in $d$ and $\lambda$ as in Cases A and B and a uniform input PDF for $g_{s,pack}$. The output sensitivities produced with a level 1 Smolyak are shown in Table 3. The distributions of each metric are dominated by the uniform uncertainty of $g_{s,pack}$, and each metric is also most sensitive to this parameter (followed closely by $\lambda$). Small $g_{s,pack}$ values form the rigid mushy zone with still enough permeability to allow significant macrosegregation-causing buoyancy induced flow, while larger values of $g_{s,pack}$ form the rigid solid with much less permeability. Also, the settling velocity of the free-floating solid is a function of the slurry fraction solid, therefore the larger $g_{s,pack}$ the stronger the settling velocity. The uncertainty of $g_{s,pack}$ is large and overwhelms the uncertainty of the other two parameters, so more information is needed about particle packing to get more certain macrosegregation predictions, also the value selected for $\lambda$ should be chosen carefully.

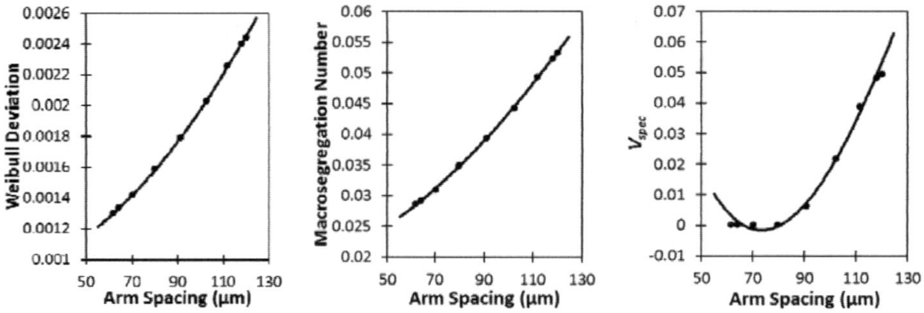


**Figure 6.** Response surfaces of $W$, $M$, and $V_{spec}$ for Case B.

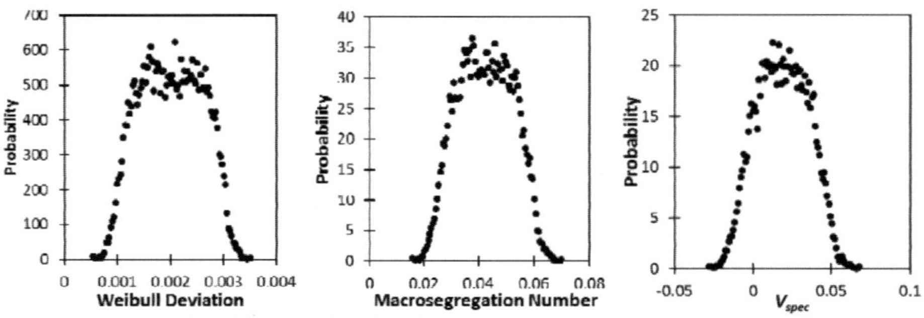


**Figure 7.** Probability distribution functions of the model outputs for Case C.

**Table 3.** Output sensitivities for Case C.

| | $W$ Sensitivities | | $M$ Sensitivities | | $V_{spec}$ Sensitivities | |
|---|---|---|---|---|---|---|
| $d$ (m) | $\mu^*$:4.18x10$^{-5}$ | $\sigma^*$:1.49x10$^{-5}$ | $\mu^*$:9.85x10$^{-4}$ | $\sigma^*$:3.60x10$^{-4}$ | $\mu^*$:1.97x10$^{-3}$ | $\sigma^*$:1.97x10$^{-3}$ |
| $\lambda$ (m) | $\mu^*$:1.14x10$^{-3}$ | $\sigma^*$:1.66x10$^{-4}$ | $\mu^*$:2.49x10$^{-2}$ | $\sigma^*$:3.18x10$^{-3}$ | $\mu^*$:4.93x10$^{-2}$ | $\sigma^*$:3.75x10$^{-2}$ |
| $g_{s,pack}$ | $\mu^*$:1.90x10$^{-3}$ | $\sigma^*$:1.21x10$^{-3}$ | $\mu^*$:3.10x10$^{-2}$ | $\sigma^*$:1.62x10$^{-2}$ | $\mu^*$:5.00x10$^{-2}$ | $\sigma^*$:3.95x10$^{-2}$ |

*3.2 Direct Chill Casting Model*

The UQ approach shown here for static casting is also applied to a transient model of direct chill casting of Al-4.5 wt. % Cu with a 50 cm diameter [16]; Figure 8 is a schematic of that process. The uncertain input parameters are the particle size and heat transfer boundary conditions. The value of the boundary heat transfer coefficients from [17] are complex and are multiplied by an uncertainty factor with a mean value of 1 and a standard deviation of 0.15. The particle size has the same uncertainty as in the previous study. The simulations were run until the solid metal height was 1.5 m and then the macrosegregation level was measured in the steady state region using the metrics described previously. Along with the macrosegregation metrics, the steady state sump depth and height above the bottom block at which the composition reaches steady state are also outputs of interest.

The resulting PDFs from the DC casting simulations are shown in Figure 9, which were produced with a level 1 Smolyak, and the sensitivities are shown in Table 4. The macrosegregation metrics are more sensitive to the thermal boundary conditions than the free-floating particle size, while the sump depth and height to steady state is equally sensitive to both uncertain inputs. A large portion of the macrosegregation in DC casting is due to negative centreline segregation caused by shrinkage driven flow. The depth of the sump does not change much even for large variations in the heat transfer, indicating that the formation of the rigid mushy zone is independent of the thermal characteristics in the range considered. The deviations of the output PDFs are relatively small; therefore the particle size and heat transfer coefficient measurements can be known within 20 and 30% respectively and still have a small influence on the results. Based on the static casting results, the critical packing fraction will likely have more of an effect on the segregation levels than the other parameters.

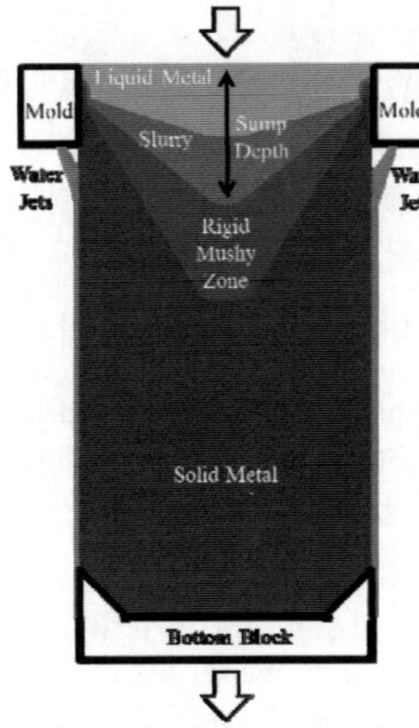


**Figure 8.** Schematic of DC casting process.

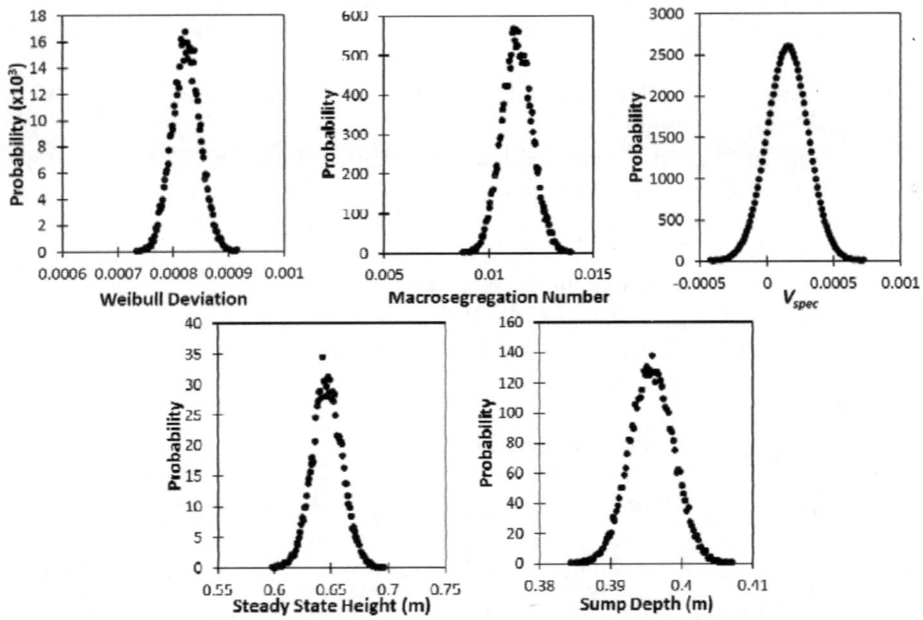


**Figure 9.** Probability distribution functions of the model outputs for DC casting.

**Table 4.** Output sensitivities for the DC casting model.

| Inputs | *W* | *M* | $V_{spec}$ | *S.S. Height (m)* | *Sump Depth (m)* |
|---|---|---|---|---|---|
| H.T.C. (W/mK) | $\mu^*$:1.42x10$^{-4}$ $\sigma^*$:1.42x10$^{-4}$ | $\mu^*$:4.91x10$^{-3}$ $\sigma^*$:4.91x10$^{-3}$ | $\mu^*$:9.43x10$^{-4}$ $\sigma^*$:9.43x10$^{-4}$ | $\mu^*$:9.00x10$^{-2}$ $\sigma^*$:9.00x10$^{-2}$ | $\mu^*$:3.16x10$^{-2}$ $\sigma^*$:3.16x10$^{-2}$ |
| *d* (m) | $\mu^*$:8.74x10$^{-5}$ $\sigma^*$:3.05x10$^{-5}$ | $\mu^*$:1.71x10$^{-3}$ $\sigma^*$:1.20x10$^{-3}$ | $\mu^*$:0.00 $\sigma^*$:0.00 | $\mu^*$:8.00x10$^{-2}$ $\sigma^*$:8.00x10$^{-2}$ | $\mu^*$:2.03x10$^{-2}$ $\sigma^*$:2.03x10$^{-2}$ |

IOP Conf. Series: Materials Science and Engineering **143** (2016) 012028 doi:10.1088/1757-899X/143/1/012028

## 4. Conclusions

Uncertain model input parameters that govern transport phenomena in grain refined alloy solidification were examined for their effect on macrosegregation using the publically available PUQ framework. Using a model of solidification in a rectangular domain, the critical packing fraction and dendrite arm spacing had more of an effect on the segregation than the free-floating particle size. More information is needed on the mechanisms of particle attachment and formation of the rigid mushy zone to lower the effects of $g_{s,pack}$ on the uncertainty of the macrosegregation model predictions. In modelling of grain refined alloys using DC casting, the thermal boundary conditions of the process and free-floating particle size do not significantly impact the depth of the sump. Also, the uncertainty of the predicted macrosegregation level was small using input parameters with 20-30% uncertainty. It is expected, as in static casting of grain refined alloys, that $g_{s,pack}$ will have the largest effect on the uncertainty of the macrosegregation predictions in DC casting.


## Acknowledgments

Financial support from Shandong Nanshan Aluminum Co., Beijing Nanshan Institute of Aeronautical Materials is gratefully acknowledged. The authors would also like to acknowledge contributions from A. Strachan and M. Hunt for their work on PUQ.


## References


[1] R.L. Wasserstein: Technometrics, 1997, vol. 39, pp. 338–338.
[2] M. Eldred: in 50th AIAA/ASME/ASCE/AHS/ASC Struct. Struct. Dyn. Mater. Conf., American Institute of Aeronautics and Astronautics, Reston, Virigina, 2009.
[3] C.J. Vreeman, M.J.M. Krane, and F.P. Incropera: Int. J. Heat Mass Transf., 2000, vol. 43, pp. 677–86.
[4] I. Vusanovic and M.J.M. Krane: 3rd Int. Conf. Adv. Solidif. Process., 2011, vol. 27.
[5] C.J. Vreeman, J.D. Schloz, and M.J.M. Krane: J. Heat Transfer, 2002, vol. 124, p. 947.
[6] M.C. Schneider and C. Beckermann: Metall. Mater. Trans. A, 1995, vol. 26, pp. 2373–88.
[7] P.J. Prescott: Ph.D Dissertation, School of Mechanical Engineering, Purdue University,1992.
[8] V.R. Voller and I. Vušanović: Int. J. Heat Mass Transf., 2014, vol. 79, pp. 468–71.
[9] S. Smolyak: Dokl. Akad. Nauk SSSR., 1963, vol. 4, pp. 240–43.
[10] M.S. Eldred and J. Burkardt: in 47th AIAA Aerosp. Sci. Meet. Incl. New Horizons Forum Aerosp. Expo., 2009, pp. 1–20.
[11] M.S. Eldred, C.G. Webster, and P.G. Constantine: in 49th AIAA/ASME/ASCE/AHS/ASC Struct. Struct. Dyn. Mater. Conf., 2008.
[12] M.D. Morris: Technometrics, 1991, vol. 33, pp. 161–74.
[13] F. Campolongo, J. Cariboni, and A. Saltelli: Environ. Model. Softw., 2007, vol. 22, pp. 1509–s18.
[14] M. Hunt: Uncertainty Quantification and Sensitivity Analysis Tool, https://github.com/c PRIMED/puq, accessed [Jan. 2015].
[15] M. Hunt, B. Haley, M. McLennan, M. Koslowski, J. Murthy, and A. Strachan: Comput. Phys. Commun., 2014, vol.194, pp. 97-107.
[16] K. Fezi, J. Coleman, and M.J.M. Krane: in Light Meals., M. Hyland, ed., Wiley, Orlando, Florida, 2015, pp. 871–75.
[17] D.C. Weckman and P. Niessan: Metall. Mater. Trans. B, 1982, vol. 13, pp. 593–602.

IOP Conf. Series: Materials Science and Engineering **143** (2016) 012029 doi:10.1088/1757-899X/143/1/012029

# New developments on optimizing properties of high-Zn aluminium cast alloys

W.K. Krajewski[1], J. Buras[1], P.K. Krajewski[1], A.L. Greer[2], P. Schumacher[3], K. Haberl[3]

[1]AGH University of Science and Technology, Faculty of Foundry Engineering, 23 Reymonta Street, 30-059 Krakow, Poland
[2]University of Cambridge, Department of Materials Science & Metallurgy, 27 Charles Babbage Road, Cambridge CB3 0FS, UK
[3]University of Leoben, Department of Metallurgy - Chair of Casting Research, Franz-Josef-Strasse 18, 8700 Leoben, Austria

[1]krajwit@agh.edu.pl

**Abstract.** Foundry alloys with Al-based matrices have a wide range of uses in today's global economy and there is a high demand for castings of Al alloys, including Al-Zn alloys. In this paper, investigations on the grain refinement of high-Zn aluminium cast alloys are presented. Aluminium alloys with relatively high zinc content have a tendency to be coarse-grained, especially in the case of castings with low cooling rates such as are found in sand moulds. The coarse-grained structure degrades the plasticity, specifically the elongation. Therefore, for aluminium alloys of high (10-30 wt.%) zinc content, inoculation is attractive, aiming to break up the primary dendrites of the α-phase solid solution of zinc in aluminium. Such dendrites are the principal microstructural component in these alloys. On the other hand, a finer grain structure usually reduces the damping (e.g. as measured by attenuation of ultrasound) in these alloys. In the present investigations, a binary sand-cast Al-20 wt.% Zn alloy was inoculated with different additions of AlTi3C0.15 (TiCAl) and ZnTi-based master alloys. The sand-cast samples were subjected to mechanical-property measurements (tensile strength and elongation), image analysis to determine grain size, and measurements of the attenuation of 1 MHz ultrasound. It is found that both of the master alloys used cause significant refinement of the α-AlZn primary dendrites and change their morphology from linear-branched to semi-globular, increase the elongation by about 40%, and decrease the attenuation coefficient by about 25% in comparison with the initial alloy without inoculation.

## 1. Introduction

Grain refinement of non-ferrous cast alloys, mainly those based on Al, is a common practice, which allows fine microstructures to be obtained with increased ductility [1–8].

Cast alloys based on Al and Zn are classified as structural materials with good damping capacity. In particular, high-Al zinc alloys, for example ZA-27, fall in the category of HiDAlloys (high-damping alloys) [9]. It has also been noted that high-Zn aluminium alloys show high damping, good tribological properties and high fatigue strength [10-12]. Grain refinement has been used for zinc-aluminium alloys with high Al content of 15-30 wt.% Al [13–18]. In fact, both groups, i.e. high-Al zinc and high-Zn aluminium alloys, solidify naturally with a coarse structure and a refinement process is necessary to permit highly refined structures to be obtained [18–22]. In practice, there are two main groups of refiners based on the Al-Ti-B and Al-Ti-C systems, commonly used in the casting of Al-alloys. A newly introduced alternative – a master alloy based on the Zn-Ti system – requires a melt temperature of only about 500°C, thus avoiding detrimental overheating, reducing the costs of energy and material, and improving the mechanical properties of the castings [17-18, 23-25]. On the other hand, many structural materials are required to have good damping properties, and unfortunately a fine grain structure is believed to reduce the damping capacity [26-28]. The present paper summarizes work on the grain refinement and damping-property measurements of high-Zn aluminium cast alloys on the basis of joint investigations performed over the last ten years at AGH University of Science and Technology – Faculty of Foundry Engineering, University of Cambridge, UK – Department of Materials Science and Metallurgy, and at University of Leoben – Department of Metallurgy, Chair of Casting Research.

## 2. Experimental

The alloys Al-(18-22)%Zn (Al-20Zn) and master alloy (Zn,Al)-4%Ti were melted from elements of minimum purity 99.9 % (all compositions in wt.%). Melting was performed in an electric furnace with an argon protective atmosphere, in a clay-graphite crucible of 2-litre capacity (the details of melting and casting are given in Refs [6–8]). During these experiments an amount of the ZnTi-based master alloy or Al-3Ti-0.15C (TiCAl) was added to the melt to introduce 25, 50, 100, 200 or 400 ppm of Ti. The inoculated melt was held for about 2–3 minutes to ensure complete dissolution of the added master alloy. Next, the melt was stirred for about 2 minutes with a silica-glass tube, and finally the alloy was cast into dried sand moulds to obtain dog-bone-shape samples (gauge section ∅12x60 mm) for tensile and structural tests, Fig. 1 (a) – (b) as well as ∅32x80 mm samples for structural and damping tests.

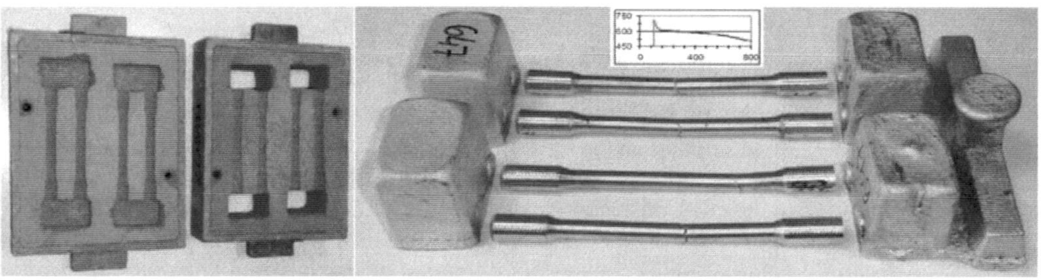


**Figure 1.** (a) Sand mould for tensile strength and elongation measurements ; (b) Machined specimens after UTS tests with risers and gating system.

Optical light metallography (LM) was performed using a Leica DM IRM microscope. Scanning electron microscopy (SEM) was performed on unetched samples with a Philips XL30 microscope equipped with an energy dispersive X-ray EDX spectrometer (Link-Isis). Damping-coefficient tests were carried out using the ultrasonic Krautkramer measurement set, model USLT 2000. The entire study was conducted for 1 MHz longitudinal waves, using the MK1S mini-transducer with head diameter of 10 mm. To determine the attenuation intensity, the echo method (*pulse-echo method*) was adopted, using the internal software of the Krautkramer USLT 2000 device. Prior to testing, the device was calibrated with a pure-zinc cast sample.

## 3. Results and Discussion

Grain refinement of the high-zinc molten alloys by the (Zn,Al)-4Ti and TiCAl master alloy is very effective, as can be clearly seen in Fig. 2 (a) – (d) and Fig. 3 (a) – (b). At the same time, the elongation significantly increases while the tensile strength remains basically preserved (Fig. 6).

As noted in the introduction, the main refiners for Al alloys are those based on the systems Al-Ti-B and Al-Ti-C. For the Al-Ti-C master alloy, TiC is the direct nucleant particle for α-Al. This is because of the similar crystal structure and lattice parameters of α-Al and TiC (Table 1).

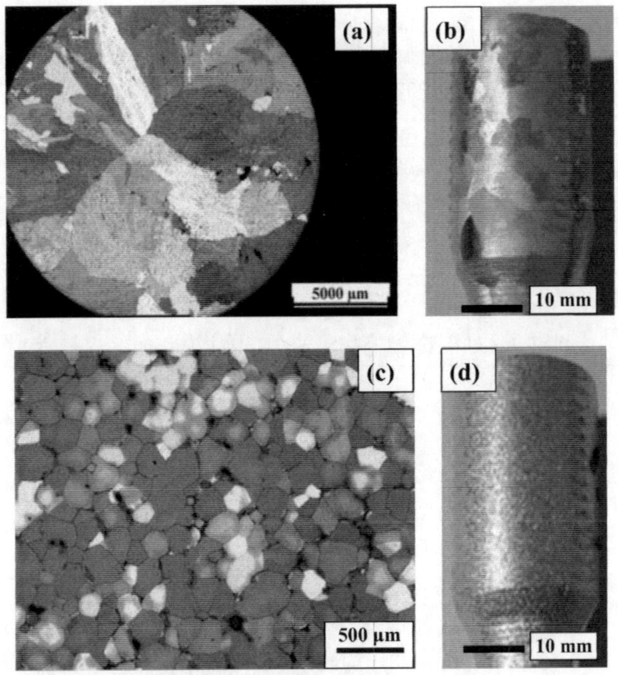


**Figure 2.** LM pictures of macrostructures and microstructures of the Al-20Zn alloy. (a) and (b) initial alloy; (c) and (d) alloy inoculated with (Zn,Al)-Ti4 MA – 0.04 wt.% Ti; (a) and (c) ground, polished and etched surfaces of a section [19-22]; (b) and (d) surfaces of the tensile samples.

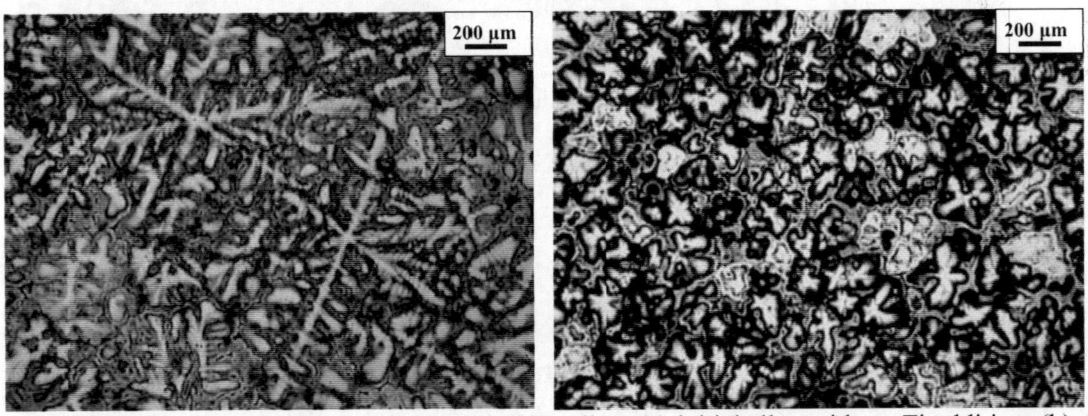


**Figure 3.** Changes of microstructure of the Al-20Zn alloy. (a) initial alloy without Ti addition, (b) the same alloy doped with 50 ppm Ti introduced with the Al-3Ti-0.15C (TiCAl) grain refiner, Light microscopy [25-26].

The new family of grain refiners presented here, based on the Zn-Ti system, contain $TiZn_3$ phase in their structure (Fig. 4), which has the same crystal symmetry as $\alpha$-Al and closely matches its lattice parameter. In an Al-Zn melt the $TiZn_3$ phase transforms into the ternary, more stable, $Ti(Al,Zn)_3$, which has also the same features as $TiZn_3$, and whose particles appear to be active centres of heterogeneous nucleation in inoculated Zn-Al alloys [21-22]. This is the most probable reason why the ZnTi-based master alloys are effective as refiners of the present Al-Zn alloys. Macrostructures and microstructures shown in Figs 2 and 3 indicate that the ternary master alloys AlTi3C0.15 and (ZnAl)-Ti4 cause strong refinement of the inoculated AlZn20 alloy. It should be noted that the ZnTi-based master alloys have a mass density higher than the inoculated Al-20Zn alloy, which simplifies their introduction into the melt. They dissolve very quickly already at temperatures of 500°C, allowing aluminium-zinc alloys to be treated without the detrimental overheating that is required when using Al-Ti based refiners.

**Table 1**. Crystallographic comparison of nucleants and the α-Al phase.

| Phase | Crystal structure | Lattice parameters ($a$), [nm] | Lattice misfit ($\delta$), [-] |
|---|---|---|---|
| α(Al) | Cubic A1 | $a$ = 0.4043 | --- |
| TiC | Cubic B1 | $a$ = 0.4328 | 1.034 |
| $TiZn_3$ | Cubic $L1_2$ ($AuCu_3$ type) | $a$ = 0.39424 | 0.9874 |
| $Ti(Al,Zn)_3$ | Cubic $L1_2$ ($AuCu_3$ type) | $a$ = 0.3958 | 0.9894 |

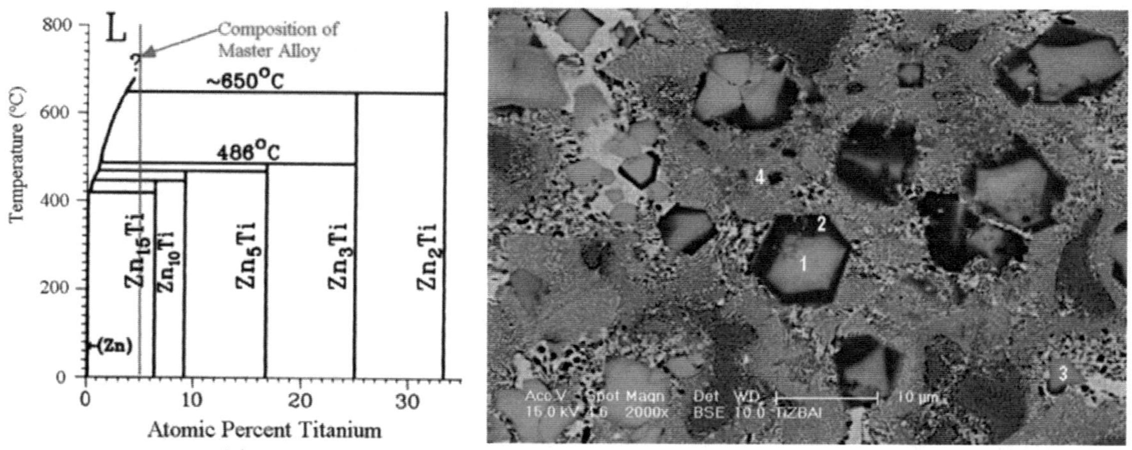


(a) (b)

**Figure 4.** (a) Zn-Ti binary system and (b) SEM (backscattered electron) image of (ZnAl)–4 wt.%Ti master alloy [18].

**Table 1.** Chemical composition of the randomly chosen particles of Ti-Al-Zn shown in Fig. 5 of ~$Ti(Al,Zn)_3$ stoichiometry. EDAX-GEMINI 4000 EDS [18].

| No. | Ti | Zn | Al |
|---|---|---|---|
| 1 | 24.4 | 66.0 | 9.5 |
| 2 | 21.9 | 16.8 | 61.3 |
| 3 | 23.8 | 67.6 | 8.6 |
| 4 | 23.3 | 16.3 | 60.4 |

From Fig. 5, it can be seen that the 2D (planar) mean grain diameter decreases from about 2300 µm for the alloy without inoculation to about 180 - 150 µm for the same alloy doped with 50 – 100 ppm Ti introduced with the TiCAl or ZnTi-based master alloys. The grain refinement that is obtained allows the elongation to be increased by about 40%, i.e. from 1.7 to 2.4%, with the tensile strength of 230-250 MPa practically preserved (increase by 5-10%) (Fig. 6).

On the other hand, in the grain-refined alloys, the attenuation coefficient of 180 – 190 dB/m is decreased by ~20-30% in comparison to the initial alloys without inoculation (220 - 240 dB/m). However, the attenuation coefficient of the refined alloys is still high, remaining on a level typical for high-damping alloys.

However, the obtained results clearly show that the level of grain refinement should be a compromise to balance good ductility with good damping, thus optimizing the combination of mechanical properties.

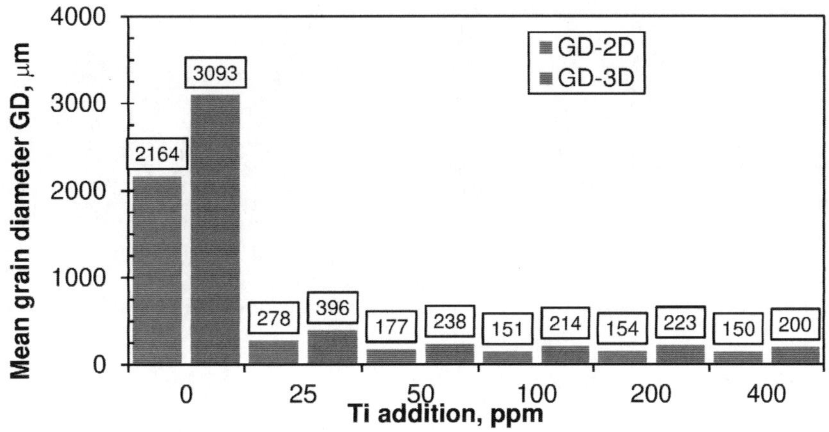


**Figure 5.** Influence of small Ti addition introduced with TiCAl master alloy on the Al-20Zn alloy mean grain diameter (GD). GD-2D is a planar grain diameter evaluated from the intercept method. GD-3D is a mean space (volumetric) grain diameter calculated from the Voronoi formula [21, 28].

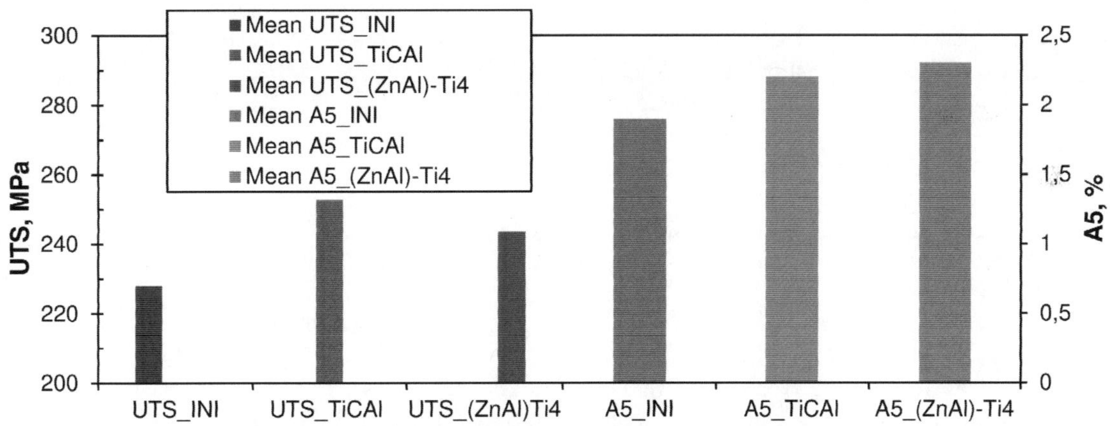


**Figure 6.** Tensile strength (UTS) and elongation (A5) of the Al-20Zn alloy inoculated with Al-3Ti-0.15C (TiCAl) and (ZnAl)-Ti4 master alloy.

## 4. Conclusions

Based on the studies described above, the following conclusions can be drawn:

For the high-Zn aluminium alloys in this study, grain refinement is a promising process leading to property improvements. At the same time, using the low-melting-point (Zn,Al)-Ti-based master alloys avoids the excessive melt overheating needed for the TiCAl or TiBAl refiners and reduces the possibility of gas pick-up and material loss.

It was also noted that grain refinement decreased the attenuation coefficient of 1 MHz ultrasound by about 25%. Taking this into account, it is concluded that further studies of other property changes, e.g. creep and tribological properties, would be desirable.

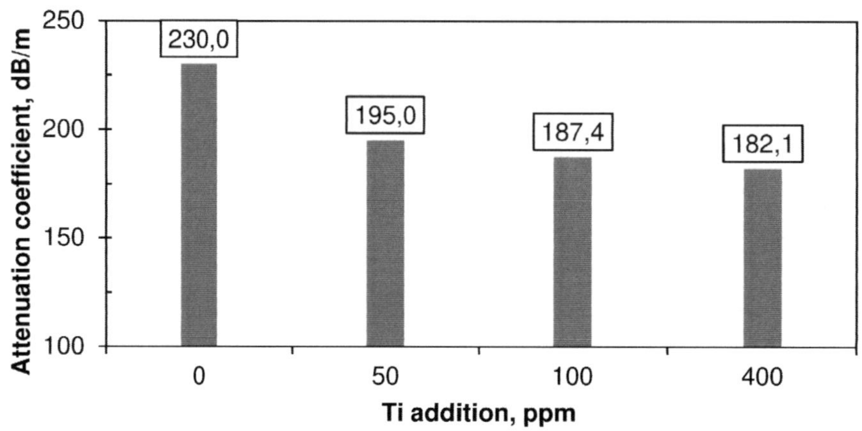


**Figure 7.** Influence of small Ti addition introduced with TiCAl master alloy on the damping properties of Al-20Zn alloy. Krautkamer USLT2000 – ultrasound of 1 MHz frequency [20-21].

**Acknowledgements**

The authors are grateful to the Polish Science National Centre for financial support under grant No. UMO-2012/05/B/ST8/01564 and to the AGH University of Science and Technology and University of Cambridge, Department of Materials Science and Metallurgy for provision of laboratory facilities.

**References**

[1] L. Backerud, Y. Shao, Aluminium, 76 (1991), 780-785
[2] J.A. Spittle, International Journal of Cast Metals Research, 19 (2006), 210-222
[3] G.K. Sigwoth, T.A. Kuhn, AFS Transactions, 115 (2007), 1-12
[4] K. Kelton, A.L. Greer, Nucleation in Condensed Matter, Pergamon Materials Series, Elsevier Ltd., Amsterdam, 2010
[5] D.H. StJohn, M. Qian, M.A. Easton, P. Cao, Acta Materialia, 59 (2011), 4907-4921
[6] W.K. Krajewski, A.L. Greer, P.K. Krajewski, Archives of Metallurgy and Materials, 58 (2013), 859-861
[7] W.K. Krajewski, A.L. Greer, P.K. Krajewski, G. Piwowarski, Proceedings of 71$^{th}$ World Foundry Congress, May 2014, Bilbao, Spain
[8] W.K. Krajewski, A.L. Greer, G. Piwowarski, P.K. Krajewski, Proceedings of ICASP-4: the fourth International Conference on Advances in Solidification Processes, Old Windsor, UK, 8-14 July 2014, Book of Abstracts, pp. 190-191
[9] I.G. Ritchie, Z-L. Pan, Metallurgical Transactions, 22A (1991), 607-616
[10] T. Savaskan, O. Bican, Y. Alemdag, Journal of Materials Science, 44 (2009), 1969-1976
[11] Y. Alemdag, T. Savaskan, Tribology Letters, 29 (2008), 221-227
[12] T. Savaskan, H.O. Tan, Materials Science and Technology, 30 (2014), 938-943
[13] W.A. Pollard, K.M. Pickwick, J.T. Jubb, R.H. Packwood, Canadian Metallurgical Quarterly, 13 (1974), 535-543
[14] A.F. Skenazi, J. Pelerin, D. Coutsouradis, B. Magnus, M. Meeus, Some Recent Developments in the Improvement of the Mechanical Properties of Zinc Foundry Alloys, Metall, 37 (1983), 898-902
[15] A.A. Abdel-Hamid, Zeitschrift für Metallkunde, 83 (1992), 314-318
[16] A.A. Abdel-Hamid, Zeitschrift für Metallkunde, 84 (1993), 40-43
[17] W. Krajewski, Archives of Metallurgy, 44 (1999), 51-61
[18] W.K. Krajewski, A.L. Greer, J. Zych, J. Buras, Foundry Trade Journal, 180 (2007), No. 3642, April 2007, 97-100

[19] W.K. Krajewski, J. Buras, M. Zurakowski, A.L. Greer, Archives of Metallurgy and Materials, 54 (2009), 335-338
[20] K. Haberl, W.K. Krajewski, P. Schumacher, Archives of Metallurgy and Materials, 55 (2010), 837-841
[21] J. Buras, PhD thesis, AGH University of Science and Technology, Krakow 2012 (in Polish)
[22] K. Haberl, Examination of AlZn20 alloyed with various grain refiners, Report Marie Curie Host Fellowship, July 2009, University of Leoben, Austria (unpublished).
[23] W.K. Krajewski, Materials Science Forum, 508 (2006), 615-620
[24] W.K. Krajewski W.K., A.L. Greer, Materials Science Forum, 508 (2006), 281-286
[25] W.K. Krajewski, Metallurgia Italiana, No. 11-12 (2006), 27-30
[26] R. Schaller, Journal of Alloys and Compounds 355(2003), 131-135.
[27] W.K. Krajewski, J. Buraś, P.K. Krajewski, G. Piwowarski, Archives of Foundry Engineering 15 (2015) Issue 2, 237-241
[28] W.K. Krajewski, J. Buraś, P.K. Krajewski, A.L. Greer, K. Faerber, P. Schumacher, Proceedings of Aluminium Two Thousand & ICEB 2015, 12 - 16 May 2015, Florence, Italy

IOP Conf. Series: Materials Science and Engineering **143** (2016) 012030 doi:10.1088/1757-899X/143/1/012030

# A numerical analysis of inverted V-segregation and its dependency on alloy properties

**T SAWADA[1], S SUZUKI[1], K KAJIKAWA[1]**

Muroran Research Laboratory, The Japan Steel Works, Ltd., 4 Chatsu-machi, Muroran, 051-8505, JAPAN

E-mail: tomoki_sawada@jsw.co.jp

**Abstract.** We ran a numerical simulation to investigate the effect of liquidus slope on the formation of inverted V-segregation in laterally solidified ingots of a virtual Fe-based binary-alloy system. The criterion used in this research is based on Suzuki's criterion. Our criterion assumed that inverted V-segregation zones form when the product of the solidification velocity and the cooling rate is below a critical value which depends on the alloy system. We studied the change in the critical value caused by the density and the liquidus slope of added element X. When different elements of varying densities were added, the results were in line with previous experimental results where large density difference causes increased formation of inverted V-segregation. Finally, the effect of the liquidus slope on the formation of the inverted V-segregation was studied. With a decrease in the liquidus slope, the critical value increased and inverted V-segregation tended to form. Furthermore, the streaks formed were more highly concentrated and sharper, and the channel population increased when the solidification temperature range was narrow.

## 1. Introduction

Thermo-solutal convection during alloy solidification transports concentrated liquid which causes microsegregation in the mushy zone. Subsequently, various types of macrosegregation form in the ingot. Inverted-V segregation is a macrosegregation that can be observed as a string-shaped region where added elements concentrate. Depending on the alloy system, unexpected phases such as intermetallic compound precipitates and fine porosity forms in the inverted-V segregation. It is hard to eliminate local concentration and precipitation of unexpected phases in the segregation region by forging and heat treatment. Hence the segregation region causes a reduction of ingot and product quality. One effectual countermeasure is to solidify the ingot under conditions where the inverted-V segregation doesn't form. To achieve such production, it is important to understand formation conditions of the inverted-V segregation.

Many researchers have investigated the formation conditions or criteria for channel-type segregations such as inverted-V segregation and freckles. One criterion is the mushy zone Rayleigh number [1-4], which is a non-dimensional number based on the kinetics of the interdendritic concentrated liquid. The kinetics-driven force of density difference and flow resistance of the interdendritic region are considered and the model provides industrial applications of prediction using numerical simulation. On the other hand, Suzuki et al. derived the formation conditions by classifying the relationships between the regions where inverted-V segregation forms and the heat transfer condition in actual ingots [5]. Although Suzuki's criterion is an empirical criterion with limited practical applications, this criterion also can predict inverted-V segregation regions in actual gravity casting ingots and ESR ingots. However, both criteria are not complete in terms of the fact that neither can predict the concentration in the channel segregation, the channel population, or the dimensions of the channels. Also, both criteria don't directly consider important alloy properties, such the liquidus slope, partition coefficient, and so on. Therefore it is necessary to investigate the basic relationship between the alloy properties and the formation of inverted-V segregation, and freckles.

Both criteria focus on density differences between the bulk-liquid and interdendritic concentrated liquid. Density difference is considered as one of parameters in the Rayleigh number and the formation condition changes according to density difference in Suzuki's criterion [6]. It is reasonable to consider density difference in the model of channel segregation because it is a directly driven force of the movement of the concentrated liquid that plays important roll. From the viewpoint of thermodynamics and the fact that channel formation is accompanied with remelting and solidification retardation, the liquidus slope, *m*, in the phase diagram can be expressed as Eq. (1), which is also important since this strongly affects the solid fraction change.

$$m_i = \frac{\partial T_L}{\partial C_{L,i}} \quad (1)$$

However it is very difficult to experimentally investigate the effect of just the liquidus slope on channel formation, since there are no two elements that have equal density and redistribution coefficients. Thus, when formation conditions in different alloy systems are studied, the change in formation conditions always contains effects of density difference, liquidus slope, and other factors.

In this work, to investigate the effect of liquidus slope on channel formation, numerical simulations were performed in a virtual binary alloy system consisting of Fe and the virtual added element X. The change in the formation conditions of inverted-V segregation in laterally-directionally solidified ingot was studied by varying the liquidus slope of the alloy system.

## 2. Procedures

### *2.1. Numerical simulation*

The laboratory-scale lateral directional solidification problem was solved as a two-dimensional problem. The height and width of the ingot are 200 mm and 150 mm, respectively. The computational domain and its boundary conditions are shown in Fig. 1. In this simulation, in order to conduct a directional solidification in the numerical simulation, it is assumed that the left wall is at a constant temperature and cool melt, and the other wall is the adiabatic boundary. The properties and parameters of the Fe-X alloy used in this simulation are shown in Table 1. These values were assumed as characteristics of typical low-alloy steels. Numerical simulations are performed with the densities of added element X and liquidus slopes shown in Table 2. Phase diagrams of virtual alloy system are shown in Fig. 2. The values in Table 2 are decided by a preliminary simulation to find the value that forms inverted-V segregation in the computational domain.

The numerical method used in this work is from a previously published model [7]. It was assumed that the melt was an incompressible fluid and that flow in the mushy zone can be expressed by Darcy's law. The thermo-solutal convection of the melt was taken into account with a density variation by temperature and liquid concentration; however, it was assumed that, divergence free, Eq. (3) is satisfied. Solutes in the liquid were moved by the fluid flow and diffusion and redistribution of the solute at the interface between the liquid and the solid was taken into account. Diffusion of the solute was neglected, because it was smaller than the numerical diffusion. The governing equations were as follows:

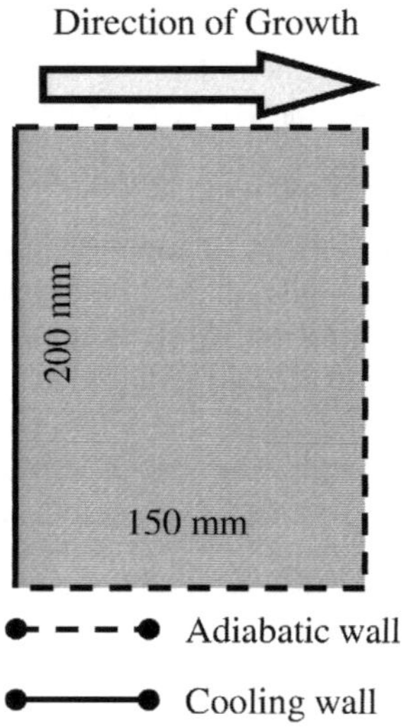


Fig. 1 Computational domain and its boundary conditions

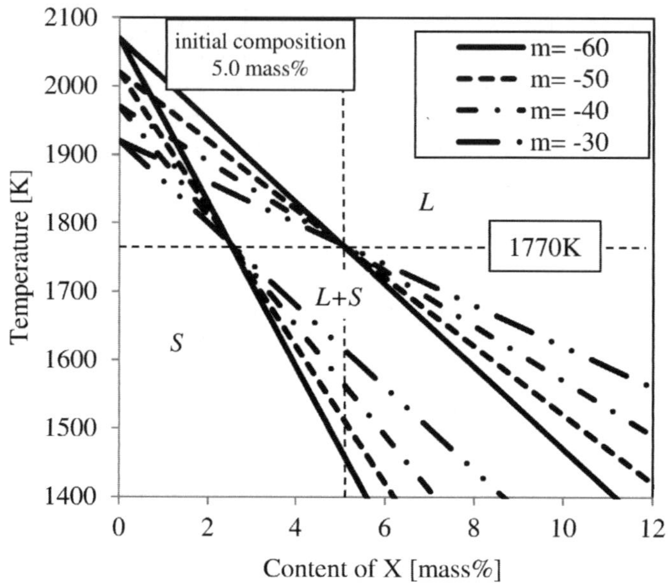


Fig. 2 Phase diagram of virtual alloy system

Table 1 Alloy properties and parameters u sed in simulation

| Properties and parameters | Value | Units |
|---|---|---|
| Thermal conductivity | 35 | W/K/m |
| Specific heat | 680 | J/K/kg |
| Density of Fe | 7000 | kg/m$^3$ |
| Latent heat | $2.6 \times 10^5$ | J/kg |
| Liquidus temperature | 1770 | K |
| Initial temperature | 1775 | K |
| Added element X content | 5 | mass% |
| Redistribution Coefficient | 0.5 | - |
| Temperature of cooling wall | 300 | K |
| Density of added element | $\rho_X$ | kg/m$^3$ |
| Liquidus slope | $m$ | K/mass% |

Table 2 Density and liquidus slope of virtual alloys

| | $\rho_X$ [kg/m$^3$] | $m$ [K/mass%] |
|---|---|---|
| Alloy 1 | 3500 | -60 |
| Alloy 2 | 3750 | -60 |
| Alloy 3 | 4000 | -60 |
| Alloy 4 | 4250 | -60 |
| Alloy 5 | 3890 | -60 |
| Alloy 6 | 3890 | -55 |
| Alloy 7 | 3890 | -50 |
| Alloy 8 | 3890 | -45 |
| Alloy 9 | 3890 | -40 |
| Alloy 10 | 3890 | -35 |
| Alloy 11 | 3890 | -30 |

$$\frac{\partial u}{\partial t} + u \cdot \nabla u = -\frac{\nabla p}{\rho} - \frac{\nu}{K} u + \nu \nabla^2 u + g \qquad (2)$$

$$\nabla \cdot u_E = 0 \qquad (3)$$

$$\frac{\partial T}{\partial t} + \nabla \cdot T u_E = \frac{\nabla \cdot q}{\rho c} + \frac{H}{c} \frac{\partial f_S}{\partial t} \qquad (4)$$

$$\frac{\partial C_{L,i}}{\partial t} + \nabla \cdot C_{L,i} u_E = D_{L,i} \nabla^2 C_{L,i} + \frac{\partial C_{L,i}}{\partial f_S} \frac{\partial f_S}{\partial t} \tag{5}$$

Eqs. (2), (3), (4) and (5) are the momentum equation, the equation of continuity, the energy equation, and the solute conservation equation in liquid, respectively. $t$ is the time, $u$ is the flow velocity, $u_E$ is the empty tower velocity, $\rho$ is the density of molten alloy, $p$ is the pressure, $\nu$ is the kinetic viscosity, $K$ is the permeability, $g$ is the gravity vector, $T$ is the temperature, $c$ is the specific heat, $q$ is the vector of the heat flux, $H$ is the latent heat in fusion, $f_S$ is the solid fraction, and $C_{L,i}$ and $D_{L,i}$ are the concentration and diffusion coefficients of a $i$-th solute element in liquid. Following equation was used to calculate density of molten alloy.

$$\rho = \frac{\sum_{i=1}^{N} M_i X_i}{\sum_{i=1}^{N} V_i X_i} = \frac{1}{\sum_{i=1}^{N} \frac{V_i}{M_i} \phi_i} \tag{6}$$

Here $V_i$, $M_i$, $X_i$ and $\phi_i$ are molar volume, molar mass, molar fraction and mass fraction, respectively. The second term on the right hand side in the solute conservation equation, *i.e.* the derivative of the liquid concentration with respect to the solid fraction, shows the micro segregation behavior. Details of the numerical method used in this work can be found in Ref. [7]

### 2.2. *Evaluation of Formation Condition of inverted-V Segregation*

Estimation of the formation condition of inverted-V segregation was based on Suzuki's criterion. In Suzuki's criterion, inverted-V segregation forms in the region where Eq. (7) is satisfied.

$$R^{1.1} V < C \tag{7}$$

Where, $R$, $V$ and $C$ denote the cooling rate [K/s], the solidification velocity [m/s], and the critical value of the inverted-V segregation formation [$K^{1.1}m/s^{2.1}$], respectively. The critical value changes according to the alloy system. As mentioned above, the value strongly depends on the density difference between the bulk liquid and the interdendritic concentrated liquid. In this work, the criterion expressed as Eq. (8) was used to simplify the description of formation conditions.

$$I = RV < S \tag{8}$$

Where, $I$ and $S$ are the segregation index and critical value [$Km/s^2$] used in this work. The cooling rate and solidification velocity were evaluated at the temperature where the solid fraction reaches 0.3, the "solid fraction" is from the solidification pass of the initial composition and not the physical value defined at each point during the macrosegregation simulation. This segregation index, $I$, was computed along the centerline at half height of the ingot. Then, the segregation index, $I$, at the position where the first inverted-V segregation arises was regarded as the critical value, $S$.

## 3. Results

### 3.1. *Relationship between segregation formation and density difference*

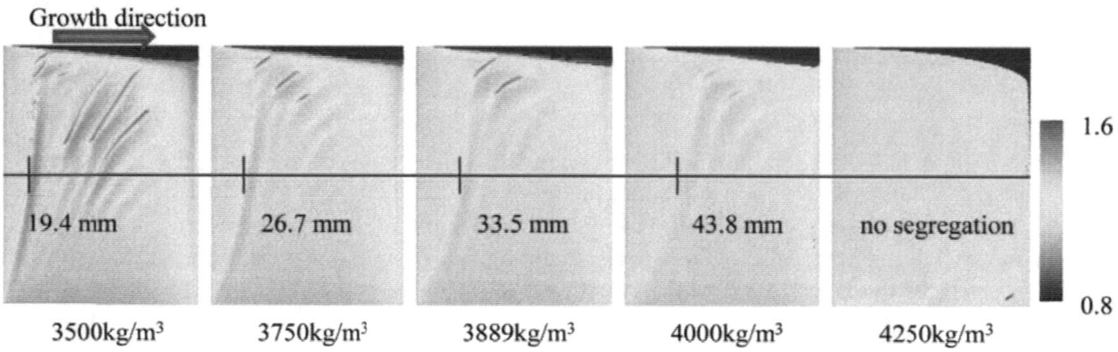


Fig. 3 Distributions of segregation ratio of added element X at fully solidified with various densities

Table 3 Relation between density of added element and critical value

| | $\rho x$ [kg/m³] | $\Delta\rho$ [kg/m³] | formation position [mm] | cooling rate [K/s] | solidification velocity [m/s] | critical value [Km/s²] |
|---|---|---|---|---|---|---|
| Alloy 1 | 3500 | -6.76 | 19.4 | 0.718 | 1.235 | 0.886 |
| Alloy 2 | 3750 | -5.87 | 26.7 | 0.514 | 1.219 | 0.627 |
| Alloy 3 | 3889 | -5.42 | 33.5 | 0.458 | 1.212 | 0.563 |
| Alloy 4 | 4000 | -5.08 | 43.8 | 0.394 | 1.121 | 0.442 |

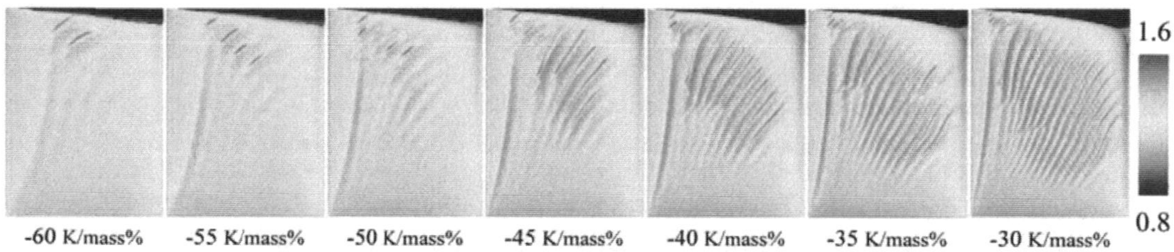


Fig. 4 Distributions of segregation ratio of added element X at fully solidified with various liquidus slopes

Table 4 relation between liquidus slope and critical value

| | liquidus slope [K/mass%] | formation position [mm] | cooling rate [K/s] | solidification velocity [m/s] | critical value [Km/s²] |
|---|---|---|---|---|---|
| Alloy 5 | -60 | 33.5 | 0.458 | 1.212 | 0.563 |
| Alloy 6 | -55 | 27.6 | 0.488 | 1.198 | 0.585 |
| Alloy 7 | -50 | 22.4 | 0.499 | 1.278 | 0.638 |
| Alloy 8 | -45 | 19.4 | 0.564 | 1.257 | 0.709 |
| Alloy 9 | -40 | 17.2 | 0.635 | 1.177 | 0.747 |
| Alloy 10 | -35 | 14.7 | 0.645 | 1.221 | 0.788 |
| Alloy 11 | -30 | 13.9 | 0.633 | 1.272 | 0.804 |

It is known that the critical value in Eq. (7) depends on the density difference between the bulk liquid and the segregated liquid and that the inverted-V segregation is more likely to occur with an increase in the density difference [6]. Numerical analyses with various alloys where the density of the added element is different were performed to validate the simulation model that we used.

The numerical results and the distributions of element X concentration are shown in Fig. 3. The position of the inverted-V segregation at half height moves toward the growth direction side with an increase in the density of the added element and the streak that arises at the bottom of the ingot doesn't

IOP Conf. Series: Materials Science and Engineering **143** (2016) 012030 doi:10.1088/1757-899X/143/1/012030

reach half height with the density of 4250 kg/m$^3$. This streak shifts to the growth direction side which means the segregation formation occurs at a lower cooling rate position. Thus, the simulation model can replicate the well-known phenomena where the inverted-V segregation formation is inhibited when the density difference between the solvent, Fe, and the added element, X, becomes small. The critical values computed by these simulations are shown in Table 3. In Table 3, $\rho_X$ and $\Delta\rho$ denote the density of the added element, X, and the density difference between the bulk liquid and the segregated liquid at solid fraction 0.3, respectively. In estimation of density difference $\Delta\rho$, scheil equation was used to compute liquid concentration.

Also, from these critical values we could confirm the tendency that the numerical results used in the present model are consistent with the knowledge that segregation tends to form with larger density differences.

### *3.2. Relationship between segregation formation and liquidus slope*

Fig. 4 and Table 4 show the computed distributions of element X concentration and estimated critical values, respectively. According to Fig. 4 and Table 3, the points at the intersection of the streak with the half-height line shift to the left wall (cold wall) as the absolute value of the liquidus slope decreases, and the critical value of segregation formation also decreases. Furthermore, the distributions show that the alloy with lower liquidus slope forms more channels and more highly concentrated streaks. Thus, the liquidus slope affects not only the formation conditions but also the concentration and numbers of segregation.

To assess the effect of the liquidus slope, the evolution equation of the solid fraction is examined in detail. The evolution equation of the solid fraction in this simulation is expressed as Eq. (9).

$$-u\cdot\nabla T+\alpha\nabla^2 T+\frac{H}{c}\frac{\partial f_S}{\partial t}=m\left(-u\cdot\nabla C_L+D_L\nabla^2 C_L+\frac{\partial C_L}{\partial f_S}\frac{\partial f_S}{\partial t}\right) \quad (9)$$

Here, $\alpha$ and $m$ are the thermal diffusivity and the liquidus slope. In binary alloy system, it is assumed that Eq. (10) was satisfied in mushy zone. When it is assumed that the local equilibrium condition is satisfied in the mushy zone, the physical temperature is equal to the liquidus temperature and the liquid concentration can be uniquely determined.

$$T=T_L\rightarrow C_L=\frac{T_L-T_M}{m} \quad (10)$$

Eq. (9) can be rewritten as Eq. (11) by Eq. (10).

$$\frac{\partial f_S}{\partial t}=\frac{(\alpha-D_L)\nabla^2 T}{m\dfrac{\partial C_L}{\partial f_S}-\dfrac{H}{c}} \quad (11)$$

Eq. (11) shows that the advection term affects variation of the solid fraction indirectly in mushy zone where local equilibrium is satisfied. Generally, flow direction in the mushy zone with lateral solidification is very close to the isothermal line since a large pressure gradient is necessary to surmount the flow resistance caused by permeability and produce a flow perpendicular to the isothermal line. Hence, the temperature and liquid concentration gradient vector is perpendicular to the flow direction and the advection term that is a dot product of the gradient vector and flow direction does not have a strong effect on the variation of these physical values. According to Eq. (10) and the relation between the gradient vector and the flow direction, the diffusion term in the temperature and

Table 5 critical values and characteristics of Ni-based Alloys

| | critical value [K mm/min$^2$] | density difference [kg/m$^3$] | liquidus temperature [K] | solidus temperature [K] | solidification temperature range[K] |
|---|---|---|---|---|---|
| Alloy A1 | 2.66 | 12.9 | 1673 | 1589 | 84 |
| Alloy A2 | 4.14 | 13.1 | 1668 | 1562 | 106 |
| Alloy B1 | 2.88 | 16.0 | 1629 | 1554 | 75 |
| Alloy B2 | 5.54 | 17.6 | 1662 | 1625 | 37 |
| Alloy C1 | 3.63 | 22.8 | 1683 | 1578 | 105 |
| Alloy C2 | 7.17 | 23.0 | 1621 | 1585 | 36 |

liquid concentration field is more significant than the advection term as long as stable solidification continues. Even though highly concentrated liquid flows from one region to another in the mushy zone, the temperature of the concentrated liquid is lower than that in the inflow region and the concentrated liquid cannot immediately melt the solid phase in the inflow region. However, the temperature is lower and the liquid concentration is higher in this inflow region than in another region that has an equal solid fraction. With these variations of temperature and concentration, and since the diffusion coefficient of the temperature is much higher than the coefficient of liquid concentration, the temperature variation is reduced by rapid diffusion and the liquidus temperature is kept low. As a result, a difference between the temperature and the liquidus temperature arises due to the difference in the two diffusion coefficients. In other words, the temperature overtakes the liquidus temperature in the region where the concentrated liquid flows. Then, the solid fraction in the region decreases to eliminate the difference between the temperature and the liquidus temperature. By reducing the solid fraction, the liquid concentration and the temperature decrease and the liquidus temperature can be equal to the temperature. It is thought that not only thermodynamic effects of the concentrated liquid but also transportation phenomena on the liquid concentration and temperature are important, in terms of remelting by the concentrated liquid. In the transportation phenomena, the intensity of the flow of concentrated liquid depends on the flow velocity and gradient of temperature. Liquid concentration and the gradients are higher with a narrower mushy zone. The effect of the diffusion cause the difference in temperature and liquidus temperature also becomes higher in a narrower mushy zone. Therefore, the inverted-V segregation tends to form in alloy systems that have lower solidification temperature ranges.

Finally, several critical values of the inverted-V segregation formation of Ni-based alloys are shown in Table 5. These values consist of three pairs. Although these three alloy pairs, Alloys A1 and A2, Alloys B1 and B2, and Alloys C1 and C2, have approximately the same density difference, each alloy has a different critical value. Excepting alloy pair A1 and A2, the critical value of one alloy in each pair has a lower solidification temperature range than that of the other. Although not all the difference in the critical values can be explained by solidification temperature range and liquidus slope, there is the possibility that solidification temperature range strongly affects the critical value and is an important factor that causes a variation in the relation between the density difference and the critical value.

## 4. Conclusion

The effect of the liquidus slope and density of added element X on the formation condition of inverted-V segregation was investigated by performing numerical simulations of lateral directional solidification with a virtual alloy Fe-X system. A criterion similar to Suzuki's criterion was used to estimate the formation condition. In investigating the effect of the density of element X, numerical simulations replicate well-known results that inverted-V segregation tends to occur when the density difference between Fe and X is larger. On the other hand, numerical simulations performed in this work suggest that inverted-V segregation also tends to form with a lower liquidus slope. Moreover, the number of inverted-V segregation and the concentration in the streaks increase with a lower liquidus slope. Examining several critical values of an actual Ni-based alloy shows that alloy pairs that have similar density difference present different critical values. Two of the pairs studied show that an alloy with a lower solidification temperature range is apt to form inverted-V segregation. One of the pairs studied did not show this effect. Thus, there is the possibility that the solidification temperature range strongly affects the formation condition of the inverted-V segregation.

## References

[1] Ramirez J C and Beckermann C, *Metall Mater. Trans.* A 2003 (34A), pp. 1525-1536
[2] Beckermann C, Gu J P, Boettinger W J, *Metall. Mater. Trans.* A 2000 (31), pp. 2545-2557
[3] Tewari S N and Tiwari R, *Metall. Mater. Trans.* A 2003 (34A), pp. 2365-2376
[4] Auburtin P, Wang T, Cockcroft S L, and Mitchell A, *Metall. Mater. Trans.* B 2000 (31B), pp. 801-811
[5] Suzuki K and Miyamoto T, *Trans. Iron Steel Inst. Jpn.* 1980, Vol. 20, pp. 377-383
[6] Yamada H, Sakuria T and Takenouchi T, *Tetsu-to-Hagané* 1989 (75), pp. 97-104
[7] Sawada T and Kajikawa K, Proceedings of the 2013 International Symposium on Liquid Metal Processing and Casting, pp. 163-170

# Liquid Steel at Low Pressure: Experimental Investigation of a Downward Water Air Flow

**Maria Thumfart**

Institute of Fluid Mechanics and Heat Transfer, Johannes Kepler University, Austria

maria.thumfart@jku.at

**Abstract.** In the continuous casting of steel controlling the steel flow rate to the mould is critical because a well-defined flow field at the mould level is essential for a good quality of the cast product. The stopper rod is a commonly used device to control this flow rate. Agglomeration of solid material near the stopper rod can lead to a reduced cross section and thus to a decreased casting speed or even total blockage ("clogging"). The mechanisms causing clogging are still not fully understood. Single phase considerations of the flow in the region of the stopper rod result in a low or even negative pressure at the smallest cross section. This can cause degassing of dissolved gases from the melt, evaporation of alloys and entrainment of air through the porous refractory material. It can be shown that the degassing process in liquid steel is taking place mainly at the stopper rod tip and its surrounding.

The steel flow around the stopper rod tip is highly turbulent. In addition refractory material has a low wettability to liquid steel. So the first step to understand the flow situation and transport phenomena which occur near the stopper is to understand the behaviour of this two phase (steel, gas) flow. To simulate the flow situation near the stopper rod tip, water experiments are conducted using a convergent divergent nozzle with three different wall materials and three different contact angles respectively. These experiments show the high impact of the wettability of the wall material on the actual flow structure at a constant gas flow rate.

## 1. Introduction

In the continuous casting process a stopper rod is a commonly used device to control the steel flow from the tundish to the mould. The stopper characteristic (throughput to stopper lift) is a critical parameter in the mould level controller design. In addition the flow field within the submerged entry nozzle (SEN) strongly influences the flow field in the mould and accordingly the quality of the cast product. The most important parameters which influence this flow field in the SEN are clogging and the structure of the two phase flow (gas - liquid).

Clogging is the result of a wide range of complex mechanisms such as chemical reactions, heat transfer, particle transport and deposition, degassing, diffusion and gas formation and transport [1], [2], which all influence one another. The chemical reactions can produce gas bubbles and inclusions. Interfaces between the liquid and the gas phase can trigger chemical reactions, collect inclusions and change global flow characteristics. Inclusions can nucleate bubbles or agglomerate at the walls and thus change their properties and shape. Accordingly the distribution of the gas phase in the SEN plays a key role.

The gas distribution and hence the flow structure depends on the source of the gas phase, the liquid flow velocity and the wetting characteristics of the bounding walls. Milan et al. [3] have shown that in case of a concurrent downward flow in a tube (8.8 mm diameter) it takes up to 130 tube diameters to

IOP Conf. Series: Materials Science and Engineering **143** (2016) 012031 doi:10.1088/1757-899X/143/1/012031

establish a developed flow regime. In addition they could show the influence of the phase distribution at the inlet on the flow structure in their whole tube with a length of 227.3 diameters. Accordingly the flow situation right at the stopper rod tip determines the phase distribution in the whole SEN.

In the SEN single phase calculation result in a negative absolute pressure in the stopper rod gap. For a more detailed discussion refer to [4] and [5]. This low pressure leads to gas nucleation at the refractory walls which defines the inlet boundary condition for the gas liquid flow in the SEN. So this paper focuses on the starting point of the two phase flow, the gap between the stopper rod tip and the bottom of the tundish.

## 2. Experimental setup

The experiments presented in this paper were done with compressed air (gas phase) and desalted water (liquid phase). Desalted water was chosen as a liquid because it is the transparent liquid with the largest surface tension. Liquid metals provide a significantly larger surface tension but they are opaque which makes the detection of the flow structure more challenging.

### *2.1. Geometry*

Figure **1** (a) shows a sketch of the water circuit. From the reservoir the water flows through a flow meter (Bio-Tech FCH-C-PA), a valve, a pump (Grundfos UPS 20-15 N 150), a flow conditioner and a transition piece to change from circular to rectangular cross section before it reaches the actual measurement section which is depicted in detail in figure **1** (b).

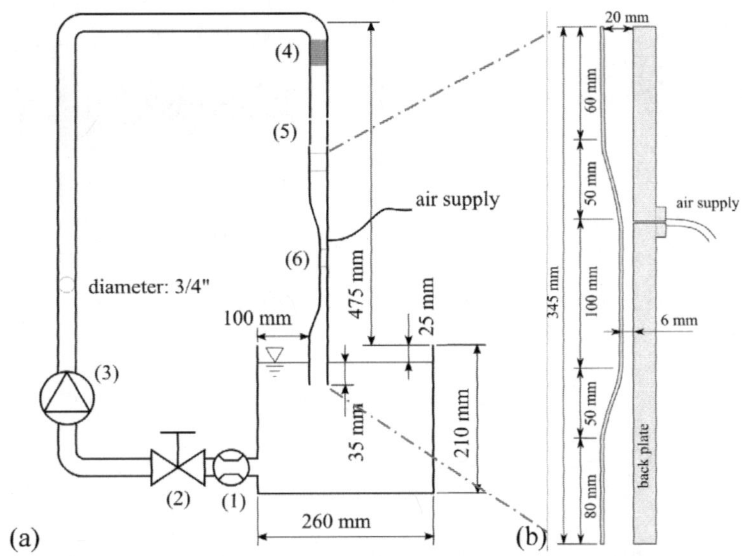


**Figure 1.** Experimental setup: (a) Water circuit: Starting in the storage reservoir the water flows through a flow meter (1) and a valve (2) to the pump (3). The right part of the circuit consists of a flow conditioner (4), followed by 100 mm of circular pipe, a transition from circular to rectangular (19 x 20 mm²) (5) and the measurement section (6). The measurement sections outflow is situated 35 mm below the free surface of the tank. (b) The measurement section consists of three Plexiglas walls and an exchangeable back plate. The channel depth is 19 mm. Its width changes from 20 mm to 6 mm and back to 20 mm. Through the hole in the back plate air is injected.

IOP Conf. Series: Materials Science and Engineering **143** (2016) 012031 doi:10.1088/1757-899X/143/1/012031

The flow conditioner consists of small pipes with a diameter of 2 mm and has a length of 50 mm. The transition from circular to rectangular has a length of 60 mm.

The measurement section (figure **1** (b)) has a constant depth of 19 mm. Its width changes from 20 mm to 6 mm and back to 20 mm. The back plate (right part in the sketch) can be exchanged. The other three sides of the channel consist of Plexiglas. Through the hole in the back plate air is injected. It is connected to a mass flow controller (Bronkhorst F-201CV -1K0) via a pipe with an inner diameter of 2 mm and a length of 1 m. The hole in the back plate has a diameter of 1 mm.

### *2.2. Contact angle*

For the experiments presented in this paper three different test plates have been used. All three of them have the same thickness. Thus the geometry of the measurement section is exactly the same. The only difference between the back plates is the surface material resulting in different contact angles (see table **1**).

**Table 1.** Static contact angles of the used wall materials.

| | Contact angle 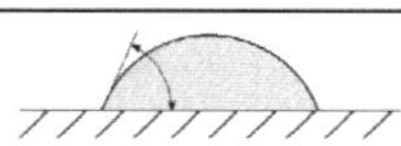 | Example picture drop |
|---|---|---|
| Plexiglas | 79° | 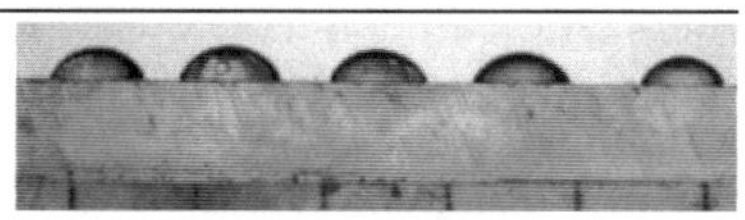 |
| PTFE oiled | 94° |  |
| PTFE clean | 105° |  |

The contact angle was measured after conducting the flow experiments using the sessile drop method. The drop diameter ranged from 2 mm to 4 mm. The values in table **1** are an average of ten drops at different positions at the surface. The contact angle between commonly used refractory material and liquid steel ranges from approximately 80° to 140° ([6], [7]). So the tested materials are well within the range of contact angles found in steel casters.

### *2.3. Measured quantities*

The measurements presented in this paper focus on the flow regime hence the special distribution of water and air in the measurement section. The measured quantities are the water flow rate $\dot{V}_w$ and the air flow rate $\dot{V}_a$.

The flow structure is recorded using a high speed camera using background light as shown in figure **2**. The camera recording rate was set to either 250 or 2000 fps depending on the dynamic of the flow. The exposure time was 1/5000 s. The resulting pictures were used to determine the flow regime.

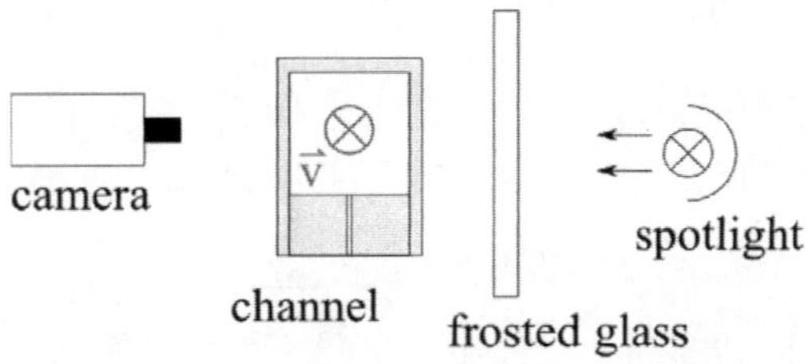


**Figure 2.** Top view of the camera arrangement. For the background light a 1000 W spotlight was used to illuminate frosted glass.

## 3. Results

Depending on the water and air flow rate ($\dot{V}_w$,$\dot{V}_a$) and on the contact angle of the back plate different flow structures were observed. The three different water flow rates used for the results presented in this paper are listed in table **2**. The Reynolds number in table **2** is calculated for the single phase flow using the hydraulic diameter of the smallest cross section of the measurement section as the length scale.

**Table 2.** Water flow rates and Reynolds numbers

| **Water flow rate $\dot{V}_w$ in l/s** | **Reynolds number** |
|---|---|
| 0.08 | 6150 |
| 0.13 | 10100 |
| 0.17 | 12800 |

### *3.1. Contact angle 79° (Plexiglas)*

Figure **4** shows a selection of the main flow structures observed using the Plexiglas back plate. Three flow regimes were observed (see figure **3**):

- Singular bubbles: From the air nozzle single bubbles detach which are significantly smaller than the channel width. So they remain almost spherical (figures **4** (a), (f)). This flow regime occurs at low air flow rates. As the water flow rate increases it can be observed that a cone shaped bubble stays attached to the air nozzle and the necking takes place slightly below the air nozzle (figures **4** (f)).
- Elongated bubbles: This flow regime occurs at a medium air flow rate. The bubble volume is too big to form a sphere in the channel so the bubbles elongate (figures **4** (b), (d), (g)). With an increased water flow rate the bubbles become longer and thinner.
- Intermediate level: In this case a large bubble is attached to the air nozzle. This bubble reaches below the water level in the reservoir. Its bottom end is horizontal. The water still completely wets the bounding walls. Small bubbles are ripped off the big bubble at the bottom and transported down the last part of the channel.

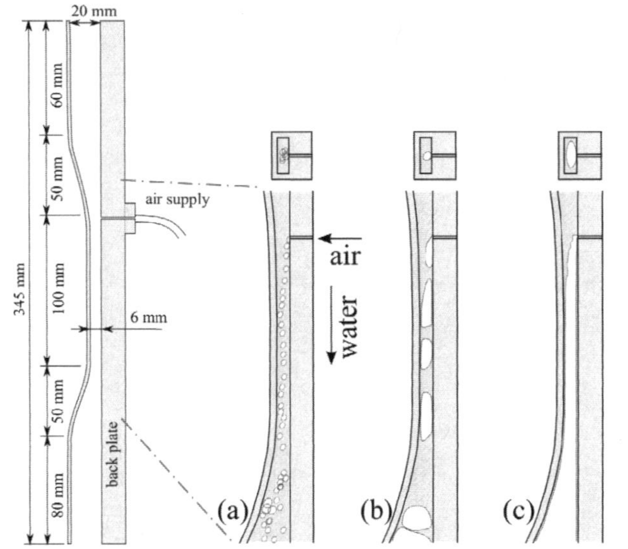


**Figure 3.** Overview of the basic flow structures observed using the Plexiglas back plate; The top shows the phase distribution in a cross section in the middle of the shown channel. The grey colour indicates water, white colour air.
(a) singular bubbles,
(b) elongated bubbles,
(c) intermediate level.

The air and water flow rates necessary for the transition between those three flow regimes are listed in table **3**. Figure **5** illustrates the regions of the different flow regimes and the relative location of the examples of figure **4**.

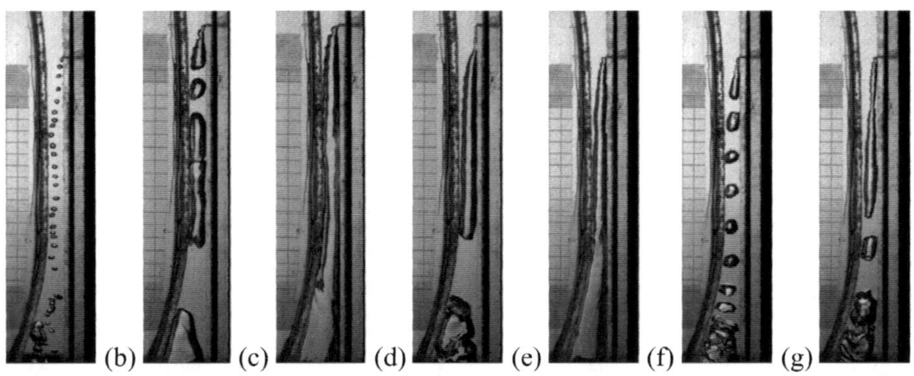

(a) (b) (c) (d) (e) (f) (g)

**Figure 4.** Results of the Plexiglas back plate.
(a) $\dot{V}_w = 0.08$ l/s, $\dot{V}_a = 0.05$ l/min, (b) $\dot{V}_w = 0.08$ l/s, $\dot{V}_a = 0.60$ l/min,
(c) $\dot{V}_w = 0.08$ l/s, $\dot{V}_a = 0.65$ l/min, (d) $\dot{V}_w = 0.13$ l/s, $\dot{V}_a = 0.80$ l/min,
(e) $\dot{V}_w = 0.13$ l/s, $\dot{V}_a = 1.00$ l/min, (f) $\dot{V}_w = 0.17$ l/s, $\dot{V}_a = 0.80$ l/min,
(g) $\dot{V}_w = 0.17$ l/s, $\dot{V}_a = 1.00$ l/min,

At a water flow rate of 0.08 l/s the bubbling regime can be observed up to an air flow rate of 0.4 l/min. At this air flow rate the bubbles do not simply detach from the nozzle. Several bubbles are formed at the nozzle, which coalesce right beneath it to form a bigger bubble. Additionally bubbles are collected in the channels diffusor section in the recirculation area. With an increase of the air flow rate these bubbles start to coalesce and form a big bubble which stays there.

At a water flow rate of 0.08 l/s and an air flow rate of 0.50 l/min there are no longer circular bubbles formed as they are constrained by the channel walls. Instead they form elongated shapes. Directly at the nozzle there stays conical bubble which grows till its lower part detaches to form a new bubble. This mode stays stable up to an air flow rate of 0.60 l/min (figure **4** (b)).

After increasing the air flow rate to 0.65 l/min the bubbles in the measurement section coalesce and form an intermediate level which is located below the storage tank water level. This single gas bubble only touches the wall at the lower left end. The rest of the walls are fully wetted.

At an increased water flow rate of 0.13 l/s the bubbles are smaller compared to the same air flow rate at a water flow rate of 0.08 l/s due to the difference in the shear forces. At this water flow rate a cone shaped bubble is attached to the air nozzle which breaks up in the middle to form a new bubble. So the bubble formation does not take place directly at the nozzle but some millimetres below it.

The transition from the formation of single bubbles to elongated bubbles is shifted to an air flow rate of 0.50 l/min. The resulting bubbles are now thinner as can be seen in figure **4** (d). This flow regime stays stable up to an air flow rate of 0.90 l/min. At an air flow rate of 0.95 l/min the transition to an intermediate level takes place (figure **4** (e)). It can be observed that especially the upper part of the gas volume is narrower compared to the water flow rate of 0.08 l/s.

Increasing the water flow rate to 0.17 l/s results in an even bigger drop shaped bubble (figure **4** (f)) attached to the air nozzle. The transition to elongated bubbles takes place at an air flow rate of 0.90 l/min. The transition to an intermediate level cannot be reached with the current setup as the air flow rate is limited to 1.00 l/min.

**Table 3.** Air flow rate $\dot{V}_a$ at transition of the flow structure using the Plexiglas back plate (contact angle 79°).

| Water flow rate in l/s | **Circular to elongated bubbles in l/min** | **Elongated bubbles to intermediate level in l/min** |
|---|---|---|
| 0.08 | 0.45 | 0.62 |
| 0.13 | 0.50 | 0.90 |
| 0.17 | 0.90 | >1.00 |

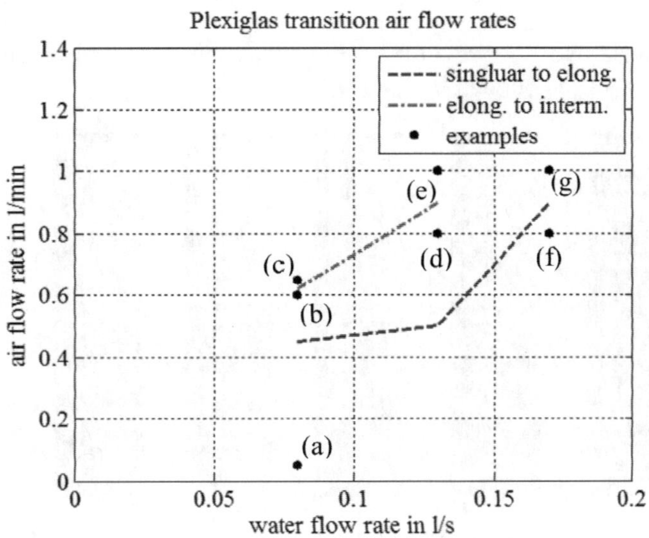


**Figure 5.** Flow chart of the Plexiglas back plate. The blue dashed line shows the transition from single bubble formation to elongated bubbles. The green dotdashed line the transition from elongated bubbles to an intermediate level. The black dots show the positions of the examples listed in figure **4**.

### *3.2. Contact angle 94° (Oil coated PTFE)*

The oiled PTFE back plate was tested using a water flow rate of 0.13 l/s and 0.17 l/s. The air flow rate was set from 0.01 l/min to 0.20 l/min. Within this range of parameters only two different flow structures were observed (see figure **6**):

- Elongated bubbles: In this case a long bubble attached to the air nozzle and to the back plate forms. Waves traveling down the bubble can be observed. They become larger as the bubble becomes longer. At some point the air water interface touches the back plate due to the surface wave, which causes the bubble to break up. The upper part stays attached to the air nozzle while the lower part accelerates and becomes shorter. While it is moving down, it stays attached to the back plate. Two examples are shown in figure **7** (a) + (b) and (d) + (e).
- Intermediate level: At higher air flow rates a large air bubble is attached to the air nozzle which reaches below the reservoir water level **7** (c). In the upper part the bubble forms a half cone which is attached to the back plate. Again the lower end of the bubble is almost horizontal and a source of small bubbles.

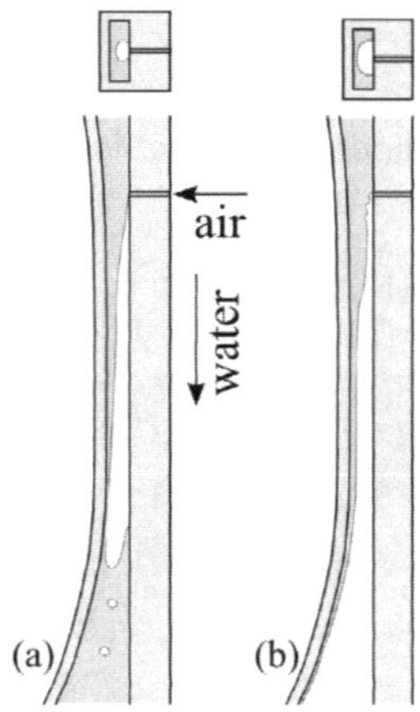


**Figure 6.** Overview of the basic flow structures observed using the oiled PTFE back plate; The top shows the phase distribution in a cross section in the middle of the shown channel. The grey colour indicates water, white colour air. (a) elongated bubbles, (b) intermediate level.

The border between those two flow regimes is illustrated in figure **8**. The air flow necessary to change to an intermediate level can be seen in table **4**.

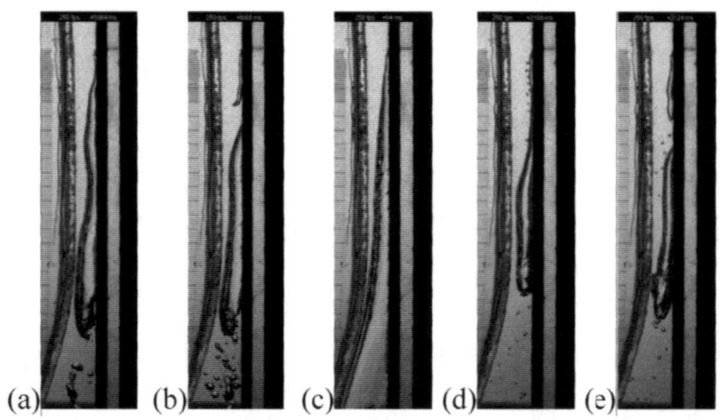


**Figure 7.** Results of the oiled PTFE back plate.
(a) $\dot{V}_w = 0.13$ l/s, $\dot{V}_a = 0.03$ l/min, t = 0 ms,
(b) $\dot{V}_w = 0.13$ l/s, $\dot{V}_a = 0.03$ l/min, t = 44 ms,
(c) $\dot{V}_w = 0.13$ l/s, $\dot{V}_a = 0.11$ l/min,
(d) $\dot{V}_w = 0.17$ l/s, $\dot{V}_a = 0.03$ l/min, t = 0 ms,
(e) $\dot{V}_w = 0.17$ l/s, $\dot{V}_a = 0.03$ l/min, t = 24 ms,

An example of elongated bubbles can be seen in figure 7 (a) and (b). Figure 7 (b) was taken 44 ms after (a). First a long bubble forms, which is attached to the air nozzle. This bubble is also attached to the wall and has an approximately conical shape. As this bubble becomes longer waves form at its surface (figure 7 (a)). When this wave touches the wall the bubble breaks apart. The upper part retreats (figure 7 (b)) while the top end of the lower part starts to accelerate down. The lower part is then transported down still attached to the wall. The upper part slowly starts growing again losing some of its air volume by forming small bubbles at its lower end. As the gas flow rate increases the length of the bubble attached to the air nozzle increases.

An intermediate level occurs if the air flow rate is increased further. For the air flow rates see table 4. As in case of the Plexiglas plate this is a very stable flow field and the intermediate level is below the water level in the storage tank. Nevertheless the air volume is now attached to the wall forming a half cone in the upper part.

An increase of the water flow rate leads to significantly smaller bubbles (figure 7 (d) and (e)). Accordingly the transition from elongated bubbles to intermediate level takes place at a higher air flow rate.

**Table 4.** Air flow rate $\dot{V}_a$ at transition of the flow structure using the oiled PTFE back plate (contact angle 94°).

| Water flow rate in l/s | **Elongated bubbles to intermediate level on l/min** |
|---|---|
| 0.13 | 0.10 |
| 0.17 | 0.20 |

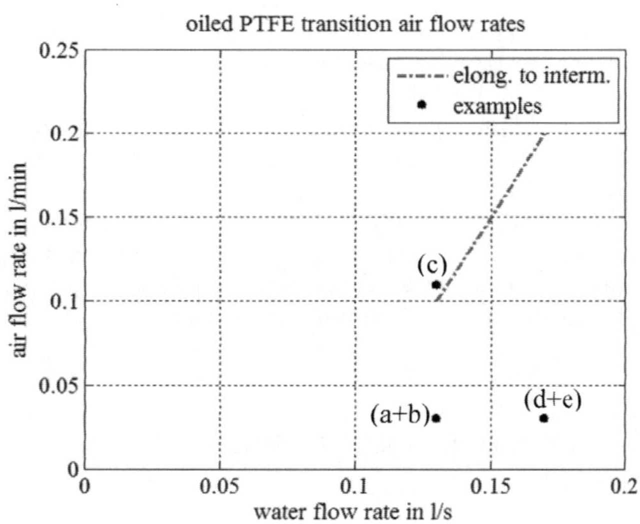


**Figure 8.** Flow chart of the oiled PTFE back plate. The green dotdashed line marks the transition from elongated bubbles to an intermediate level. The black dots show the positions of the examples listed in figure 7.

### *3.3. Contact angle 105° (PTFE)*

The clean PTFE back plate was tested using water flow rates of 0.08 l/s, 0.13 l/s and 0.17 l/s. The air flow rate was chosen between 0.01 l/min and 0.25 l/min. Within this range three different flow regimes could be observed (figure **9**):

- Singular bubbles: They are no longer spherical as in case of the Plexiglas back plate. After separating from the air nozzle they slide down the wall (figure **10** (a)). These bubbles can detach from the wall if the receding contact angle[1] reaches 0°. This results in flow structures where the bigger bubbles depart half way down the channel while the smaller bubbles slide down to feed the recirculation bubble. This big bubble in the recirculation area is in this case attached to the PTFE wall (figure **10** (b), (c) and (e)).
- Elongated bubbles: As the air flow rate increases elongated bubbles form (figure **10** (d)). This flow structure is quite similar to the case of the oiled PTFE back plate but the bubble diameter is smaller now. Accordingly the transition to the intermediate level takes place at a slightly higher air flow rate compared to the oiled PTFE back plate.
- Intermediate level: This flow structure is quite similar to the oil coated PTFE back plate. Again the gas volume is attached to the back plate.

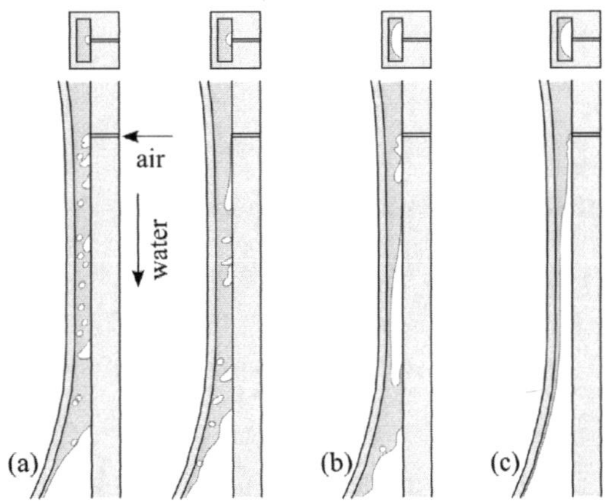


**Figure 9.** Overview of the basic flow structures observed using the clean PTFE back plate; The top shows the phase distribution in a cross section in the middle of the shown channel. The grey colour indicates water, white colour air. (a) single bubbles, (b) elongated bubbles, (c) intermediate level.

The dividing lines between the observed flow regimes are listed in table **5** and illustrated in figure **11**. Note that the air flow rates for both transitions are a factor of six smaller compared to the experiments with the Plexiglas back plate.

The single bubble regime at high water flow rates shows a long bubble attached to the air nozzle which breaks up approximately in the middle. The lower part slides down or even detaches from the back plate while the upper part stays attached to the air nozzle. This behavior is quite similar to the Plexiglas back plate but now the feeding bubble is also attached to the wall.

---

[1] The receding contact angle can be observed if the liquid phase moves away from the gas phase, for example if a solid object is pulled out of a water pool.

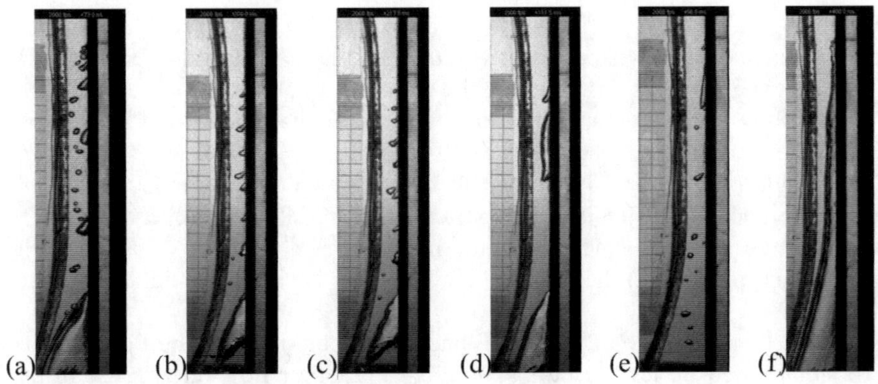


**Figure 10.** Results of the clean PTFE back plate.
(a) $\dot{V}_w = 0.08$ l/s, $\dot{V}_a = 0.05$ l/min, (b) $\dot{V}_w = 0.13$ l/s, $\dot{V}_a = 0.05$ l/min, t = 0 ms, (c) $\dot{V}_w = 0.13$ l/s, $\dot{V}_a = 0.05$ l/min, t = 8.5 ms, (d) $\dot{V}_w = 0.13$ l/s, $\dot{V}_a = 0.10$ l/min, (e) $\dot{V}_w = 0.17$ l/s, $\dot{V}_a = 0.05$ l/min, (f) $\dot{V}_w = 0.17$ l/s, $\dot{V}_a = 0.25$ l/min,

**Table 5.** Air flow rate $\dot{V}_a$ at transition of the flow structure using the PTFE back plate (contact angle 105°).

| Water flow rate in l/s | **Singular bubbles to elongated bubbles in l/min** | **Elongated bubbles to intermediate level in l/min** |
|---|---|---|
| 0.08 | 0.07 | 0.09 |
| 0.13 | 0.08 | 0.14 |
| 0.17 | 0.13 | 0.24 |

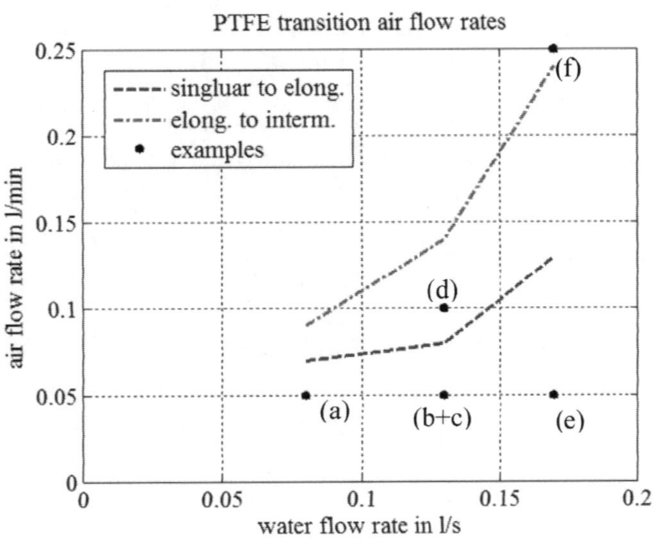


**Figure 11.** Flow chart of the PTFE back plate. The blue dashed line shows the transition from single bubble formation to elongated bubbles. The green dotdashed line the transition from elongated bubbles to an intermediate level. The black dots show the positions of the examples listed in figure **10**.

## 4. Conclusion

The experiments presented in this paper clearly show that the contact angle strongly influences the flow regime. The major differences observed are the distribution of the air phase in the cross section of the channel, the air flow rate necessary to sustain an intermediate level and the bubble size in the bubbling regime at the same air and water flow rate.

With a contact angle of 94° and 105° the air to moves down along the back plate while in better wetting conditions (contact angle 79°) it is transported in the middle of the channel.

The air flow rate necessary to generate an intermediate level depends on the wetting conditions. With the Plexiglas back plate (79°) the required air flow rate is six times higher as in the case of the PTFE back plate (105°).

The bubble size is larger in case of the PTFE back plate compared to the Plexiglas back plate. As the transition to an intermediate level occurs at a much smaller air flow rate in case of the PTFE back plate, there is a limited number of water and air flow rates where both the Plexiglas and the PTFE back plates both show bubbling.

These differences in the flow regime will clearly influence all other transport phenomena mentioned in section 1. Inclusions can only agglomerate at the refractory walls if the liquid steel wets the wall. The amount of inclusions collected by the gas liquid interface depends on its size. The bubble size distribution generated strongly depends on the contact angle which affects the bubble size distribution and the velocity field in the SEN and the mold.

## 5. Acknowledgements

This work was kindly supported by the Austrian Center of Competence in Mechatronics (ACCM), a K2-Center of the COMET/K2 program, which is aided by funds of the Austrian Republic and the Provincial Government of Upper Austria. The authors thank all involved partners for their support.

## References

[1] Kojola N, Ekerot S, Andersson M and Jönsson P G 2011 *Ironmaking and Steelmaking* **38** 1 1-11
[2] Rachers K G and Thomas B G 1995 *78th Steel-making Conference Proceedings* **78** 732–734
[3] Milan M, Borhani N and Thome J R 2013 *Int.Journalof Multiphas Flow* **56** 126-137
[4] Thumfart M and Javurek M 2014 *Proc. 8th ECCC* **8**
[5] Thumfart M and Javurek M 2015 *Steel research int.* **86** 25-32
[6] Karasangabo A 2009 *Wetting and Interfacial Reaction Investigations of Iron-Based Alloys / Alumina Systems by Means of the Sessile Drop Shape Analysis Method* Leoben 116
[7] Kawai Y 1988 *Handbook of Physico-chemical Properties at High Temperatures* Tokyo The Iron and Steel Institute of Japan Special Issue No. 41

IOP Conf. Series: Materials Science and Engineering **143** (2016) 012032 doi:10.1088/1757-899X/143/1/012032

# Simulation of macrosegregation in a large vertical continuous casting of steel

**Y Zheng[1], M Wu[1,2], A Kharicha[1,2], A Ludwig[1]**
[1]Chair of Modeling and Simulation of Metallurgical Processes, Montanuniversitaet Leoben, Austria
[2]Christian Doppler Lab for Advanced Simulation of Solidification and Melting, Montanuniversitaet Leoben, Austria

E-mail: menghuai.wu@unileoben.ac.at

**Abstract:** A three-phase mixed columnar-equiaxed solidification model considering fluid flow, heat and solute transport is applied to simulate the solidification in a vertical continuous casting. The key features of solidification phenomena in this process, such as evolution of columnar phase, evolution and floatation/sedimentation of equiaxed crystals, thermal solutal convection of the melt and the flow caused by crystal sedimentation, development of as-cast structure, the columnar-to-equiaxed transition (CET), and formation of macrosegregation, are simulated. It is predicted that there is an equiaxed zone in the central part of the strand, and the rest section is filled with columnar phase (or dominant with columnar phase). A relatively strong negative segregation in the equiaxed zone and a mostly neutral concentration in the columnar region are found. Near the CET, there is a so-called middle radius positive segregation band. Formation mechanisms of this segregation pattern are discussed.

## 1. Introduction

Vertical continuous casting technique is recently applied to produce large round steel strands and to replace some of the conventional ingot castings [1-3]. As shown in Figure 1 (a), the molten steel is conducted into the mold through a submerged entry nozzle (SEN) and the melt solidifies against the water-cooled copper mold. Below the copper mold, water and air mist sprays cool the strand continuously to maintain solidification of the melt core until it becomes fully solid. One problem is that the as-cast product is prone to macrosegregation, when the diameter of the strand increases. A special segregation profile across the section is developed, which differs from those of conventional continuously cast round products (small diameter) and conventional strands. No simple experience can be referred to control the macrosegregation directly, and experimental investigation would be extremely costly. Therefore, a numerical study of the flow phenomena and the formation mechanisms of macrosegregation during solidification is performed. In present paper, a three phase mixed columnar-equiaxed solidification model by Wu [4] is applied to analyze flow, solidification, columnar equiaxed phase transfer and evolution of macrosegregation in the vertical continuous casting. The aim of this paper is to simulate the as-cast structure including macrosegregation.

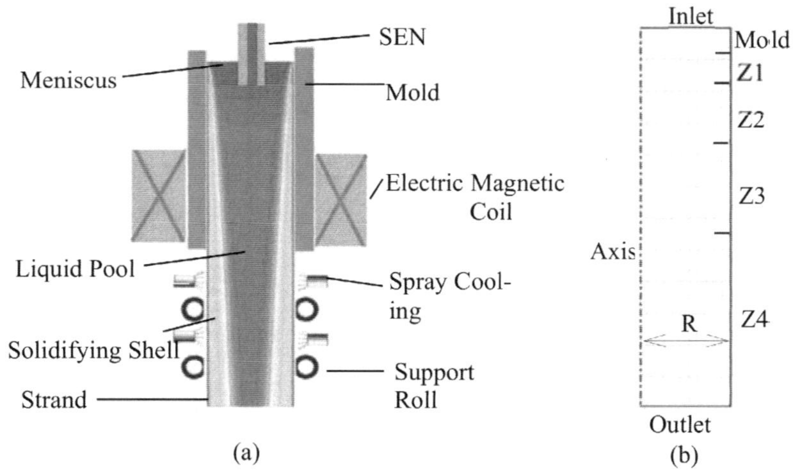


**Figure 1.** Schematic of vertical steel continuous casting process (a) and the calculation domain in 2D axial symmetry (b).

## 2. Numerical Model

A volume averaged Eulerian-Eulerian three phase mixed columnar-equiaxed solidification model is applied. Detailed descriptions of the model can be found in previous studies [4-7]. In the following a brief description and further simplifications of the numerical model are given:

a) Three phases are considered: primary liquid melt ($\ell$), equiaxed phase (e), columnar phase (c). Ideal crystal morphologies are assumed: spheres for equiaxed (globular) grains, and cylinders for columnar (cellular) dendrite trunks.
b) Both liquid and equiaxed phases are moving phases for which the corresponding velocity fields ($\vec{u}_\ell$) and ($\vec{u}_e$) are solved. The motion of columnar phase is set equal to the casting speed ($\vec{u}_{pull}$).
c) Enthalpy equations for all 3 phases are solved. A large inter-phase volume heat transfer coefficient between the phases is applied to balance the temperatures among the phases. Therefore only one temperature field is necessarily used to represent all three phases.
d) Volume-averaged concentrations for each phases ($c_\ell$, $c_e$, $c_c$) are solved by global species conservation equations. To evaluate macrosegregation, a mixture concentration is defined: $c_{mix} = (c_\ell \rho_\ell f_\ell + c_e \rho_e f_e + c_c \rho_c f_c)/(\rho_\ell f_\ell + \rho_e f_e + \rho_c f_c)$. Or it is sometimes evaluated with a so-called macrosegregation index: $\Gamma = 100 \times (c_{mix} - c_0)/c_0$.
e) A constant number density ($n_e$) of equiaxed crystals, predefined to mimic the origin of crystals by electro-magnetic stirring, is assumed to enter the domain from the inlet.
f) Growth of equiaxed crystals and columnar crystals are explicitly calculated according to diffusion-governed growth kinetics. Interactions between equiaxed crystals and columnar crystals are considered. Therefore, the columnar-to-equiaxed transition (CET) can be predicted.
g) No shrinkage porosity is considered. The Boussinesq approach is employed to model thermo-solutal convection, grain sedimentation, and sedimentation-induced melt convection.

Due to the axis symmetry, the calculation domain is set up in Figure 1(b). The influence of the submerged entry nozzle (SEN) is ignored and the electromagnetic force is not taken into account. The liquid melt with superheat is assumed to be conducted into the mold continuously from the meniscus with constant velocity, and to be cooled down by the primary cooling (mold) and second cooling zone (from Z1 to Z4). Gravity acts in the axial direction. The model domain is discretized by a mesh of $20 \times 40$ mm$^2$. The physical properties and boundary parameters are listed in Table 1.

The model is realized numerically with a control-volume based finite difference method through the ANSYS FLUENT software version 14.5, run on a Cluster paralleled with 12 cores. Both the liquid and solid share a single pressure field $p$. The pressure correction equation is obtained from the sum of the normalized mass continuity equations using an extended SIMPLE algorithm [8]. Transient calculations were performed in order to estimate steady-state. For each time step, 60 iterations were adopted to decrease the normalized residual of concentration, flow quantities and pressure below $10^{-4}$ and enthalpy below $10^{-6}$. It took about 15 days for a single simulation.

**Table 1.** Simulation settings and material properties.

| Contents | | Value |
|---|---|---|
| Steel composition (wt.% C) | | 0.432 |
| Latent heat (J kg$^{-1}$) | | $2.71\times10^{5}$ |
| Heat conductivity (W m$^{-1}$ k$^{-1}$) | | 34.0 |
| Specific heat (J kg$^{-1}$ k$^{-1}$) | | 500.0 |
| Initial equiaxed number density (m$^{-3}$) | | $1.0\times10^{8}$ |
| Density (kg m$^{-3}$) | | 6990.0 |
| Density diff. between liquid and equiaxed (kg m$^{-3}$) | | 150.0 |
| Viscosity (kg m$^{-1}$ s$^{-1}$) | | $4.2\times10^{-3}$ |
| Strand diameter (m) | | 0.6 |
| Strand length (m) | | 20.0 |
| Casting speed (m s$^{-1}$) | | $3.3\times10^{-3}$ |
| Pouring temperature (K) | | 1795.0 |
| Mold | Mold length (m) | 0.7 |
| | Heat transfer coefficient (W m$^{-2}$ K$^{-1}$) | 1000.0 |
| Z1 | Length (m) | 0.53 |
| | Heat transfer coefficient (W m$^{-2}$ K$^{-1}$) | 333.0 |
| Z2 | Length (m) | 1.52 |
| | Heat transfer coefficient (W m$^{-2}$ K$^{-1}$) | 195.0 |
| Z3 | Length (m) | 2.65 |
| | Heat transfer coefficient (W m$^{-2}$ K$^{-1}$) | 111.0 |
| Z4 | Length (m) | 14.6 |
| | Heat transfer coefficient (W m$^{-2}$ K$^{-1}$) | 10.0 |

## 3. Results

The results analyzed below include distributions of the temperature ($T$), liquid volume fraction ($f_\ell$), equiaxed volume fraction ($f_e$), columnar volume fraction ($f_c$), velocity of liquid melt ($\vec{u}_\ell$), velocity of equiaxed crystals ($\vec{u}_e$) and macrosegregation index ($\Gamma$). Here only the steady-state results are evaluated.

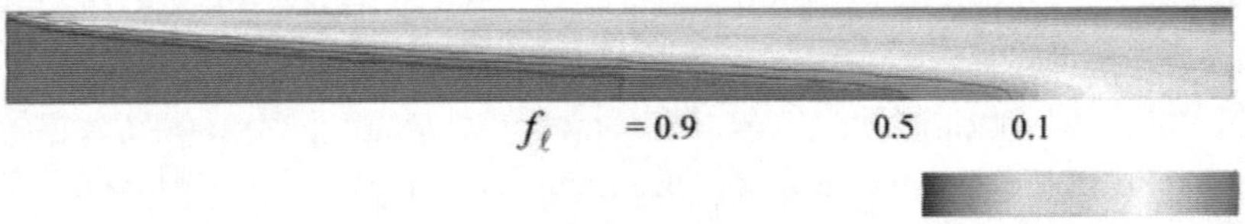


**Figure 2.** Temperature distribution overlapped with the liquid volume fraction isolines ($f_\ell$ = 0.1, 0.5, and 0.9). All of contours are scaled by a factor of 1/4 in the longitudinal direction.

### *3.1 Temperature distribution*

The calculated temperature field (Figure 2) is quite symmetrical with a relatively high temperature in the upper center of the strand due to the superheat (25K) of the steel melt, a lower temperature in the outer part of the strand because of the heat extraction on the strand surface thorough water and air cooling. The pool depth, the so-called metallurgical length, is mainly governed by the thermal field. Cooling curves along the centerline and strand surface are plotted in Figure 3. The cooling curve along the strand surface clearly indicates different cooling zones from those discontinuity points. The temperature along the stand surface is much lower than that along the center. No obvious change in $T$ occurs along the axis initially from the meniscus until reaching a turning point, at which the cooling rate increases dramatically. This turning point indicates the end of solidification at the casting center. The positon of metallurgy length is about 16 m.

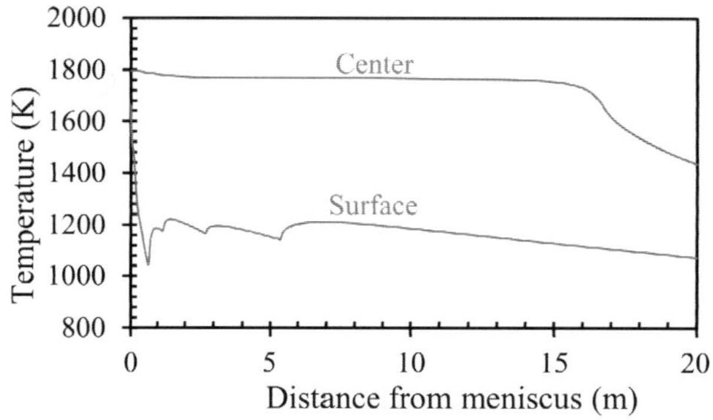


**Figure 3.** Temperature profiles along centerline and casting surface.

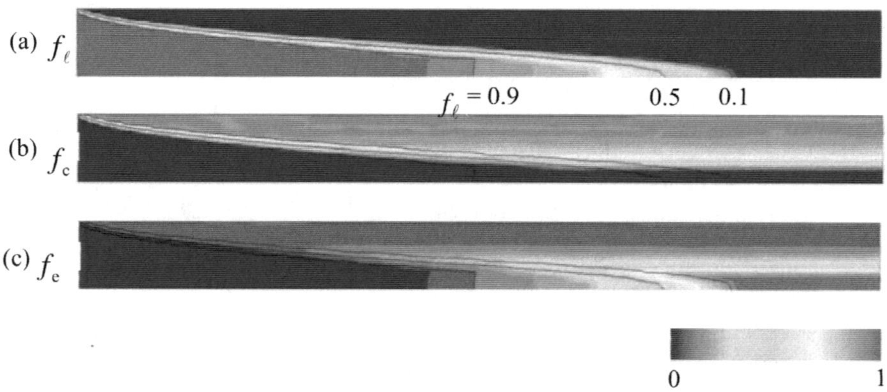


**Figure 4.** Phase distributions overlapped with isolines ($f_\ell$ =0.1, 0.5 and 0.9): (a) liquid volume fraction $f_\ell$; (b) columnar volume fraction $f_c$; (c) equiaxed volume fraction $f_e$.

*3.2 As-cast structure and columnar-to-equiaxed phase transition (CET)*

The as-cast structure of the strand is described by the distributions of the three phases, which are shown in Figure 4. The liquid phase dominates the upper center part of the domain. The rest of the domain is occupied by the two solid phases: the columnar phase in the outer part and the equiaxed phase in the center. Growth of equiaxed crystals and columnar crystals are explicitly calculated according to diffusion-governed growth kinetics. Interaction between equiaxed and columnar phase is considered. When the volume fraction of the equiaxed phase ahead of the columnar tip front is bigger than 0.49, the progress of the columnar tip front is terminated mechanically by the so-called hard block mechanism. Moreover, solutal blocking, which refers to the growth of columnar being stopped by the solute-enriched melt, is also introduced [9]. Therefore, the columnar-to-equiaxed transition (CET) is predicted.

The phase distribution along the centerline and in radial direction at the outlet are analyzed qualitatively (Figure 5). Along the centerline, $f_\ell$ decreases while $f_e$ increases, and both phases add up to 1, giving no space for the columnar phase, Figure 5(a). The columnar zone is predicted in the outer radial region. If we define the position of CET (columnar-to-equiaxed transition) at $f_e = 0.5$, the equiaxed zone takes 37% of the strand radius. That is ca. 13% of whole casting cross section.

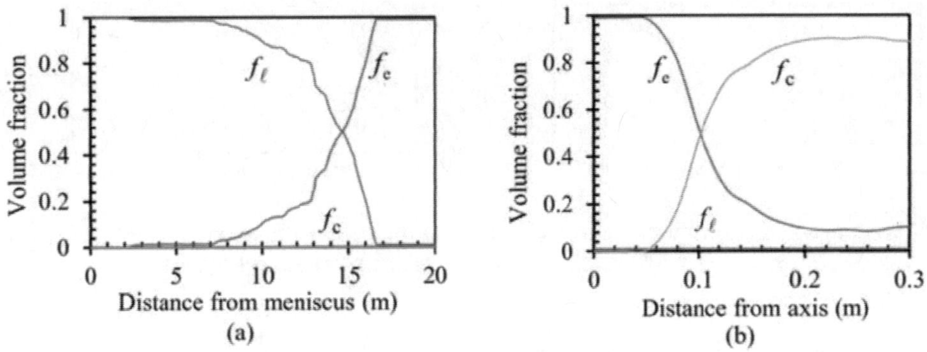


**Figure 5.** Phase distribution (a) along the centerline and (b) along the radial direction at the outlet.

*3.3 Macrosegregation*

Figure 6 shows the contour of the macrosegregation index, $\Gamma$. In the mold region near the meniscus, $\Gamma$ is zero (neutral concentration in the original melt). In the liquid core region, where $f_\ell$ is still larger than 0.9, the melt has a positive value of $\Gamma$ indicating the enrichment of solute elements. When $f_\ell$ in the casting center drops to about 0.5, strong negative segregation develops. Between the two distinguish areas is the location where velocity of the equiaxed crystals becomes equal to the casting speed (no relative velocity between equiaxed and the moving strand). The rapid development of negative segregation at the transition point corresponds to the pile-up of the equiaxed crystals due to settlement. On surface of the strand a tiny negative segregation zone is predicted.

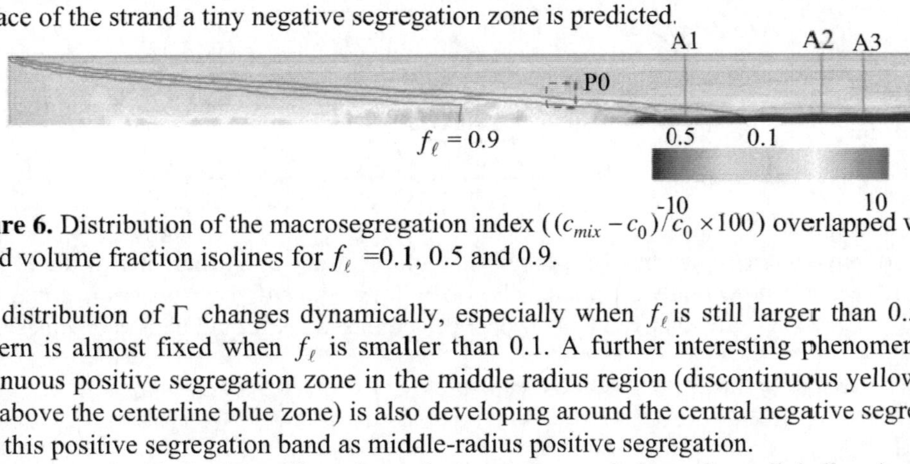


**Figure 6.** Distribution of the macrosegregation index ($(c_{mix} - c_0)/c_0 \times 100$) overlapped with the liquid volume fraction isolines for $f_\ell$ =0.1, 0.5 and 0.9.

The distribution of $\Gamma$ changes dynamically, especially when $f_\ell$ is still larger than 0.5. However, the pattern is almost fixed when $f_\ell$ is smaller than 0.1. A further interesting phenomenon is that a discontinuous positive segregation zone in the middle radius region (discontinuous yellow zones, just located above the centerline blue zone) is also developing around the central negative segregation zone. We call this positive segregation band as middle-radius positive segregation.

Macrosegregation index $\Gamma$ profiles along the centerline and along the radial direction at different positions are presented in Figure 7. Along the centerline (Figure 7(a)), $\Gamma$ near the meniscus is zero indicating the initial concentration in the original melt. Below that $\Gamma$ becomes positive and then followed by a positive-to-negative transition at about 13 m from the meniscus. The negative $\Gamma$ in the casting center reaches an extreme of -13%. After the positive-to-negative transition, the $\Gamma$ curve fluctuated slightly. The final segregation profiles at different strand sections are analyzed along the radial direction (A1, A2 and A3), as plotted in Figure 7 (b). The segregation profile along the section A3 represents the final segregation profile. Near the casting surface, a slightly negative segregation is predicted. Nearby, there is a slightly positive segregation zone. The most significant segregation zones are the mid-radius positive segregation and the center negative segregation. The positive segregation index at the mid-radius region (ca. 80 mm from axis) reaches the extreme of about 3%.

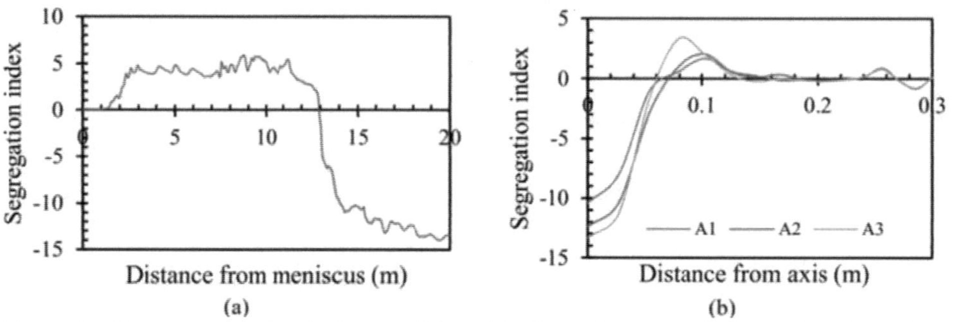


**Figure 7.** Macrosegregation index profiles: (a) along the centerline and (b) along the radial direction at three different sections A1-A3 as marked in Figure 6.

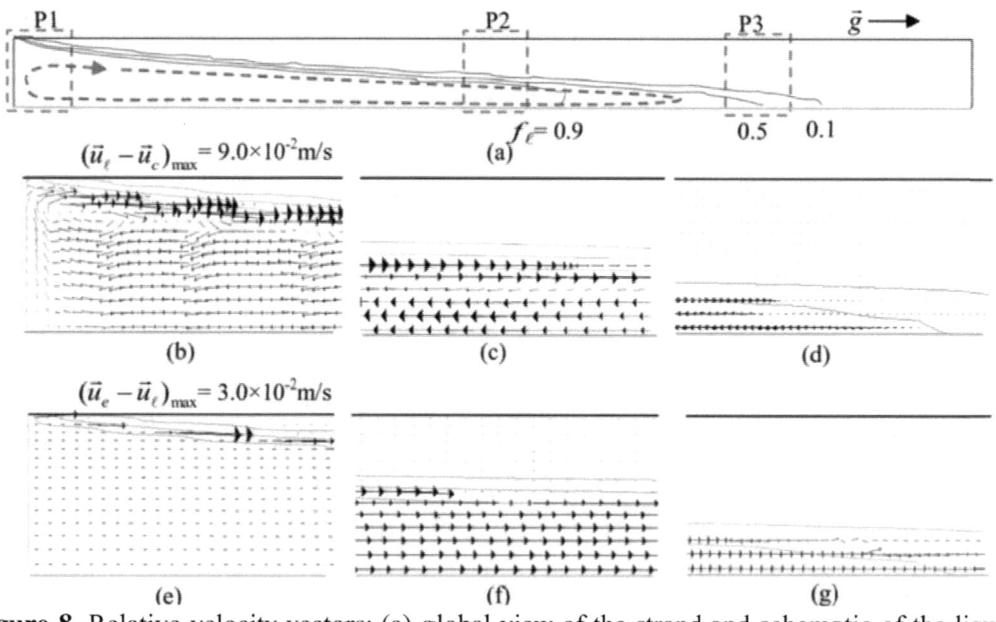


**Figure 8.** Relative velocity vectors: (a) global view of the strand and schematic of the liquid flow; (b), (c) and (d) show relative velocities between liquid and columnar phases in zones P1, P2 and P3; (d), (e), (f) show relative velocity between equiaxed and liquid phases in zones P1, P2 and P3.

## 4. Discussions

### *4.1 Macrosegregation mechanisms*

Macrosegregation occurs due to the relative movement between phases. Both thermal-solutal convection and equiaxed crystal sedimentation are dynamic and unstable. Figure 8(a) gives a global view of the strand. Figure 8 (b)-(d) are zoom-in views of the relative velocities between liquid and columnar phases in zones P1, P2 and P3, respectively. Similarly, Figure 8 (e)-(f) are zoom-in views of the relative velocities between equiaxed and liquid phases in zones P1, P2 and P3. In the upper part of the strand near the mold wall, many vortexes form. In the middle of the domain, a large circulation loop occur. The melt sinks along the columnar front and rises at the casting center. At the bottom of the pool, the rising melt transports and mixes the rejected carbon solute into the liquid pool, leading to a solute enriched area in the upper part of the domain (Figure 8 (b)-(d)). The relative equiaxed-liquid velocity is always downwards due to the density difference between liquid and equiaxed. Therefore the equiaxed phase would sediment and pile up at the bottom of the liquid pool, leading to a negative segregated center (Figure 8 (e)-(f)).

The relative motion between phases is also responsible for other macrosegregation phenomena, such as the above-mentioned mid-radius positive macrosegregation, whose formation mechanism is demonstrated in Figure 9. The background color fluctuation in Figure 9(a) indicates the formation of the mid-radius positive macrosegregation (red color for positive). The black vectors show the motion of equiaxed relative to columnar crystals. The white vectors illustrate the motion of liquid melt relative to the columnar crystals. The formation mechanism for this mid-radius positive segregation is mainly due to the motion of equiaxed crystals, which tend to leave the mush zone and sediment downwards at the casting center. In the mush zone near the columnar tip front the space of the leaving equiaxed crystals (solute-depleted) is filled by the solute-enriched melt (Figure 9(b)). This kind of relative motion causes the local increase of the mixture concentration, *i.e.* the formation of a positive macrosegregation.

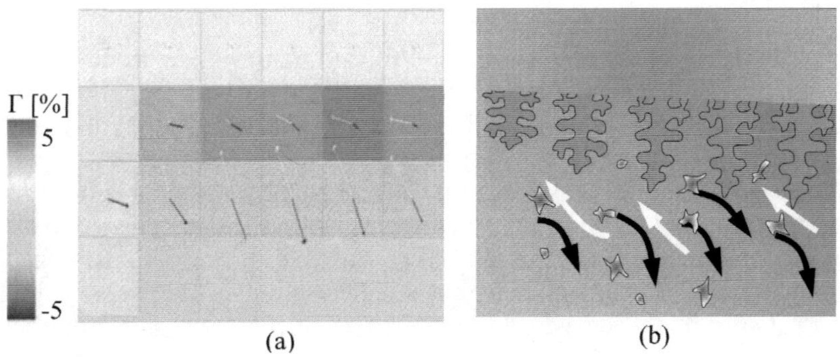


**Figure 9.** Analysis of the formation mechanism of the mid-radius positive segregation: (a) zoom-in of P0 as marked in Figure 6; (b) schematic of the relative motion between phases in a representative volume element.

*4.2 Evaluation of the segregation result*

The three-phase mixed columnar-equiaxed solidification model used to simulate macrosegregation in a vertical continuous casting is able to successfully explain the segregation pattern from the relative motion between liquid, columnar and equiaxed phases. The predicted segregation agrees qualitatively well with the industry praxis (Primetals): central negative segregation zone surrounded by a slightly positive segregation ring. It is worth mentioning that the electromagnetic stirring (EMS) is currently not modeled explicitly. The most important function of EMS is to promote the equiaxed zone in the casting center. In the current model, the effect of EMS is considered by setting a predefined initial number density of equiaxed crystals, depending on the intensity of the EMS-induced forced convection. Here the initial number density of equiaxed crystals is determined by a parameter study to fulfill the condition that the calculated equiaxed zone takes around 30~40% of the strand radius (industry experience).

One may notice that different segregation patterns were observed in continuous casting [10-16]. In conventional small bloom and billet castings without EMS, the columnar crystals grow typically towards the casting center, leading to the so-called bridging and resulting in positive macrosegregation in the casting center [10, 11]. Nowadays the utilization of EMS is popular. The EMS can promote significant amount of equiaxed crystals ahead of the columnar tip front. With the increase of the equiaxed zone in the casting center, segregation pattern can change from aforementioned positive segregation to negative segregation [12-13]. Buoro *et al.* [14] reported that the ∅ 500 mm steel strand cross section had a negative segregation in the center. These references support the current modeling result.

Some other industry results show different segregation patterns, conflicting with the current modeling result. Sun *et al.* [15] reported that there is a remarkable positive macrosegregation in the center of a round steel bloom (350 mm diameter) with F-EMS. Li *et al.* [16] reported the similar results with positive macrosegregation at the center of a 380×280 mm$^2$ steel bloom under complex EMS condition. The major difference is that they have used F-EMS just before end of solidification, which is not considered in the current vertical casting process. The F-EMS can modify the flow pattern and the motion of the settling equiaxed crystals dramatically.

Although the first attempt to simulate macrosegregation in the large vertical continuous casting gives some promising results and helps to interpret the formation mechanisms of some segregation phenomena, the current model is still subject to further improvement. The origin of equiaxed crystals by heterogeneous nucleation and fragmentation due to the effect of EMS are not properly modeled. The flow pattern influenced by SEN in the mold is also not calculated. Full 3D simulation is needed to give a more comprehensive understanding of the transport phenomena. Additionally, quantitative experimental validation is desired.

## 5. Summary

1) A first attempt to simulate the macrosegregation in a large vertical continuous casting (∅ 600 mm) by using a three-phase mixed columnar-equiaxed solidification model was made. The modeling results give valuable insight into the solidification process of the large vertical continuous casting.
2) The solidification process in such a large vertically cast strand includes: evolution of columnar dendrites, evolution and flotation/sedimentation of equiaxed crystals, thermal-solutal convection and flow caused by crystal drag, development of a columnar-to-equiaxed transition, and formation of macrosegregation.
3) The calculated segregation pattern agrees qualitatively with the industry praxis: negative segregation in the center equiaxed zone, surrounded by a middle-radius positive segregation ring in the CET region. The relative motion between different phases successfully explains the segregation pattern.
4) Further modeling effort, especially for the origin of equiaxed crystals by heterogeneous nucleation and fragmentation due to electromagnetic stirring, and the corresponding experimental evaluation are required in future.

## 6. Acknowledgements

The authors acknowledge the financial support from Austrian Research Promotion Agency (FFG) through the project of Bridge Early Stage (No. 842441), as well as the technical support from the industrial partner Primetals Technologies (former Siemens VAI).

## 7. References

[1] Cherukuri H and Johnson R 2001 *Int.J.Mech. Sci.* **43** 1243
[2] Lopez P, Jalali P N, Björkvall J, Sjöström U and Nilsson C 2014 *ISIJ Int.* **54** 342
[3] Wimmer F, Thöne H and Pennerstorfer P 2015 *Metall. Res. & Techn.* **112** 1
[4] Wu M and Ludwig A 2006 *Metall. Mater. Trans. A* **37** 1613
[5] Wu M, Li J, Ludwig A and Kharicha A 2014 *Comp. Mater. Sci.* **92** 267
[6] Wu M, Li J, Ludwig A and Kharicha A 2013 *Comp. Mater. Sci.* **79** 830
[7] Li J, Wu M, Ludwig A and Kharicha A 2014 *Int. J. of Heat and Mass Transfer* **72** 668
[8] Patankar S V 1980 *Numerical Heat Transfer and Fluid Flow* (Carlsbad: Hemisphere Pub. Corp.)
[9] Ludwig A and Wu M 2005 *Mater. Sci. and Eng. A* **413** 109
[10] Bode O, Schwerdtfeger K, Geck H and Höfer F 2008 *Ironmaking & Steelmaking* **35** 137
[11] Moore J 1980 *Iron and Steelmaker* **10** 8
[12] Ayata K, Nakata H, Miyazaki J, Nozaki T and Mori T 1985 Steelmaking Proceedings **68** 463
[13] Zhou D, Jie F, Ping W X, Junhao, Yaqing X and Zheng L 2000 *J. Mater. Sci. Techn.* **16** 273
[14] Buoro S and Romanelli G 2012 *Iron and Steel Technology* **9** 39
[15] Sun H and Li L 2015 *Ironmaking and Steelmaking* D **10**.1179/1743281215Y.0000000018
[16] Li J, Wang B, Ma Y and Cui J 2006 *Mater. Sci. Eng. A* **425** 201

IOP Conf. Series: Materials Science and Engineering **143** (2016) 012033 doi:10.1088/1757-899X/143/1/012033

# Experimental und numerical investigations on cooling efficiency of Air-Mist nozzles on steel during continuous casting

**G. Arth[1], M. Taferner[1], C. Bernhard[1], S. Michelic[2]**
[1] Chair of Ferrous Metallurgy, Montanuniversität Leoben, Franz-Josef-Straße 18, 8700 Leoben, Austria
[2] now with INTECO special melting technologies GmbH, Wienerstrasse 25, 8600 Bruck an der Mur, Austria

G. Arth, Chair of Ferrous Metallurgy, Montanuniversität Leoben, Franz-Josef-Straße 18, A-8700 Leoben, Austria.

E-mail: gregor.arth@unileoben.ac.at

**Abstract.** Cooling strategies in continuous casting of steel can vary from rapid cooling to slow cooling, mainly controlled by adjusting the amount of water sprayed onto the surface of the product. Inadequate adjustment however can lead to local surface undercooling or reheating, leading to surface and inner defects. This paper focuses on cooling efficiency of Air-Mist nozzles on casted steel and the experimental and numerical prediction of surface temperature distributions over the product width. The first part explains the determination of heat transfer coefficients (HTC) on laboratory scale, using a so called nozzle measuring stand (NMS). Based on measured water distributions and determined HTC's for air-mist nozzles using the NMS, surface temperatures are calculated by a transient 2D-model on a simple steel plate, explained in the second part of this paper. Simulations are carried out varying water impact density and spray water distribution, consequently influencing the local HTC distribution over the plate width. Furthermore, these results will be interpreted with regard to their consequence for surface and internal quality of the cast product. The results reveal the difficulty of correct adjustment of the amount of sprayed water, concurrent influencing water distribution and thus changing HTC distribution and surface temperature.

## 1. Introduction

Secondary cooling strategies of continuous casting processes can vary from rapid to slow cooling. Water or air-mist nozzles with a wide range of possible spray water distributions and/or impact densities aim on uniform heat removal during the casting process. High effort is done in research in homogeneous water distribution over a wide range of operation parameters (varying water flow rate, air pressure e.g.) [1–3], nevertheless heat transfer coefficients (HTC) can still differ over spray width, leading to uneven heat removal during casting. [4–7]
Mean heat transfer coefficients based on adjusted water impact densities [8–11] are commonly used to describe the heat removal in the secondary cooling zone in continuous casting process simulations, aiming on simplicity and calculation speed. More complex approaches separate these mean values into local mean HTC values over spray width [12–14] to reach a more detailed look on temperature

distribution below the nozzle. However, different researchers illustrate beside surface temperature parameters like spray water temperature [15], water/air ratios [16], casting speed [7,17], droplet size and velocity [5,7,9,11,18,19], and more [2,17,19–25] as important influence factors on heat removal. The demand on more detailed information on heat removal thus lead to the development of several test assemblies on laboratory scale to measure local heat transfer coefficients. [10,26–28]

## 2. Experimental equipment

A so called nozzle-measuring-stand (NMS) was developed as prototype within the frame of a K1-MET-Phase1 project in cooperation with voestalpine Stahl Donawitz GmbH, at the Chair of Ferrous Metallurgy (CoFM), Montanuniversität Leoben (MUL). The NMS is used to determine local HTC's with the option on variation of a number of nozzle- and casting parameters. Figure 1 presents a schematic drawing of the HTC measurement set-up. The NMS can be separated into 3 levels:

- Level 1contains a cylindrical steel sample, the induction heating unit and the sample support stand (positioning unit on linear axes). Maximum casting speed currently tested is 6 m/min.
- Level 2 is called the "wet-zone", where the spray nozzles are fixed in definite positions, spraying upwards for HTC measurements, spraying downwards to measure water distribution.
- Level 3 stores technical equipment like air and water supply and the cooling unit of the induction device.

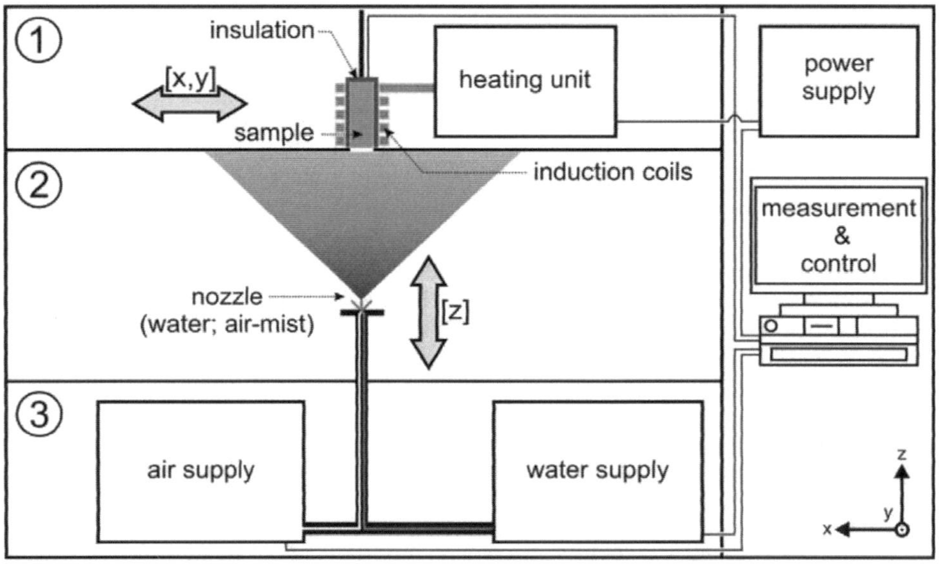


**Figure 1.** NMS build-up at the CoFM to determine local HTC.

### 2.1. Determination of local heat transfer coefficients

HTC's measurements are carried out using a cylindrical sample made of scale resisting austenitic steel (1.4841) with diameter to height ratio of 1:1.8. Three thermocouples are positioned along the center axis of the steel sample at different distances to the cooled surface (bottom of the sample), where an additional thermocouple is fixed. The lateral and the top of the steel surface are insulated to ensure one-dimensional heat flow towards the bottom during cooling.

Figure 2 displays the principle to characterize a flat fan nozzle regarding local HTC values:

i. The spray water distribution has to be known (or determined using the NMS) to identify the ideal positions for HTC measurement and to avoid incomplete nozzle characterization.
ii. The first HTC measurement typically starts at the center of the nozzle. The sample is positioned at a certain distance to the water spray to ensure no contact between water and steel during heating.

iii. Heating of the cylindrical steel sample to the aimed surface temperature ($T_{surface,max}$ = 1100°C) is done by inductive heating under normal atmosphere.

iv. When reaching constant volumetric flow of water (or water-air mixture), inductive heating is switched off and the sample is moved at constant velocity (casting speed) through the water spray ("a" in figure 2).

v. Temperatures are measured inside the steel sample during this movement, using type S or type K thermocouples, depending on the desired maximum surface temperature.

vi. After this first measurement in the center of the spray water distribution, the water (water-air mixture) is switched off and the steel sample is moved to the next position (C+1/5L in figure 2). The cycle continues at step "iii".

Local HTC are computed by inverse calculation of measured temperatures inside the steel sample.

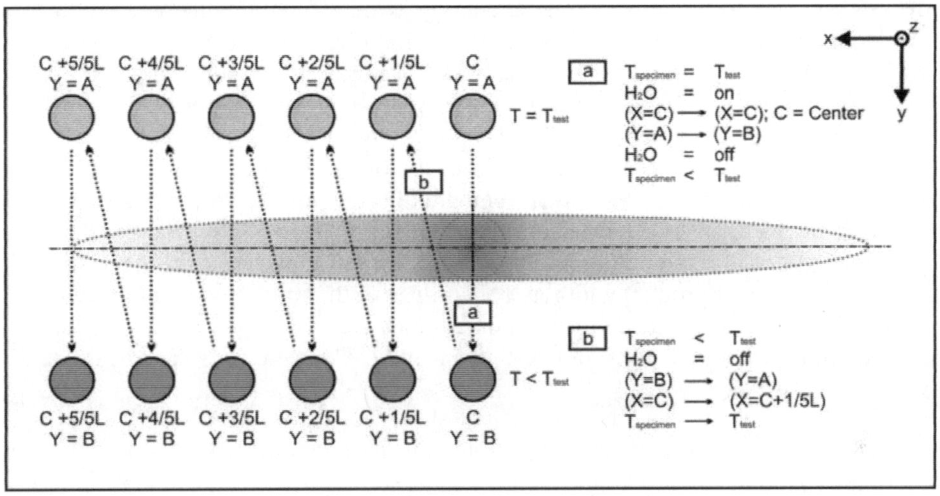


**Figure 2.** Scheme of nozzle characterization of a flat fan nozzle by the NMS; No variation of temperature, velocity or water impact density

### *2.2. Determination and results of spray water distribution*

Water distribution is likewise measured using the NMS, details concerning the measurement assembly and the nozzle results can be taken out of [6].

### *2.3. Test schedule*

Both investigated nozzles are typically used in parallel arrangement in a conventional slab casting machine. Table 1 presents the layout and mounting information; the test schedule with measurement parameters can be taken out of table 2.

HTC measurements were carried out for combinations of minimum water flow rate at minimum air pressure (further stated as "case a"), and maximum water flow rate and maximum air pressure ("case b") for each nozzle.

**Table 1.** Nozzle layout and mounting

| Nozzle | Spray Angle | Distance to surface |
|---|---|---|
| | [°] | [mm] |
| A | 90 | 396 |
| B | 60 | 346 |

**Table 2.** Testing parameters.

| Nozzle | $H_2Omax$ [l/min] | $H_2Omin$ [l/min] | $Air_{max}$ [bar] | $Air_{min}$ [bar] | $T_{surf}$. [°C] | $v_{cast}$ [m/min] |
|---|---|---|---|---|---|---|
| A | 7 | 3.0 | 2.5 | 1.1 | 900 ; 1100 | 0.8 ; 1.2 |
| B | 3.5 | 1.5 | 2.5 | 1.1 | 900 ; 1000 | 0.8 ; 1.2 |

## 3. Results

### *3.1. Simulation*

Measured local heat transfer coefficients at definite positions, for case "a" and "b" of both nozzles, were used to describe the heat removal over the spray width (X-direction). Measurements of local HTC were only done at half length of the nozzle due to the clear symmetry of the spray pattern, thus the results were mirrored for the simulation of the plate cooling. Limits to heat removal in spray length (Y-direction) and width were defined according to the water distribution, a HTC value of 200 W/m²K was adjusted for radiation at the remaining regions on the surface. Plate width was fixed as 0.95m with adiabatic boundaries. The same steel grade that is used for HTC measurements was taken as plate material; the thickness of the plate was set as 40mm with adiabatic regime in the bottom. The plate was moved through the different HTC distributions at (casting) speed of 0.8 [6] and 1.2m/min.
Simulation results are printed as contour plots with additional color bars for better imaging of the temperature distributions over product width and according to the initial temperature of the steel plate. Images of the results at a casting speed of 0.8m/min can be taken out of [6]. Details to these results are contemporary explained with the results of a casting speed of 1.2m/min for sake of comparability.

#### *3.1.1. Nozzle A:*

Using a minimum amount of water with concurrent low air pressure (case "a") leads to a uniform water distribution over roundabout 400mm spray width, see also [6]. The simulation also predicts a nearly uniform heat removal in this region, independent on 0.8 or 1.2m/min casting speed (figure 3, top).
Increasing the water impact density (case "b") shifts the distribution to a more pyramidal appearance, leading to a higher heat transfer coefficient in the center of the nozzle with decreasing values towards the edge of the spray width (figure 3, bottom). Increasing the casting speed to 1.2m/min results in a more uniform temperature distribution as for 0.8m/min.
This first example predicts a higher heat removal when increasing the water impact density, and is in good accordance to the literature. However, even if the homogeneous water distribution leads to the assumption of a uniform heat removal in the spray width (case "a"), the temperature contour plot in the top of figure 3 already reveals small differences and local temperature drops at the edges of ~10°C. These effects cannot directly be explained using only the determined spray water distribution.
In case "b" of nozzle A, the heat transport from the bottom of the plate towards the top leads to a nearly homogeneous temperature distribution on the surface of the plate, after leaving the region of water impact. This example reveals the severity to detect local maxima of heat removal during casting, due to the surface reheating by heat transport from the inner side of the product. Nevertheless, temperature gradients were present while passing the water distribution, owing to the determined heat transfer distribution over the spray width.
Decreasing the initial surface temperature of the steel plate to 910°C, and applying the measured local heat transfer coefficients at a casting speed of 1.2m/min, leads to results depicted in figure 4. Even if the water impact density was held constant during the tests, while varying only the initial surface temperature, the distribution differs to the simulations at 1110°C.

- For low but nearly uniform water impact densities (case "a"), the surface temperature distribution becomes more inhomogeneous at a casting speed of 0.8m/min [6].

- For case "b" a nearly uniform surface temperature can be achieved, in contrast to tests at 1100°C, casting with 0.8m/min, where high temperature drops were detected [6].
- At a casting speed of 1.2m/min the results at 910°C nearly mirror those of 1110°C.

Slight differences can be visualized using snapshots of the temperature distribution at the surface of the plate, depicted in figure 7 in chapter 4. Higher values of the local heat transfer coefficients can lead to temperature differences, reaching a maximum of 18°C for case "b" at $T_{surface,start}$ = 1110°C. Decreasing the initial temperature to 910°C however predicts a difference of roughly 22°C.
At low water impact densities (case "a") no considerable differences can be found, when changing the casting speed from 0.8m/min to 1.2m/min. Higher water impact densities (case "b") on the other hand results at 1.2m/min in a minor temperature difference at 1110°C, and a higher value at 910°C.

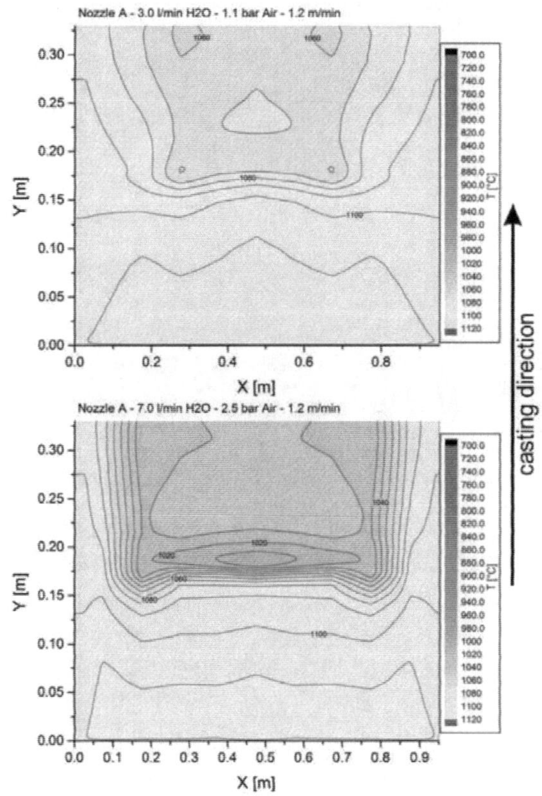


**Figure 3.** Temperature distribution for case "a" (top) and case "b" (bottom), nozzle A; v = 1.2m/min; $T_{init.}$ = 1110°C.

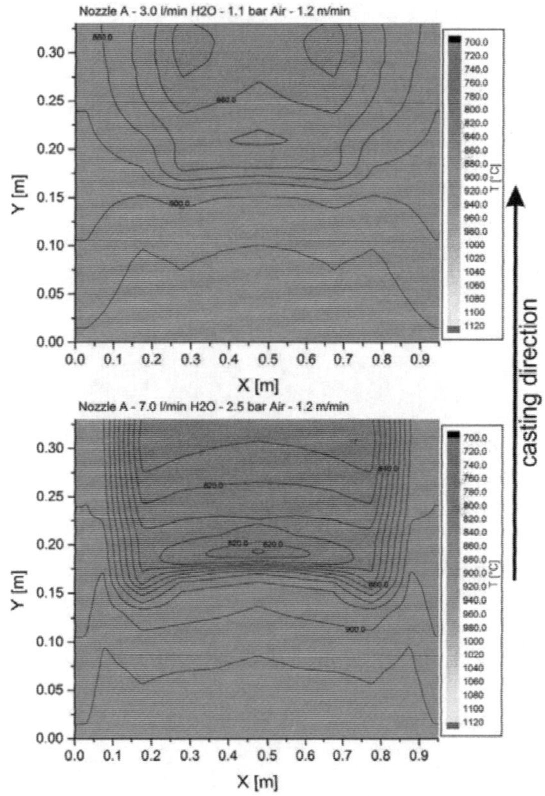


**Figure 4.** Temperature distribution for case "a" (top) and case "b" (bottom), nozzle A; v = 1.2m/min; $T_{init.}$ = 910°C.

*3.1.2. Nozzle B:*
Independent on the water impact density of nozzle B (case "a" or "b"), two local maxima are present at the edges of the spray pattern, resulting in higher local heat transfer coefficients at these positions. Figure 5 and 6 present the contour plots of the simulations, carried out for nozzle B at different temperatures and constant casting speed of 1.2m/min.
Temperature profiles at Y = 0.15m in cast direction are depicted in figure 8 for all parameter variations of nozzle B at 1.2m/min. At low water impact densities (case "a"), temperature differences already appear at the surface, reaching values of roughly 25°C. High water impact densities (case "b") have a

IOP Conf. Series: Materials Science and Engineering **143** (2016) 012033 doi:10.1088/1757-899X/143/1/012033

tremendous effect on the heat removal, located mainly on the spray pattern edges, resulting in temperature differences up to 100°C at 1.2m/min.

Heat transport towards the surface from the inside of the plate seems only to be sufficient for case "a" at 910°C and 1.2m/min, to nearly equalize the temperature distribution after leaving the area of water impact. Surface defect formation owing to thermal stresses in these areas for all other cases is thus highly probable.

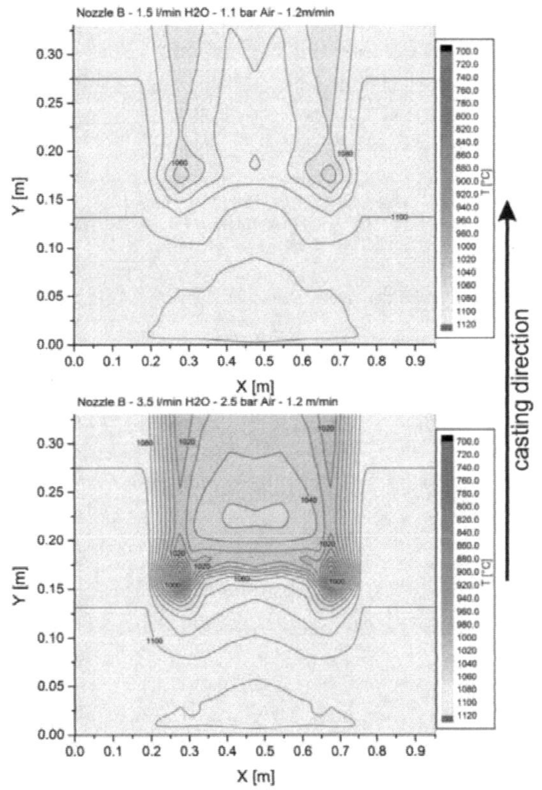


**Figure 5.** Temperature distribution for case "a" (top) and case "b" (bottom), nozzle B; v = 1.2m/min; $T_{init.}$ = 1110°C.

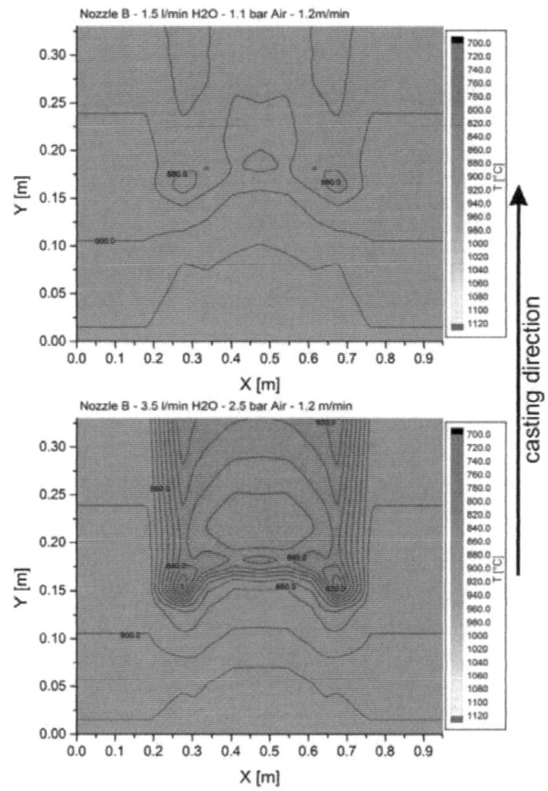


**Figure 6.** Temperature distribution for case "a" (top) and case "b" (bottom), nozzle B; v = 1.2m/min; $T_{init.}$ = 910°C.

## 4. Summary

Measurements to determine local heat transfer coefficients were carried out at the Chair of Ferrous Metallurgy, Montanuniversität Leoben, using a so called nozzle measuring stand. Spray parameters of two different nozzles as well as surface temperatures were varied; the determined local HTCs were used as input parameter for simulation. A steel plate was moved through the local HTC distribution at two different casting speeds. The results can be summarized as follows:

- Water impact density does not simply correspond to heat transfer coefficient. More detailed information about droplet size, velocity and impact distribution is required.
- The influence of casting speed on HTC seems to be dependent on water impact density.
- Uniform spray water distributions can lead to inhomogeneous heat withdrawal (case "a", nozzle A) and vice versa (case "b", nozzle A).

- However, local maxima in water impact density can result in high temperature gradients (100°C, figure 8) over spray width, increasing the probability of surface defect formation during casting.
- Detection of local undercooling of the surface may be difficult, if the heat transport from the inside of the product causes a reheating of the surface (figure 3, nozzle A, case "b"). Only the surface defects may remain, but the underlying phenomenon is no longer detectable at the end of the casting.

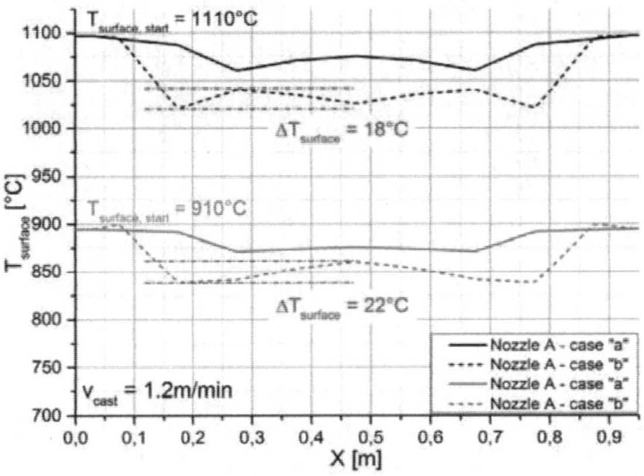


**Figure 7.** Temperature distribution nozzle A; v=1.2m/min; Y=0.17m.

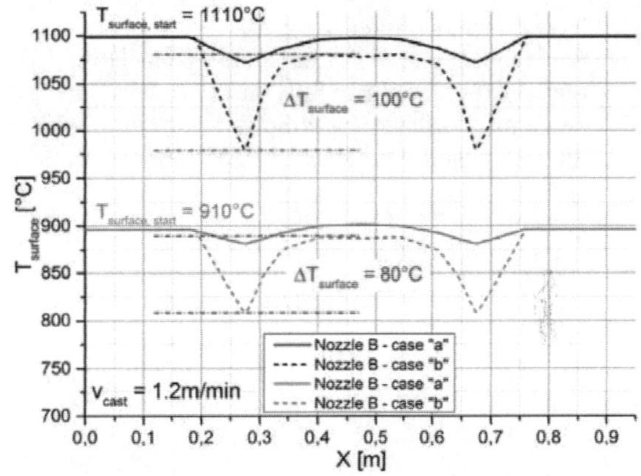


**Figure 8.** Temperature distribution nozzle B; v=1.2m/min; Y=0.15m.

## 5. Conclusion

The measurement of local heat transfer coefficients for better characterization of the heat withdrawal during casting is vital. Using a mean value based on water impact density for simulation, in order to keep calculation times low, leads consequently to information losses and makes it difficult to reveal the origins of possible defects in the casted product.

## References

[1] Pavlova A A, Otani K and Amitay M 2008 *Int. Journal of Heat and Fluid Flow* **29** 985.
[2] Frick J and Boyle R 2008 *Proc. of the 6$^{th}$ ECCC*, (Riccione).
[3] Weyer A and Frick J 2013 *MPT International* 36.
[4] Puschmann F 2003 *Experimentelle Untersuchung der Spraykühlung zur Qualitätsverbesserung durch definierte Einstellung des Wärmeübergangs.* Dissertation (Magdeburg).
[5] Puschmann F, Specht E and Schmidt J 2000 *Continuous Casting* (Weinheim, FRG: Wiley-VCH Verlag GmbH & Co. KGaA) pp 101–108.
[6] Arth G, Taferner M and Bernhard C 2015 *METEC and 2$^{nd}$ ESTAD Proc.*, (Düsseldorf).
[7] Horský J and Raudensky M 2005 *Proc. of the Metal 2005 Conf., 14 $^{th}$ Int. Metallurgical & Materials Conf.*(Hradec n/Moravicí).
[8] Schwerdtfeger K J ed 1991 *Stranggießen.* (Düsseldorf: Verlag Stahleisen mbH).
[9] Müller H and Jeschar R 1973 *Arch. Eisenhüttenwesen* **8** 589.
[10]Wendelstorf J, Wendelstorf R and Spitzer K-H 2009 *steel research international* **80**.
[11]Ramstorfer F, Roland J, Chimani C and Mörwald K 2009 *Int. Journal of Cast Metals Research* **22** 39.
[12]Reiners U, Jeschar R, Scholz R, Zebrowski D and Reichelt W 1985 *Steel Research* **56** 239.
[13]Moravec R, Horský J, Blazek K, Graham C, Fiegle S, Dombovic T and Kaurich T 2011. *METEC 2011 Proc.*, (Düsseldorf).
[14]Horský J, Ondrouskova J, Moravec R, Blazek K and Yin H 2013 *AISTech 2013 Proc.*,(Pittsburgh)
[15]Chabičovský M, Raudensky M and Hnizdil M 2012 *Proc. of the 18$^{th}$ Int. Conf. on Engineering Mechanics 2012*(Prague) 499.
[16]Bendig L *Wärmeübertragungseigenschaften verschiedener Zweistoffdüsen bei der Sprühkühlung.*

[17]Raudensky M and Horsky J 2005 *Ironmaking and Steelmaking* **32** 159.
[18]Bendig L, Raudensky M and Horsky J 1994 *Spray 94 Conf. Proc.*, (Nürnberg).
[19]Heidt V and Jeschar R 1993 *Steel Research* **64** 157.
[20]Chen J, Zhu J Q, Rogall C-M and Kopp R 1989 *Steel Research* **60** 550.
[21]Sommerhofer H and Sommerhofer P 2010 *BHM* 12.
[22]Ondrouskova J, Luks T and Horsky J 2012 *EPJ Web of Conferences* **25**.
[23]Viscorova R 2007 *Untersuchung des Wärmeübergangs bei der Spritzwasserkühlung unter Berücksichtigung des Einflusses der Verzunderung.* Dissertation (Clausthal).
[24]Hrabovský J, Chabičovský M and Horský J 2015 *Int. Journal of Mechanical, Aerospace, Industrial and Mechatronics Engineering* **9** 81.
[25]Wendelstorf R, Spitzer K-H and Wendelstorf J 2008 *Int. Journal of Heat and Mass Transfer* **51** 4892.
[26]Wendelstorf J, Spitzer K-H and Wendelstorf R 2008 *Int. Journal of Heat and Mass Transfer* **51** 4902.
[27]Reiners U, Jeschar R and Scholz R 1989 *Steel Research* **60** 442.
[28]Hnizdil M and Raudensky M 2012 *Metal 2012 Conf. Proc.* (Brno).

IOP Conf. Series: Materials Science and Engineering **143** (2016) 012034 doi:10.1088/1757-899X/143/1/012034

# On the importance of thermodynamic investigations for the re-assessment of selected ternary Fe-base systems

**P Presoly and C Bernhard**
Chair of Ferrour Metallurgy, Montanuniversitaet Leoben,
Franz-Josef-Straße 18, A-8700 Leoben, Austria

Email: peter.presoly@unileoben.ac.at

**Abstract**. Reliable thermodynamic data are essential for the design and the production of new alloying systems. Particularly, the knowledge of the high-temperature phase transformations ($T_{Liquid}$, $T_{Solid}$, $T_{Perit}$, $T_{\gamma\rightarrow\delta}$) are important for the solidification and the further processing. Investigations of selected commercial Dual-Phase, TRIP and high-Mn TWIP steels by DTA/DSC measurements show that the experimental results differ significantly from the calculation results of thermodynamic databases with respect to the phase transformation temperature and sequence. Based on these findings, it is very important to identify the defective subsystems of complex alloys in order to optimise the thermodynamic databases. In order to verify a quaternary system, e.g. the Fe-C-Si-Mn system, it is important to check the corresponding ternary subsystems. This was performed by DSC measurements of selected model alloys. By doing so, it was found that in Si- and Mn-alloyed Dual-Phase steels the thermodynamic description of the Fe-Si-Mn system is currently inadequate. This is a very important result, since all new designed steel grades for the automotive industry are based on a Fe-C-Si-Mn matrix.

## 1. Introduction

Steel, the main metallic construction material, is subjected to a constant development. In the last decades high strength and ductile steels were developed for the automotive industry. Three of these alloying concepts are Dual-Phase (DP), TRansformation Induced Plasticity (TRIP) and Twinning Induced Plasticity (TWIP) steels. These new steel grades are considered as a new generation of automotive steels due to the advantages of weight reduction, energy saving and safety performance for automobiles. All these steel grades base on the iron-carbon-manganese system with additions of ferrite former such as silicon and alumina, followed by micro-alloying elements. This means, that the alloy matrix consists of the Fe-C-Si-Al-Mn system (in different variations, often with or without Al, or with very high Mn contents up to 25 wt.-% as in TWIP steels). The main differences compared to previous steel grades are the significantly higher levels of the ferrite formers Si and Al.

Apart from research on the material and product properties of these new steels, the knowledge about the production process is particularly important. In this regard, the high-temperature phase transformation temperatures and the thermodynamic properties play an important role. As visualized in Figure 1a. four different transformation sequences (I-IV) can be distinguished in the equilibrium Fe-C system, characterize by the points $C_A$, $C_B$, and $C_C$. Special features are the two different high-temperature phases: δ- and γ-iron, showing significant differences in their density due to the diverse crystal structure (δ = face centred cubic / γ = body centred cubic) [1]. Steels which solidify according to the transformation sequence of range II are of special interest because in this case the transformation of δ-Fe to γ-Fe (i.e. the peritectic phase transition L+δ → δ+γ → γ) starts in the solid-liquid two phase region and ends in the solid. Particularly, this specific peritectic transformation range II is responsible for an increased appearance of defects (such as hot tears, surface defects, depression formation, and in the worst case breakouts) in the continuous casting process [2,3]. As many new developed steel grades are higher alloyed multi-component alloys, the strong influence of alloying elements also needs to be taken into account, as visualized in the pseudo binary phase diagram in Figure 1b. The effect of the alloying elements can be distinguished between austenite formers (like Mn, Ni, Co, Cu, N...) and ferrite formers (like Cr, Mo, W, Al, Si, Ti, V....).

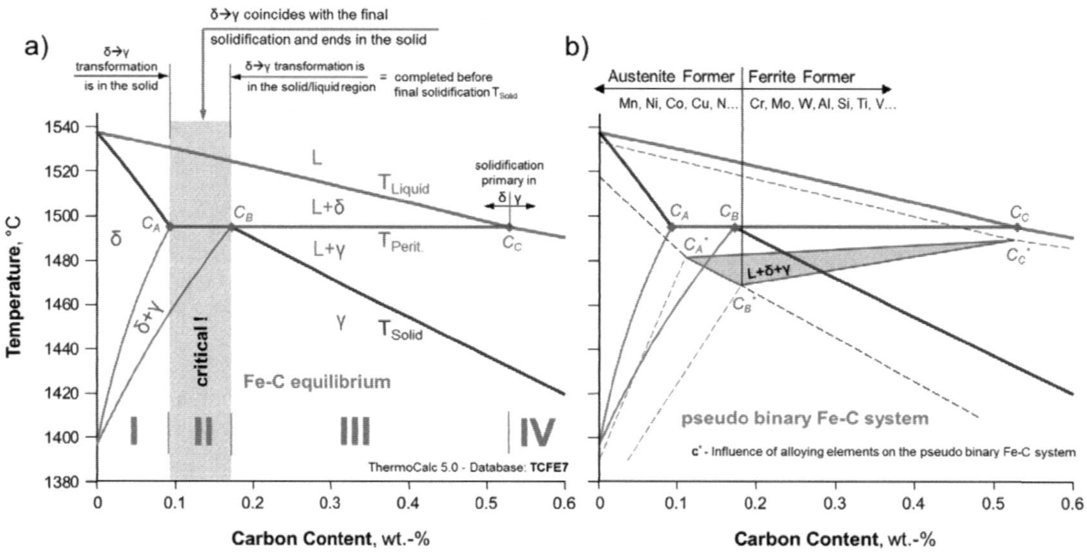


**Figure 1.** a) Fe-C equilibrium diagram with the critical peritectic range and b) influence of alloying elements on the Fe-C system in the high temperature range.

## 2. Considerations on the Fe-C-Mn-Si-Al system

In the following section, four examples of new steel grades are presented to illustrate the strong influence of alloying elements and to demonstrate the need of a thermodynamic database validation and development. From existing steel grades, small samples were used for DSC measurements to investigate the high temperature phase transformations. DSC measurements, carried out under equilibrium conditions as described in [3,4], are an excellent method to record all transitions associated with an exo- or endothermic effect. The high-Mn TWIP steels (Mn >20 wt.-%) were measured with the more robust DTA method, since manganese evaporates very strong at high temperatures and will destroy the platinum DSC sensor. DSC/DTA measurements allow to determine the high temperature transformation temperatures and sequences (range I, II, III, or IV).

In addition CALPHAD calculations of the investigated steels were carried out with two different databases. Therefore the ThermoCalc steel database [TCFE7] and Scientific Group ThermoData Europe - Alloy Database 2014 [SGTE2014] were used. All calculation of multicomponent phase diagrams using the CALPHAD approach (ThermoCalc, FactSage, MTData, Pandat...) are only as good as the underlying thermodynamic databases. It has to be noted again, that the following four alloys are only selected points of the Fe-C-Si-Al-Mn system and that they are insufficient for serious and comprehensive considerations.

The first investigated alloy is a Mn- and Si- alloyed Dual-Phase steel, without Al but with 1,1 wt.-% Si. The DSC measurement in Figure 2 shows clearly a γ→δ+γ transformation in the solid and a small peritectic peak, which is typical for alloys between $C_A^*$ and $C_B^*$ (=transformation sequences II). Both calculated phase diagrams illustrate nearly the same behaviour. A view in the database description shows, that both products use the same references for the ternary subsystems Fe-C-Si, Fe-C-Mn and Fe-Si-Mn (no Fe-C-Si-Mn description is available). Therefore, the calculation results are also identical, but they are in contrast to the measurement, as visualized in Figure 2. The measurement shows clearly a total different phase transformation behaviour compared to the thermodynamic calculations. Only the liquidus temperature is calculated correct with a small deviation of -4°C.

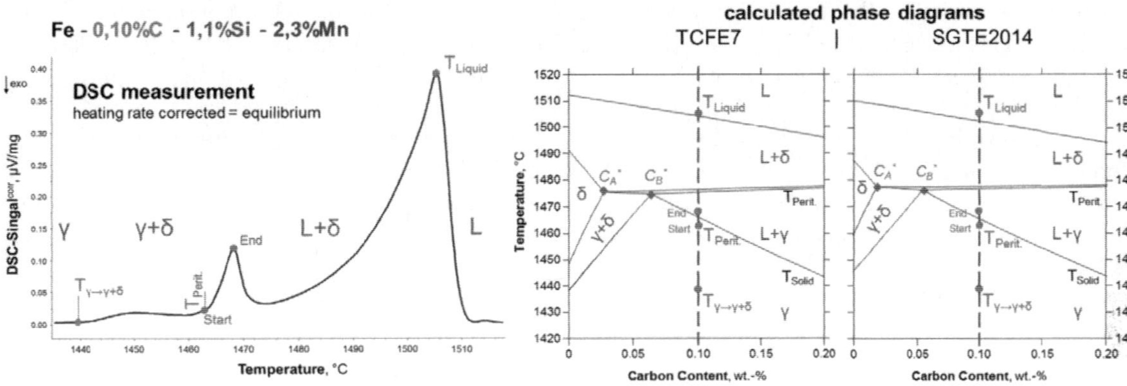


**Figure 2.** Investigation of a Dual-Phase steel (DP).

The second investigated alloy is a medium alloyed TRansformation Induced Plasticity steel with 0,77 wt.-% Al. The DSC measurement in Figure 3 shows clearly a large peritectic peak between the solidus and liquidus temperature, which is typical for alloys between $C_B^*$ and $C_C^*$ (=transformation sequences III). The measured steel grade is quite good described by the TCFE7 database, only the peritetic temperature is calculated a bit too high. In comparison, the SGTE2014 database calculates solidus and peritectic more than 20°C too low and shows a conspicuous wide L+δ+γ ternary region. A view in the database description shows, that only the TCFE7 database has a thermodynamic description of the ternary Fe-C-Al system from Connetable *et al.* [5] which based mainly on ab initio calculations. The missing description of the Fe-C-Al system in the SGTE2014 databases (at the moment, an implementation seems suitable) seems to be the reason for the different calculated phase diagrams in Figure 3.

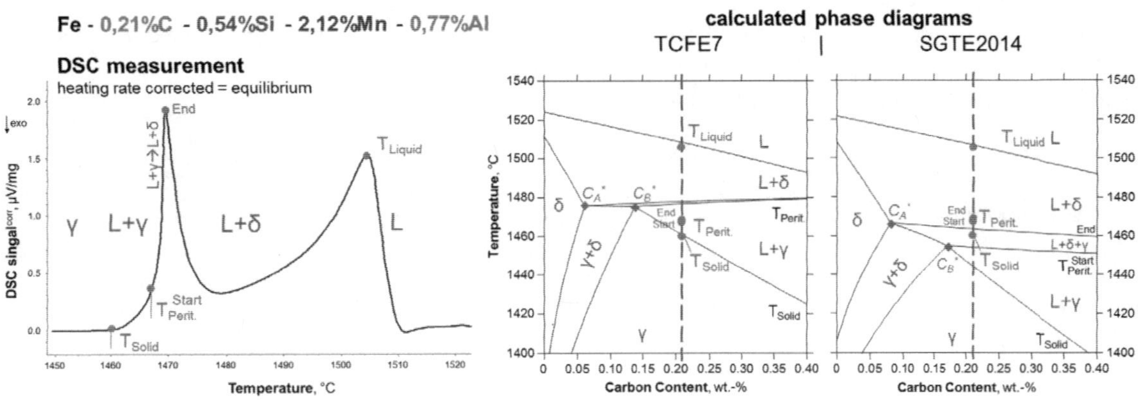


**Figure 3.** Investigation of a TRansformation Induced Plasticity steels (TRIP).

Compared to the medium alloyed DP- and TRIP-steels, the high-Mn alloyed TWIP steels show even larger discrepancies between measurements and calculations. It is also difficult to measure these steels, due to their high reactivity and the risk of evaporating manganese, which easily destroy platinum-thermocouples. Therefore only DTA measurements in special crucibles are possible (slightly lower accuracy than DSC measurements).

The third analysed steel is a high-Mn alloyed TWIP with high Al and low C. The DTA measurement in Figure 4 shows the melting of the alloy, whereat only the solidus and liquidus temperature can be observed, without a peritectic or gamma-delta transformation, which is typical for alloys left of $C_C^*$ (=transformation sequences IV, primary $\gamma$). Since DTA investigations as a single method are not sufficient to determine the type of the solid phases (primary $\gamma$ or $\delta$), a high temperature X-ray diffraction measurement would be helpful (because also a pure $\delta \rightarrow L$ transitions might be possible).

Even without these investigations, the calculated phase diagrams and temperatures differ significantly from the DTA measurement (4 DTA measurements with different heating rates were carried out in order to determine the equilibrium temperatures). Both calculated phase diagrams show a total different construction. On the first view the calculation with the SGTE2014 databases seems to be appropriate, but on the second view the calculated phase diagram shows a totally unrealistic behaviour with increasing carbon content, as visualized on the right side of Figure 4. Furthermore, the SGTE 2014 database does not include a description of the Fe-C-Al and of the Fe-Mn-Al system, which is essential for this alloy. With this single analysis, no explanations for the large discrepancies between measurement and calculation (TCFE7 and SGTE2014) can be ascertained.

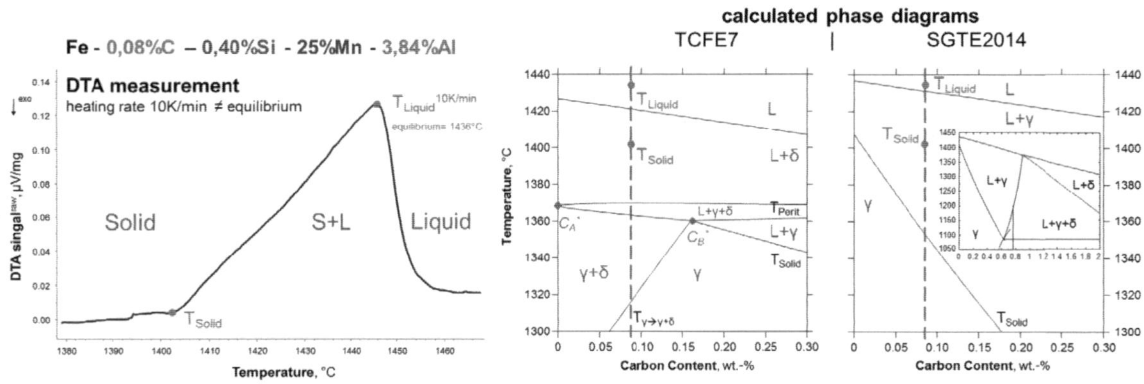


**Figure 4.** Investigation of a Twinning Induced Plasticity steel (TWIP) with low C.

The fourth analysed steel is a high-Mn alloyed TWIP with high Al, low Si and medium C. The DTA measurement in Figure 5 shows the melting of the alloy, with a large peritectic peak between the solidus and liquidus temperature, which is typical for alloys between $C_B^*$ and $C_C^*$ (=transformation sequences III).

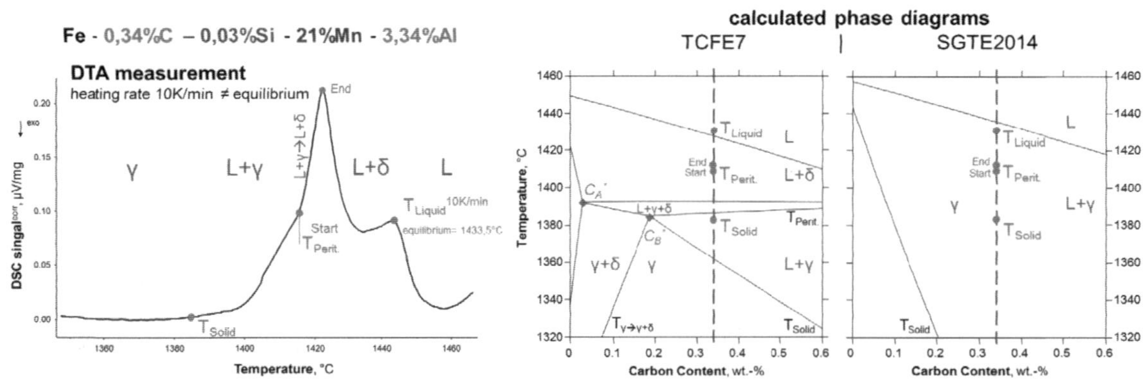


**Figure 5.** Investigation of a Twinning Induced Plasticity steel (TWIP) with medium C.

Even though the TCFE7 database calculates the solidus and peritectic temperature by more than 20 °C too low, the phase transformation behaviour coincides with the measurement which is pretty good for this exotic alloy. Similar to the TWIP steel from Figure 4, the SGTE2014 database cannot handle this alloy, due to the missing thermodynamic description of relevant subsystems.

## 3. Approach for optimizing the Fe-C-Mn-Si-Al system

As these four selected examples of new steel grades show, there is a great need to validate and optimize thermodynamic databases. Especially higher concentrations of alloying elements and their interactions with each other significantly influence the Fe-C system. In order to calculate multicomponent systems, the thermodynamic description of the subsystems needs to be secured. Figure 6 gives an overview of the methodology of the CALPHAD method and the way how to asses higher ordered systems [6]. The Gibbs Energy $G^{\varphi}$ of each phase $\varphi$ consists of three terms; ${}^{0}G^{\varphi}$ describes the standard Gibbs energy of a mechanical mixture of the constituents of the phase, ${}^{ideal}G^{\varphi}$ describes the entropy of mixing an ideal solution and ${}^{E}G^{\varphi}$ describes the so-called excess term.
Based on experimental dates, the excess term ${}^{E}G^{\varphi}$ is assessed to describe the interactions of the elements with each other and therefore the "real behaviour". Figure 6 shows also the systematic to describe higher order systems. In order to calculate a 5-component system, the knowledge of 5 quaternary subsystems, 10 ternary subsystems and 10 binary subsystems are needed - which is quite a large amount of thermodynamic data. Especially the systems from the Fe-corner are the most important one, which needs to be verified.

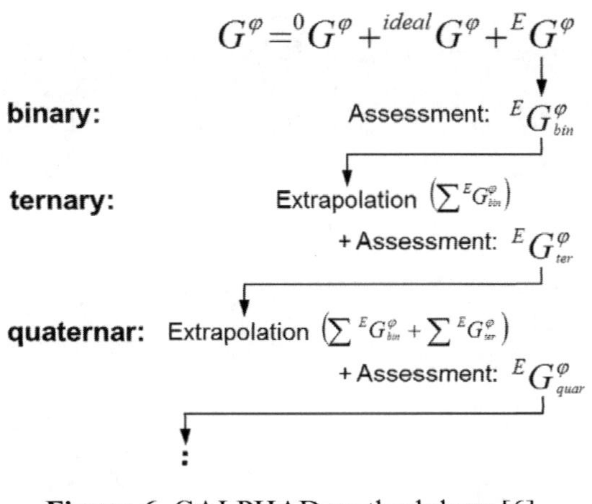


**Figure 6.** CALPHAD methodology [6].

Regarding the investigations represented in Figure 2 to 5 and the motivation to improve the thermodynamic calculations, the most important step is the identification of the affected subsystem, which could be wrong. This is quite difficult for a 5-component Fe-C-Si-Al-Mn system. In this case, the investigation of the Dual-Phase steel in Figure 2 is quite helpful, as the measurement seems to show an error in the quaternary Fe-C-Si-Mn system. As Silicon and Manganese are one of the most important alloying elements, this finding needs to be investigated in detail.

## 4. Methodology to identify the defective subsystems of complex alloys

In order to investigate the findings from Figure 2, an examination with selected model alloys was performed. For this purpose, a Fe-C-Si-Mn alloy with an increased content of 2%Mn and 2%Si (stronger effect of the Mn x Si interaction) and the ternary subsystems Fe-C-2%Mn and Fe-C-2%Si were produced. For all alloys, a carbon content in the peritectic range between $c_A$ and $c_B$ with of 0,11 to 0,12 mass-% was selected. As a reference, a pure Fe-0,11%C alloy was also prepared.

The model alloys (about 60 g) were melted from high purity raw materials in an alumina crucible with a high-frequency remelting and spin-casting machine under argon atmosphere 5.0. The samples were very homogeneous due to the inductive melting, the spin casting and the rapid solidification in a copper mould. The alloy composition was analysed by emission-spectroscopy and is summarized in Table I. The equipment used in the study was a NETZSCH DSC404F1 with a platinum DSC sensor, type-S thermocouples and $Al_2O_3$ sample cups. In order to achieve the best equilibrium conditions, 50 mg samples and a heating rate of 10 K/minutes were used. The performed DSC measurements are compiled in Figure 7 and the measured equilibrium temperatures are summarized in Table I.

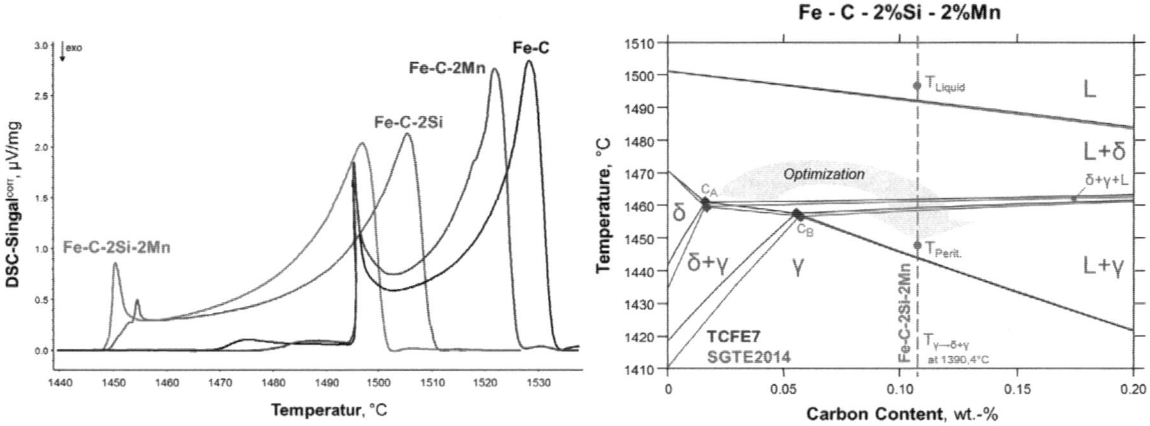


**Figure 7.** Systematic investigation of selected subsystems of the Fe-C-Si-Mn systems.

**Table 1.** Chemical analysis, measured phase transformation temperatures and deviations from the thermodynamic calculations.

| Alloy | | Fe-C | | Fe-C-2Mn | | Fe-C-2Si | | Fe-C-2Si-2Mn | |
|---|---|---|---|---|---|---|---|---|---|
| Analysis | Carbon | 0,11 | | 0,11 | | 0,12 | | 0,11 | |
| | Silicon | - | | - | | 2,04 | | 2,00 | |
| | Manganese | 0,05 | | 2,00 | | 0,04 | | 1,94 | |
| | Position | peritectic between $c_A$ and $c_B$ | | peritectic between $c_A$ and $c_B$ | | peritectic between $c_A$ and $c_B$ | | peritectic between $c_A$ and $c_B$ | |
| Transformation | deviation | measured | ΔT | measured | ΔT | measured | ΔT | measured | ΔT |
| $T_{\gamma\rightarrow\delta}^{Start}$ | ΔT DSC - TCFE7: | 1467,8 | 3,5 | 1476,8 | -5,6 | 1358,3 | -16,4 | 1390,4 | xxx |
| | ΔT DSC - SGTE2014: | | 3,4 | | -5,6 | | -18,9 | | xxx |
| $T_{Solidus}$ | ΔT DSC - TCFE7: | - | - | - | - | - | - | - | xxx |
| | ΔT DSC - SGTE2014: | | - | | - | | - | | xxx |
| $T_{Perit.}^{Start}$ | ΔT DSC - TCFE7: | 1494,0 | -0,5 | 1494,1 | 3,3 | 1449,3 | 9,5 | 1448,2 | xxx |
| | ΔT DSC - SGTE2014: | | 3,1 | | 5,8 | | 8,2 | | xxx |
| $T_{Liquidus}$ | ΔT DSC - TCFE7: | 1528,3 | -0,9 | 1521,4 | 1,9 | 1506,6 | 2,6 | 1497,2 | 5,2 |
| | ΔT DSC - SGTE2014: | | -0,7 | | 2,1 | | 3,1 | | 5,7 |

The analysis of the DSC measurements and the comparison with the thermodynamic calculations demonstrates the following findings. The measurements of the Fe-0,11%C alloy clearly shows a $\gamma\rightarrow\delta+\gamma$ transformation in the solid and a sharp peritectic peak, which is typical for alloys between $C_A^*$ and $C_B^*$ (=transformation sequences II). Both thermodynamic databases (FCFE7 and SGTE2014) describe this binary alloy very well. The addition of 2%Mn (Fe-0,11%C-2%Mn) shows no effect on the peritectic temperature, but the liquidus is lowered by -7°C and the $\gamma\rightarrow\delta$ temperature is raised to +9°C (=enlargement of the gamma region). Both thermodynamic databases describe this ternary alloy similar and quite well. The measurement of the ternary Fe-0,12%C-2%Si alloy shows the strong effect of 2%Si based on the significant reduction of $T_{Liquid}$ up to -21,7°C, $T_{Perit.}$ up to -44,7°C and $T_{\gamma\rightarrow\delta+\gamma}$ up to -109,5°C, in comparison to the binary Fe-0,11%C alloy. For this alloy, the very wide two-phase $\gamma$+$\delta$-region of 91°C is very difficult to measure by the DSC method, due to the very low enthalpy change of the $\gamma\rightarrow\delta$ phase transformation. Both thermodynamic databases describe this ternary alloy similar, but show an increased deviation for $T_{Perit.}$ and even larger deviation for $T_{\gamma\rightarrow\delta+\gamma}$. Therefore, further investigations of the ternary Fe-C-Si system are recommended.

The combination of 2%Mn and 2%Si leads to a strong decrease of $T_{Liquid}$ up to -31,1°C, $T_{Perit.}$ up to -45,8°C and $T_{\gamma\rightarrow\delta+\gamma}$ up to -77,4°C in comparison to the binary Fe-0,11%C alloy. The liquidus temperature is the lowest one of all tested alloys, which was to be expected, since Mn and Si both reduce $T_{Liquid}$. The $T_{Liquid}$ decrease of 2%Mn (-7°C) and 2%Si (-21,7°C) can be added and this simple calculations correspond nearly with the real measured $T_{Liquid}$ of the quaternary alloy. This simple correlation applies only for the liquidus temperature and not for the other phase transformations, which follow more complex interactions. The measurement of the Fe-0,11%C-2%Mn-2%Si alloy clearly shows, that the 2%Mn has no effect on the peritectic temperature, same as for the Fe-0,11%C-2%Mn alloy. The peritectic temperature depends in this case, only on the Fe-C system and the silicon content. The measurements also show that +2%Mn, compared to the ternary Fe-0,12%C-2%Si alloy, leads to a smaller δ+γ two-phase region at a slightly larger peritectic peak – this increase of gamma region is caused by manganese (austenite former). Both thermodynamic databases describe this ternary alloy similar but wrong, as visualized in Figure 7 and compiled in Table I. The liquidus temperature is calculated 5 °C too deep and the residual phase diagram is definitely wrong, since the alloy is clearly located between $C_A^*$ and $C_B^*$ (= transformation sequences II) and not between $C_B^*$ and $C_C^*$ (= sequences III) as calculated.

As both thermodynamic databases (TCFE7 and SGTE2014) do not include a quaternary description of the Fe-C-Si-Mn system, the reason must be searched in the relevant subsystems. The ternary Fe-C-Mn system (source [7]) appears well described, however, the ternary Fe-C-Si (source [8,9]) system shows a certain need for improvement. But this deviation in the Fe-C-Si system is too small, to explain the large discrepancy of the quaternary system. Therefore, the ternary Fe-Si-Mn system [Fe-Mn-Si:*Va*] (source [10]) has to be considered, since this system is particularly responsible for the low carbon region of the Fe-C-Si-Mn system, as visualized in Figure 7. As publications about measurements of the high temperature area of the Fe-Si-Mn and Fe-C-Si-Mn system are hardly available, it appears highly recommendable to investigate these systems systematically. One further possible reason for the large deviations in the Fe-C-Si-Mn system could also come from binary Si-Mn subsystem, which describes the interaction of this important ferrite and austenite formers. This binary system was recently reassessed by Paek [11] and Zheng [12] visualized in Figure 8 (for information, these two publications are not yet implemented in the TCFE7 and SGTE14 database). Both authors show similar phase diagrams, wherein the data used for the description of the high manganese site are only from the publication of Voegel [13]. It should be noted, that these experimental study is already more than 80 years old and it would be interesting and useful to repeat and verify these results, especially the δ⇔γ transition and the peritectic range, as visualized in the red circle in Figure 8.

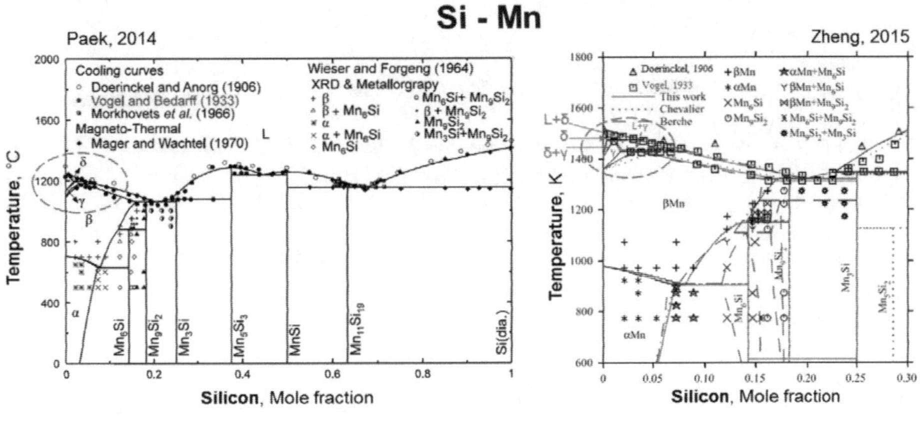


**Figure 8.** Reassessments of the binary Mn-Si system [11,12].

The performed investigation in this study clearly shows in Figure 2 and Figure 7 that the actual thermodynamic calculations of the quaternary Fe-C-Si-Mn system are definitive wrong. As visualized in the phase diagram in Figure 7, the position of the peritectic range should be at higher carbon content and the δ-region should be larger (bcc more stable). The investigations clearly identify, that the relevant major fault is in the ternary Fe-Si-Mn system and not in the Fe-C-Mn, and even not in the Fe-Si-C system; therefore this should also be checked. The main focus is therefore on the Fe-Si-Mn system and in this regard also on the binary Si-Mn subsystem should be checked, since their original experimental investigation is very old. Only with this systematic approach and the use of model alloys, it is possible to identify relevant subsystems and their interactions. With this knowledge, the selected systems can be further investigated in detail.

## 5. Summary and Outlook

The current study illustrates the importance of thermodynamic investigations for the re-assessment of selected ternary Fe-base systems. At first DSC/DTA measurements were performed from existing DP, TRIP and TWIP steel grades to investigate the high temperature phase transformations and the transformation behaviour regarding the peritectic range. These measurements showed that even "simple" DP steels are incorrectly calculated with commercial thermodynamic databases (TCFE7 and SGTE2014). In order to examine this finding in detail, an investigation of the Fe-C-Si-Mn system was performed using selected model alloys and DSC measurements. The investigations clearly identify, that the relevant major fault is in the ternary Fe-Si-Mn system and not in the Fe-C-Mn, and even not in the Fe-Si-C system. This is a very important finding, since all new designed steel grades for the automotive industry based on a Fe-C-Si-Mn matrix.

The next steps should be to verify the aluminium alloyed ternary and quaternary subsystems with the same systematics. With the methodology to identify the defective subsystems of complex alloys, a powerful method is available to focus on the relevant systems. Only if these basic binary, ternary and quaternary subsystems are well investigated and thermodynamically described, then the CALPHAD-approach has the potential to calculate complex higher alloyed systems. Especially the measurements of the TWIP steels demonstrate, that there exists a huge potential for further investigations. Also the low number of published measurement and the sometimes very old raw data are motivations, to perform further intensive investigations.

## 6. Acknowledgement

Financial support by the Austrian Federal Government (in particular from Bundesministerium für Verkehr, Innovation und Technologie and Bundesministerium für Wirtschaft, Familie und Jugend) represented by Österreichische Forschungsförderungsgesellschaft mbH and the Styrian and the Tyrolean Provincial Government, represented by Steirische Wirtschaftsförderungsgesellschaft mbH and Standortagentur Tirol, within the framework of the COMET Funding Programme is gratefully acknowledged.

## 7. References

[1] Jablonka A, Harste H and Schwerdtfeger K 1991 *Steel Research* **62** 24.
[2] Xia G, Bernhard C, Ilie S and Fürst W 2008 *6th ECCC* (Riccione, Italy).
[3] Presoly P, Xia G, Reisinger P and Bernhard C 2014 *BHM* **159** 430.
[4] Presoly P, Pierer R Bernhard C 2013 *Metallurgical and Materials Transactions* **44A** 5377.
[5] Connetable D, Lacaze J, Maugis P and Sundman B 2008 *CALPHAD* **32** 361.
[6] Lukas H L, Fries S G and Sundman B 2007 *Computational Thermodynamics – The Calphad Method* (Cambridge University Press).
[7] Huang W 1990 *Metallurgical and Materials Transactions* **A21** 2115.
[8] Lacaze J and Sundman B 1991 *Metallurgical and Materials Transactions* **22A** 2211.
[9] Miettinen J and Hallstedt B 1998 *Calphad* **22** 231.
[10] Forsberg A and Agren J 1993 *Journal of Phase Equilibria* **14** 354.
[11] Paek M K, Pak J J and Kang Y B 2014 *Calphad* **46** 92.
[12] Zheng W, Lu X G, He Y, Cui Y and Li L 2015 *Journal of Alloys and Compounds* **632** 661.
[13] Vogel R and Bedarff H 1933 *Arch. Eisenhüttenwesen* **7** 423.

IOP Conf. Series: Materials Science and Engineering **143** (2016) 012035 doi:10.1088/1757-899X/143/1/012035

# Formation of shrinkage porosity during solidification of steel: Numerical simulation and experimental validation

**M. Riedler[1], S. Michelic[2], C. Bernhard[1]**
[1] Chair of Ferrous Metallurgy, Montanuniversitaet Leoben, Austria
[2] INTECO TBR casting technologies GmbH, Austria

E-mail: michael.riedler@unileoben.ac.at

**Abstract**. The phase transformations in solidification of steel are accompanied by shrinkage and sudden changes in the solubility of alloying elements, resulting in negative side effects as micro- and macrosegregation and the formation of gas and shrinkage porosities. This paper deals with the numerical and experimental simulation of the formation of shrinkage porosity during the solidification of steel.
First the physical basics for the mechanism of shrinkage pore formation will be discussed. The main reason for this type of porosity is the restraint of fluid flow in the mushy zone which leads to a pressure drop. The pressure decreases from the dendrite tip to the root. When the pressure falls below a critical value, a pore can form.
The second part of the paper deals with different approaches for the prediction of the formation of shrinkage porosity. The most common one according to these models is the usage of a simple criterion function, like the Niyama criterion. For the computation of the porosity criterion the thermal gradient, cooling rate and solidification rate must be known, easily to determine from numerical simulation. More complex simulation tools like ProCAST include higher sophisticated models, which allow further calculations of the shrinkage cavity.
Finally, the different approaches will be applied to a benchmark laboratory experiment. The presented results deal with an ingot casting experiment under variation of taper. The dominant influence of mould taper on the formation of shrinkage porosities can both be demonstrated by the lab experiment as well as numerical simulations. These results serve for the optimization of all ingot layouts for lab castings at the Chair of Ferrous Metallurgy.

## 1. Introduction

The solidification process for steel is coupled with volume shrinkage due to an increase in the density with decreasing temperature. This volume shrinkage can be compensated by feeding with liquid steel. As long as feeding is sufficient, porosity will not occur. At the end of solidification the resistance against the flow increases leading to a higher pressure drop and therefore to the formation of shrinkage porosity.

The aim of this paper is to compare experimental investigations regarding shrinkage porosity with numerical result. Therefore ingots with different geometry were cast and cut along the ingot length axis to make the shrinkage porosity apparent. The numerical simulation of the ingots was performed with the software ProCAST. Different porosity criteria and their respective porosity models were used to predict the formation of shrinkage porosity in the ingot. Finally, a comparison between the different numerical and experimental results was made.

## 2. Theoretical aspects of the formation of porosities

The formation of porosity can be divided into two main categories: gas porosity and shrinkage porosity. The formation mechanism and the porosity shape of both porosity types exhibit substantial differences.

Investigations of gas porosity show that the shape of this porosity type is the most time spherical or sometimes ellipsoid. The formation of gas porosity is based on the fact that the saturation for the important gases [H], [N] and [O] (formation of CO) in liquid steel decreases as the temperature drops, which is depicted in figure 1. The heat withdrawal of the mould lowers the temperature of the liquid steel and leads to an enrichment of gaseous elements at the front or between the dendrite arms. The reason for the formation of gas porosity is due to the gas exceeding its saturation point. The spherical shape of this porosity type implies that the formation occurs in the liquid state or during the early stage of solidification. At the end, it is of note that the nucleation of gas porosity is not homogeneous as the internal pore pressure would be very high. Therefore, heterogeneous nucleation is preferable, and, hence, inclusions or uneven surfaces will be used as a nucleation site. [1], [2], [3]

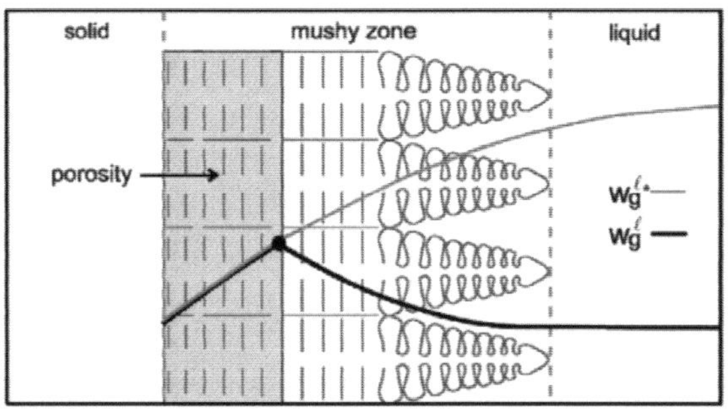


**Figure 1**. Schematic illustration of the raising gas concentration $w^l_g$ in the mushy region based on gas microsegregation and the diminution of the solubility limit $w^{l*}_g(T,p_l)$ [2], [3].

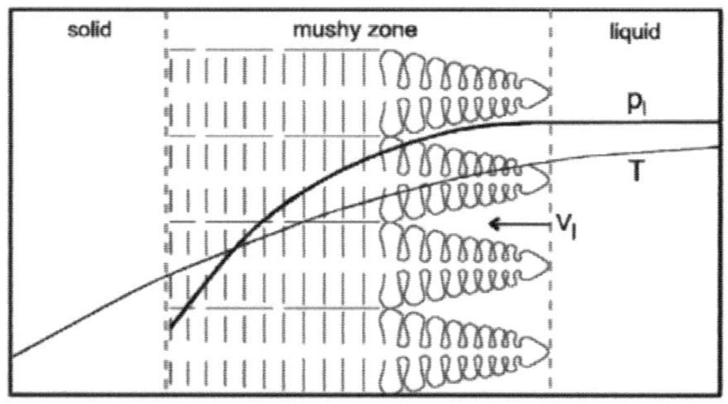


**Figure 2.** Schematic illustration of the liquid pressure drop in the mushy region based on solidification shrinkage[2], [3].

In contrast to gas porosity the shape of shrinkage porosity is very tortuous and is dendritic solidifying alloys limited by the primary dendrites and dendrite arms. Shrinkage porosity appears at the end of solidification range and so the shape of this type is adjusted by the remaining space between the dendrites. The shrinkage effect of the steel due to decreasing temperature can be compensated in the mushy zone by feeding liquid steel. This can be a problem for wide range mushy zones as the fluid flow of the liquid steel must be sufficient down to the root of the dendrite. The restraint of the fluid flow in the mushy zone leads to a pressure drop, as illustrated in figure 2. An inadequate development of the interdendritic flow reduces the pressure to a certain limit, which leads to an insufficient feeding, and, to the formation of shrinkage porosities. Thus, it is clear that the formation of this porosity type occurs at the end of the solidification range and furthermore, the shape is related to the remaining space of the dendrites. Finally, a 2D metallographic image shows only a small part of this porosity type (size is around a few tens of micrometers), as it normally interconnects in a complex manner and only a 3D illustration can capture the full extension. [1], [2], [3]

## 3. Prediction of the formation of shrinkage porosity

Thus far, many methods for the prediction of the appearance of shrinkage porosity exist. These methods range from very simple methods, which use only material properties or solidification parameters, to highly complex methods, which are the mostly self-programmed and try to explain the formation. To calculate the shrinkage porosity for an ingot, the Niyama criterion and the porosity algorithm of ProCAST are used in this work. These methods are explained in detail below.

### *3.1. Porosity Criterion*

Niyama et al. [4] established a criterion to determine the appearance of shrinkage porosities. Therefore they considered different porosity criterion which can be calculated easily and compared these values with experimental results. As a result a porosity criterion was investigated, which is now widely used and known as Niyama criterion. The criterion can be calculated with the following equation (1)

$$Ny = \frac{G}{\sqrt{\dot{T}}} < 1 \tag{1}$$

where $G$ is the thermal gradient and $\dot{T}$ is the cooling rate. Regarding steel, the critical value is around 1.0 [(K min)$^{0.5}$/cm]. A Niyama value below this critical value leads to the occurrence of shrinkage porosity. The advantage of this criterion is the independence of the casting geometry. Nevertheless, the critical value depends on the corresponding casting material and must be determined for every casting material. [5], [6], [7], [8]

The Niyama criterion was further developed by Carlson et al. [9]. The aim of this improvement was to predict not only the position of the shrinkage porosity in the casting, but also the amount of the pore volume fraction. The advanced Niyama criterion is also called the dimensionless Niyama criterion and can be calculated according to equation (2)

$$Ny^* = \frac{G\,\lambda_2\sqrt{\Delta P_{cr}}}{\sqrt{\mu_l \beta \Delta T_f \dot{T}}} = Ny\frac{\lambda_2\sqrt{\Delta P_{cr}}}{\sqrt{\mu_l \beta \Delta T_f}} = \sqrt{I(g_{l,cr})} \tag{2}$$

where $\lambda_2$ is the secondary dendrite arm space, $\Delta P_{cr}$ is the critical pressure drop; $\mu_l$ is the liquid dynamic viscosity; $\beta = (\rho_s - \rho_l)/\rho_l$ is the solidification shrinkage; $\Delta T_f = T_{liq} - T_{sol}$ is the freezing range and $I(g_{l,cr})$ is an integral, which depends on the solid fraction-temperature curve. For simple curves the integral can be calculated analytically with equation (3)

$$I(g_{l,cr}) = \int_{g_{l,cr}}^{1} 180\frac{(1-g_l)^2}{g_l^2}\frac{d\theta}{dg_l}dg_l \tag{3}$$

where $\theta = (T - T_{sol})/\Delta T_f$ is the dimensionless temperature. Further details regarding the dimensionless Niyama criterion can be found in [9]. Carlson et al. uses the dimensionless Niyama criterion to understand the formation of shrinkage porosity for an aluminium alloy, a magnesium alloy and mild steel. Further verification using casting experiment with steel was made

### *3.2. Porosity model of ProCAST*

ESI Group has implemented a model to predicting shrinkage porosity into the software ProCAST. Two important parameters are needed to calculate the porosity. The first parameter is MacroFs which corresponds to a solid fraction and defines the starting point for the porosity calculation. The second parameter is called FeedLen and can be seen as an interdendritic feed length. The porosity calculation is performed for regions with a solid fraction between MacroFs and 1. For these areas two possible situations can occur:

1. The mushy zone contains regions with a solid fraction lower than MacroFs

In this case the distance between the solidus temperature isotherm and the MacroFs isotherm is calculated. The distance between these isotherms must be larger than FeedLen for the formation of porosity. Hence, a high temperature gradient leads to a smaller distance and vice versa, which is schematically illustrated in figure 3. The temperature gradient in region A is low and therefore the distance between the two isotherms is greater than FeedLen. Porosity will form in this area. In zone B, the distance is smaller since the gradient is much higher and as a consequence no porosity will occur. [10]

2. All elements in the mushy zone have a solid fraction higher or equal than MacroFs
The appearing shrinkage between MacroFs and solidus temperature is compensated through the formation of porosity in the whole region. [10]

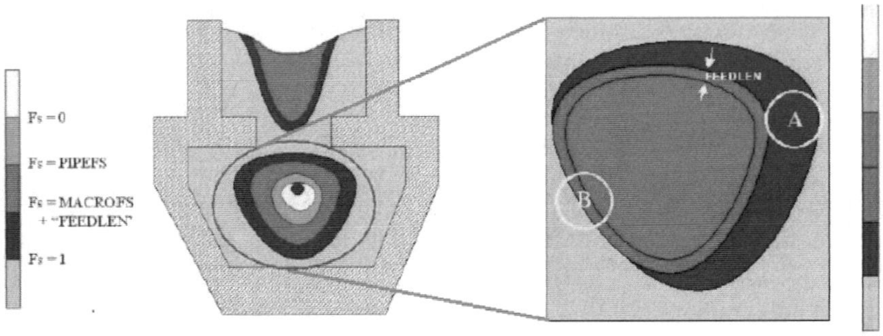


**Figure 3.** Schematic illustration of the internal shrinkage porosity calculation from ProCAST [10].

## 4. Development of the experimental casting trials

Casting trials were performed in the melt shop of the Chair of Ferrous Metallurgy in Leoben. Thereby the geometries of the mould were varied to see whether there is an influence on porosity formation.

At the top of the ingot a feeding head is installed to influence the solidification behaviour in this region. Furthermore, please note that the mould is not completely filled with liquid steel. The meniscus level of the liquid steel is around 30 mm beneath the top edge of the mould. Thus the ingot has dimensions of 100 mm x 100 mm x 370 mm and a weight around 27 kg. The inner surface of the mould was covered with release agents for a better separation. For a better stability the mould was embedded into sand. Three experimental trials were cast with varying taper which changed from 0° to 5° and 10° only for the narrow mould wall. The broader mould side is parallel. Furthermore, the mould was not heated up before casting. At the end of casting experiment the top of the ingot were covered with an insulation powder. The steel was cast into the mould by use of the top pouring method.

From the literature it can be seen that the formation of shrinkage porosity depends on the material. A material with a wide-spread mushy zone tends to form porosities as a material with a narrow mushy zone. Thus, steel with a high carbon content was used as cast material. The composition of the cast steel is shown in table 1. For the mould a common structural steel (S 335) was used.

**Table 1.** Steel composition of the ingot

| | C [wt.-%] | Si [wt.-%] | Mn [wt.-%] | Al [wt.-%] |
|---|---|---|---|---|
| Cast Material | 0.85 | 0.30 | 1.00 | 0.04 |

An important part of the metallographic examinations is the detection of shrinkage porosity. For this purpose, a 10 mm slice was cut out at the middle of the ingot. A macroscopic image was made to identify the porosity in the ingot centre.

## 5. Modelling of the experiment

The simulation of the solidification process was performed with the finite element software ProCAST from ESI Group. The software has the ability to solve the transient behaviour of a three dimensional thermal and fluid flow problem. ESI Group introduced a new discretization method called edge-based finite element method (EBFEM) in the version 2013.0. EBFEM uses an up-winding scheme with limiters to discretize the advection operator, thereby leading to a higher-order of accuracy and rendering an improved mould filling behaviour. Furthermore, ESI Group have implemented an

adapted solver into ProCAST 2013.0. The solver based on the algebraic multigrid method and is a robust and highly powerful linear solver. [10], [11], [12]
ProCAST solves following partial differential equations:

- Momentum balance for the velocities U, V and W
- Mass balance for the pressure P
- Energy balance

### *5.1. General simulation scheme*

The ingot dimensions necessitate a 3D simulation as the third dimension has an influence on the results. There were several constraints regarding the setup of the model:

- mould filling process is not calculated
- formation of the pipe is not considered
- only natural convection is simulated

For the discretization of the geometry, a structured mesh grid was used. The average size of the element was 5 mm, which leads to a total number of elements of around 85000. A simple permeability model, which is widely used, was applied at these simulations. The Carmen-Kozeny model is an isotropic permeability model and based on the pressure drop during the flow of a fluid through a porous medium. Equation (4) describes this model.

$$K = K_0 \frac{g_l^3}{(1-g_l)^2} \tag{4}$$

$K_0$ represents the microstructure of the steel and can be calculated with $K_0 = \lambda_2{}^2/180$; $\lambda_2$ is the secondary dendrite arm space. For the calculation of the permeability a secondary dendrite arm space $\lambda_2$ of 50 µm was estimated.

### *5.2. Material properties*

ProCAST can handle temperature-dependent material data which are used for these simulations. The material data are calculated for cast and mould material with the software IDS [13]. The software was developed from Miettinen et al. to determine the material data for low alloyed steel and consider further a microsegregation model. The insulation at the top of the ingot is a typical ceramic material, thus typical properties for such a material are used. [14], [15]

### *5.3. Initial and boundary condition*

For the simulation a superheat of 15 °C was used. The initial temperature of the mould was assumed to have an elevated room temperature of 40 °C. The majority of the ingot's heat and also mould will be removed through heat conductivity as the mould is embedded into sand. Hence, the applied boundary conditions for the mould surface is a heat transfer coefficient of 200 [W/(m²K)] at an ambient temperature of 25 °C. It is assumed that no heat radiation between mould and sand occurs. Therefore the emissivity was set to 0. The top of ingot is covered with an insulation powder. For the sake of simplicity, this powder was not explicitly defined at the model. However, the influence of the powder is considered since an adiabatic boundary condition was used. The interface heat transfer coefficient between mould and ingot was assumed to be 150 [W/(m²K)] and between insulation and mould respectively ingot a value of 100 [W/(m²K)] was set.

### *5.4. Possibilities to calculate the porosity formation*

ProCAST offers many possibilities to calculate porosity criteria, especially the Niyama criterion. It is possible to compute this criterion during the simulation or afterwards in the post processing menu. In this case the Niyama criterion was calculated during the simulation and a solid fraction of 0.9 was used. To calculate the dimensionless criterion, a critical pressure drop $\Delta P_{cr}$, and also the secondary dendrite arm space $\lambda_2$ must be defined. In this case $\Delta P_{cr}$ is assumed at 1.01 bar, which corresponds to a liquid pressure of 1.01 bar and no capillary pressure; $\lambda_2$ is set to 200 μm which should be a good assumption for the ingot centre. The other parameter can be calculated from the material properties. Furthermore, it is possible to use the internal model from ProCAST. Therefore the parameter MacroFs and FeedLen must be defined. For theses simulations the parameter MacroFs is set to 0.85 and FeedLen is 5 mm. [10]

## 6. Results

The primary structure of the cast ingots with a mould taper of 0°, 5° and 10° are depicted at figure 4. The shrinkage porosity is marked with a green colour in this picture.

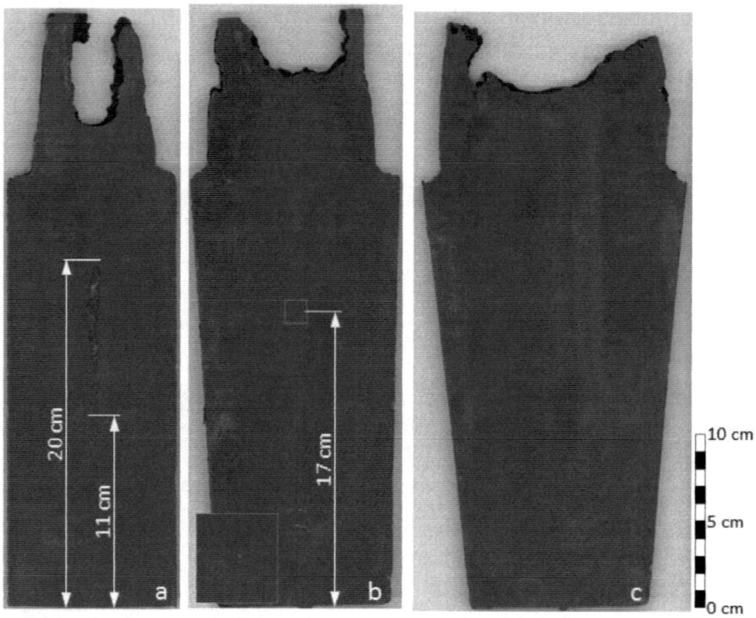


**Figure 4.** Experimental results of the cast ingots (a: 0° taper; b: 5° taper; c: 10° taper).

On the left, the ingot with no taper is shown. As can be seen from the image, shrinkage porosity appears at the centre of the ingot and extends from 11 cm to 20 cm ingot height. For an ingot with a 5° taper, a small pinhole at 17 cm is detectable. The ingot with a 10° taper is free of shrinkage porosity. Furthermore, it is clear to see how the pipe geometry changes with the mould taper. Looking at the ingot with no taper, the pipe is very deep. However, this changes with a higher mould taper to a flatter and wider pipe.

The Niyama criterion for the ingot with no taper is illustrated in figure 5. As mentioned above, shrinkage porosity will appear if the Niyama criterion falls below 1 [$(K\ min)^{0.5}$/cm]. The criteria have low values at the centre of the ingot an also in the top. A 2D plot of the ingot centre axis is shown in figure 6 for different mould tapers. The difference between the behaviour of the Niyama value seems to be minor, but it is clear to see that the minimum Niyama value for an ingot with no taper falls below 1 [$(K\ min)^{0.5}$/cm] and for an ingot with 10° taper it exceeds 1 [$(K\ min)^{0.5}$/cm]. The simulation describes the experimental results very well. The position of the minimum Niyama values matches quite well with the experimental results.

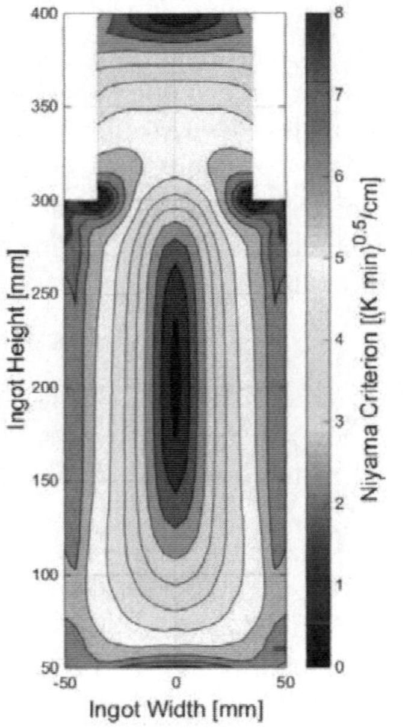


**Figure 5.** Calculated Niyama criterion for cast ingot with no taper.

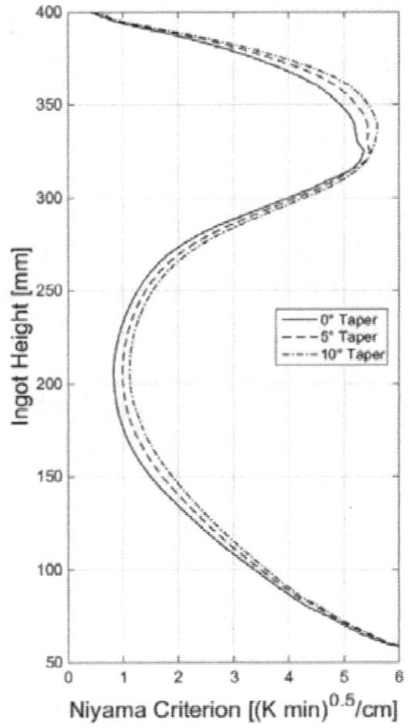


**Figure 6.** Comparison of the Niyama criterion for three ingots with different taper.

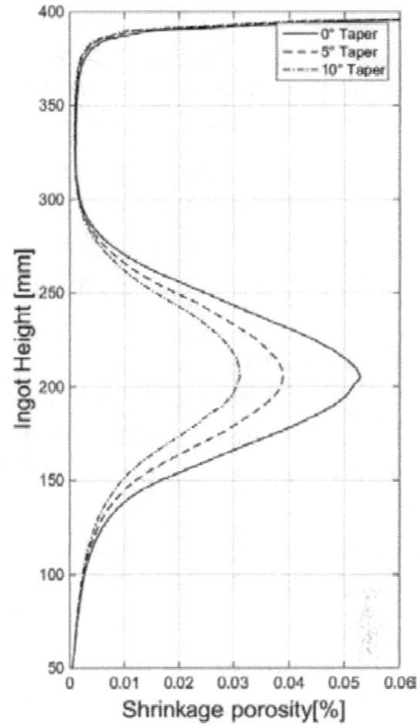


**Figure 7.** Comparison of the shrinkage porosity for three ingots with different taper.

Using the dimensionless Niyama criterion it is possible to predict the shrinkage porosity amount for the cast ingots. The linkage between the Niyama value and the shrinkage porosity percentage is depicted in figure 8. The calculated distribution of the pores is centred on the middle of the ingot and decreases rapidly outside the centre.

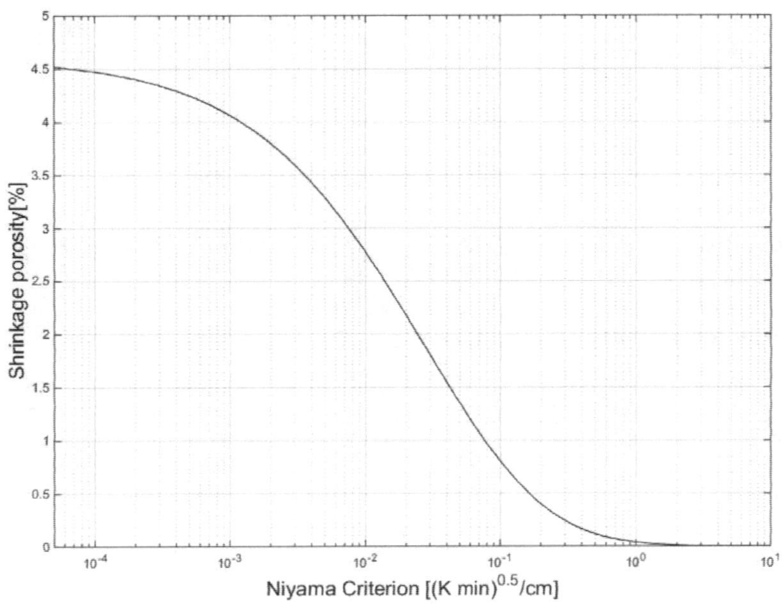


**Figure 8.** Relation between Niyama criterion and shrinkage porosity.

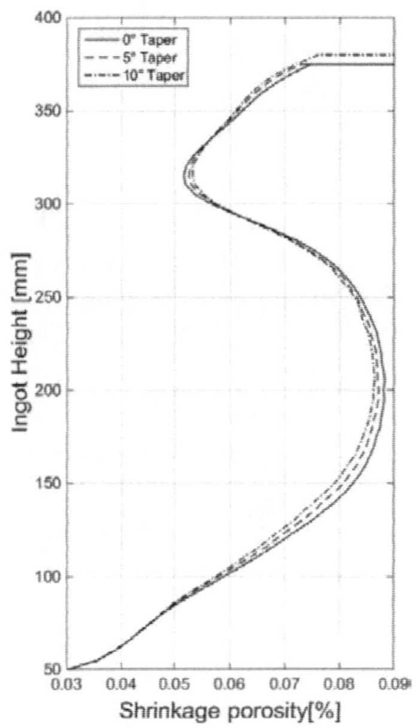


**Figure 9.** Comparison of porosity amount using the ProCAST internal method

Figure 7 shows the amount of shrinkage porosity in the ingot centre for different mould taper. It is obvious that the amount decreases with increasing taper. The amount of shrinkage porosity calculated with ProCAST internal model is depicted in figure 9. The position of maximum shrinkage porosity is the same for both calculation methods. Comparing the amount of the occurring porosity the method after Carlson et al. is about 40 % lower than with the use of the ProCAST internal method. The amount and also distribution of the pores calculated with the ProCAST internal model is higher and more uniform than the dimensionless Niyama criterion. Hence, the distinction of the shrinkage porosity between different taper dimensions is tiny when using the internal method.

## 7. Conclusion

Experimental investigations were made regarding the correlation between mould taper and shrinkage porosity. For this purpose, three ingots with different tapers (0°, 5°and 10°) were cast. The amount of the centre porosity decreases with increasing mould taper, which reflects the literature. The ingot with a 10° taper has no visible porosity. Further efforts were made to model the cast ingots and implement the Niyama criterion to obtain information about porosity formation. The Niyama criterion for the ingot with a 10° taper does not fall below a critical value of 1 [(K min)$^{0.5}$/cm], and, consequently no porosity appears. This is in good agreement with the experimental results. The Niyama value for the ingot with no taper is below 1. Therefore, shrinkage porosity will occur. The position of this formation regarding the ingot height is slightly higher compared to the experimental results.

Using the dimensionless Niyama criterion it is not only possible to predict the formation of shrinkage porosity, but also the amount of porosity can also be determined. It can be seen that the amount of shrinkage porosity decreases from 0.05% to 0.03% with increasing mould taper. ProCAST has also the ability to calculate the amount of shrinkage porosity. The calculated porosity percentage (by use of the ProCAST internal method) is approximately two-thirds higher when compared to the dimensionless Niyama criterion. Moreover, the distribution of the shrinkage porosity is more uniform.

## 8. References

[1] Dantzig J A and Rappaz M 2009 Solidification (Lausanne: EPLF Press)
[2] Couturier G and Rappaz M 2006 *Model. Simul. Mater. Sc.* 253–71
[3] Couturier G and Rappaz M 2006 Modeling of porosity formation in multicomponent alloys in the presence of several dissolved gases and volatile solute elements *Symp. on Simulation of Aluminum Shape Casting Processing (San Antonio, Texas, USA)* pp 143-52
[4] Niyama E, Uchida T, Morikawa M and Saito S 1982 *AFS Int. Cast Met. J.* **7** 52–63
[5] Stefanescu D M 2005 *Int. J. Cast Metal. Res.* **18** 129–43
[6] Stefanescu D M 2009 Science and Engineering of Casting Solidification (New York, NY: Springer Science + Business Media)
[7] Carlson K D, Ou S, Hardin R A and Beckermann C 2002 *Metall. Mater. Trans. A* **33** 731–40
[8] Sigworth G K and Wang C 1993 *Metall. Mater. Trans. B* **24** 349–64
[9] Carlson K D and Beckermann C 2009 *Metall. Mater. Trans. A* **40** 163–75
[10] ESI Group 2013 ProCAST 2013.5: *User's Manual* (Paris: ESI Group)
[11] Sang B, Kang X and Li D 2010 *J. Mater. Process. Tech.* **210** 703–11
[12] Orgeas L, Gabathuler J P, Imwinkelried T, Paradies C and Rappaz M 2003 *Model. Simul. Mater. Sc.* **11** 553–74
[13] Miettinen J, Lauhenkiipi S and Laine J 1997 IDS - Solidification Analysis Package (Helsinki: Helsinki University of Technology)
[14] Miettinen J 1997 *Metall. Mater. Trans. B* **28** 281–97
[15] Miettinen J, Louhenkilpi S, Kytönen H and Laine J 2010 *Math. Comput. Simulat.* **80** 1536–50

# Free-surface flow in horizontally rotating cylinder: experiment and simulation

**J Bohacek[1], A Kharicha[2, 4], A Ludwig[1], M Wu[2], A Paar[3], M Brandner[3], L Elizondo[3] and T Trickl[3]**
[1]Dept. of Metallurgy, Montanuniversitaet Leoben, Austria
[2]Christian Doppler Lab for Advanced Simulation of Solidification and Melting, Montanuniversitaet Leoben, Austria
[3]Eisenwerk Sulzau-Werfen, R. & E. Weinberger AG, Austria

E-mail: abdellah.kharicha@unileoben.ac.at

**Abstract.** The horizontal centrifugal casting process targets on a liquid layer with a uniform thickness. To achieve this, the rotations of the mold have to be large enough so that the liquid can pick up the speed of the mold. In the present paper, an experiment was conducted using a laboratory plexi-glass mold with water as a working fluid. Starting with an initial volume fraction of liquid resting in the bottom of the mold, the mold rotations were gradually increased from 0 rpm to max rpm and a new position of the contact line was recorded. In addition, first critical rpm was recorded, at which the transition from the liquid pool to a uniform liquid layer occurred. While gradually going back from max rpm to 0 rpm, second critical rpm was recorded, at which the uniform liquid layer collapsed. The experiment was compared with the numerical simulation solving the modified shallow water equations using the Newton-Raphson method with the Wallington filter.

## 1. Introduction

The horizontal centrifugal casting (HCC) is a process at which it is among other things necessary to use high enough rotations of the cylindrical mold, so that the liquid picks up the speed of the rotating mold and creates a layer of a uniform thickness around the mold circumference. In fact, also upper limit exists such that too high rotations can lead to excessive vibrations of the system, which are commonly related to undesired hoop stresses in the product [1]. An experimental and numerical study of the first is a main target of the present paper. The experiment and the numerical simulation are discussed in detail in section 2 and section 3, respectively.

The survey of related experimental articles showed that most researchers concentrated on free-surface patterns corresponding to mold rotations at which the liquid was already moving with the speed of the mold. For example in [2], Martinez identified several different wave patterns on the free surface of the liquid layer and also stated that some of the wave patterns are related to mold vibrations and deformations. In [3], the effort of Esaka moved from observing free surface patterns to investigation of what is happening inside the transparent solidifying liquid. By attaching a high-speed camera to the rotating mold, a relative movement of a selected equiaxed crystal could be analysed. During each rotation of the mold the position of the equiaxed crystal oscillates due to the pumping effect caused by the gravity. Moreover, the selected equiaxed crystal slowly moves in the anti-

[4] To whom any correspondence should be addressed.

rotational direction due to a certain slip between the liquid bulk and the mold. The pumping effect as a result of interaction between the inertia force and the gravity is also responsible for having a thicker layer in the top than in the bottom. In another experimental study [4], Prasad searched for minimal mold rotations required to form a uniform liquid layer. Liquids with higher viscosities and lower volume fractions naturally require lower rpm. He also pointed out that minimal rpm is influenced by the formation of the Ekman vortex near the mold extremities. The liquid in a longer mold will pick up the speed easier than in a shorter mold of the same diameter. Other investigators (e.g. [5]) were more focused on examination of coating flows inside a horizontally rotating cylinder generally containing a small amount of a very viscous liquid. A rich variety of flow patterns has been observed.
Unlike flows in the horizontal centrifugal casting, the coating flows were intensively investigated using the thin-film lubrication equations often in dimensionless form. For example in [6], Hosoi used the lubrication theory to capture so-called shark-teeth pattern, pseudo-stationary pendants, and periodic sloshing. Numerical tools used in the modelling of the horizontal centrifugal casting generally rely on commercial CFD packages namely the VOF free-surface tracking [7]–[9]. In our previous work we applied so-called shallow water equations especially for the purpose of major computational time savings [10]. In the present paper, in addition to the experiment discussed later we introduced modified shallow water equations in dimensionless form solved by the iterative Newton-Raphson method [11]. The objective of the present numerical study is to demonstrate the capability of capturing the main flow features. The numerical model is compared with the experiment.

## 2. Experimental set-up

Here, the experiment focuses on analysis of fluid flow in the horizontal centrifugal casting process namely on estimation of the contact line position within a given range of Froude numbers. The dimensionless Froude number $Fr$ is defined as ratio between the centrifugal inertia and the gravity $\left(\Omega\sqrt{R/g}\right)$, where $\Omega$ is the angular frequency of the mold in rad/s, $R$ is the inner mold radius, and g is the gravitational acceleration. The definition of the contact line is pictured in figure 1. In addition, we looked for a critical Froude number $\uparrow Fr_c$ at which transition from the liquid pool to a uniform layer occurred. We also looked for other critical Froude number $\downarrow Fr_c$ at which raining [12] appeared and the originally uniform layer collapsed. For this experiment, a laboratory plexi-glass mold was used with the length of 705 mm and the inner diameter of 285 mm. On one side, the mold is connected through a shaft to an AC motor, whose speed can be varied via a frequency changer from 0 rpm to 500 rpm. On the other side, the mold is opened so that the mold filling can be easily done. The schematic of the whole device is illustratively shown in figure 2. Snapshots of the free surface were taken by the digital camera CASIO EX-F1 (F2.7, 1/1000) mounted approximately coaxially with the mold axis, placed in a laboratory (non-rotating) reference frame. The amount of 1 liter of water at room temperature was used as a working liquid. The experiment involved the following steps:

- Pour a specific amount of liquid.
- For the post-processing namely automatic axis insertion, take a picture of the calm free surface.
- Very slowly increase mold rotations and at defined rpm take a picture of the actual free surface.
- During increasing mold rotations check for $\uparrow Fr_c$.
- After exceeding $\uparrow Fr_c$ very slowly decrease mold rotations and check for $\downarrow Fr_c$.

The experiment was followed by a simple image post-processing. The snapshot with the calm free surface was used to transform cylindrical coordinate system in perspective projection and insert tangential ticks (figure 3) eventually copied into all other pictures. From each picture, the contact line position was extracted and plotted against the corresponding Froude number $Fr$ (figure 5). On horizontal axis we also marked the location of both, $\uparrow Fr_c$ and $\downarrow Fr_c$. Note that $\uparrow Fr_c$ required to form a uniform layer is generally much higher than $\downarrow Fr_c$ corresponding to the moment when the uniform layer collapses. The magnitude of $\downarrow Fr_c$ is approaching unity ($\downarrow Fr_c \rightarrow 1$) indicating an exact balance between the centrifugal force and the gravity. The theoretical $\downarrow Fr_c$ ($\downarrow Fr_c = 1$) is shown as a dot-and-dashed line on horizontal axis in figure 5.

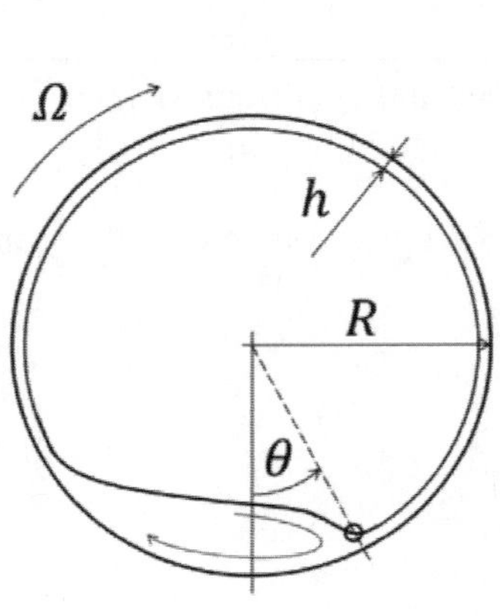


**Figure 1.** Schematic diagram of a cross section of the rotating plexiglass mold. Angle $\theta$ signifies tangential position. (The actual magnitude of angle $\theta$ corresponds to the contact line position.

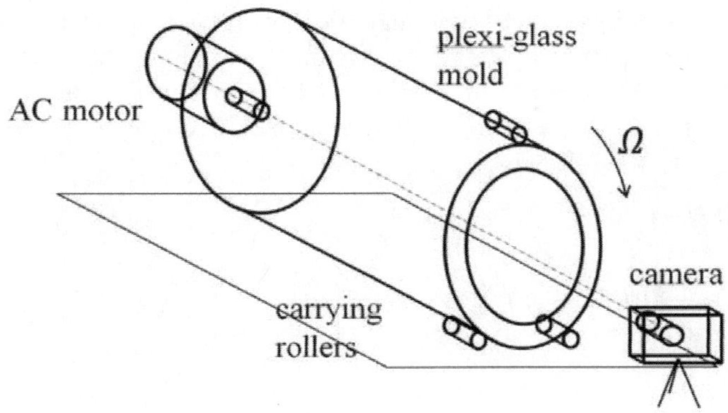


**Figure 2.** Sketch of the experimental set-up.

In the reality, $\downarrow Fr_c$ is always greater than unity due to the fact that $Fr$ is defined using the mold radius $R$ and not the radius measured from the free surface to the mold center. Next, it is also the relative velocity field generated by the pumping effect and inherent vibrations that increase $\downarrow Fr_c$. A different story applies to $\uparrow Fr_c$ where it is the ratio between the viscous forces and the gravity $\left(\alpha = \sqrt{\nu\Omega \mathrm{R}/gh^2}\right)$ responsible for picking up the liquid, where $h$ and $\upsilon$ are the thickness of the liquid film and the kinematic viscosity. $\uparrow Fr_c$ can be estimated from setting $\alpha = 1$ and substituting $Fr$ into $\alpha$. Having the Reynolds number defined as $Re = \Omega h^2/\nu$, $\uparrow Fr_c$ becomes simply $\sqrt{Re}$. -However, when the turbulence is present the effective viscosity should be used instead. Hence, calculation of- $\uparrow Fr_c$ is not trivial. In addition to $Fr$ analysis, in figure 4 one can notice so-called shark-teeth pattern studied earlier in [14]. The range of occurrence is also demarcated in the figure 5. In the next section we attempt to simulate the aforementioned scenario using modified shallow water equations with more details given for example in [15]–[17].

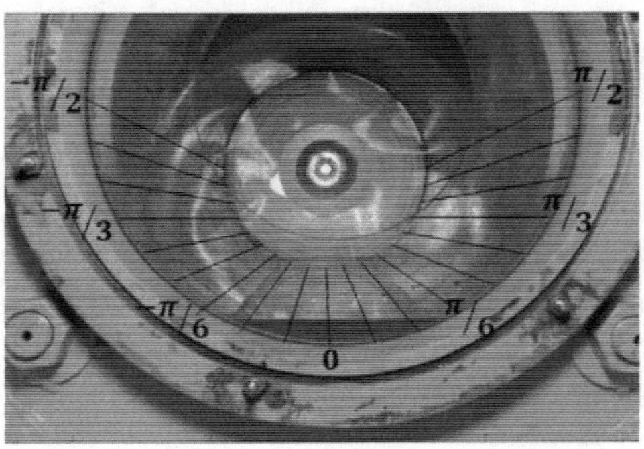


**Figure 3.** Snapshot of the calm free surface. The liquid volume fraction of 1 liter.

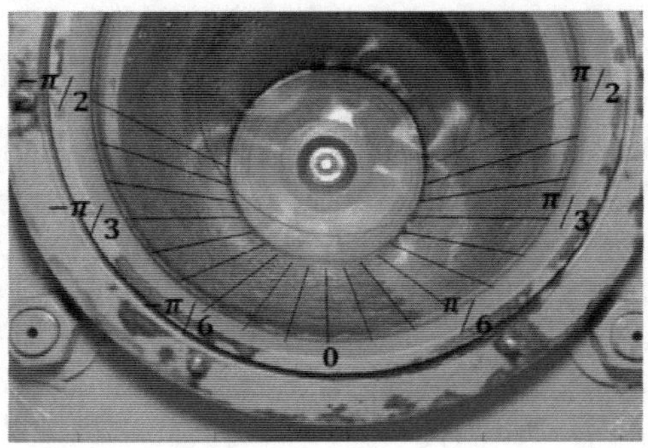


**Figure 4.** Shark-teeth pattern at $Fr \cong 1.3$ (105 rpm).

## 3. Numerical model

Here, the objective of the numerical model is to examine whether the modified shallow water equations (SWE) can capture the main flow features such as the recirculation region in the bottom of the mold, correct position of the contact line, $\uparrow Fr_c$ and $\downarrow Fr_c$. The SWE belong to the group of hydrostatic models [18]. For the cylindrical geometry, the original SWE were modified and can be written as the following:

$$h_t + (hu)_x = 0 \tag{1}$$

$$(hu)_t + \left(hu^2 + \frac{1}{2}gh^2\cos\theta + \frac{1}{2R}(hu)^2\right)_x = -gh\sin\theta - 3v\frac{u - \Omega R}{h} \tag{2}$$

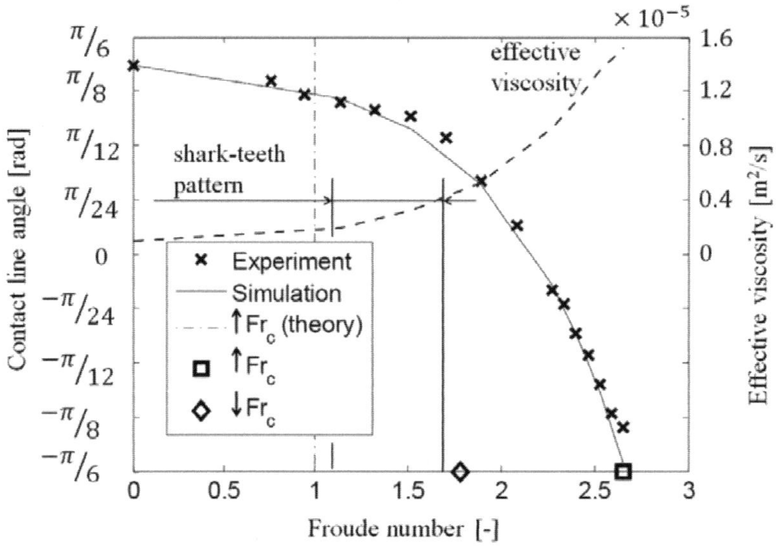


**Figure 5.** Contact line vs. Froude number for the liquid volume fraction of 1 liter obtained from the experiment and the simulation. The occurrence range of shark-teeth pattern is demarcated. $\uparrow Fr_c$ and $\downarrow Fr_c$ are marked on the horizontal axis. Effective viscosity used in simulation is shown in dashed line.

In the present study, the axial momentum was omitted from the calculation. Equations (1)–(2) are the continuity equation and the momentum equation for the tangential direction. Subscripts $t$ and $x$ denote temporal and spatial derivatives, respectively. Since $h \ll R$, the Cartesian coordinate system is used and the tangential coordinate $x$ is defined as $x = R\theta$ with $\theta = 0$ rad in the bottom of the mold. In (2), the second and the third term in the fluxes $(\ldots)_x$ represent the gradient of the hydrostatic pressure integrated over the liquid height $h$, resulting from the radial component of the gravity and the motion in a curved geometry, respectively. On the rhs, the first term corresponds to the tangential component of the gravity, while the last term denotes the bed shear stress with the moving mold wall assuming a parabolic velocity profile [19]. Then, equations (1)–(2) were further rewritten in the dimensionless form. Dimensionless variables in the problem become

$$\tilde{h} = \frac{1}{R}h \qquad \tilde{u} = \frac{1}{\Omega R}u \quad \tilde{t} = \frac{1}{\Omega}t$$
$$Fr = \Omega\left(\frac{R}{g}\right)^{1/2} \qquad Re = \frac{\Omega R^2}{v} \tag{3}$$

IOP Conf. Series: Materials Science and Engineering **143** (2016) 012036 doi:10.1088/1757-899X/143/1/012036

Equations (1)–(2) in the dimensionless form become (Note that the tilde ~ was dropped)

$$h_t + (hu)_x = 0 \quad (4)$$

$$(hu)_t + \left(hu^2 + \frac{1}{2}Fr^{-2}h^2\cos\theta + \frac{1}{2}(hu)^2\right)_x = -Fr^{-2}h\sin\theta - 3Re^{-1}\frac{u-1}{h} \quad (5)$$

Equation (4)–(5) can be categorized as the non-linear conditionally hyperbolic equations. The hyperbolicity can be lost exclusively due to the negative second term in the fluxes $(\ldots)_x$ of (5) in the upper part of the mold. To solve (4)–(5) an implicit scheme for the time differencing was used with a timestep $\Delta t$ chosen by step doubling. The resulting equations were solved simultaneously via the Newton-Raphson method using <5 iterations. Handling of dry cells was remedied by adding an artificial viscosity term $Dh_{xx}$ to (4) with the diffusivity term $D$, acting only when $h \to 0$. In addition, dispersive errors were damped via the Wallington filter [20]. In figure 6, free surface patterns are shown at different Froude numbers $Fr$. The corresponding position of the contact line is plotted along with the experimental data in figure 5. As already mentioned, the kinematic viscosity needs to be replaced with the effective viscosity due to turbulent effects. Since the turbulent effects are not accounted for by (4)–(5), the effective turbulent viscosity is obtained by solving the inverse task with the objective function equal to the difference between the computed and measured position of the contact line. The calculated effective viscosity is shown on the second $y$-axis in figure 5. The higher rotations of the mold will produce a more turbulent flow with a higher effective turbulent viscosity.

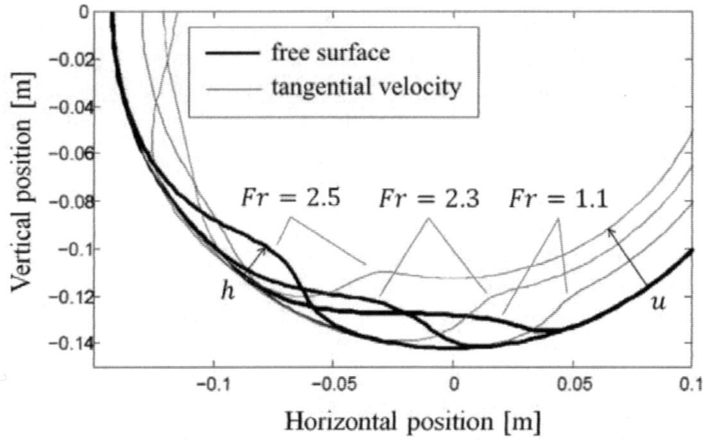


**Figure 6.** Calculated free-surface and magnitudes of average tangential velocity shown at different Froude numbers $Fr$.

In figure 7, we show a detail of the recirculation zone at $Fr = 2.3$. Note that although the 2D velocity vector field is not directly calculated from (4)–(5), it can be reconstructed with the help of the continuity equation, 1D field of tangential velocities $u$, and the assumption of a parabolic velocity profile.

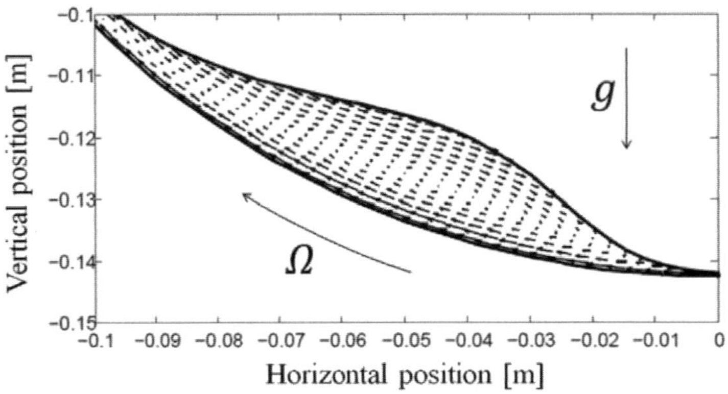


**Figure 7.** Recirculation zone with a reconstructed velocity vector field at $Fr = 2.3$.

## 4. Conclusions

Using a laboratory plexi-glass mold we performed experiments with water at room temperature as a working fluid. Here, only a single liquid volume fraction was tested (1 liter). During the experiment the rotations were gradually increased and the position of the contact line was recorded. In addition, $\uparrow Fr_c$ required for the liquid to pick up the speed of the mold and $\downarrow Fr_c$ necessary to prevent raining and consequent collapse of the uniform liquid layer were examined. The experiment showed that $\uparrow Fr_c$ is generally much higher than $\downarrow Fr_c$. While it is possible to estimate $\downarrow Fr_c$ quite reliably using the condition of the hydrostatic balance, it is hard to draw any quantitative judgement on $\uparrow Fr_c$ due to due to strong turbulent effects adjusting the effective viscosity. During the industrial centrifugal casting process namely during the pouring stage, it is necessary to keep $Fr \geq \uparrow Fr_c$ in order to pick up the speed of the mold. On the other hand, once the pouring is finished and the liquid forms a uniform layer, it is apparently sufficient to keep $Fr$ slightly larger than 1 ($\downarrow Fr_c$). The positive difference is needed to compensate the pumping effect and mold vibrations.

In addition to the experiment, we investigated whether the modified shallow water equations are able to capture the recirculation region formed close to the bottom of the mold. We concluded that the mathematical model possess this ability, however; requires an estimation of the effective viscosity. In fact, this relatively simple mathematical model can also successfully predict the pumping effect caused by the gravity. This topic is not covered in the present study. For the future, we plan to test wide variety of liquid volume fractions and liquids with different viscosities.

## 5. Acknowledgement

Financial support by the Austrian Federal Government (in particular from the Bundesministerium fuer Verkehr, Innovation und Technologie and the Bundesministerium fuer Wirtschaft, Familie und Jugend) and the Styrian Provincial Government, represented by Oesterreichische Forschungsfoerderungsgesellschaft mbH and by Steirische Wirtschaftsfoerderungsgesellschaft mbH, within the research activities of the K2 Competence Centre on "Integrated Research in Materials, Processing and Product Engineering", operated by the Materials Center Leoben Forschung GmbH in the framework of the Austrian COMET Competence Centre Programme, is gratefully acknowledged. This work is also financially supported by the Eisenwerk Sulzau-Werfen R. & E. Weinberger AG.

IOP Conf. Series: Materials Science and Engineering **143** (2016) 012036 doi:10.1088/1757-899X/143/1/012036

## 6. References

[1] Mandal D 2008 Special Metal Casting and Forming (CAFP) pp 1-19
[2] Martinez G, Garnier M and Durand F 1987 *Appl. Sci. Res.* **44** pp 225–239
[3] Esaka H, Kawai K, Kaneko H and Shinozuka K 2012 *Mater. Sci. Eng.* **33** 012041
[4] Keerthiprasad K S, Murali M S, Mukunda P G and Majumdar S 2010, *Front. Mater. Sci. China* **4** pp 103–110
[5] Thoroddsen S T, Mahadevan L 1997 *Exp. Fluids* **23** pp 1–13
[6] Hosoi A E, Mahadevan L 1999 *Phys. Fluids* **11** pp 97–106
[7] Xu Z, Song N, Tol R V, Luan Y and Li D 2012 *Mater. Sci. Eng.* **33** 012030
[8] Keerthiprasad K S, Murali M S, Mukunda P G and Majumdar S 2010 *Metall. Mater. Trans.* B **42** pp 144–155
[9] Kaschnitz E 2012 *Mater. Sci. Eng.* **33** 012031
[10] Bohacek J, Kharicha A, Ludwig A and Wu M 2014 *ISIJ Int.* **54**(2) pp 266–274
[11] Tjalling J Y 1995 *SIAM Rev.* **37**(4) pp 531–551
[12] Lebeau T C 2008 *MSc thesis* University of Alabama p 16
[13] Bagheri J, Das S K 2013 J. Appl. Computat. Math.
[14] Krasnopol'skaya T S, van Heijst G J F, Voskamp J H and Trigger S A 2001 *Int. Appl. Mech.* **37**(7) pp 929–934
[15] Bohacek J, Kharicha A, Ludwig A and Wu M 2013 *Mater. Sci. Eng.* **33** 012032
[16] Bohacek J, Kharicha A, Ludwig A and Wu M 2015 *Appl. Math. Comput.* **267** pp 179–194
[17] Kharicha A, Bohacek J, Ludwig A and Wu M 2015 *J. Fluids Eng.* **137**(11) 111105
[18] George D. L. 2006 *PhD thesis* University of Washington, p 36
[19] Murillo J, Gracía-Navarro P 2012 *J. Comput. Phys.* **231**, pp 1963–2001
[20] Wallington C E 1962 *J. Roy. Meteor. Soc.* **88** pp 437–484

IOP Conf. Series: Materials Science and Engineering **143** (2016) 012037 doi:10.1088/1757-899X/143/1/012037

# Effect of high-intensity ultrasonic treatment on microstructure, hardness and wear behaviour of the hypereutectic Mg-5Si alloy

**M E Moussa**[a], **M A Waly**[a], **A M El-Sheikh**[b]

[a]Department of Manufacturing Technology, Laboratory of Foundry, Central Metallurgical Research and Development Institute (CMRDI), P.O. 87, Halwan, Egypt

[b]Department of Mining, Petroleum and Metallurgical Engineering, Faculty of Engineering, Cairo University, P.O. 12613, Giza, Egypt

E-mail address: mwaly@cmrdi.sci.eg

**Abstract**. The effect of high-intensity ultrasonic treatment (HIUST) on microstructure, hardness and wear behavior in Mg–5wt.%Si hypereutectic alloy has been investigated. The results showed clearly that without HIUST, most of primary $Mg_2Si$ appeared as coarse dendritic morphology with average size of about 200 µm. With HIUST, the average size of primary $Mg_2Si$ decreased significantly to about 33 µm and their morphologies changed to polyhedral shape. The modification mechanism is mainly attributed conjugation of two mechanisms: cavitation-enhanced heterogeneous nucleation and cavitation-induced dendrite fragmentation. The alloy treated with HIUST has higher hardness and wear resistance than that untreated with HIUST. The wear mechanism of investigated alloys at low applied load (10 N) and low sliding speed (0.3 m/s) is a mild abrasive oxidative wear with little adhesion. However, the wear mechanism due to the applied high loads (30, 50 N) at low sliding speed (0.3 m/s) and/or to the applied high sliding speeds (0.6, 0.9 m/s) under low load (10 N), could be described as delamination mechanism. The microstructures of the specimens were analyzed by optical microscope (OM) (model OPTIKA M–790, Italy). Energy dispersion spectrum (EDS) affiliated to field emission scanning electron microscopy (FESEM) (model Quanta FEG, The Netherlands) were performed to reveal the concentration of alloying elements in selected areas of the microstructure.

## 1. Introduction

Magnesium alloys are widely applied in the fields which are strongly driven towards weight-reduction, such as automotive, aeronautic and astronautic industries [1]. Meanwhile, improving the elevated temperature properties of magnesium alloys has become a critical issue for the further application [2]. In recent years, the fascinating properties and promising application of hypereutectic Mg–Si alloys have attracted particular interest due to the formation of thermally stable $Mg_2Si$ phase [3]. It is known that the intermetallic compound of $Mg_2Si$ exhibits an excellent combination of superior properties, such as high melting temperature (1085 °C), low density ($1.99 \times 10^3$kg m$^{-3}$), high hardness ($4.5 \times 10^9$ Nm$^{-2}$), low thermal expansion coefficient ($7.5 \times 10^{-6}$ K$^{-1}$) and reasonably high elastic modulus (120 GPa). Furthermore, the $Mg_2Si$ phase is exceptionally stable and therefore could effectively impede grain boundary sliding at elevated temperatures. However, the hypereutectic Mg–Si alloys prepared by ordinary ingot metallurgy process showed very low ductility and strength due to the large primary $Mg_2Si$ particle size and the brittle eutectic phase [3, 4].

Much work has been focused on the modification effects of Ca [1], P [2], Sr–Sb [3], Y [4], $KBF_4$ [5] and Bi [6] on the primary and eutectic $Mg_2Si$ in magnesium alloys. However, certain internal defects could be existed. For example, although $KBF_4$ has a good effect on the modification of $Mg_2Si$ phase in Mg–5Si alloys, a large quantity of smoke and sputter is produced, which is harmful to the surrounding environment [5].Y [4] and Bi [6] can modify and refine the primary $Mg_2Si$, but the $Mg_2Si$ phase becomes coarse again when the adding amount exceeds a certain limit. Therefore, an alternative method for the development of hypereutectic Mg–Si alloys with modification of $Mg_2Si$ phase and high mechanical properties is desired.

HIUST is a simple and effective physical method for solidification control. Several investigations were carried out between 1960 and 1990 [7, 8], mainly in the former Soviet Union countries. They clearly demonstrated that the grain–refinement effect on magnesium alloys improvement of mechanical properties. The renewed interest in magnesium materials in recent years has resulted in increased attention in the grain-refinement potential of HIUST for magnesium alloys [9–11], particularly for Mg–Al based alloys.

Recent research [12, 13] has shown that the primary $Mg_2Si$ could be effectively refined and modified by the application of HIUST on the hypereutectic Mg–5 wt.%Si alloy during solidification process. However, the effect of HIUST on wear behavior of hypereutectic Mg–Si alloys has not been reported. Therefore, the main aims of this work are to investigate the effect of HIUST on hardness and wear behavior at different applied loads (10, 30, 50 N) and different sliding speeds (0.3, 0.6, 0.9 m/s) of Mg–5wt.%Si hypereutectic alloy and to explore the wear mechanisms of resultant samples.

## 2. Experimental procedures

### *2.1. Materials and processing*

Commercial pure Mg (99.8 wt.% purity) and Si (99.9 wt.% purity) were used as starting materials. Charges of about 20 kg with the nominal composition of hypereutectic Mg–5wt.%Si alloy were prepared as the base material in the present study. The charge was melted in a graphite crucible by a 200 kW medium frequency induction furnace under the mixed gas protection consists of tetrafluoroethane ($CF_3CH_2F$, HFC-134a, 1 vol.%) and carbon dioxide ($CO_2$, Bal.). Firstly commercial Mg ingots were melted to above 650 °C, and then silicon was added into Mg melt. After that the melt were heated to above 800 °C and kept about 30 min to ensure Mg and Si fully reacted and formed $Mg_2Si$. Finally the melt was poured into a cast iron mold. The hypereutectic Mg–5wt.%Si alloy ingots were sliced for subsequent experiments.

The ultrasonic treated alloy were prepared as following. The hypereutectic Mg–5 wt.%Si alloy of about 1 kg was remelted at 800 °C in a mild steel crucible using an electric resistance furnace under the mixed gas protection consists of tetrafluoroethane ($CF_3CH_2F$, HFC-134a, 1vol.%) and carbon dioxide ($CO_2$, Bal.). The melt was manually stirred for 2 min using a stainless steel rod, and then was held for additional 10 min in order to get full homogenization. After that the slag was removed, and then the melt was poured into a cylindrical resin-bonded sand mold with dimensions of outer diameter (Ø 100 mm), inner diameter (Ø 42 mm) and length (250 mm) which mounted on the ultrasonic sonotrode of diameter (Ø 40 mm) as shown in Figure 1. The reason for using a cylindrical resin-bonded sand mold is to reduce the cooling rate effect on the resulted microstructures of the prepared samples. Therefore, the difference in the morphology and size of primary $Mg_2Si$ in the microstructures of investigated samples were obtained mainly as a result of the difference in the application of HIUST conditions. HIUST which was generated by using ultrasonic generator (model TS6MD1, Russia) and magnetostrict transducer (model PMS-15-22, Russia) with the maximum output power of 5 KW and the fixed frequency of 21.4 KHz was applied right before pouring the melt at pre-determined optimum temperature of about 800 °C for the pre-determined optimum vibration time of about 90 s based on recorded data [20]. At the end of planed HIUST application time, the ultrasonic source was switched off and the melt was left to room temperature. Casting temperature was controlled within an accuracy of ±2 °C. The ultrasonic waves emitted from the transducer and passed through the acoustic sonotrode were propagated directly into the melt during solidification. The poured melt became a part of acoustic sonotrode, so the action of ultrasonic energy on the melt was raised remarkably. For comparison, alloy without the application of HIUST was prepared at pouring temperature of 800 °C. The chemical composition of the prepared alloy without and with HIUST was measured with X-ray fluorescence analyzer (XRF) (model Axios advanced-PANALYTCAL, The Netherlands) as shown in Table 1.

**Table 1:** The chemical composition of prepared alloys (wt.%).

| Prepared alloys | Mg | Si | Fe | Cu |
|---|---|---|---|---|
| Mg-5Si | Bal. | 5.02 | 0.033 | 0.041 |

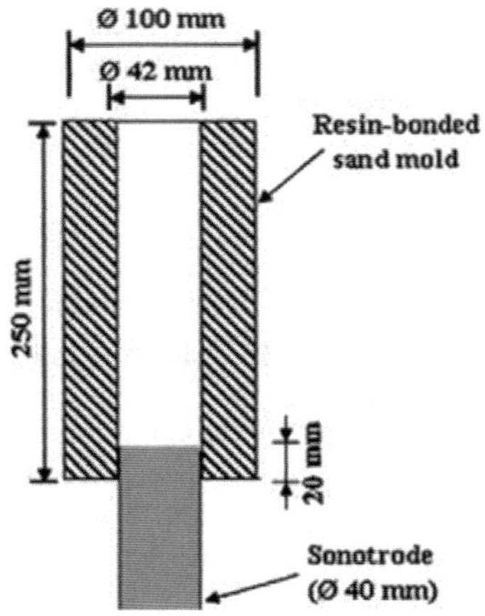


**Figure 1:** Schematic of the experimental setup used in this study.

*2.2. Materials characterization*

All metallographic specimens were cut at the same position of 10 mm from the bottom of castings and prepare according to usual procedures developed for magnesium alloys. Grinding of specimens is carried out on silicon carbide papers wet with water, down to grit size 1200 then followed by fine polishing with diamond, particle size from 6 to 1 μm, on nylon cloth, rinsing with water and ethyl alcohol. For best results, the polished surface must be mirror like, i.e., free from any residual tarnishing layer. After that, they are etched by solution with 10 ml nitric acid, 30 ml acetic acid, 40 ml water, and 120 ml ethanol for 2–3 min. They are then rinsed in anhydrous ethyl alcohol and dried in a blast of dried air.

The microstructures of the specimens were analyzed by optical microscope (OM) (model OPTIKA M–790, Italy). In the present study, the average length of primary $Mg_2Si$ was measured as the size of $Mg_2Si$. Six OM micrographs were taken for each sample from the optical observed area at magnification of 100x. The average size of the primary $Mg_2Si$ was measured by ImageJ1.44 software. All $Mg_2Si$ existed in one picture taken from the observed area were measured. Energy dispersion spectrum (EDS) affiliated to field emission scanning electron microscopy (FESEM) (model Quanta FEG, The Netherlands) were performed to reveal the concentration of alloying elements in selected areas of the microstructure.

Phase constituents of samples were analyzed by X–ray diffraction (XRD) (model X'PERT PRO, The Netherlands) using Cu Kα radiation in step scan of 2θ from 20° to 120° with an increment of 0.02° and a scanning speed of 4°/min.

*2.3. Hardness and wear tests*

The Vickers hardness test of the prepared alloys using the metallographic specimens, at room temperature, was carried out in a Vickers hardness tester (model INSTRON WOLPERT GMBH–930/250, England) according to ASTM E92-82 with a normal load of 3 Kgf (designated as $HV_3$). The mean of ten successful measurements was taken to establish the hardness values.

Dry sliding wear test without lubricant was conducted using a pin–on–disc type apparatus (model TNO TRIBOMETER, The Netherlands) in accordance to the ASTM G99-05 standard. The cylindrical pin specimens having diameter (Ø7 mm) and length (12 mm) machined out from the same position of the prepared castings were used as test samples. Hardened ball bearing steel disc (HRC 63) of outer diameter (Ø73 mm), inner diameter (Ø65 mm) and thickness (25 mm) was used as the counterpart surface. Specimens and counterpart surfaces were ground with different emery papers up to 1200 grit and cleaned ultrasonically in acetone to avoid the presence of humidity and non–desirable deposits. During testing, a jet of compressed air was pointed at the edge of the disc to avoid accumulation of wearing particles on the disc. All tests were performed under ambient atmosphere. The wear test conditions were divided into three groups. In the first group, the wear test of samples was carried out with three different normal loads (10, 30, 50 N) at a constant sliding speed of 0.3 m/s for 10 min. In the second group, the wear test was carried out with two different sliding speeds (0.6 and 0.9 m/s) at a constant load of 10 N for 10 min. The weights were measured before and after the experiment using electronic scales with 0.1 mg accuracy, after which the results of the experiment were evaluated according to the loss in weight. Worn surfaces and wear debris of the specimens were examined and analyzed using FESEM equipped with EDS in order to determine the post–experimental wear mechanisms.

## 3. Results

### *3.1. Materials investigation*

XRD results reveal that the constituents of the obtained microstructures for both conditions without and with HIUST are only $Mg_2Si$ and Mg phases, as shown in Figure 2. Therefore, no change of phase constituents obtained due to HIUST.

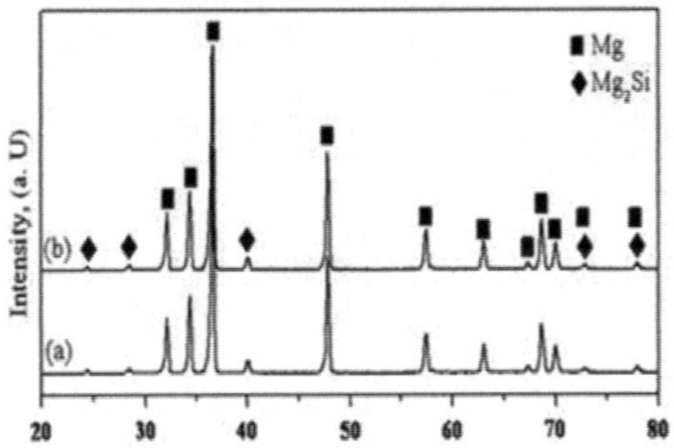


**Figure 2:** XRD patterns of the investigated hypereutectic Mg–5wt.%Si alloys (a) without and (b) with HIUST.

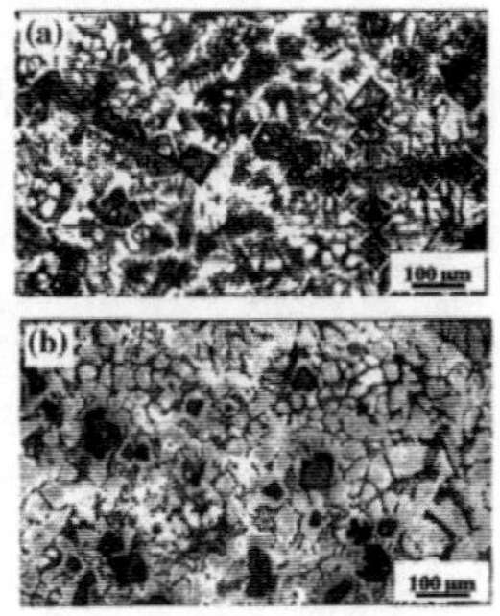


**Figure 3:** Optical images of the as-cast Mg–5wt.%Si alloy: (a) Without treatment and with (b) HIUST.

Figure 3 shows the optical images of the investigated alloy without and with HIUST. Furthermore, Figure 4 shows the FESEM micrographs with EDS line scan across primary $Mg_2Si$ and EDS elemental mappings for the elements Mg, Si and Ca for the investigated alloy without and with HIUST. The as-cast microstructures of the prepared Mg–5wt.%Si hypereutectic alloy reveal the presence of primary $Mg_2Si$, Mg halos and eutectic Mg–$Mg_2Si$ (Figure 3). Moreover, all the primary $Mg_2Si$ are surrounded by Mg halos, followed by eutectic structure. Without any treatment, most of primary $Mg_2Si$ are coarse dendritic morphology (Figure 3a) with average size of about 200±2 μm. Also, it can be seen that the coarse primary $Mg_2Si$ is formed in the preferred growth direction that occurs at the tips of branches (Figure 4a), resulting in a complex regular with sharp-angled shapes and dendritic morphologies with a nonuniform distribution of Mg in the interdendritic regions (Figure 4b, c). With HIUST, most of primary $Mg_2Si$ become polyhedral shape (Figures 3b, 4d) containing a network of segregated Mg along the grain boundaries (Figure 4e, f) and their average size are reduced to about 33±1μm.

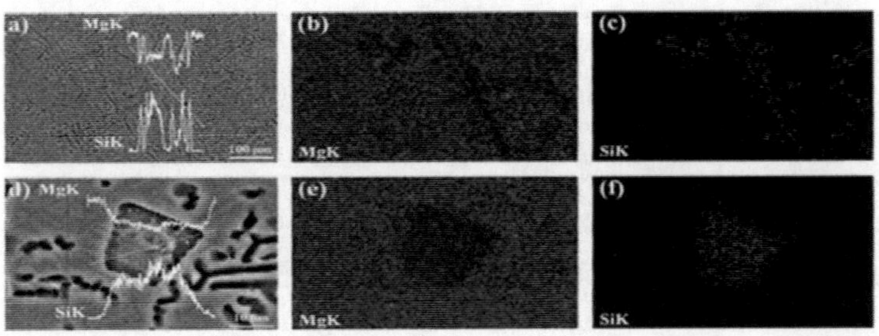


**Figure 4:** The FESEM micrographs with EDS line scan across primary $Mg_2Si$ and EDS elemental mappings of Mg, Si and Ca for the prepared Mg–5 wt.%Si alloy: (a), (b) and (c) without HIUST and (d), (e) and (f) with HIUST.

### *3.2. Hardness and wear behavior*

The average hardness values of the investigated alloy significantly increases from about 50±5.5 $HV_3$ without HIUST to about 69±1 $HV_3$ with HIUST (i.e. a 38% increase) , implying that HIUST can improve the hardness of the investigated alloy.

Figures 5 and 6 show the variation of weight loss of the investigated alloy without and with HIUST for various loads (10, 30, 50 N) at constant sliding speed (0.3 m/s) and for various sliding speeds (0.3, 0.6, 0.9 m/s) at constant load (10 N). The investigated alloy with HIUST has better wear resistance than without HIUST due to its lowest weight loss at the same load level (Figure 5) and at the same load sliding speed level (Figure 6). However, with the application of high normal loads or high sliding speeds for the same process, the weight loss increases drastically. However, the investigated alloy due to HIUST still has better wear resistance than without HIUST.

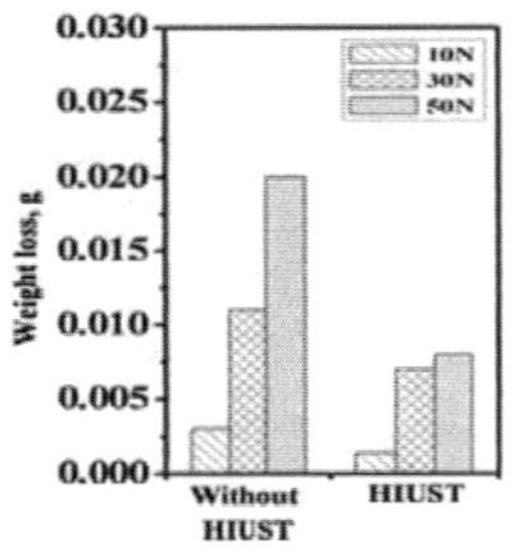


**Figure 5:** The variation of weight loss for various loads at constant sliding speed (0.3 m/s) of investigated hypereutectic Mg-5wt%Si alloy without and with HIUST.

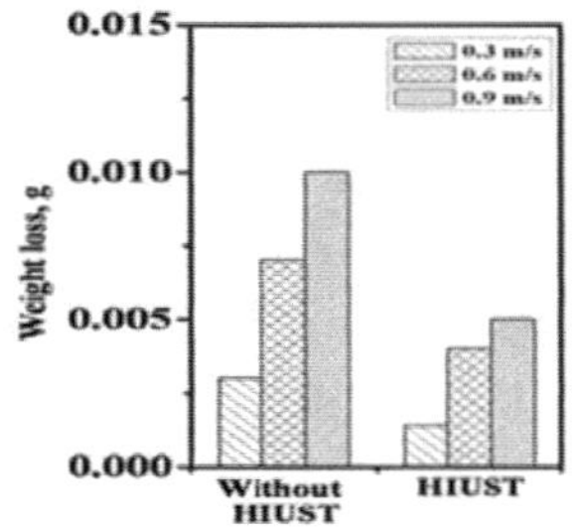


**Figure 6:** The variation of weight loss for various sliding speeds at constant load (10 N) of investigated hypereutectic Mg-5wt%Si alloy without and with HIUST.

Figure 7 shows FESEM micrographs of the worn surfaces and their wear debris with their EDX analysis at different wear conditions of the investigated alloy without and with HIUST. It is found that the worn surfaces after wear test exhibit different morphologies. Traces of parallel grooves and ridges can be observed on the surface of the sample without HIUST (Figure 7a, g, m). Moreover, there are both delamination of large plateaus and also some cavities or craters on the worn surfaces of alloy at wear conditions of high load (50 N) (Figure 7g, j) or high sliding speed (0.9 m/s) (Figure 7m, p) as compared with that at wear conditions of low load (10N) and low sliding speed (0.3 m/s) (Figure 7a, d). Furthermore, the surface of sample without HIUST is rougher than that of HIUST-sample as shown in Figure 7a, d, g, j, m, p. The wear debris of sample without HIUST are large size as shown in Figure 7b, h, n which gives rise to a relatively high weight loss. The wear debris of HIUST-sample are smaller in size as shown in Figure 7e, k, q, which gives rise to a relatively low weight loss. The size of wear debris in worn surfaces under wear conditions of high load (50 N) (Figure 7h, k) or high sliding speed (0.9 m/s) (Figure 7n, q) reveal numerous large flakes or sheets as compared with that of low load (10N) and low sliding speed (0.3 m/s) (Figure 7b, e) give rise to a relatively high weight loss with increasing applied load or sliding speed. For the samples without and with HIUST, analysis on the debris confirms the presence of Mg, Si, Fe and $O_2$ as shown in Figure 7c, f, i, l, o, r.

## 4. Discussion

### *4.1. Material investigation*

Under the present experimental conditions, the effect of the cooling rate on the modification of primary $Mg_2Si$ in hypereutectic Mg–5 wt.%Si could be neglected. Therefore, the difference in the morphology and size of primary $Mg_2Si$ were resulted almost exclusively from the difference in the application of HIUST. Usually, the microstructure of materials depends on the nucleation process and growth conditions. For the $Mg_2Si$, its structure belongs to face centered cube (FCC) and its dendrite arm should grow along the preferential [100] crystallographic directions [12]. As a result, the morphologies of primary $Mg_2Si$ in the sample without HIUST are mainly characterized by dendrites with complex morphologies, as shown in Figures 3a and 4a.

The refinement and modification of primary $Mg_2Si$ (Figures 3b, 4d) can be mainly attributed to HIUST during the solidification process. This can be explained as conjugation of two mechanisms: cavitation-enhanced heterogeneous nucleation and cavitation-induced dendrite fragmentation.

IOP Conf. Series: Materials Science and Engineering **143** (2016) 012037 doi:10.1088/1757-899X/143/1/012037

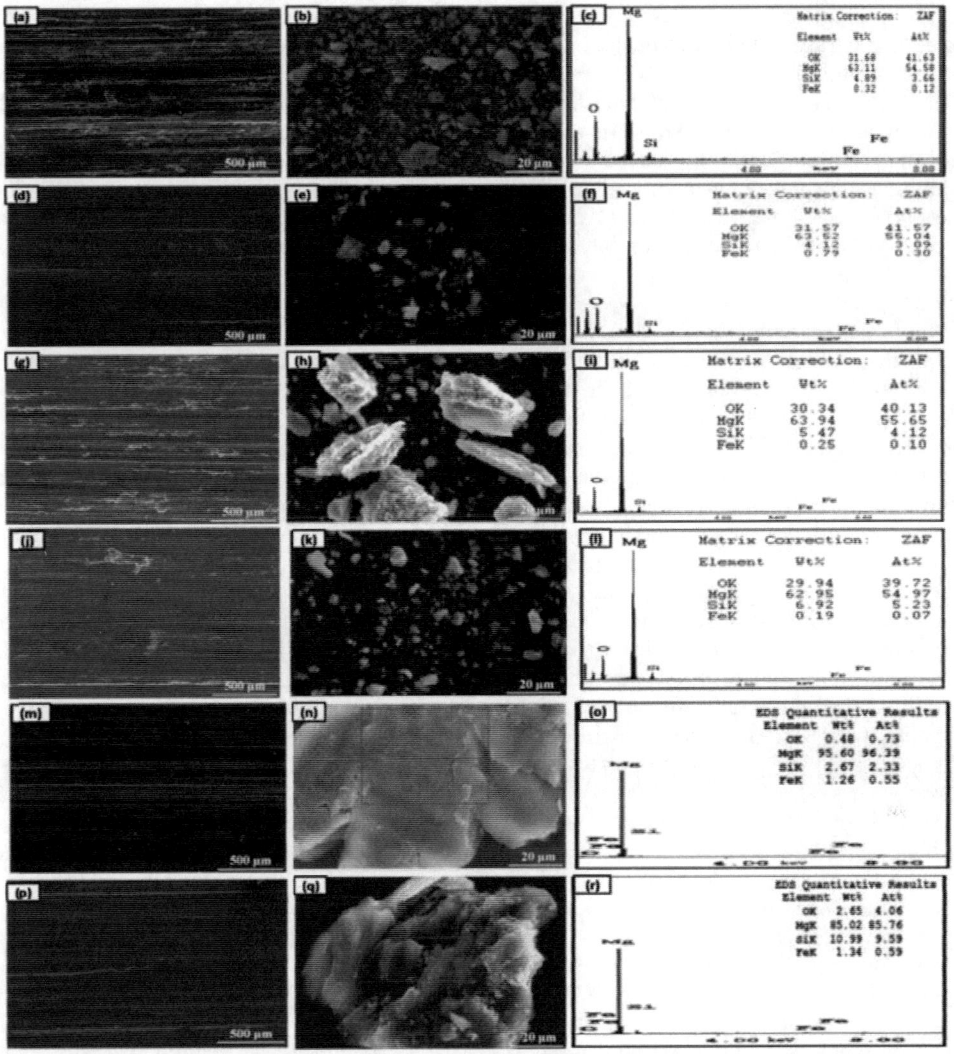


**Figure 7:** FESEM micrographs of the worn surfaces and their wear debris with their EDX analysis of the investigated Mg–5wt.%Si: (a), (b) and (c) without HIUST and (d), (e) and (f) with HIUST at wear conditions of 10 N load and 0.3 m/s sliding speed, (g), (h) and (i) without HIUST and (j), (k) and (l) with HIUST at wear conditions of 50 N load and 0.3 m/s sliding speed and (m), (n) and (o) without HIUST and (p), (q) and (r) with HIUST at wear conditions of 10 N load and 0.9 m/s sliding speed.

During the first stage of solidification between pouring temperatures and the liquidus temperature of the investigated alloy, HIUST can only be attributed to cavitation-enhanced heterogeneous nucleation. This because dendrite fragmentation will not be possible at those temperatures since solidification has not started yet. The cavitation-enhanced heterogeneous nucleation is further explained by three different mechanisms. The first is based on the pressure pulse-melting point ($T_m$) mechanism [12, 13], where the pressure pulse arising from the collapse of bubbles alters $T_m$ according to the Clausius–Clapeyron equation as $dT_m/dP = T_m(V_L - V_S)/\Delta H$, where $T_m$ is the freezing point in K, P is the pressure in MPa, $V_L$ and $V_S$ are the specific volume of the liquid and the solid phase in $cm^3/g$, respectively, and $\Delta H$ is the latent heat of freezing in J. An increase in $T_m$ is equivalent to increasing the undercooling so that an enhanced nucleation event is expected. The second is based on cavitation-enhanced wetting [8], which assumes that cavities and cracks on the substrate surfaces and insoluble non-metallic inclusions that either pre-exist in the melt or form on cooling during solidification, can be wetted by the melt under the pressure pulse from the collapse of the bubbles. Consequently, this enables these substrates to act as effective nucleation sites. The third mechanism [12, 13] assumes that rapid adiabatic expansion of gas inside the bubbles created during cavitation undercools the liquid at the bubble–liquid interfaces resulting in

IOP Conf. Series: Materials Science and Engineering **143** (2016) 012037 doi:10.1088/1757-899X/143/1/012037

nucleation on the bubble surfaces. Collapsing of these bubbles will distribute the nuclei into the surrounding liquid producing a significant number of nuclei in the liquid. Below the liquidus temperature, corresponding to the formation of the first solid metal, the cavitation-induced dendrite fragmentation is more effective and dominant. This mechanism assumes that the shock waves generated from the collapse of bubbles lead to fragmentation of dendrites [8, 12, 13], which are redistributed through acoustic streaming, thereby increasing the number of crystals. The coexistence of these two mechanisms seems to promote a high density of nuclei in the melt, thus leading to the development of a large number of fine primary $Mg_2Si$ crystals which exhibit an octahedral morphologies with a variety of polygonal outlines in the two-dimensional view, such as triangle, square, trapezoid and hexagon because of the random cutting angles polishing process of Metallographic samples.

*4.2. Hardness and wear behavior*

It is well known that the fine microstructure is usually beneficial to the mechanical properties of engineering alloys. In the present work, the coarse dendritic primary $Mg_2Si$ will give a detrimental effect on the hardness because of its sharp tips. Therefore, it can be inferred that the improvement of the hardness through HIUST is possibly related to the modification and refinement of the hard primary $Mg_2Si$. In the present work, the coarse dendritic primary $Mg_2Si$ without HIUST will give a detrimental effect on the wear resistance because of its sharp tips. Furthermore, at wear conditions of low load 10 N and low sliding speed 0.3 m/s , the high weight loss (Figure 5), the surface scratches (Figure 7a) and the large wear debris (Figure 7b) of the sample without HIUST addition may be attributed to coarse dendritic primary $Mg_2Si$ detached from the contact face between pin and disc. The wear resistance of an alloy can generally be enhanced by introducing secondary phase(s) to the matrix material [14]. Nevertheless, microstructural characteristics like morphology, size and distribution of hard particles have a great influence on the sliding wear properties of alloys [15]. In the present study, since the hard primary $Mg_2Si$ is almost uniform distributed in the alloy with HIUST, the abrasion degrees of contact face on the sample should be homogeneous. Furthermore, the fine $Mg_2Si$ also reduces the extent of abrasion wear. Moreover, the alloy with HIUST has higher hardness than that without HIUST which leads to the highest wear resistance. As cited in [16], the wear law states that the materials with higher hardness will exhibit better resistance to wear. Thus, all of these lead to the lowest weight loss (Figure 5), the smooth surface (Figure 7d) and the small wear debris (Figure 7e) of the alloy with HIUST and therefore its high wear resistance. However, it can be found that with the application of wear conditions at higher loads or higher sliding speeds for the same process, the weight loss increases drastically (Figures 5, 6). This is obvious that at higher loads and/or higher sliding speeds, the soft Mg matrix smears off leading to higher wear in spite of the presence of hard $Mg_2Si$ particles [17]. However, the sample due to HIUST still has the best wear resistance.

The presence of Fe in wear debris of investigated alloy (Figure 7c, f, i, l, o, r) is due to the ploughing by hard primary $Mg_2Si$ on the rotating steel disc [15, 17]. Also, the presence of $O_2$ was considered to have arisen in reactions with the environment, indicating that the mode of wear is mildly oxidative [15]. At very low load and/or sliding speed, an oxidative mechanism controls the wear process, generating debris comprising predominantly oxides. The oxidation was considered to be taken place due to the ability of the metal to oxidize under ambient conditions. On the other hand, material loss under abrasive wear conditions was removed by ploughing and micro-cutting. These mechanisms required penetration by hard abrasive particles which in turn are controlled by the hardness of material [15]. It has also reported [16] presence of abrasive wear in the low speed regime. Numerous grooves and shallow scratches running parallel to the sliding direction generally characterize abrasive wear. In the present work, abrasive wear is favor at wear conditions of a low sliding speed of 0.3 m/s under the low applied load of 10 N. The presence of grooves and ridges on the surfaces of the sample without HIUST (Figure 7a) characterize abrasive wear. Adhesion is a relevant wear mechanism in many lightweight alloys [18]. The adhesion wear mechanism appears due to the formation of micro-joints between the pin and the disc. As a consequence of their relative movement, the softer material breaks, leaving a small void in the Mg alloy and transferring some material to the steel disc. Figure 7a, d present some evidence of the existence of adhesion wear mechanism in the worn surface of investigated alloy. Therefore, it can be concluded that these as–cast alloys had experienced a mild abrasive oxidative wear with little adhesion.

However, examinations of the wear debris due to the applied high load (50 N) (Figure 7h, k) at low sliding speed (0.3 m/s) and/or the applied high sliding speed (0.9 m/s) (Figure 7n, q) at low load (10 N) reveal numerous flakes or sheets linking to the process of delamination. Since, at relatively slower speed 0.3m/s , if the load increases considerably (50 N), mixing of wear debris and counterface material could take place which due to higher temperature rise, gets oxidized. The oxidized layer may cover the surface and increase the weight loss. The formation and removal of this layer determines the overall wear rate of the material [19]. At higher load, removal the oxide film through sliding become greater that its rate of formation, and a transition to metallic wear occurred that caused delaminating in the bulk material and increased the wear rate [17, 19]. This is termed as delamination mechanism. Delamination is observed more extensively under the high loads [17]. Furthermore, at higher sliding speed (0.9 m/s), relatively at lower applied load of 10 N, delamination mechanism is also prevailing. This may be due to more adiabatic type of heating and cause more adhesive action between the two surfaces. At a critical applied load and sliding velocity, temperature rises so high and because of severe degree of delaminaiton, mixed layer becomes discontinuous and naked material comes in contact with counter surface. This also leads to a rise in surface temperature to a critical value i.e. flashing temperature at which strong adhesion between counter surface and specimen takes place.

**Acknowledgement**
The authors would like to appreciate the financial support from Academy of Scientific Research and Technology in the framework of the scholarship for support of Ph.D study, Egypt.

**References**
[1] Cong M, Li Z, Liu J, Yan M, Chen K, Sun Y, Huang M, Wang C, Ding B and Wang S 2012 Effect of Ca on the microstructure and tensile properties of Mg–Zn–Si alloys at ambient and elevated temperature J. Alloys Compd. **539** 168
[2] Hou J, Li C and Liu X 2011Nucleating role of an effective in situ $Mg_3P_2$ on $Mg_2Si$ in Mg–Al–Si alloys J. Alloys Compd. **509** 735
[3] Wang H Y, Chen L, Liu B., Li X R, Wang J G and Jiang Q C 2012 Heterogeneous nucleation of $Mg_2Si$ on $Sr_{11}Sb_{10}$ nucleus in Mg-x(3.5, 5 wt.%) Si-1Al alloys Mater. Chem. Phys. **135** 358
[4] Jiang Q C, Wang H Y, Wang Y, Ma B X and Wang J G 2005 Modification of Mg2Si in Mg–Si alloys with yttrium Mater. Sci. Eng. A **392** 130
[5] Zheng N, Wang H Y, Gu Z H, Wang W and Jiang Q C 2008 Development of an effective modifier for hypereutectic Mg–Si alloys J. Alloys Compd. **463** L1
[6] Guo E J, Ma B X, Wang L P 2008 Modification of $Mg_2Si$ morphology in Mg-Si alloys with Bi J. Mater. Process. Technol. **206** 161
[7] Abramov O V 1998 High-Intensity Ultrasonics: Theory and Industrial applications (First ed., Gordon and Breach Science Publishers, Amsterdam)
[8] Eskin G. I 1998 Ultrasonic Treatment of Light Alloy Melts (First ed., Gordon and Breach Science Publishers, Amsterdam)
[9] Bhingole P P and Chaudhari G P 2012 Synergy of nano carbon black inoculation and high intensity ultrasonic processing in cast magnesium alloys Mater. Sci. Eng. **A 556** 954
[10] Moussa M E, Waly M A, Nofal A A and El-Sheikh A M 2011 Influence of ultrasonic vibration on the grain refinement and mechanical properties of AZ91 magnesium alloy (in: Proceedings of 1st InterQuadrennial Conference in Middle East and Africa (IQCMEA-ICF): Processing, Performance and failure analysis of Engineering Materials, Egypt) 103
[11] Moussa M E, Waly M A, Nofal A A and El-Sheikh A M 2010 Effect of ultrasonic vibration and boron addition on microstructure and mechanical properties of AZ91 magnesium alloy (in: Proceedings of 8th Arab Foundry symposium (ARABCAST 2010) , Egypt) 1
[12] Moussa M E, Waly M A and El-Sheikh A M 2013 Effect of high-intensity ultrasonic treatment on modification of primary $Mg_2Si$ in the hypereutectic Mg–Si alloys J. Alloys Comp. **577** 693
[13] Moussa M E, Waly M A and El-Sheikh A M 2014 Combined effect of high-intensity ultrasonic treatment and Ca addition on modification of primary $Mg_2Si$ and wear resistance in hypereutectic Mg–Si alloys J. Alloys Comp. **615** 576
[14] Asl K M, Masoudi A and Khomamizadeh F 2010 The effect of different rare earth elements content on microstructure, mechanical and wear behavior of Mg–Al–Zn alloy Mater. Sci. Eng. **A 527** 2027
[15] Anasyida A S, Daud A R and Ghazali M J 2010 Dry sliding wear behaviour of Al–12Si–4Mg alloy with cerium addition Mater. Des. **31** 365
[16] Muley S V, Singh S P, Sinha P, Bhingole P P and Chaudhari GP 2014Microstructural evolution in ultrasonically processed in situ AZ91 matrix composites and their mechanical and wear behavior Mater. Des. **53** 475
[17] Kumar K K A, Pillai U T S, Pai B C and Chakraborty M. 2013 Dry sliding wear behavior of Mg–Si alloys Wear **303** 56
[18] Aung N N, Zhou W and Lim L E N 2008 Wear behavior of AZ91D alloy at low sliding speeds, Wear **265** 780
[19] Lim C Y H, Lim S C and Gupta M 2003 Wear behaviour of SiCp-reinforced magnesium matrix composites Wear **255** 629

IOP Conf. Series: Materials Science and Engineering **143** (2016) 012038 doi:10.1088/1757-899X/143/1/012038

# Pressurized metallurgy for high performance special steels and alloys

**Z H Jiang[1, 2], H C Zhu[1], H B Li[1], Y L1[1] and F B Liu[1]**
[1] Dept. of Ferrous Metallurgy, School of Materials and Metallurgy, Northeastern University, Shenyang, 110819, P.R. China

Email: jiangzh@smm.neu.edu.cn

**Abstract.** The pressure is one of the basic parameters which greatly influences the metallurgical reaction process and solidification of steels and alloys. In this paper the history and present situation of research and application of pressurized metallurgy, especially pressurized metallurgy for special steels and alloys have been briefly reviewed. In the following part the physical chemistry of pressurized metallurgy is summarized. It is shown that pressurizing may change the conditions of chemical reaction in thermodynamics and kinetics due to the pressure effect on gas volume, solubility of gas and volatile element in metal melt, activity or activity coefficient of components, and change the physical and chemical properties of metal melt, heat transfer coefficient between mould and ingot, thus greatly influencing phase transformation during the solidification process and the solidification structure, such as increasing the solidification nucleation rate, reducing the critical nucleation radius, accelerating the solidification speed and significant macro/micro-structure refinement, and eliminating shrinkage, porosity and segregation and other casting defects. In the third part the research works of pressured metallurgy performed by the Northeastern University including establishment of pressurized induction melting (PIM) and pressurized electroslag remelting (PESR) equipments and development of high nitrogen steels under pressure are described in detail. Finally, it is considered in the paper that application of pressurized metallurgy in manufacture of high performance special steels and alloys is a relatively new research area, and its application prospects will be very broad and bright.

## 1. Introduction

There are three thermodynamic parameters in the metallurgical reaction and the metal solidification process, namely chemical composition, temperature and pressure of the system. The metallurgical reaction and solidification process, especially in the metallurgical reaction and solidification process with gas phase is greatly influenced by pressure.

The metallurgical process at present, especially ferrous metallurgy process is carried out mostly at atmospheric pressure or under the pressure of less than 0.1MPa (vacuum). High pressure hydrometallurgy-leaching has been applied in non-ferrous metallurgical industry [1-5]. The earliest research work-"Pressure NaOH leaching aluminum mineral" was done by a chemist, K. J. Bayer since 1887. Since the 1940s, high pressure hydrometallurgy had been used in production of heavy non-

[2] Corresponding author. Tel.: +86-24-83686453; fax: +86-24-23890559.
Address: Mailbox 241, School of Materials and Metallurgy, Northeastern University, No. 3-11, Wenhua Road, Heping District, Shenyang, P. R. China. Postal code: 110819.

IOP Conf. Series: Materials Science and Engineering **143** (2016) 012038 doi:10.1088/1757-899X/143/1/012038

ferrous metals including Ni, Cu and Co etc.. In the 1970s, the greatest advances were made in acid leaching in zinc concentrate treatment under pressure. In the 1980s, the most important progress in the pressure leaching was the use of pressure pre oxidation of refractory gold ore to replace the roasting. Application of pressure technology in ferrous metallurgy process is mainly concerning two aspects. One is for the process with gas in the reaction, using pressure to increase reaction speed, shorten the reaction time, such as bath smelting reduction, BOF, etc. [6-7]. However, up to now there is not a real application in industry. The other is to increase the solubility of gases and volatile elements in liquid steel in the steelmaking temperature under the pressure, such as "dephosphorization by Ca reduction", "high nitrogen steels" etc.. Bulgarian people developed pressurized metallurgy method of pilot and industrial IMS-BAS method in the 70's of last century, with large molten pool melting method which can produce ingot from 500Kg, 1300Kg, 2500Kg [8-9]. Since 1980's, a series of methods for manufacture of high nitrogen steels have been developed, such as hot isostatic pressing melting(HIP), pressurized induction furnace melting(PIM), pressurized electroslag remelting (PESR), pressurized ladle nitrogen purge(PLBN), pressurized plasma arc melting(PARP), pressurized arc slag remelting (PASR) etc. [10-11].

There are two ways that pressure affects casting solidification and structure formation: mechanical effects related to the physical phenomena at macro level (elastic and/or plastic deformation, intensification of heat exchange between casting and mould, variation of cooling rate, control of mould filling, etc.) and thermodynamic and transport phenomena at micro level (changes in phase diagrams, physical properties, Gibbs free energy, chemical potentials, specific heat, surface tension, diffusion coefficients, etc) [12]. The former is effective at the pressure of several MPa, while the latter is valid at the pressure of more than 10MPa or even more than 0.1GPa. The earliest work concerning the effect of pressure on the structure and properties of the material was done by Nobel Prize winner P. W. Bridgman in 20-30s of the last century. He developed techniques to achieve high pressures (~10 GPa) and investigated pressure effects on thermal and electrical conductivities, melting, reaction kinetics, viscosity, compressibility, tensile strength and a variety of other properties and phenomena.[13] Bridgman's work demonstrated the importance of pressure for studying continuous and discontinuous changes in the structure and properties of matter. His scientific results and engineering solutions have found a very large number of practical applications.

There are many research works regarding casting technology under high pressure for the light metals and alloys [14-19]. G. Zhang studied 515GPa high pressure on the microstructure of Al2Si alloy [14]. R. Li performed the microstructure and properties of ZnAl alloy under the action of high pressure [15-16]. The results show that the microstructure of ZA43 alloy was obviously refined without the formation of coarse dendrite, replaced by a small, granular structure on the cross section with random distribution, and the hardness of ZA27 alloy solidified at high pressure of 2.5 GPa was 55.8% higher than atmospheric pressure. A few works on casting of iron and steel under high pressure were made because it is relatively difficult to apply a high pressure at high temperature due to high melting points of such materials [12, 20].

The aim of this work is to understand the physical chemistry during melting and solidification of special steels and alloys at pressurized condition in order to improve the existed pressurized processes or to find new methods of pressurized metallurgy for manufacture of high performance special steels and alloys.

## 2. Physical chemistry of pressurized metallurgy

### *2.1. Gas equation*

Ideal gas equation can be expressed by the relation

$$PV = n\mathrm{R}T \tag{1}$$

where $P$, $V$, $T$ are the pressure, volume and temperature of ideal gas, respectively; R is gas constant, $n$ the number of mole. For a real gas, Van der Waals equation can be used.

$$\left(P+\frac{an^2}{V^2}\right)(V-nb)=nRT \tag{2}$$

Where a and b are constants. Pressure is a basic thermodynamic parameter. The change of $P$ makes the change of $T$ and $V$.

### *2.2. Gas solubility in liquid and solid steel*

Gas solubility in liquid steel can be expressed by Sievert's law.

$$[G\%]=K\sqrt{P_G} \tag{3}$$

It is shown that the solubility of gas in molten steel of [G%] is proportional to the square root of gas partial pressure. However, the solubility will deviate from the Sievert's law with the increase of pressure. Therefore, a modified equation at pressure above 0.1MPa is applied as follows:

$$[G\%]=K\left(\sqrt{P_G}\right)^{1-\delta} \quad (0\le\delta<1) \tag{4}$$

Where $\delta$ is the correction factor of pressure. A new model for calculation of nitrogen solubility in molten stainless steel in a wide range of the alloy concentrations, temperatures, and pressures has been successfully established by introducing a new term of pressure-effect on nitrogen activity coefficient which is 0.06 [21].

$$\begin{aligned}\lg[\%N]=&\frac{1}{2}\lg\left(p_{N_2}/p^\theta\right)-\frac{188}{T}-1.17-\left(\frac{3280}{T}-0.75\right)\Big(0.13[\%N]+1.118[\%C]+0.043[\%Si]\\&+0.011[\%Ni]+3.5\times10^{-5}[\%Ni]^2-0.024[\%Mn]+3.2\times10^{-5}[\%Mn]^2\\&-0.01[\%Mo]+7.9\times10^{-5}[\%Mo]^2-0.048[\%Cr]+3.5\times10^{-4}[\%Cr]^2+\delta_N^p\log\sqrt{p_{N_2}/p^\theta}\Big)\end{aligned} \tag{5}$$

where $\delta_N^p$=0.06, when $p_{N_2}/p^\theta>1.0$; $\delta_N^p$=0, when $p_{N_2}/p^\theta\le1.0$.

Figure 1 gives the nitrogen solubility of the austenitic stainless steels at different nitrogen partial pressure, which means that the high nitrogen austenitic stainless steels can be produced under the high pressure condition. As shown in figure 2, the nitrogen solubility of Fe-18Cr-18Mn alloy increases 0.3% when the temperature decreases from 2000K to 1750K, which means that low temperature is favorable to increase the nitrogen solubility of stainless steel.

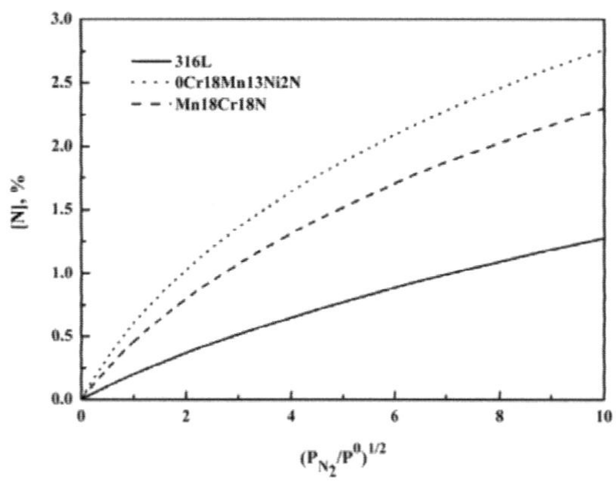


**Figure 1.** Effect of nitrogen partial pressure on nitrogen solubility of austenitic stainless steels at 1873K.

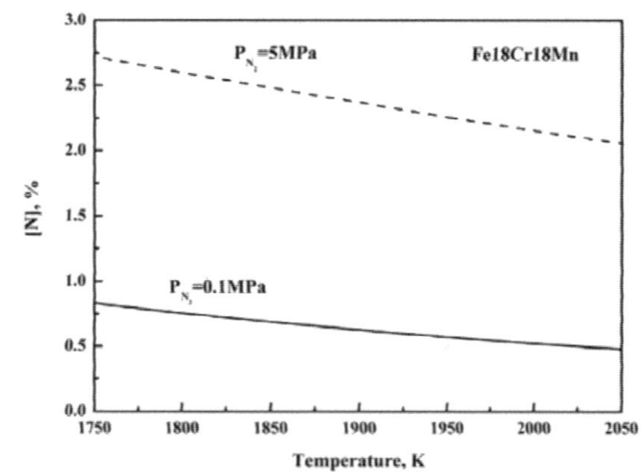


**Figure 2.** Effect of temperature on nitrogen solubility in liquid Fe-18Cr-18Mn alloy at 0.1MPa and 5MPa.

We also developed a model for calculation of nitrogen solubility in δ and γ phase of stainless steel [22]. As shown in figure 3, with the increase of nitrogen partial pressure, δ phase region gradually decreased, when nitrogen pressure up to 0.6MPa, δ phase completely disappeared. The solidification process directly entering from the liquid phase into the γ phase austenite region, which effectively suppressed the precipitation of nitrogen in solidification process

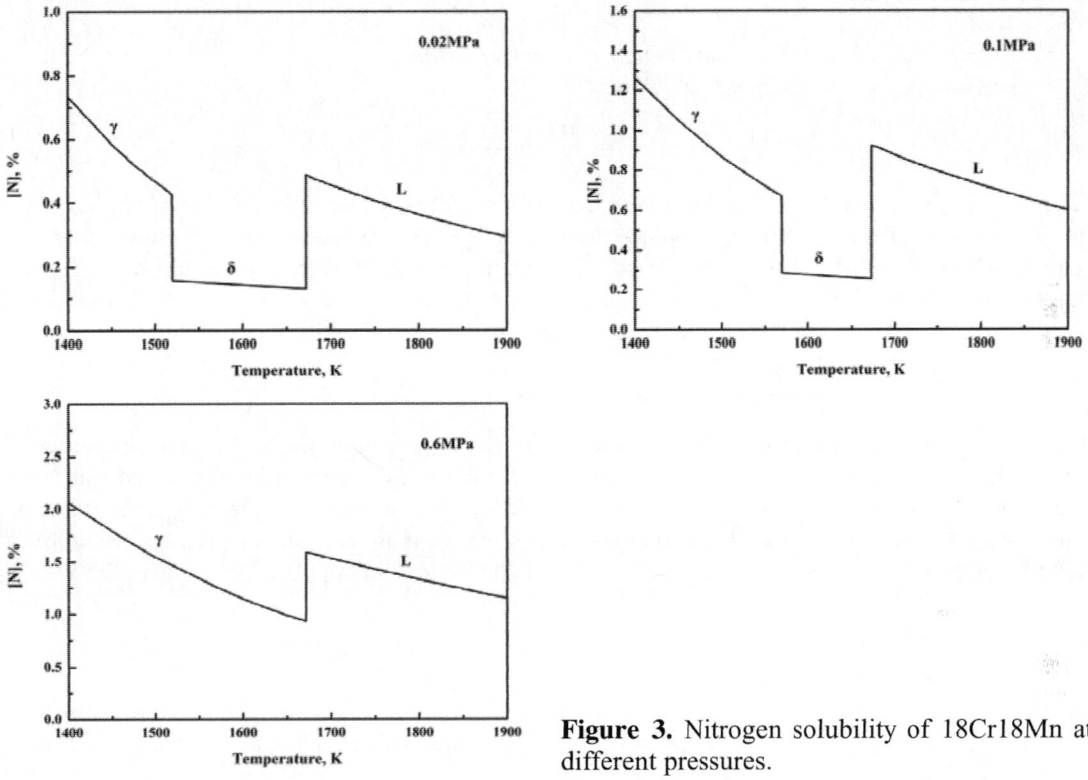


**Figure 3.** Nitrogen solubility of 18Cr18Mn at different pressures.

In fact, in addition to N, H, O, many of the elements in the steelmaking temperature (for example temperature of 1873 K) exist in gaseous state, such as Ca, Mg, K, Na, Pb, Sn, Zn, Li, Bi, which has very high vapor pressure and are easily volatilized under normal pressure. The increase in pressure can significantly improve the solubility of these elements in molten steel, and thus play a role in that cannot be achieved under normal pressure. For example, increasing the pressure of Ca, Mg can significantly improve the treatment effect of liquid steel, greatly increase the utilization rate.

### *2.3. Chemical reactions with gas*

The chemical reaction will take place in the direction in which gas generates or the volume increases with the decrease of pressure. Vacuum (by reducing pressure) will promote the reaction: [C]+[O] ={CO}, for example. The chemical reaction will take place in the direction in which the volume decreases with the increase of pressure (by pressurization), for example, $N_2$=2[N].

### *2.4. Effect of pressure on activity and activity coefficient*

$$\left(\frac{\partial \ln a_i}{\partial P}\right)_{T,C_j} = \frac{\overline{V}_i - \overline{V}_i^0}{RT} \tag{6}$$

$$\left(\frac{\partial \ln \gamma_i}{\partial P}\right)_{T,C_j} = \left(\frac{\partial \ln f_i}{\partial P}\right)_{T,C_j} = \frac{\overline{V}_i - \overline{V}_i^0}{RT} \tag{7}$$

Where $a_i$, $\gamma_i$, $f_i$ - activity, activity coefficients of component i in the standard state of pure substance and Henry's law, respectively; $\overline{V}_i$, $\overline{V}_i^0$ partial molal volumes of component i at present pressure and at standard pressure, respectively.

### 2.5. *Effect of pressure on melting point or phase transformations*

The Clausius–Clapeyron relation can be expressed

$$\frac{dT}{dP} = \frac{T\Delta v}{\Delta H} \tag{8}$$

The equation relates the slope of the temperature–pressure boundary for coexisting phases at some point ($T$, $P$) to the specific enthalpy of phase transformation $\Delta H$ and the specific volume change $\Delta v$. Equation (8) is very useful because it permits quantitative estimates to be obtained from the thermodynamic parameters for phase transformation. Using the Clausius-Clapeyron equation, changes in phase transformation temperature with pressure can be calculated with following formula.

$$T_{eq}(P_0 + \Delta P) = T_{eq}(P_0) + \Delta P \frac{dT_{eq}}{dP} = T_{eq,0} + \Delta P \frac{T_{eq,0}\Delta v}{\Delta H} \tag{9}$$

Where $T_{eq}$, $T_{eq,0}$ are the equilibrium temperatures for phase transformation, $\Delta P$ pressure increment, Pa. Figure 4 shows the correlations between melting point and pressure for aluminium and the eutectic temperature and pressure for the Al-Si system[12]. However, the pressure effecting on melting point or phase transformations becomes obvious when the order of magnitude of pressure is more than 0.1GPa.

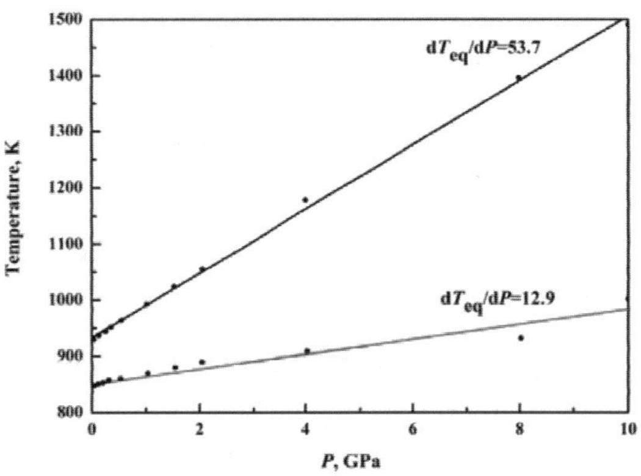


**Figure 4.** Melting point of aluminium $T_m$ and eutectic temperature of Al-Si system $T_{eut}$ as function of pressure.

### 2.6. *Pressure effect on interfacial free energy*

$$\left(\frac{\partial \sigma}{\partial P}\right)_T = \Delta P \left(\frac{\partial V_A}{\partial P}\right)_T \tag{10}$$

As the volume of every closed system with constant phase composition decreases with pressure, i.e. $(\partial V_A/\partial P)_T < 0$, an increase in pressure will decrease the interfacial free energy. The interfacial free energy determines the critical radius for solid phase nucleation. Since an increase in pressure decreases the interfacial free energy (surface tension) of the crystal/melt interface, the critical radius for nucleation will decrease proportionally. This in turn will increase the number of nucleation sites and consequently refine the solidified structure.

### 2.7. *Effect of pressure on nucleation*

Pressure effects on critical radius for nucleation can be expressed.

$$r = \frac{2\sigma dT}{[\Delta T(V_2 - V_1)dP]} \tag{11}$$

where σ is the surface tension, J $m^{-2}$.

Pressure effects on nucleation rate $I$ can be determined by equation

$$\ln I = \ln a - \frac{b}{(d+P)^2} - cP \tag{12}$$

Where a, b, c, d are constants.The increase of pressure will decrease the critical radius for nucleation, and will increase nucleation rate**,** and then refine the solid's structure. Figure 5 presents the simulation results of nucleation in Fe-0.6%C alloy at various solidification times under atmospheric pressure and pressure of 10 MPa [23].

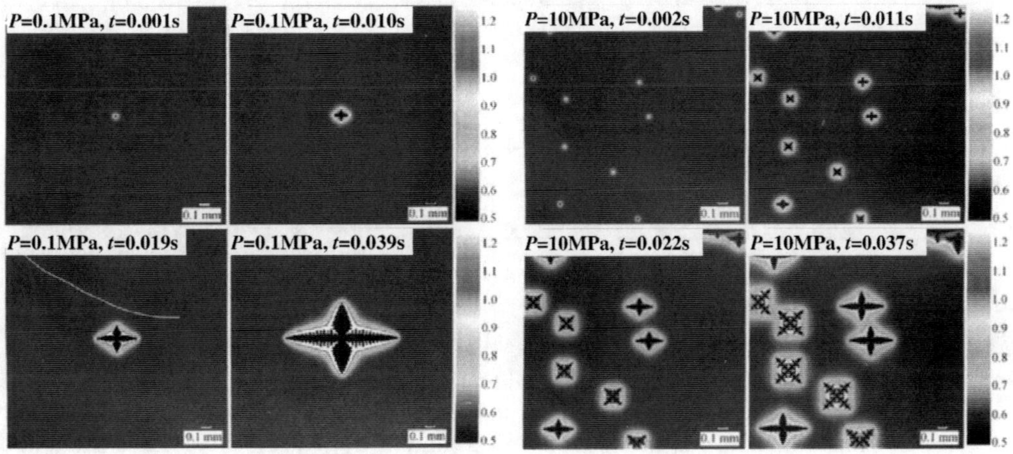


**Figure 5.** Dendrite morphologies of Fe-0.6%C alloy at various solidification time under atmospheric pressure and pressure of 10 MPa (simulation).

*2.8. Effect of pressure on heat transfer between casting and mould*

External pressure during solidification increases the metal/mould contact and raises the heat transfer coefficient. Theoretical and experimental research on the solidification of Al-Si eutectic alloy against steel dies have shown that a pressure of 196 MPa increases the heat transfer coefficient $h$ by a factor of 15 from about $3.4\times10^3$ to $5.25\times10^4$ W $m^{-2}$ $K^{-1}$. The structures as shown in figure 6 [24] are formed in high nitrogen steel ingots solidified in a metal mould under various gas pressures: $P_1$=0.1 MPa, $P_2$=1.7 MPa and $P_3$=3.6 MPa. Owing to the plasticity of semisolid and solid regions of solidifying ingot, the heat exchange between mould and casting is forced by an external pressure, which results in a reduction in total solidification time, decrease in local solidification time (LST), the increase in cooling rate $V_{cool}$ and, finally, the refinement of the dendritic structure. There results indicated that refinement of casting structure can be realized even at relatively small increases of pressure. Figure 7 shows that the heat transfer coefficient between casting and mould is greatly influenced by pressure [25].

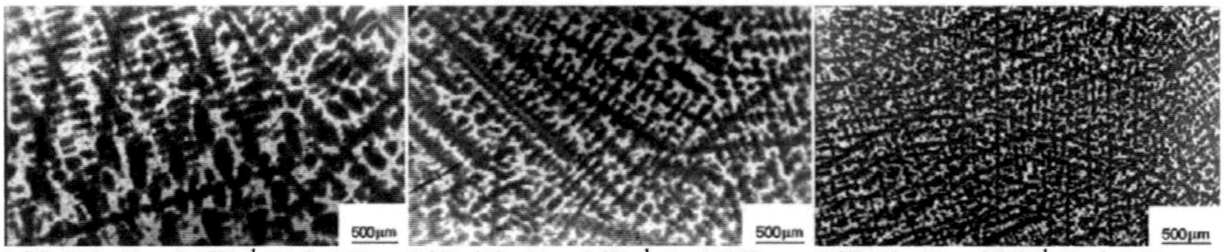


**LST=715s, $V_{cool}$=0.17Ks$^{-1}$, $P_1$=0.1MPa** **LST=418s, $V_{cool}$=0.3Ks$^{-1}$, $P_1$=1.7MPa** **LST=60s, $V_{cool}$=2.0Ks$^{-1}$, $P_1$=3.6MPa**

**Figure 6.** Structure of high nitrogen steel solidified in metal mould under different gas pressures: all structures are observed in very same region of ingots.

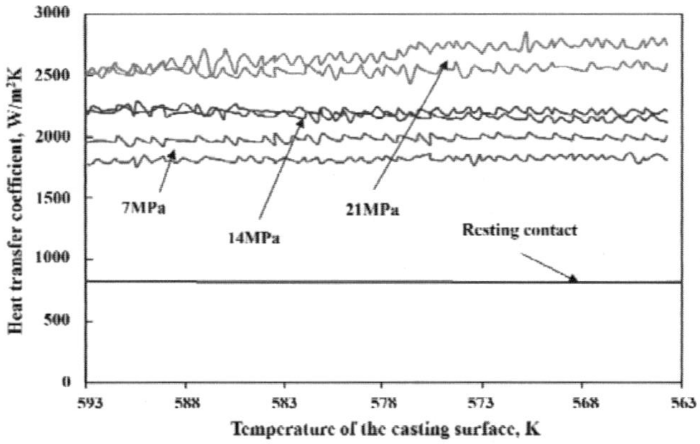


**Figure 7.** Effect of pressure on heat transfer coefficient between casting and mould.

## 3. Manufacture and application of special steels and alloys by pressurized metallurgy

According to the above fundamental consideration, there are two useful effects of pressure on special steels and alloys in manufacture process. One is to increase the solubility of gases or volatile elements in steel or alloy by pressurization, for example, to obtain high nitrogen steels that have been made by many research works. To increase the solubility of volatile elements (Ca, Mg …) in steel or alloy will be the future works. The other is to refine the solidification structure of steel or alloy by improvement of heat transfer between casting and mould at relatively small increases of pressure (≤10MPa). However, more high pressure (10-100MPa, even more) will be difficult to be realized in steel industry.

### *3.1. Manufacture of high nitrogen steels (HNS) by pressurized metallurgy*

There are different approaches to manufacture of high nitrogen steels, as shown in figure 8. Except for EAF and AOD process, other methods including gas purging, induction furnace, plasma arc melting, arc slag remelting (ASR) and PESR can work either in normal pressure or increased pressure. Of course, the pressurized method is more effective to obtain high nitrogen content for steel or alloy.

In our laboratory a 25kg pressurized induction furnace with max pressure of 6MPa and a 50kg pressurized ESR furnace with max pressure of 7MPa have been established a few years ago, as shown in figure 9. Several types of high nitrogen steels in including austenitic stainless steels and martensitic stainless steels have been manufactured by PIM, as shown in table 1. The working pressure during the melting and casting can be controlled separately. The pressures are normally controlled at 0.2-0.5MPa at melting stage and at 1.0-1.5MPa at casting stage. Nitrogen alloying in steel was realized by adding chromium nitride and nitrogen gas absorption. The percentage of gas alloying could be from 20%-100% which depends on the nitrogen pressure and aimed nitrogen content in steel. The rate of nitrogen absorption from gas into liquid steel is quite fast due to strong magnetic stirring of liquid steel and slag-free on top of bath. For example, Cronidur30 with 0.55% nitrogen in table 1 was prepared by gas alloying without addition of chromium nitride. Figure 10 shows the macrostructure and photo by γ-ray detection of HNS-A ingot, and nitrogen and oxygen distribution in ingot. It can be seen from the figure that except significant shrinkage in the upper part of ingot, the rest of the macrostructure is very uniform, and compact without holes. Table 2 gives the N, O distribution at different positions of ingot HNS-A made by PIM. The results indicate that there is not macrosegregation of nitrogen in ingot. Non-metallic inclusions in the HNS-A ingot mainly are $Al_2O_3$, MnS and $Al_2O_3$-TiN with the sizes of less than 5μm. After homogeneous solid solution treatment of ingot at 1100℃ for 1hour and water cooling, the microstructure with uniform austenite is shown in figure 11.

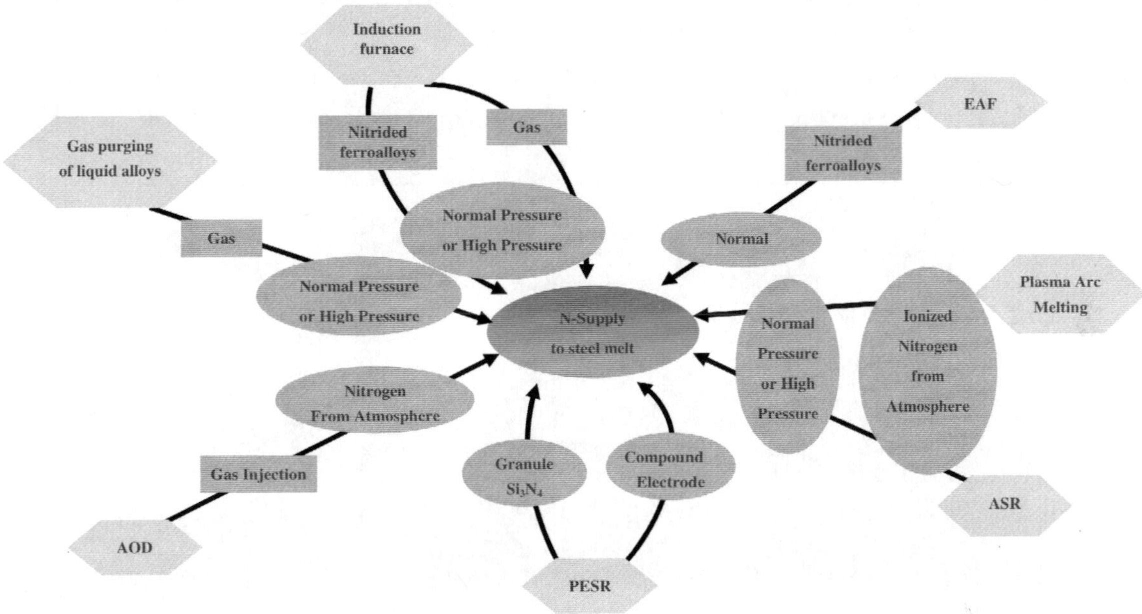


**Figure 8.** Manufacture of high nitrogen steels (HNS) by different methods.

**Figure 9.** The lab-scale PIM and PESR equipments.

**Table 1.** Chemical composition of high nitrogen steels made by PIM, wt%.

| No. | Steel grade | Cr | Mn | Mo | C | N | O | Al | P | S |
|---|---|---|---|---|---|---|---|---|---|---|
| HNS-A | 18.0Cr16.5Mn2.0Mo0.8N | 18.02 | 16.10 | 1.96 | 0.046 | 1.08 | 0.0033 | 0.01 | 0.014 | 0.0080 |
| HNS-B | 18.0Cr16.5Mn2.0Mo1.0N | 18.56 | 15.29 | 1.98 | 0.051 | 1.20 | 0.0130 | 0.01 | 0.013 | 0.0080 |
| P900N | 18Cr18MnN | 18.27 | 18.47 | -- | 0.100 | 1.20 | 0.0023 | 0.02 | 0.010 | 0.0029 |
| P900NMo | 18Cr18Mn2MoN | 18.26 | 17.87 | 2.02 | 0.110 | 1.23 | 0.0054 | 0.03 | 0.011 | 0.0028 |
| P2000 | 18Cr14Mn3MoN | 17.88 | 14.17 | 2.97 | 0.110 | 1.15 | 0.0026 | 0.02 | 0.012 | 0.0029 |
| Cronidur30 | 15Cr1MoN | 15.33 | 0.52 | 0.99 | 0.310 | 0.40 | 0.0026 | 0.03 | 0.013 | 0.0026 |

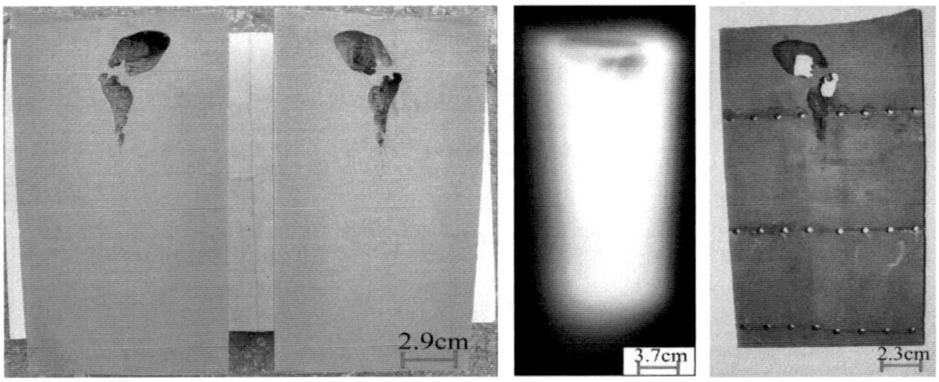


**Figure 10.** Macrostructure of HNS ingot prepared by PIM.

**Table 2.** N, O distribution at different positions of ingot HNS-A made by PIM, wt%.

| | No.1 | No.2 | No.3 | No.4 | No.5 | No.6 | No.7 | No.8 |
|---|---|---|---|---|---|---|---|---|
| N | 1.10 | 1.10 | 1.10 | 1.10 | 1.08 | 1.08 | 1.08 | 1.08 |
| O | 0.0033 | 0.0041 | 0.0035 | 0.0051 | 0.0055 | 0.0034 | 0.0061 | 0.0030 |

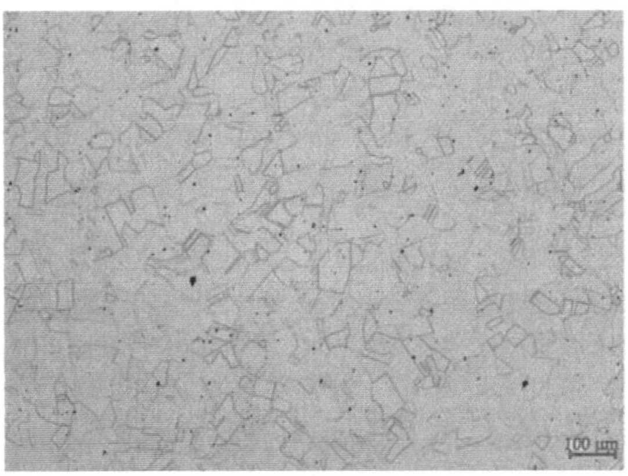

**Figure 11.** The microstructure of HNS-A after solid solution treatment.

The high nitrogen austenitic stainless steels with very uniform and compact macrostructure have been made by 50kg pressurized electroslag remelting furnace with the pressure from 2.0-3.2MPa by using composite electrode with mixture of nitrides and deoxidizers, as shown in figure 12. Table 3 gives the chemical composition of HNS manufactured by PESR. The precipitation behaviour, mechanical properties, corrosion resistance properties and biological properties of high nitrogen austenitic stainless steels have been systematically investigated associated with the nitrogen concentration and the other influence factors [26-32]. The results show that HNS manufactured by PESR exhibits good mechanical properties, excellent corrosion resistance properties and the excellent blood compatibility.

**Figure 12.** Composite electrode and Macrostructure of P2000 ingot made by PESR.

**Table 3.** Chemical composition of HNS manufactured by PESR, wt%.

| Grades | MPa | Cr | Mn | C | Si | P | Mo | Ni | N |
|---|---|---|---|---|---|---|---|---|---|
| A | 2.1 | 21.22 | 15.92 | 0.120 | 0.49 | 0.023 | -- | -- | 1.21 |
| P900N | 2.0 | 20.13 | 16.51 | 0.100 | 0.54 | 0.020 | -- | -- | 1.00 |
| P900NMo | 2.1~3.0 | 18.34 | 18.36 | 0.068 | 0.50 | 0.024 | 2.13 | -- | 0.93 |
| P2000 | 3.0 | 17.06 | 13.18 | 0.042 | 0.75 | 0.021 | 3.37 | -- | 0.79 |
| B | 3.2 | 21.33 | -- | 0.026 | 0.72 | 0.019 | 2.05 | 4.00 | 0.88 |

### *3.2. Application of special steels and alloys by pressurized metallurgy*

The family of high nitrogen steels consists of not only austenitic, martensitic stainless steel, and also ferritic stainless steel, high speed tool steel and die steel. As shown in table 4, there are many performance characteristics for future applications in industry. For example, nitrides uniformly distributed in tempered high nitrogen martensitic stainless steel can improve the strength and hardness of steel. The excellent resistance against wear as well as pitting corrosion of CRONIDUR 30 (15% Cr, 1% Mo, 0.4% N), promotes this steel for application in aviation. Typical applications are bearings in aviation turbines and ball screw gearing shafts, also used for rams in water cooled turbo drills and screws, fasteners for the chemical, car, and construction industry [33]. Figure 13 gives examples of application of high nitrogen austenitic stainless steels in manufacture of human medical materials, precision thin wall tube and retaining ring forging for large thermal power unit (P900N) by Chinese companies based on our research results.

With the progress of manufacturing technology, the cost of manufacturing will continue to be reduced together with further improvement of the performance; the scope of application of high nitrogen stainless steel will continue to expand. The highest tensile strength of high nitrogen stainless steel is currently 3600MPa, may be more than 4000MPa in the future, which still maintain a good ductility and high corrosion resistance. Therefore, it can be expected that high nitrogen stainless steels will be widely used in transportation (car, train, and ship), construction (such as ultra-high strength rebar), aerospace industry, atomic energy, ocean engineering and military industry in the future.

IOP Conf. Series: Materials Science and Engineering **143** (2016) 012038 doi:10.1088/1757-899X/143/1/012038

**Figure 13**. Examples of application of high nitrogen austenitic stainless steels.

**Table 4.** Performance characteristics of different high nitrogen steels.

| Type | N, wt% | Grades | Performance characteristics |
|---|---|---|---|
| Austenitic Stainless Steel | ≤1.20-2.80 | Cr18Mn18N<br>Cr18MnN)<br>16Cr14Mn3MoN<br>Cr25Mn11Si3N | The strength is increased, the low temperature impact toughness improved obviously;<br>Excellent corrosion resistance and resistance to stress corrosion;<br>Stabilization of austenite and non-magnetic stability |
| Ferritic Stainless | 0.08-0.60 | Cr12Ni2Mo2VN<br>1Cr12Ni3Mo2VN | Nitride dispersion distribution<br>The high temperature creep properties Improved |
| Martensitic stainless | 0.08-1.60 | 4Cr13MoN<br>X30CrMoN15-1 | The strength and wear resistance is increased greatly, the low temperature impact toughness improved obviously |
| High speed tool steel | 0.08-0.60 | W6Cr5V2N<br>W2Cr6V2N | The refinement of solidification structure<br>Nitride dispersion distribution |
| Die steel | 0.10-0.60 | 55NiCrMoV7N<br>55CrVMoN | The refinement of solidification structure<br>The strength and toughness improved obviously |

## 4. Summary and outlook

Pressurization may change the conditions of chemical reaction in thermodynamics and kinetics, and make the volatile element increase solubility in molten metals, not only to raise the utilization rate of the elements, but also to improve the roles of these elements in metal melt, such as purifying and alloying.

Pressurization can also enhance heat transfer and the cooling effect between casting and mould, reduce the local solidification time, promote the refinement of solidification structure and eliminate shrinkage, porosity and segregation and other casting defects. This phenomenon can appear even at relatively small increases of pressure (at several MPa), which means that this technology could be easily realized in industry in the near future.

The high pressure may change the physical and chemical properties of metal melt, thus greatly influencing phase transformation during the solidification process and the solidification structure, such as increasing the solidification nucleation rate, reducing the critical nucleation radius, accelerating the solidification speed and significant microstructure refinement. The pressure has effect on solute distribution coefficient and diffusion coefficient of elements during solidification, and affects the segregation behavior of elements during solidification process.

The pressurized technology used in ferrous metallurgy mainly concerns the high nitrogen steels, especially high nitrogen stainless steels. The behaviours of Ca, Mg, K, Na, Pb, Sn, Zn, Li, Bi and other volatile elements in pressurized steelmaking process is still in the embryonic stage of research. Application of pressurized metallurgy in manufacture of special steel and alloy is a relatively new research direction, so its application prospects will be very broad and bright.

**Acknowledgements**
This project supported by the National Nature Science Foundation of China (No.51434004, U1435205, 51304041) and the National High Technology Research and Development Program of China (863 Program) (No. 2012AA03A502). In addition, the work is supported by Program for Liaoning Innovative Research Team in University with the grant No. LT20120008.

**References**
[1] Qiu D F 1994 *Min. Metall.* **4** 55-67
[2] Ke J J 1996 *Hydrometall. Chin.* **15** 165-6
[3] Song F L 2001 *Hydrometall. Chin.* **20** 165-6
[4] Derry R 1972 *Miner. Sci. Eng.* **4** 3-24
[5] Habashi F 1971 *Eng. Min. J.* **172** 96-100
[6] Lu L M, Ding W Z, Jiang G C, Song Y B, Zhang M F and Shi H Z 1997 *J. Iron Steel Res. Int.* **9** 9-12
[7] Chen J X, Jiang G C, Lu L M, Ding W Z, Lu X G, and Chen J Y 1990 *J. Baotou Univ. Iron Steel Technol.* **18** 246-51
[8] Andreev C, Rashev T 1999 *Mater. Sci. Forum.* **318** 255-60
[9] Rashev T 2001 *Annual Meeting Pro.* (Beijing) pp 541-549
[10] Li H B, Jiang Z H, Shen M H You X M and CaoY 2006 *Chin. Metal.* **16** 9-13
[11] Jiang Z H, Li H B, Dong Y W, Yang Y and CaoY. 2006 *J. Iron Steel Res. Int.* **16** 1-6
[12] Sobczak J, Drenchev L and Asthana R. 2012 *Int. J. Cast Met. Res.* **25** 1-14
[13] Bridgman P W 1931 *The Physics of High Pressures* (London: G. Bell & Sons, Ltd.)
[14] Zhang G Z, Xin Q B and Wang X Y 1998 *J. Northeast. Univ., Nat. Sci. (China)* **19**(5): 479-81
[15] Li R D, Huang Z P and Zi Y H 2003 *Spec. Cast. & Nonferrous Alloy.* **24** 16-17
[16] Li R D, Huang Z P and Zi Y H 2003 *Foundry* **52** 92-94
[17] Wang Z L, Wang H W, Wei Z J and Cao L 2007 *Chin. J. Nonferrous Met.* **17** 383-9
[18] Zhao Z, Fan Z T, Cheng P, Jiang W M and Dong W P 2010 *Chin. J. Nonferrous Met.* **20** 1520-5
[19] Suarez B, Asensiolozano J and Vandervoort G 2007 *Microsc. Microanal.* **13** 1030-1
[20] Han B and Wu S J. 2012 *Mater. Lett.* **70** 7–10
[21] Jiang Z H, Li H B, Chen Z P and Huang Z Z 2005 *Steel Res. Int.* **76** 740-5
[22] Li H B 2008 *PHD thsis* (Shenyang: Northeastern University)
[23] Li Q, Guo Q Yand Li R D 2006 *Miner. Metall. Mater.* **13** 516-22
[24] Gavrilova R, Manolov V, Yotova A, Popov Sand Saraivanov L 2004 *Proc. 4th Int. Cong. on 'Mechanical engineering technologies'*(Varna: Bulgaria) pp135-7
[25] Griffiths W D and Kawai K. 2010 *J. Mater. Sci.* **45** 2330-9
[26] Li H B, Jiang Z H, Fan H, Zhu H C and Zhang Z R 2013 *J. Cent. S. Univ. Technol.* **20** 3345-53
[27] Li H B, Jiang Z H, Fan H, Zhu H C, Ma Q F and Zhan D P 2012 *J. Iron Steel Res. Int.* **19** 43-51
[28] Li H B, Jiang Z H, Cao Y and Zhang Z R 2009 *Int. J. Mine. Metall. Mater.* **16** 387-92
[29] Jiang Z H, Zhang Z R, Li H B, Li Z and Ma Q F 2010 *Int. J. Mine. Metall. Mater.* **17** 729-736
[30] Fu Y, Wu X Q, Han E H, Ke W, Yang K, Jiang Z H 2009 *Electrochim. Acta* **54** 4005-14
[31] Fu Y, Wu X Q, Han E H, Ke W, Yang K, Jiang Z H 2009 *Electrochim. Acta* **54** 1618-29
[32] Li H B, Jiang Z H, Zhang Z R, CaoY and Yang Y 2009 *Int. J. Mine. Metall. Mater.* **16** 654-60
[33] Stein G and Hucklenbroich I. 2004 *Mater. Manuf. Process.* **19** 7-17

**Institute of Physics**
Dirac House, Temple Back
Bristol BS1 6BE UK

ISSN: 1757-8981
ISBN 978-1-5108-2821-6

9 781510 828216

CURRAN ASSOCIATES INC.
proceedings.com